Software and Hardware Engineering

Assembly and C Programming for the Freescale HCS12 Microcontroller

Fredrick M. Cady

Department of Electrical and Computer Engineering
Montana State University

New York   Oxford
Oxford University Press
2008

Oxford University Press, Inc., publishes works that further Oxford University's objective of excellence in research, scholarship, and education.

Oxford   New York
Auckland   Cape Town   Dar es Salaam   Hong Kong   Karachi
Kuala Lumpur   Madrid   Melbourne   Mexico City   Nairobi
New Delhi   Shanghai   Taipei   Toronto

With offices in
Argentina   Austria   Brazil   Chile   Czech Republic   France   Greece
Guatemala   Hungary   Italy   Japan   Poland   Portugal   Singapore
South Korea   Switzerland   Thailand   Turkey   Ukraine   Vietnam

Published by Oxford University Press, Inc.
198 Madison Avenue, New York, New York 10016
http://www.oup.com

Library of Congress Cataloging-in-Publication Data
Cady, Fredrick M., 1942-
    Software and hardware engineering: assembly and C programming for the freescale HCS12
    microcontroller / Frederick M. Cady.—2nd ed.
        p. cm.
    Includes bibliographical references and index.
    ISBN-13: 978-0-19-530826-6
        1. Programmable controllers.   2. Software engineering.   I. Title.

TJ223. P76C336 2007
629.8'95—dc22   2006049595

Printing number: 9 8 7 6 5 4 3 2 1

Printed in the United States of America
on acid-free paper

"To embedded systems students everywhere, with thanks to Douglas Adams,

DON'T PANIC"

Contents

Preface

Software and Hardware Engineering: Assembly and C Programming for the Freescale HCS12 Microcontroller is a merger of *Software and Hardware Engineering: Motorola M68HC12* and *Microcontrollers and Microcomputers: Principles of Software and Hardware Engineering*. The latter was written to give a general-purpose view of software and hardware engineering in microcontroller and microprocessor systems. It was meant to be accompanied by a technical reference for a particular processor being used in a class and laboratory. *Software and Hardware Engineering: Motorola M68HC12* and its predecessor, *Software and Hardware Engineering: Motorola M68HC11*, filled that role for these two processors. In the ten years since the publication of the first pair, we have found that students and instructors like to have both general principles and specific applications in a single text. That is what we have accomplished with *Software and Hardware Engineering: Assembly and C Programming for the Freescale HCS12 Microcontroller.* The text has many application examples as well as general information to help the student understand what is behind today's microcontroller embedded applications.

As semiconductor technology advanced, Freescale (spun off as a separate company from Motorola in 2004) created a completely new version of the M68HC12 called the HCS12. New features were added to the microcontroller and most internal registers changed. We have chosen to tell you about one member of this family of processors. The MC9S12C32 microcontroller is representative of all the HCS12 microcontrollers. Learning about it will serve you well in using and applying any of the other HCS12 family members.

As the hardware has improved over the years, so has the software development environment. In the M68HC11 and M68HC12 era, software was written in assembly (or C) and downloaded to RAM for testing. The D-Bug12 ROM-based debugging monitor program allowed the student to run the program, set breakpoints, trace and inspect registers and memory to verify the program was running correctly, and debug it if it was not. Sometimes simulators were used but these were limited in how they simulate input and output procedures. Software development tools are now significantly advanced in comparison to these earlier tools. We use the CodeWarrior® Integrated Development Environment. This powerful tool supports both assembly and C programming, has a simulator with some I/O simulation capabilities, and supports the Freescale background debugging mode. In this mode, software can be loaded into the microcontroller's flash memory and debugging operations can be carried out in the background.

Organization and Features

This new book includes many elements of *Microcontrollers and Microcomputers: Principles of Software and Hardware Engineering.* Some material has been added as separate chapters. Because we write so much software for embedded systems in C, a chapter on C programming and features needed for embedded systems is included. We have not included introductory C programming details, assuming the student has had a C or other high-level programming class before studying embedded microcontrollers. However, we have included many C programming examples showing how to replicate assembly language examples in C.

1. **Introduction:** The introduction contains definitions used throughout the book and a brief history of the microprocessor/microcontroller development at Motorola and Freescale.

2. **General Principles of Microcontrollers:** Much of *Microcontrollers and Microcomputers* is included in this chapter to introduce the architecture of a computer system. In addition to the hardware descriptions, the software development process and tools used are described.

3. **Structured Program Design:** The principles of top-down software design covered in Chapter 6 in *Microcontrollers and Microcomputers* are included in this chapter. Students should be exposed to these ideas early, before they start writing programs to learn about the MC9S12C32.

4. **Introduction to the HCS12 Hardware:** The CPU programmer's model and addressing modes are covered in this chapter.

5. **An Assembler Program:** The CodeWarrior assembler syntax and operation is covered. Students programming only in C or using another assembler can skip this chapter.

6. **The Linker:** All modern software development, except for very small applications, should use modular programming techniques. Functional modules are developed separately, assembled or compiled separately, and then linked together to form an executable program. This chapter describes the operation of the CodeWarrior linker that does this.

7. **The HCS12 Instruction Set:** For those of you going to learn assembly language programming, here is your chapter. All CPU instructions are covered with many examples of use.

8. **Assembly Language Programs for the HCS12:** A good way to learn programming techniques is to look at examples. This chapter shows a good structure for creating readable and understandable assembly language programs. It shows the student how to write structured programming assembly language and includes a section of hints for assembly programmers.

9. **Debugging HCS12 Programs:** Almost all of us download our programs, push the run button, and expect the program to work the first time. Almost all of us are disappointed. Programs rarely work the first time. This chapter presents some debugging strategies and techniques and shows how to use the CodeWarrior debugger. It also shows some of the common mistakes beginning assembly language programmers make and gives hints on how to find them. For instructors continuing to use D-Bug12, Chapter 5 from *Software and Hardware Engineering: Motorola M68HC12* will continue to be available on the Instructor's CD.

10. **Program Development Using C:** This chapter shows some of the things that are different about programming in C for an embedded system as compared to a desktop computer application. It is not an introductory chapter to learn C programming. We assume the student has learned C in another programming course.

In each of the following chapters the reader will find many programming examples in assembly language and in C.

11. **HCS12 Parallel I/O:** This chapter starts the description of the hardware features of the HCS12 specializing in the MC9S12C32. The Port Interface Module—PIM—is covered and all digital input and outputs are described.

12. **HCS12 Interrupts:** The general principles of interrupts are covered along with the specifics of interrupts in the HCS12. Hints on how to avoid problems when writing interrupt service routines or interrupt handlers are given.

13. **HCS12 Memories:** The internal memory features of the MC9S12C32 are covered.

14. **HCS12 Timer:** Perhaps the most complex of the I/O features of the HCS12 is the timer. Fortunately, it is virtually identical to the timer in the M68HC12 family. Although not strictly part of the timer subsystem, we include a discussion of the computer operating properly (COP), real-time interrupt (RTI), and the pulse-width modulation (PWM) modules in this chapter.

15. **HCS12 Serial I/O—SCI and SPI:** The asynchronous (SCI) and synchronous (SPI) serial interfaces are described.

16. **HCS12 Serial I/O—MSCAN:** The Freescale implementation of the controller area network (CAN) bus is explained.

17. **HCS12 Analog Input:** The 10-bit analog-to-digital converter is described.

18. **Single-Chip Microcontroller Interfacing Techniques:** A variety of interfaces to the real world are described in this chapter. We have chosen to group all these application examples in this one chapter instead of spreading them around in other chapters to give the student one place to find this useful information. If desired, an instructor can easily tie specific sections of this chapter to the relevant earlier chapters. Included are parallel and serial I/O, switch and display interfaces including an LCD display driver, digital-to-analog output using the Serial Peripheral Interface (SPI), analog electronics for the A/D converter, and a discussion of I/O software timing and handshaking.

19. **HCS12 Fuzzy Logic:** The fuzzy logic instruction set is given with a complete programming example.

20. **Debugging Systems:** Advanced debugging techniques are covered in this chapter. Included is a description of the background debug module (BDM), which in the MC9S12C32 includes an on-chip in-circuit emulator. This allows the user to capture and trace bus activity while the program is running at full speed. Two examples showing how to use this feature are given.

21. **Advanced HCS12 Hardware:** This last chapter covers a variety of topics not covered in previous chapters, including the clock generator circuits.

Appendixes: Material in the appendixes includes a review of binary codes and binary arithmetic, the ASCII character table, a reference to syntax for the CodeWarrior assembler, the complete HCS12 instruction set in an easy to use table, compact reference listings of all I/O registers, and a listing of the include files from the assembler examples in the text.

Chapter Problem Sets: Most chapters have end-of-chapter problems organized into basic, intermediate, and advanced difficulty. Each chapter problem is cross-referenced to ABET accreditation outcomes a–k. Instructors may use student performance on these problems to quantify the students' compliance with the a–k outcomes.

Acknowledgment

Jim Sibigtroth, of Freescale, Inc., was a coauthor of the M68HC12 book. Although Jim was not able to participate as much in this text, his contributions are spread throughout. Jim badgered me into changing to the CodeWarrior development system. For that, I am extremely grateful. It has saved a good two weeks of laboratory time for the students and shows them what a modern development environment is like. When students leave this course they hit the ground running and are able to write complex applications in assembly or C. Jim is the author of Chapter 20, Debugging Systems. His description of the in-circuit emulator shows us not only how to use the trace features to debug a program but also how to program the flash memory on the fly. This very useful technique can be used to program the flash memory for nonvolatile variables normally stored in EEPROM. A thank you also goes to Eduardo Montañez for his patient help with my CodeWarrior questions.

Thanks also go to the team extraordinaire at Oxford University Press who were helpful and encouraging throughout the publication process. Karen Shapiro (Managing Editor) handled all the production and copyediting details with flair, Dawn Stapleton (Associate Editor, Engineering) helped organize reviews and developed the instructor's support materials, and Adriana Hurtado (Editorial Assistant, Engineering) kept the project on schedule. Trent Haywood (Senior Copywriter) and Liz Cosgrove (Art Director) made the cover look great. Danielle Christensen, Engineering Editor shares my love of the whimsy and approved the Douglas Adams quote "Don't Panic" among all the other tasks of bringing this text to completion. Thanks Danielle.

Thanks to Freescale Semiconductor for offering the CodeWarrior™ Development Studio on the in-text CD to all readers.

Finally, the reviewers who commented on this book throughout its development contributed enormously to its ideas and its accuracy. Thanks are particularly due to the students of Jon Bredeson (Texas Tech University) and Tim Mohr (Grove City College) who class-tested early versions of the manuscript and gave feedback.

Massood Atashbar	Western Michigan University
Jon Bredeson	Texas Tech University
Ji Chen	University of Houston
Robert Gao	University of Massachusetts, Amherst
Xiaojun Geng	California State University-Northridge
Sarah Harris	Harvey Mudd College
Russell Kraft	Rensselaer Polytechnic University
Rob Maher	Montana State University
Aleksandar Milenkovic	University of Alabama-Huntsville
Tim Mohr	Grove City College
William Murray	California Polytechnic State University, San Luis Obispo

Karen Panetta	Tufts University
Albert Richardson	California State University, Chico
John Ridgely	California Polytechnic State University, San Luis Obispo
Fernando Rios Gutierrez	University of Minnesota-Duluth
Eric Schwartz	University of Florida
Greg Semeraro	Rochester Institute of Technology
Mukul Shirvalkar	University of Texas at Tyler
Gerald Sobelman	University of Minnesota
David Lowell Smith	Glendale Community College
Jianjian Song	Rose-Hulman Institute of Technology
Girma Tewolde	Kettering University
Nathan VanderHorn	Iowa State University
Fan Yang	University of Mississippi

1 Introduction

1.1 Introduction

This text gives specific information and examples for the Freescale HCS12 microcontroller. Our goal is not simply to repeat the information that is in Freescale's *MC9S12C Family Device User Guide* or the various *Block User Guides* for any of the various components of the HCS12 family. Instead, we want to give you the extra information needed to become proficient at using an HCS12 family device by giving examples to explain the many details found in the *Block User Guides*.

1.2 Some Basic Definitions

Throughout this text we use the following digital logic terminology.

Logic High: The higher of the two voltages defining logic true and logic false. The value of a logic high depends on the logic family. For example, in the high-density complementary metal-oxide semiconductor (HCMOS) family, logic high (at the input of a gate) is signified by a voltage greater than 3.15 volts. This voltage is known as V_{IHMIN}.

Logic Low: The lower of the two voltages defining logic true and false. In HCMOS, a logic low (at the input of a gate, V_{ILMAX}) is signified by a voltage less than 1.35 volts.

Tristate™ or Three-State: A logic signal that can neither source nor sink current. It presents a high impedance load to any other logic device to which it is connected.

Assert: Logic signals, particularly signals that control a part of the system, are *asserted* when the control, or action named by the signal, is being done. A signal may be low or high when it is asserted. For example, the signal WRITE means that it is asserted when the signal is logic high.

Active Low: Used to define a signal whose assertion level is logic low. For example, the signal READ_L is asserted low. Although many data sheets and schematic diagrams make use of an overbar or some other notation, in this text we will denote active-low signals by adding the _L suffix to the signal name.

Active High: Used to define a signal whose assertion level is logic high.

Mixed Polarity Notation: The notation used by most manufacturers of microcomputer components defines a signal by using a name, such as WRITE, to indicate an *action*, and a polarity indicator to show the *assertion level* for the signal. Thus, the signal WRITE indicates that the CPU is doing a write operation when the signal is high. READ_L denotes a read operation is going on when the signal is low.

Logical Complement: The complement of a logical signal is an operator. We will use the overbar to donate the complementation. Thus, $\overline{\text{PUMP\_ON}}$ is the complement of the active high signal PUMP_ON.

RAM: Random access memory. This memory can be read from and written to and is used in the microcontroller for variable data storage. The memory contents are lost when the power is removed. Therefore the memory is said to be volatile.

ROM: Read only memory. The contents of this memory are programmed one time when manufactured and are nonvolatile. That is, the memory contents persist when the power is removed. ROM is used in microcontrollers for program storage.

PROM: Programmable read only memory; memory that can be programmed by the user instead of at the factory as must be done for ROM.

EPROM: Erasable programmable read only memory. First introduced by Intel in 1971, this PROM could be erased by exposing it to ultraviolet (UV) light. These PROMs have a quartz window to allow the UV light into the package.

OTP EPROM: One-time programmable EPROM. This is an EPROM without the quartz window and thus cannot be erased after it is programmed.

EEPROM: Electrically erasable programmable read only memory—pronounced *double e* prom. This is an EPROM that can be erased by an electrical signal, eliminating the need to remove the chip from its circuit and exposing it to UV light as is the case for EPROM.

Flash: EEPROM may be erased and written to one byte at a time. Flash allows data to be erased and written in blocks and is thus faster than EEPROM. Flash is used mostly for program memory and EEPROM for variable data that must be retained when the power is removed. Note that Flash is sometimes called Flash EEPROM.

Byte: A byte is 8 bits.

Nibble: A nibble is 4 bits. There are 2 nibbles for each byte.

Word: A word is 16 bits.

1.3 Computers, Microprocessors, Microcomputers, Microcontrollers

A computer system is shown in Figure 1-1. We see a *CPU* or *central processor unit, memory*, containing the program and data, an *I/O interface* with associated *input and output devices*, and three *buses* connecting the elements of the system together. The organization of the program and data into a single memory block is called a *von Neumann* architecture, after John von Neumann who described this general purpose, stored-program computer in 1945. In Figure 1-1 the data, address, and control buses consist of many wires, for example, 8, 16, 32, or more, that carry binary signals from one place to another in the computer system. The HCS12 microcontroller is a von Neumann architecture machine.

There is another major computer architecture type called the *Harvard* architecture in which two completely separate memories are used—one for the program and one for the data. This architecture is often found in digital signal processing (DSP) chips and some other microcontroller chips such as Microchip Technology PIC microcontrollers.

Figure 1-2 shows a timeline for the development of a few of the computers, minicomputers, microprocessors, microcomputers, and microcontrollers leading to the Freescale HCS12. In the beginning computers were very large, in size if not in capacity and capability. In 1958 the Air Force SAGE computer required 40,000 square feet and 3 megawatts of power and had 30,000 vacuum tubes with a 4K × 32-bit word magnetic core memory. In 1964 the Digital Equipment Corporation's PDP-8 was the first mass-produced minicomputer. This was the start of a trend toward less expensive, smaller computers suitable for use in nontraditional, non-data-processing applications. In 1971 Intel introduced the first microprocessor, the Intel 4004. This was constructed with only four chips consisting of a CPU, 256-byte program memory (ROM), and 40-byte data RAM (it was a Harvard architecture machine) and a shift register chip to do I/O. Intel's great contribution was to integrate the functions of the many element CPU into one (or at most a few) integrated circuits. The term *microprocessor* first came into

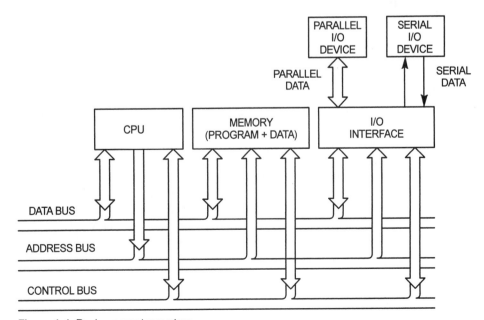

Figure 1-1 Basic computer system.

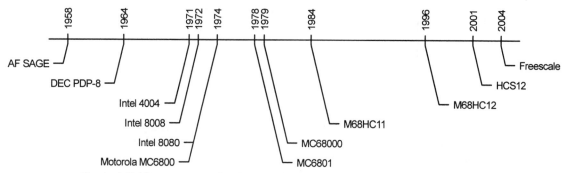

Figure 1-2 Microcomputer development timeline.

A *microcomputer* is a microprocessor with added memory and I/O.

use at Intel in 1972[1] and, generally, refers to the implementation of the central processor unit functions of a computer in a single, large scale integrated (LSI) circuit. A *microcomputer*, then, is a computer built using a microprocessor and a few other components for the memory and I/O.

The Intel 4004 was a 4-bit microprocessor and led the way to the development of the Intel 8008, the first 8-bit microprocessor, which was introduced in 1972. This processor had 45 instructions, had a 30-microsecond average instruction time, and could address 16K bytes of memory. Today, of course, we have advanced far beyond these first microcomputers.

1.4 Motorola/Freescale Developments Leading to the HCS12

Motorola introduced their first 8-bit microprocessor, the MC6800, in 1974.[2] This was the first microprocessor to operate from a single 5-volt power supply. The MC6800 included only the central processing unit (CPU). Even the oscillator, which is used for generating the clock timing to synchronize the CPU operation, was implemented outside the MC6800. A series of peripheral integrated circuits including a 128-byte RAM, a ROM chip, a parallel I/O interface (two 8-bit ports), a programmable timer, and an asynchronous serial I/O interface were combined to form usable systems. About 1978 Motorola introduced the MC6801, which integrated an 8-bit CPU with 128 bytes of RAM, 2K bytes of ROM, a 16-bit timer, and an asynchronous serial communications interface (SCI). This was the first "single-chip microcomputer." The CPU was a source code and object code compatible upgrade from the MC6800, which allowed system developers to easily upgrade to the new chip.

From the MC6801, Motorola went in three new directions. In 1979 one group developed the MC68000 16-bit microprocessor, which has since developed into the current 32-bit ColdFire family. A second group of software experts developed the MC6809. A third group stripped the MC6801 CPU down and added peripheral modules to form the M6805 family of very low cost microcontrollers, which led to the current M68HC08 and HCS08 families. Meanwhile, the 2K ROM in the MC6801 was replaced with 4K of UV erasable EPROM to create a reprogrammable version of the M6801 family called the MC680144.

[1] R. N. Noyce and M. E. Hoff, Jr., "A History of Microprocessor Development at Intel," *IEEE Micro* **1,1**: 8–21 (Feb. 1981).

[2] R. Gary Daniels, "A Participant's Perspective," *IEEE Micro* **16** (6): 21–31, (Dec. 1996).

In 1984 Motorola introduced the CMOS MC68HC11,[3] which was also known to General Motors as the GMSCM (General Motors Single-Chip Microcontroller). The M68HC11 CPU was a compatible upgrade path from the M6800 and M6801. It added more registers, bit manipulation instructions, two 16-bit by 16-bit divide instructions, and many other miscellaneous instructions. In all, more than 90 new instruction opcodes were added to the earlier M6801 instruction set. The M68HC11 also represented a major advance in the level of integration by including an 8-channel A-to-D converter, synchronous (SPI) and asynchronous (SCI) serial I/O systems, more complex timers, and electrically erasable and programmable read only memory (EEPROM). The MC68HC711E9 was the first microcontroller to include both EPROM and EEPROM on the same silicon die.

In 1996 Motorola introduced the M68HC12 to further extend the performance and integration of the M6800–M6801–M6811 lineage. The new silicon processes allowed multiple metal layers so they were able to implement full 16-bit buses throughout the M68HC12. They used Flash memory to replace the EPROM program memory of the M68HC11. This made it possible to erase and reprogram the main program memory without removing the microcontroller from the application system. The M68HC12 instruction set remained compatible with earlier M6800, M6801, and M68HC11 instruction sets while adding memory-to-memory moves and several 16- and 32-bit math operations. This instruction set remains one of the most efficient available for high-level languages such as C.

In 2001 Motorola released the first HCS12 devices. Both the M68HC12 and the HCS12 share the same instruction set and peripheral functions but the HCS12 uses newer 0.25-μm silicon processing and a more automated model-based design methodology. This results in better utilization of the available silicon and gives more "bang for the buck."

The newer 0.25-μm process also improved the speed so the HCS12 can now operate at 25-MHz bus frequency as compared to the 8-MHz frequency in the MC68HC12. A few HCS12 family devices can operate even faster than this. This drive to smaller silicon geometry will continue into the future and experimental microcontrollers similar to the HCS12 already have been built in a 0.18-μm process.

In 2004 Motorola spun off the semiconductor businesses to form Freescale Semiconductor. Although the company name has changed, the same engineers continue to design a full range of microprocessors, microcontrollers, and many analog and sensor devices. The HCS12 family of microcontrollers is the latest incarnation of the architecture that was first introduced in 1974 as the MC6800. At 30+ years, this architecture has been continuously available longer than any other microprocessor architecture.

1.5 Embedded Systems

The HCS12 is used primarily in applications where the system is dedicated to performing a single task or a single group of tasks. These are called *embedded* systems because the microcontroller's presence is transparent (nonobvious) to the user. Examples of embedded applications are found almost everywhere in products such as microwave ovens, toasters, and automobiles. These are often *control* applications and make use of microcontrollers.

---

[3] James M. Sibigtroth, "Motorola's MC68HC11: Definition and Design of a VLSI Microprocessor," *IEEE Micro* **4**(1): 54–65 (Feb. 1984).

TABLE 1-1 Notation

$	Hexadecimal numbers are denoted by a leading $; for example, $FFFF is the hexadecimal number FFFF. When two memory locations are to be identified, the starting and ending addresses are given as $FFFE:FFFF.
%	Binary numbers are denoted by a leading %; for example, $F may be written %1111.
@	A base-8 or octal number is preceded by @. $F = @17. Base 10 is the default base and has no base indicator like hexadecimal, binary, or octal. $F = 15.
0x	In C program examples found in later chapters, the C notation, 0xFFFF, is used for hexadecimal numbers.
0b	In C programs, the 0b prefix is used to signify a binary number.
#	A # indicates immediate addressing mode. Be *very* careful about this because it is *very* easy to forget this symbol when writing assembly language programs.
x	An x indicates a don't care bit—that is, the bit may be zero or one.
*	The * indicates a pointer in a C program.
_L	A signal whose assertion level is low.

A *microcontroller* is a computer with *CPU, memory*, and *I/O* in one integrated circuit chip.

A *microcontroller* is a microcomputer with its memory and I/O integrated into a single chip. The number of microcontrollers used in products is mind boggling. In 1991 the chip manufacturers delivered over 750 million 8-bit microcontrollers[4] and by 1997 Motorola had delivered over 2-billion MC6805 microcontrollers. In 2004 the industry delivered 6.8 billion microcontroller units.[5]

1.6 Notation

Throughout this text, the notation given in Table 1-1 is used.

1.7 Bibliography and Further Reading

Freescale provides a device user guide for the MC9S12C and each subsystem in the basic processor has its own block user guide.

---

[4] *EDN*, Jan. 21, 1993.

[5] http://www.instat.com/press.asp?ID=1445&sku=IN0502457SI.

2 General Principles of Microcontrollers

OBJECTIVES

This chapter introduces the principles of a stored program computer and shows how we develop the software for an embedded microcontroller system. When finished, you should understand the hardware of a typical system. You will see the importance of the instruction fetch, how the sequence controller works, and how to determine system timing. You will understand how memory operates and how it affects the design of the computer. We also consider the software needed and introduce the idea of a tool set to produce the code that ultimately resides in the microcomputer's read only memory.

2.1 Introduction

In this chapter we investigate the operation of a typical microcontroller. Our goal is to have you see that a computer is *not* a mysterious box but rather is a collection of basic digital logic components that you could design. By the end of this chapter you will appreciate that a computer works in a predictable way and that you have complete and absolute control over what it does at all times.

2.2 A Typical Microcontroller

A typical microcontroller is shown in Figure 2-1. It consists of the following elements:

- A *central processor unit* (*CPU*) that contains registers, an arithmetic and logic unit (*ALU*), and a sequence controller to control all activities of the microcontroller.
- *Read only memory* (*ROM*) to hold our program and any constant data. Modern microcontrollers have reprogrammable types of read only memory such as Flash memory, which is a particular type of electrically erasable programmable read only memory (EEPROM).
- *Random access memory* (*RAM*) to store variable data.
- An *input/output* (*I/O*) *interface* to connect the microcontroller to the real world. The I/O interface in most microcontrollers contains other useful functions such as timers, pulse-width modulators, and other special I/O functions.

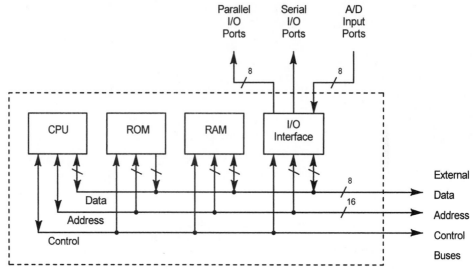

Figure 2-1 Typical microcontroller.

- Connecting these blocks are three buses—the *data bus,* the *address bus,* and the *control bus.* Often these buses are available outside the microcontroller to allow additional memory and I/O to be used.

The Program

Any program in an embedded system, such as the famous C program that prints the message "Hello World!" (Example 2-1), must be in the memory (normally ROM) and looks something like that shown in Table 2-1. This example illustrates that even though you may write a program in a high-level language like C, it is converted to bytes representing the *operation* that must be done and *operands* that are being operated upon. The memory addresses shown correspond to the ROM in the embedded system. Note that all addresses and instruction code bytes are in hexadecimal. The right-hand column shows each instruction in the program in assembly language. Don't worry about what these instructions are at this stage because we will be covering all that in a later chapter.

> A computer *instruction* is an *opcode* plus *operands.*

Look closely at the Instruction Code Bytes in Table 2-1. In each case, the first byte (CF, CC, 16, etc.) is a unique code for each *operation* to be *executed* by the microcontroller. For example, *CF* is a code for the LDS (*immediate load stack pointer register*) operation. This is the *opcode* byte.[1] The following two bytes (0A 00) are the bytes for the *operands* for this operation. A computer *instruction* is the combination of an operation (what the computer is to do) and zero, one, or more operands (what the computer is going to do it to). For this instruction the microcontroller will load, or initialize, the stack pointer register with the value $0A00.

Constant data also may be stored in the ROM. The data for this program is the string "Hello World!" that is to be printed. Data constants may be defined in an assembly

---

[1] Some computer operations may need to be specified by more than one byte.

Example 2-1 Hello World! Program

```
Metrowerks HC12-Assembler
(c) COPYRIGHT METROWERKS 1987-2003
Abs. Loc    Obj. code Source line
--- -----   ------  --------.

    1             ; Example program to print
    2             ; "Hello World"
    3             ; Constant equates
    4   0000  000D CR:        EQU  $0d   ; Carriage return
    5   0000  000A LF:        EQU  $0a   ; Line feed
    6   0000  0000 EOS:       EQU  0     ; End of string
    7             ; Memory map equates
    8   0000  8000 PROG:      EQU  $8000 ; Flash memory
    9   0000  0A00 STACK:     EQU  $0a00 ; Stack pointer
   10                         ORG  PROG  ; Locate program
   11             Entry:
   12             ; Initialize stack pointer
   13 008000 CF0A 00          lds  #STACK
   14             loop:
   15             ; Print Hello World! string
   16 008003 CC80 0B          ldd  #HELLO
   17 008006 1680 18          jsr  printf
   18             ; Do it forever
   19 008009 20F8             bra  loop
   20             ; Define the string to print
   21 00800B 4865 6C6C HELLO: DC.B 'Hello World!',EOS
      00800F 6F20 576F
      008013 726C 6421
      008017 00
```

TABLE 2-1 Contents of Memory for Hello World!

Memory Addresses	Instruction Code Bytes	Instruction
8000 – 8002	CF 0A 00	LDS #$0A00
8003 – 8005	CC 80 0B	LDD #$800B
8006 – 8008	16 80 18	JSR $8018
8009 – 800A	20 F8	BRA $8003
800B – 8017	48 65 6C 6C 6F 20	Constant data for
	57 6F	the Hello
	72 6C 64 21 00	World! string

Example 2-2

For each of the instructions in Table 2-1, give the memory locations and the hexadecimal value for each opcode.

Solution:

Memory	Op Code Byte
8000	CF
8003	CC
8006	16
8009	20

Example 2-3

For each of the instructions in Table 2-1, give the memory locations and the hexadecimal value for each operand code.

Solution:

Memory	Operand Code Bytes
8001, 8002	0A, 00
0804, 0805	80, 0B
0807, 0808	80, 18
800A	F8

Example 2-4

For each of the instructions in Table 2-1, give the memory locations and the hexadecimal value for each constant data byte.

Solution:

Memory	Constant Data Byte
800B	48
800C	65
800D	6C
800E	6C
800F	6F
8010	20
8011	57
8012	6F
8013	72
8014	6C
8015	64
8016	21
8017	00

language program like you have done in other programming languages. Line 21 in Example 2-1 shows how this is done in the assembly program. In Table 2-1 the memory locations $800B–$8017 contain the *constant* data used by the program. These bytes are the ASCII codes for the characters in the message that will be printed on the screen by a *printf* routine jumped to, or called in line 17.

Memory

There are two types of memory shown in Figure 2-1. The read only memory, ROM, is memory whose contents are not lost when the power is removed from the microcontroller. This is *nonvolatile* memory. There are a variety of types of ROM including factory programmed (normally just called ROM) and field programmable ROM, or PROM. PROM comes in three varieties including UV erasable EPROM, electrically erasable EEPROM, and Flash (sometimes called Flash EEPROM). EEPROM and Flash are very similar in that the memory can be erased and then reprogrammed without removing it from the circuit to place in a UV PROM eraser. They are different in that EEPROM can be erased and programmed a byte at a time. It is used to store program variables that must remain after the power has been cycled off and then on. Flash EEPROM uses similar integrated circuit technology but is organized so that it can be erased and then programmed in large blocks. This makes it faster to erase and program and so Flash is used to store our programs.

The RAM is *random access memory*. This terminology is somewhat misleading because ROM can be randomly accessed too. RAM is memory that can be read from and written to, making it useful for program variable data. Most RAM is volatile and so the program must initialize it after the power has been turned on.

The memory organization in a computer is shown using a *memory map*, which shows what memory addresses are used for what type of memory. A typical microcontroller's memory map may contain RAM, EEPROM, Flash, and even spaces without memory as shown in Figure 2-2.

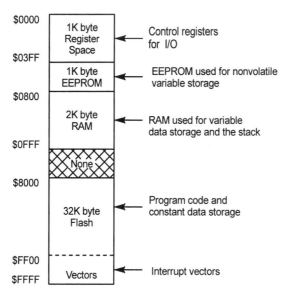

Figure 2-2 A microcontroller's memory map.

TABLE 2-2 Memory Contents

Memory Address	Memory Contents (Binary)	Memory Contents (Hexadecimal)
8000:	1100 1111	CF
8001:	0000 1010	0A
8002:	0000 0000	00
8003:	1100 1100	CC
8004:	1000 0000	80
8005:	0000 1011	0B
. . .		. . .
8016:	0010 0001	21
8017:	0000 0000	00

ROM OPERATION

ROM is *nonvolatile* and is used for program code and constants.

The ROM contains the opcode, operand, and constant data bytes. We will not worry about how they get there at present, leaving that discussion to a later chapter. The microcontroller CPU needs to *fetch,* or read, the opcode and operand bytes from the ROM to be able to execute the program. To see how this is done, let us first consider how the ROM, and for that matter the RAM, works.

We can visualize the ROM as a sequence of storage locations, each containing a byte of information as shown in Table 2-2.[2] Figure 2-3 shows a ROM. The bus labeled *address bus* consists of *address* bits coming from the CPU. The number of bits depends on the number of memory storage locations. For example, a 64K-byte memory (65,536 storage locations) must have 16 address bits to specify each location uniquely. The CPU provides this address, as we will describe later. The line labeled READ_L is a *read control* signal, and it is also asserted by the CPU. When the memory receives an address and READ_L is asserted, the ROM places the data byte stored in that address onto the *data bus*.

RAM OPERATION

RAM is *volatile* and is used for variable data storage and the stack.

Figure 2-4 shows us a picture of RAM, and it is very similar to the ROM shown in Figure 2-3. Unlike ROM, the data connection to a RAM is bidirectional so data from the data bus can be written into the selected address in RAM in response to the new WRITE signal. The most notable difference between ROM and RAM, however, is

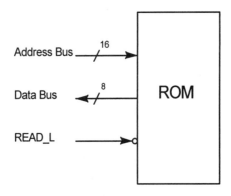

Figure 2-3 Read only memory.

---

[2] This is a *byte-wide* memory. Some applications may have 16 or more bits per location.

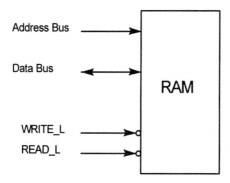

Figure 2-4 Random access memory.

that the data in RAM is *volatile*. If the power to the RAM is turned off at any time, the data are lost. There are two uses for RAM in our embedded microcontroller systems. The first is for variable data storage. For example, if we were to read a value from an analog-to-digital converter and wanted to save it for later reference, we would write that value into the RAM. The second use is for the stack. The *stack* is an area of RAM that is used for temporary variable data storage and for return addresses for subroutines. We will learn much more about the stack in later chapters.

The Central Processor Unit

The *central processor unit* is the heart and lungs of the microcontroller.

Figure 2-5 shows a representation of a central processor unit, or CPU. Although the operation of the HCS12 is somewhat more involved than this general description of CPU operation, it is sufficient to let us see, in principle, how a typical microcontroller CPU functions.

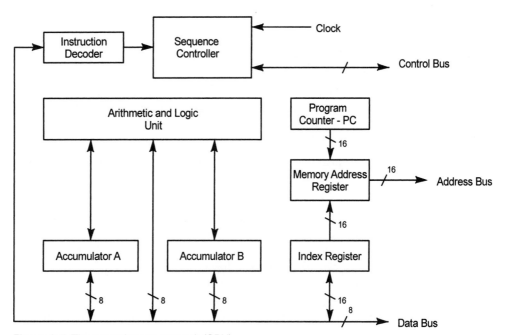

Figure 2-5 The central processor unit (CPU).

THE CPU REGISTERS

The *programmer's model* is the set of registers in the CPU that you will use in your programs.

The CPU contains registers that you will use extensively in your assembly language programs. In general, there are accumulator registers and registers used to access memory.

Accumulators A and B: The two 8-bit accumulators, *A* and *B*, may be a source or destination operand for 8-bit instructions. For example, if we were to retrieve a byte of data from memory we may put it into accumulator A or B. The registers are called accumulators because the results of an arithmetic or logic operation may accumulate there.

Program Counter: Although the program counter is usually shown in the programmer's model, the programmer does not have direct control over it like the other registers. The number of bits in the program counter shows how much memory can be directly addressed. In this example, a 16-bit address bus is needed for a 64K-byte memory.

Index Register: The program may use an *index* register to access memory. As we will see when we find out more about the HCS12 microcontroller, there are a variety of instructions that access memory using this type of register.

THE INSTRUCTION EXECUTION CYCLE

The process by which the microcontroller executes each instruction in our program is called the *instruction execution cycle.* When an instruction opcode is to be fetched from ROM, the memory must be supplied with the address of the opcode and the read control signal asserted.

Each computer instruction is completed during the *instruction execution cycle*, which consists of one or more steps.

This is how the *instruction cycle* starts and it continues by fetching the rest of the instruction bytes, doing whatever is required by the instruction, incrementing the program counter to point to the next opcode, and then repeats. We can describe the full instruction execution cycle in the following way (refer to Figure 2-5):

- The CPU's *program counter* contains the address of the first byte of the instruction to be executed. We say the program counter *points* to the opcode. The CPU places that address on the address bus.
- The *sequence controller* asserts the *Read* control signal on the *control bus.*
- After a small delay, called the *memory access time*, the ROM places the contents of the addressed memory location on the *data bus.*
- The sequence controller writes this byte into the *instruction decoder.*
- The *instruction decoder* holds the opcode byte and decodes it for the sequence controller.
- The decoded instruction causes the sequence controller to go through a sequence of actions that complete the execution of the instruction. These include fetching operands from memory, loading registers, performing an arithmetic or logical operation on a pair of operands, and incrementing the program counter.
- When the instruction execution is complete, the program counter is pointing to the next opcode to be fetched and executed. The instruction execution cycle then repeats.

A microcontroller is *always* fetching and executing instructions.

The instruction execution cycle continues forever, or at least until the power is turned off or a special instruction that stops it is encountered. Remember, while power is turned on, the microcontroller is *always* fetching and executing instructions.

THE SEQUENCE CONTROLLER

The sequence controller and instruction decoder shown in Figure 2-5 are, in combination, a sequential state machine, much like you may have designed in an earlier digital logic course. It is the "brains" of the microcontroller. All computers have some kind of sequence controller. There are several ways to design one, but the basic function and purpose remain the same. The sequence controller generates control signals required by the currently executing instruction, at the correct time, to accomplish the information transfer or other operation.

The sequence controller is designed to allow different instructions to be executed in different amounts of time. Take, for example, an 8-bit immediate addressing instruction that transfers a byte from the memory immediately following the opcode byte to some register in the CPU. In the Freescale HCS12 instruction set this might be the

```
ldaa    #$22
```

instruction that loads accumulator A with the data $22. Figure 2-6(a) shows how the opcode and the data are located in the memory. At the start of the instruction execution cycle, the program counter is pointing at the opcode, and the execution cycle starts by fetching the opcode.

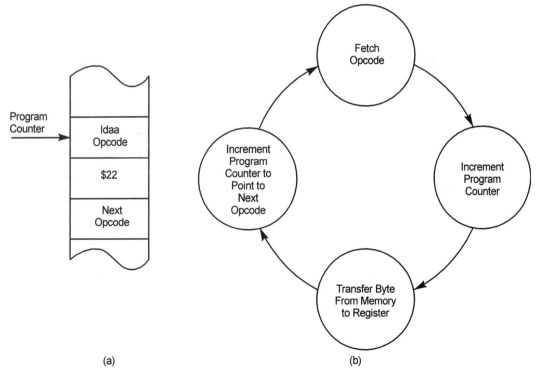

(a)   (b)

Figure 2-6 The `ldaa #$22` instruction: (a) contents of memory and (b) sequence of states.

Figure 2-6(b) shows a sequential state transition diagram illustrating that four states are needed to accomplish this instruction.

Another instruction, which loads accumulator A with data from somewhere else in memory, is

`ldaa     $1234`

This loads accumulator A with a byte from memory location $1234. The memory map for this instruction is given by Figure 2-7(a) and the sequential state transition diagram in Figure 2-7(b) shows that seven states are needed.

These explanations of the immediate and extended addressing instructions are not completely true for the HCS12 microcontroller, although the general principles apply. In the HCS12 an instruction pipeline, or cache memory, is kept filled with instruction opcode and operand bytes by a mechanism that accesses memory while the CPU is doing other operations. This results in fewer CPU clock cycles needed to execute each instruction, depending on what is in the pipeline. Chapter 20 explains this hardware architecture in more detail.

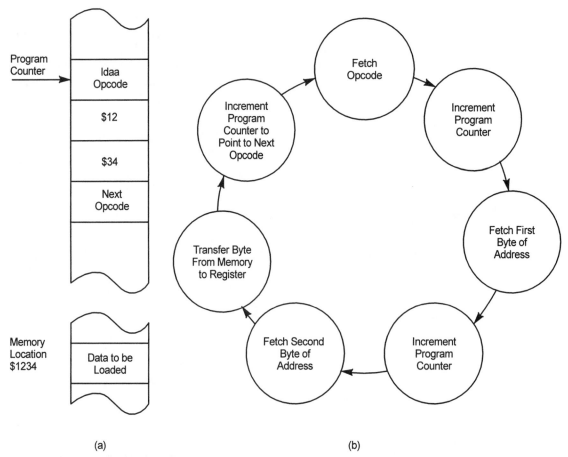

(a)   (b)

Figure 2-7 The `ldaa $1234` instruction: (a) opcode and data in memory and (b) instruction execution sequence.

ARITHMETIC AND LOGIC UNIT (ALU)

If the sequence controller is the brain of the microcontroller, the ALU in Figure 2-5 does the work. It contains the digital logic to operate on the operands in the way specified by the opcode. It does arithmetic (add, subtract, etc.), logic (and, or, etc.), and other operations such as shifts, rotates, increments, and decrements. As you can see in Figure 2-5, the ALU receives its inputs from, and places it outputs to, accumulator registers and the data bus.

The I/O Interface

The I/O interface shown in Figure 2-8 has two components—one to *input* data into the micro-controller and one to *output* data from it.

The input interface connects an input device, such as a bank of switches, to the data bus through a set of three-state gates. The input three-state gates are activated when the *address* of the input device is placed on the address bus, generating an ADR_OK_L signal, and the READ_L control signal is asserted.

The I/O interface controls the input and output of data.

The output interface consists of a set of latches to capture data from the data bus. Like the input interface, the correct address on the address bus asserts the address decoder output. The CPU then asserts the WRITE_L control signal to latch the data.

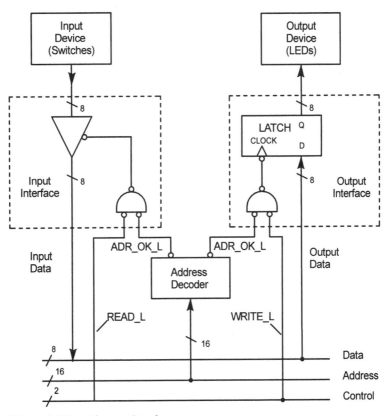

Figure 2-8 Input/output interface.

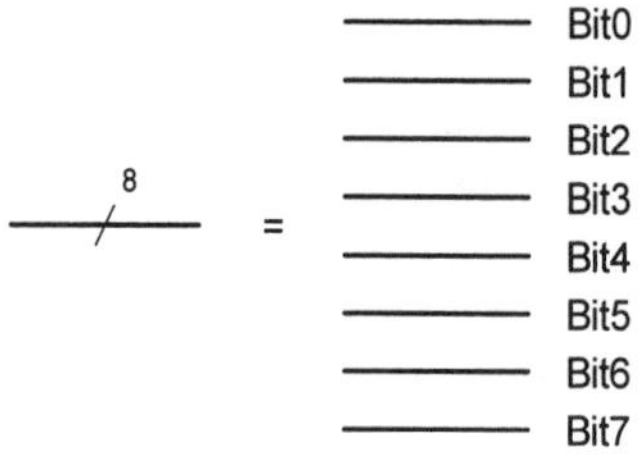

Figure 2-9 Computer bus notation.

The Address, Data, and Control Buses

Figure 2-1 shows three structures, called buses, connecting the CPU, ROM, RAM, and I/O interface together. A bus can be defined as follows:

> *A bus is a multiple wire, information pathway with multiple sources and destinations for the information.*

A source places information onto a bus and a destination takes information from it. Although the bus has many wires, it is normally drawn on schematic diagrams as one wire with an indicator showing how many wires are used. See Figure 2-9.

Address Bus: The address bus carries the address from the CPU to the ROM, RAM, or I/O interface to select one particular byte location in the 2^{16} locations in the memory map.

Data Bus: The data bus carries information to and from the CPU and the ROM, RAM, and I/O interface.

Control Bus: The control bus has a variable number of wires depending on the particular system. At a minimum, at least for this example, it contains the READ_L and WRITE_L memory read and write control signals. The control signals provide direction information (reading or writing) and control the timing of the data transfer as described in the next section.

Timing

We mentioned in our discussion on the operation of the ROM that there is a short delay when reading the data out of the ROM. Also, we see in Figures 2-6 and 2-7 that there is a sequence of steps taken when fetching and executing an instruction. The clock we see in Figure 2-5 controls the timing of all this.

> Data must be taken from the bus or placed onto the bus at the correct time. The CPU controls this timing.

Two fundamental processes of the microcontroller when executing programs are the writing to and reading from the data bus. These operations are called the *read cycle* and the *write cycle* and they are used for both memory and I/O access.

WRITE CYCLE

The CPU is the bus master and controls all information transfer timing. Consider transferring data from a CPU register to an output data latch. The CPU's timing is controlled by its clock, and this output operation is called a *write cycle*. Figure 2-10 shows a typical CPU write cycle.

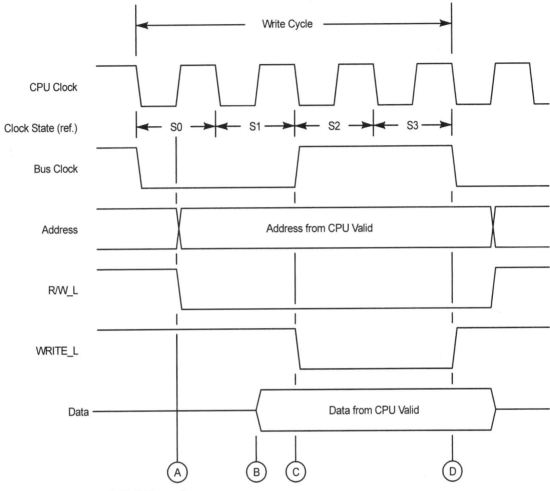

Figure 2-10 Write cycle.

The CPU places the address on the address bus at point A. The data bits are supplied at point B, and the WRITE_L control signal is asserted low a short time later at point C, when the read/write (R/W) signal from the CPU is low and the bus clock is high. This signal is used by the output device interface (or RAM) to latch the data at the correct time (after data is stable on the data bus). The data may be captured by the output latch or the memory on the falling edge (C) or rising edge (D) of WRITE_L, depending on the type of latch.

READ CYCLE

Transferring information from an external source or from ROM or RAM to the CPU is called a *read cycle*. A typical CPU read cycle is shown in Figure 2-11. Again, an address is supplied by the CPU at point A. The READ_L control signal is asserted at B (when the read/write signal from the CPU is high and the bus clock is high) to enable the input interface three-state gates. The input data becomes valid to the CPU a short time later at point C. The CPU actually

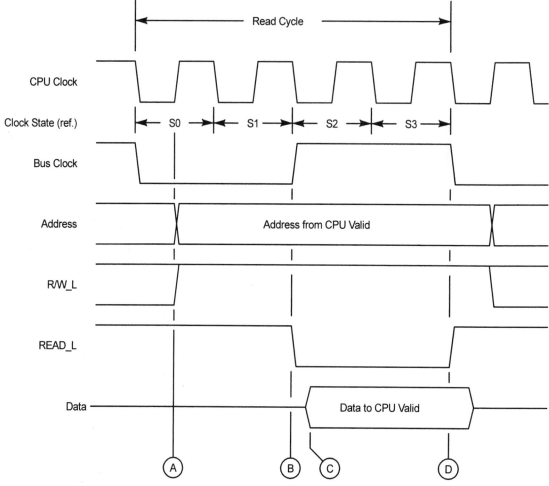

Figure 2-11 Read cycle.

latches (reads) this data at the falling edge of the bus clock at D. An important point to mention at this time is that the CPU reads the data bus at this time whether or not the input device has it ready. If it is not, we need some form of I/O synchronization or some way to extend the CPU read cycle.

INSTRUCTION TIMING

The time taken to execute an instruction depends on the number of read and write cycles needed and the time taken to complete the operation. All instructions need at least one read cycle to fetch the opcode byte.[3] The microcontroller data sheets give the number of clock cycles each instruction takes to execute. See Example 2-5.

---

[3] Some CPUs use a pipeline architecture that allows some of the read and write cycles to be done while other actions are going on. For example, while the CPU may be completing some internal operation, such as an ALU operation, the fetch of the next instruction could be taking place. This, of course, speeds up the overall operation of the processor, which is a good thing.

Example 2-5

For each of the instructions below, describe the instruction execution cycle in terms of the read and write cycles needed assuming each read or write cycle operates on one byte assuming there is no pipeline for instruction bytes.

```
LDAA    #$12      ; Load the 8-bit A register with the data $12
STD     $1234     ; Store the 16-bit D register in memory location
                    $1234
```

Solution:

LDAA#$12   First read cycle to fetch the opcode

Second read cycle to read the data ($12) from the next memory location

Two memory cycles total.

STD $1234   First read cycle to fetch the opcode

Second read cycle to fetch the high byte of the memory address ($12)

Third read cycle to fetch the low byte of the memory address ($34)

First write cycle to write the first byte of data into memory address $1234

Second write cycle to write the second byte of the data into memory address $1235

Five memory cycles total.

2.3 Software/Firmware Development

The software developed for embedded systems is often called *firmware* because, unlike programs you might have written for your computer science classes that are loaded into RAM on a PC or other desktop system, an embedded system requires its program to be in *read only memory*. Thus, the "software" is more "firm" because it is retained in the computer memory even while the power is removed from the system.[4] As "software" developers we must know something about the hardware upon which our software is installed. This hardware is called the *target system* and we must know the addresses used for the various kinds of memory in the system. Table 2-3 shows a typical *memory map* with both random access memory (RAM) and read only memory (ROM). The ROM may be of several types including *programmable ROM (PROM)* such as *Flash electrically erasable PROM*. Flash is used in many microcontrollers and allows us to create our software using a development tool set as described later and then to convert it to firmware by programming the EEPROM. Our program development process must take into account the *location* of each type of memory and must *locate* the various parts of the program correctly.

Table 2-4 shows where the various parts of our program must be located to work in an embedded system. Whether we are writing in assembly language or a high-level language such as C, our software development tools must allow us to control this code location process.

> Embedded system software is called *firmware* because it is in ROM and is not so easily changed as programs in RAM.

---

[4] An unknown author, critical of many computer programs being written, once referred to these programs as *mushware*!

TABLE 2-3 Memory Map with Both RAM and ROM

$0000	
	I/O interface registers
$01FF	
$0200	
	None
$07FF	
$0800	
	2K bytes RAM
$0FFF	
$8000	
	32K bytes Flash
$FFFF	

TABLE 2-4 RAM and ROM Used in an Embedded Application

Memory Type	Program Use
Flash	1. All program code.
	2. Constants such as messages and lookup tables.
	3. Any other information that does not change.
RAM	1. Program variables and data.
	2. Stack data storage.

2.4 The Software Development Tool Set

A software development tool set, such as CodeWarrior®, includes a variety of tools to allow you to develop, and sometimes debug, the program for your embedded system. In the following chapters we will cover some of these tools in detail, but here we discuss how to generate and then locate the code in the appropriate memory.

In an embedded application, code is located in ROM and variable data in RAM.

The code location question is tied to the hardware's memory map. Various parts of the program must be allocated to the two different kinds of memory. Table 2-4 shows how to locate different parts of the program in an embedded, ROM-based system where the program must exist in the computer after the power has been turned off and then on. You will be writing your programs in assembly language or a high-level language such as C, or maybe even both. An *assembler* (program) converts the assembly language application program to the opcode and operand bytes in the embedded system's memory. The C program is similarly converted by a *compiler*, usually to an intermediate file called an *object* file. In Chapter 5 we discuss in detail the

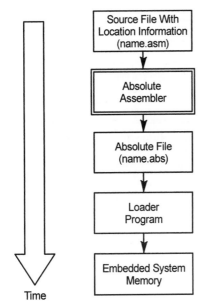

Time

Figure 2-12 Absolute assembler operation.

operation of the assembler included in the CodeWarrior tool set. Chapter 10 discusses C programming of our microcontrollers. Let us now consider two types of assemblers and how the code and data are located using each.

Absolute Assemblers

When writing programs, we know from our hardware design where the code is to be located in memory. A special directive called *ORG* provides this information to the assembler. All code is located, *absolutely* at a *specific* memory address, from this information. This is the simplest form of assembler. It takes the source code file and produces an executable file that is transferred (called *downloading*) to the target system. Figure 2-12 shows an absolute assembler in use.

Downloading transfers an executable file from the computer that created it to the computer that executes it.

A major disadvantage of the absolute assembler is that the source file must contain *all* of the source code to be in the program. This means that when large programs are being written, all code must be assembled whenever any change is made. Furthermore, the project cannot be split easily into elements that can be written and debugged by different project engineers.

Relocatable Assemblers

A *relocatable* assembler can overcome the disadvantages of the absolute assembler. As shown in Figure 2-13, the assembler accepts a program, or a program segment, as a source file. The source file does not need to be the complete program, nor does it need to contain location information or ORG directives. The assembler produces an output file, called the *object* file, which contains the binary codes for the operations and as many operands as the assembler can evaluate. When an operand, such as a branch address, cannot be evaluated, the assembler adds this fact (that an address needs to be resolved) to the object file so a *linker* program can provide the final addresses. Notice that the program can be split into multiple source files and assembled at different times.

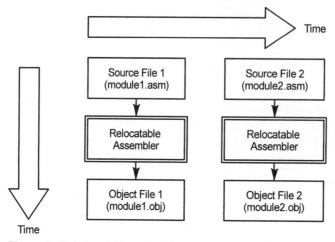

Figure 2-13 Relocatable assembler.

Compilers

Compilers allow us to write our programs in high-level languages much more efficiently than in assembly languages. One high-level C program statement can replace 10 or more assembly language program steps. Nevertheless, in the final analysis, the microcontroller's memory must have the operation code and operand bytes as we described in the previous sections. To accomplish this, the compiler *compiles* the source program, often to an intermediate assembly language program, which is then assembled into an object file. The assembler to do this may be hidden within the compiler or may be a separate program. See Figure 2-14. Some compilers, such as the CodeWarrior compiler, can produce a listing of the assembly language code they produce. This can be very useful during the program debugging stages.

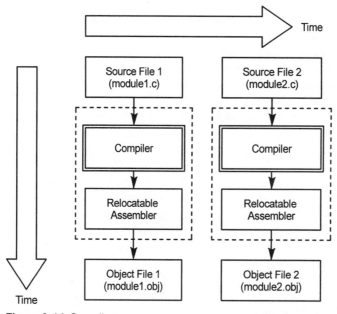

Figure 2-14 Compiler.

The Linker

A *linker* program takes object modules that have been assembled by a relocatable assembler or a compiler, *links* them together, and *locates* all addresses. Figure 2-15 shows two source files, module1.asm and module2.c, which are separately assembled and compiled by a relocatable assembler and C compiler. The linker combines the object files to produce the executable file. You can see in Figure 2-15 that the location information for the code and data parts of the program is given to the linker by a *linker parameter* file (*.prm*). Figure 2-15 also shows that object files can be linked from a library.

Creating a Relocatable Program

The beauty of creating firmware using the relocatable method is that the project can be partitioned using top-down design techniques and allocated to separate programmers. Each programmer is responsible for developing *modules* that ultimately fit into the whole program. The modules are assembled separately by the relocatable assembler to produce object files. In addition, C program modules may be compiled. These object modules are put together by the linker, and any addresses or operands that the assembler or compiler was not able to create are generated at this time. Notice in Figure 2-15 that previously assembled and compiled object modules may be maintained in a library. A *librarian* program is included in the tool set to manage the libraries used by various projects.

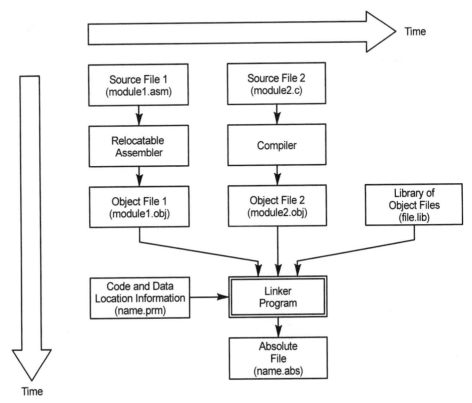

Figure 2-15 Linker program.

2.5 Remaining Questions

In this chapter we have covered a lot of ground to explain some of the basic principles of the operation of a microcontroller. There are still many questions and topics remaining for further discussion. These include the following topics:

- How does the microcontroller start executing a program when it is first turned on? *When the starting address of your program is known, that address is placed into a special place in ROM. When the microcontroller is powered up, or the reset signal asserted, it goes to that special address to find the starting address of your program.*

- How do interrupts work? *That is a big question! We will defer answering it to Chapter 12.*

- How does the program get into the read only memory? *There are a variety of read only memory types and we will cover those in Chapter 13. A very common type is EEPROM. Most microcontrollers have some sort of interface that allows you to program this kind of ROM.*

- What kind of I/O features do microcontrollers have? *There are a wide variety of microcontrollers and families of microcontrollers, each with different features and capabilities. In this book we will cover the Freescale HCS12 family and in addition to the parallel I/O capabilities we will cover the timer, serial I/O, analog input, and background debugging capabilities.*

- How do I learn the assembly language? *Stand by for Chapters 5 and 7.*

2.6 Conclusion and Chapter Summary Points

In this chapter we discussed, from the central processor unit's point of view, how a microcontroller works. Our goal is for you to see that it is not a mysterious beast at all, but one whose basic operation is understandable.

- The microcontroller has RAM, ROM, I/O interfaces, and data, address, and control buses within a single integrated circuit.
- Embedded system programs are in the ROM.
- Types of ROM include electrically erasable programmable ROM (EEPROM) and Flash EEPROM.
- RAM is used for variable data storage and the stack.
- An instruction is an operation plus zero, one, or more operands.
- To retrieve or read data from ROM or RAM, you must supply it with the address and a Read control signal.
- To write data into RAM you must supply it with the address, the data, and a Write control signal.
- The instruction execution cycle repeats continuously.
- An input interface is a set of three-state gates connecting an input device to the data bus.
- An output interface is a set of latches into which data from the data bus is latched during an output operation.
- The CPU controls the timing of all read and write operations.
- Embedded system software is often called "firmware."
- A relocatable assembler allows you to develop the application software in modules that are linked together to create the program to go into the ROM.

2.7 Bibliography and Further Reading

Cady, F. M., *Microcontrollers and Microprocessors*, Oxford University Press, New York, 1997.

2.8 Problems

Basic

2.1 What is the difference between an assembler and a compiler? **[a, c]**

2.2 What is the advantage of a relocatable assembler compared to an absolute assembler? **[a]**

2.3 What is a microcontroller memory map? **[a]**

2.4 What does a sequence controller do? **[a]**

2.5 Give short answers to the following: **[a,g]**

   a. What is a data bus?

   b. Why is an address decoder used in I/O interfaces?

   c. How is an information source, such as a set of switches, interfaced to a data bus?

   d. What control signals are needed to latch data from the data bus into an output interface at the correct time?

   e. Give the sequence of events that occur when a CPU does an input (or read) cycle.

Intermediate

2.6 Discuss the difference between an absolute and a relocatable assembler. **[a, k]**

2.7 How do most microcomputer systems solve the problem of multiple sources of information present on a data bus? **[g]**

Advanced

2.8 Why must a tristate gate be used to interface an input device to the data bus? **[a, c]**

2.9 Why must a latch be used to interface an output device to the data bus? **[a, c]**

2.10 For a CPU performing a write cycle, why does the CPU place the data on the data bus before asserting the WR_L control signal? **[a]**

2.11 A microcontroller memory map shows 16K bytes of Flash EEPROM (ROM) in memory space $C000–$FFFF and 1K bytes of RAM in memory space $1000–$13FF. **[c, k]**

   a. Give a range of addresses (in hex) suitable for locating code.

   b. Give a range of addresses (in hex) suitable for allocating variable data storage.

2.12 Draw a timing diagram relative to the CPU clock shown in the figure, which includes the address and data buses, R/W_L and the write control signal (WR_L = active low) and which shows a write cycle. **[a]**

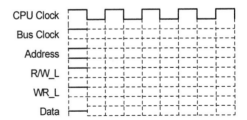

2.13 Draw a timing diagram relative to the system CPU clock shown in the figure, which includes the address and data buses, R/W_L, and the read control signal (RD_L = active low) and which shows a read cycle. **[a]**

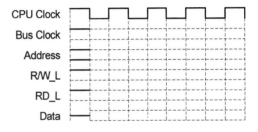

3 Structured Program Design

OBJECTIVES

In this chapter we present a design procedure, called top-down design, suitable for both hardware and software projects. You will learn to use tools to design programs following the top-down design procedure and the principles of structured programming.

3.1 The Need for Software Design

In the design and development of many systems, the cost of producing software is higher, often much higher, than the cost of the hardware. Frederick Brooks, in *The Mythical Man-Month*,[1] compares large-system programming that does not use good design techniques with the tar pits that swallowed saber-toothed tigers, dinosaurs, and mammoths. Few of these systems meet their goals in terms of schedules and costs. Designing the software before writing the code is vital to control costs and to meet requirements and schedules.

Software design means *designing* the software *before* writing the code. When beginning your studies of any processor, or any programming language, designing before writing is difficult. You are wrapped up in just learning the details of the processor and its instruction set or the syntax of the programming language. Soon, however, the problems get more complicated and, with your newfound mastery of the language, your efforts should be spent in designing the solution to the problem instead of just programming the solution.

In this chapter we assume you are about to learn the instruction set of a microcontroller and the operation of the assembler or high-level language compiler. To prepare for this task we would like you to learn how to design software properly instead of just writing it. We will look at various design philosophies and at tools used to design software.

3.2 The Software Development Process

There are several steps to take when developing fully designed, coded, debugged, and documented software for any real system. These are (1) design, (2) coding of modules, (3) testing and debugging of modules, (4) system testing and verification, and (5) documentation.

---

[1] Frederick P. Brooks, Jr. *The Mythical Man-Month*, Addison-Wesley Publishing, Reading, MA, 1982.

Design: The design for any complex system might well take 50% or more of the total effort required for a project. In the sections that follow we will distinguish between design methodologies and design tools. A design methodology is a philosophy of how we do design. Design tools are the mechanics used to do the design. The goal of the design phase is to understand completely the problem and to propose a solution broken down into modules or functional elements that can be coded, tested, and documented.

All software development starts with a *design* phase.

Coding: Coding means writing the program in the chosen programming language. We would hope to use a high-level language for most of the code, but often, especially in time-critical applications, we must use assembly language.

Module Testing: If the design is done properly, we will have coded modules that can be tested and prove to work correctly. The testing and debugging tools used depend on how we have done the coding. Fortunately, many high-level languages have very powerful debuggers that allow us to test and debug our software.

System Testing: This step follows subsystem or module testing and is necessary to prove that the software and hardware works as a whole.

Documentation: Although mentioned last in the list of steps, *each step of software development is accompanied by documentation*. The design documentation specifies what the system is to do and how it is implemented; typically, this work will form the basis of user manuals. Documentation effort is never wasted. Documentation begins in the design step and various types of design documentation are discussed in later sections of this chapter. The documentation produced in the coding phase is the code itself and includes features of the design as comments. Code testing phases are documented with test plans and results. These become templates to show that the system meets the specifications and allow future modifications of the software to be tested to the same standard. Documentation efforts also include the installation and user manuals.

Documentation is so important that it accompanies each step in the process.

3.3 Top-Down Design

A design methodology is a stepwise procedure for doing the design. We contrast this with design tools, which are the mechanical things we use (e.g., pseudocode or flow charts) to produce the design. The top-down design (TDD) method is the design procedure of choice. If we follow the steps given next, we can almost assure ourselves that we will have a good design in the end.

Understand the Problem Completely

This is the first principle of TDD. Unfortunately, many programmers violate this right away because it is so much fun to program that they start before they fully understand the problem. For example, consider designing the hardware and software for a digital voltmeter. Questions that should be asked (and answered) before proceeding with the design might include the following: What is the range of input voltages? What is the resolution needed for the display? How are the analog voltages coded? How does the analog-to-digital converter work?

Understand what is required of the system before starting to program.

Understanding the problem means we must specify exactly what the software is required to do. It is not necessary to understand (at least in the initial stages of the design) how elements of the proposed solution work in detail. For example, when designing the digital voltmeter, we do not need to know how the output display works. We just have to know what we need for the output.

A student recently suggested that we call this part of the design process "outside-in design" to emphasize that the specifications for the software often come from an outside customer. The specifications must be written so that both the end user and the engineer of the system know exactly what the system is to be.

> The *requirements specifications* tell exactly what the program is supposed to do.

A document that is produced during this phase of the design is called a *require-ments specification*. This bit of jargon simply means that you specify (write it down) what the system is required to do. We are not specifying *how* something is to be done, just *what* is to be done.

The design process should consider potential error conditions and allow for them in the rest of the design. Often when customers supply specifications, they fail to consider all error conditions. You should make it your responsibility to think about errors and error handling requirements.

A statement that summarizes this first principle of top-down design is: "Think first, program later."

Design in Levels

After specifying the requirements, it is time to start designing a system to meet them. This is the "how" part of the design process. It is natural to feel overwhelmed by the complexity of the problem. Often one cannot see a way to the end. Do not worry. The design procedures will help us through to the end.

> Upper levels of the design are more general; lower levels are more detailed.

Designing in levels means that we recognize that the whole solution to the problem cannot be seen at once. We start at an upper level and propose a solution to the problem. As we learn more about the problem and how to solve it, levels that are more detailed can be added to the design. When designing in levels, a tree structure as shown in Figure 3-1 is developed to represent the design. The upper levels of the tree are more general statements of the problem solution, and, as one progresses down the tree, more detailed information is shown.

Let us look at an example. Consider designing the software for the digital voltmeter. The requirements are the following:

The input voltage ranges from 0 to 5 volts.

Use an analog-to-digital converter producing an 8-bit unsigned binary code.

Display the voltage on a two-digit, seven-segment LED display to a resolution of 0.1 volt.

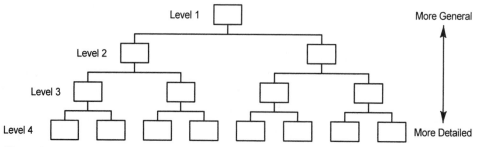

Figure 3-1 Tree structure that results from designing in levels.

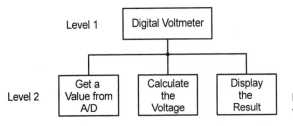

Level 1

Level 2

Figure 3-2 Two-level design for digital voltmeter.

We will not complete this design to the final level of details needed in a real-world project. Our goal is to show how to start a top-down design. The first two levels of the design are shown in Figure 3-2.

The top level is a simple statement of the problem, with the next level providing some details of how that top block is to be done. This level starts to focus our thoughts as we consider what should be done to program the digital voltmeter. The design may not be correct or complete at this stage, but it is at least a start, and starting is often the hardest part of any project. Notice that the blocks in level 2 are algorithmic. That is, if we read them from left to right, they describe a sequence of things done to input the voltage and display the result.

Ensure Correctness at Each Level

The design started in Figure 3-2 is not necessarily correct or complete after our first pass. Before going on to lower levels, we should make sure the algorithm is correct at this level. In going back over the design, try to think if anything else should be done. For example, we might remember that we need to initialize some of the I/O devices in the system. It is easy at this stage to add another block to the design as shown in Figure 3-3.

Postpone Details

There will be unknown and unresolved details at all upper levels of the design. Postpone thinking about the details until you reach the lower levels later in the design process. For example, when working at level 2 of the digital voltmeter, we do not need to know the details

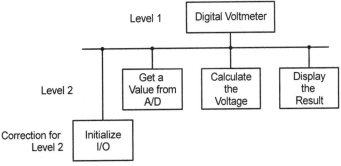

Figure 3-3 A more correct design for level 2.

TABLE 3-1 Pseudocode Design for a Digital Voltmeter

Initialize I/O devices
Get a value from the A/D
Calculate the voltage
Display the results

of how to get data from the A/D converter. Nor do we need to know the details of the algorithm to convert the 8-bit unsigned binary code to a voltage value. Thinking about and designing for these details can be postponed until later. It is only necessary at level 2 to know that this conversion needs to be made, not the details of how to do it.

Successively Refine Your Design

As we progress through the lower levels, we learn more about the details of what is required. Inevitably, as these lower level details become apparent, we think of something that could be done at an upper level to make the lower level design easier. That is OK. We have not invested any time in programming so it is easy to change the design. Go back to the upper level, change it, make sure it is now correct at that level, and continue to work at the lower levels.

Design Without Using a Programming Language

The initial design should propose solutions to the problems that are independent of any programming language. It should make no difference to the design how the machine code in the memory of the computer is generated. We are now beginning to talk about design tools—the tools and techniques used to write down the design. One widely used design tool is pseudocode. This is a programming-like language used for design. For example, a pseudocode design for the digital voltmeter at level 2 is shown in Table 3-1.

3.4 Design Partitioning

The top-down design method allows us to partition the design into easily handled pieces. At the upper levels, we can concentrate on more general ideas, leaving the detailed design until later. Also, when working at the upper levels, it is usually easy to see where work can be divided among different people working on the project. In the digital voltmeter design, it

> Most programming problems can be partitioned into elements that are divided among the programmers working on the job.

would be easy to split the design at level 2 into two parts. One engineer could work on the I/O initialization and getting data from the analog-to-digital converter and another could be assigned to convert the unsigned binary data to the voltage display. Partitioning the design and allocating work to different people is part of managing a software development project.

3.5 Bottom-Up Design

Bottom-up is another design philosophy that some people use. They think they are doing top-down design but they really are not. Here is how a designer could fall into a bottom-up design. We begin with a top-down design for the first levels. For example, the digital voltmeter design

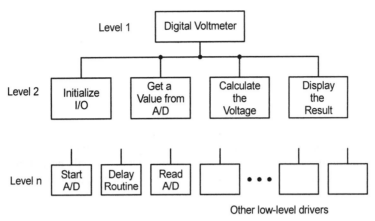

Figure 3-4 Bottom-up design.

In *bottom-up design*, low-level functions are designed, coded, and tested before the upper levels of the design are completed.

could be started just as before. So far, so good; but the time comes when we start looking ahead to doing some coding. After all, we are programmers aren't we? We can see that there will be some low-level drivers required, such as a routine that reads the A/D. Why not, we argue, take a break from this design stuff and do some programming for a change? Our design starts to look like Figure 3-4.

What is wrong with this? First, by writing programs before doing the complete design, we cast in code[2] how things are being done at lower levels before understanding the upper levels of the design. This violates the principle of postponing details. Ideally, we would like to *design* the lower levels based on a well thought out upper-level design. If we design and code the lower levels first, we may make decisions that could make the upper levels harder to implement. This also violates the principle of successive refinement of design. We do not want to invest time in writing code until all design levels have been completed. By coding the low levels first, we do not get a chance to optimize lower levels of the design based on decisions made for the higher levels. If later we do decide to change the low-level design, the work put into that coding is wasted.

Another problem with the bottom-up approach is that when code is written for the low-level drivers, extra code has to be written to test them. This means extra work for the programmer. The top-down approach, on the other hand, gives a testing structure that can test low-level programs. Top-down testing and debugging is discussed in Section 3.9.

Bottom-up design is not all bad, however. Bottom-up can be a tool building exercise. In any system one can see functional elements that are needed. If the tool building phase is approached so that the new tools are not application specific, and they do not have a great impact on upper levels of the design, they can be used in several applications. This may save work in the end.

3.6 The Real-World Approach

Rarely in the real world do we have the opportunity to follow the ideals of the top-down philosophy and complete the design to all levels of detail before doing any coding. Often low-level functions are available that have been coded and tested; you can use them in your design.

---

[2] Sometimes very much like concrete!

Top-down design combined with judicious use of functions already programmed works best in the real world.

It makes sense to use these working functions and not have to redesign, code, and test them again. Most high-level languages come with libraries of functions, and your company or co-workers may have useful libraries too. Using these functions violates the principles of top-down design but this is OK, providing you understand why the principles are being compromised and what the consequences may be. Using previously written low-level functions may impose constraints on the higher levels of the design. The time saved by using already working functions can offset the disadvantages of these constraints. Note that this is different from the bottom-up philosophy. In bottom-up, one sets out to write the low-level drivers, putting effort into their design, coding, and testing.

In summary, the real-world approach is one in which we recognize the power of the top-down method and attempt to do as much design as possible before coding, but we use previously developed and debugged functions where possible.

3.7 Types of Design Activity

Functional design is the more general activity where the required functional elements are defined.

The top-down design philosophy supports two types of design activity found in any software development project. The first is oriented toward defining the *functionality* required in the software. We do not care *how* the software does its thing as much as *what* function has to be provided. The design is to be refined to a level where we have components that are manageable by one person. That function is then assigned to a person to program.

Detailed design specifies the details necessary for each function called for in the functional design stage.

The second activity is the *detailed design* necessary to produce the functionality required in a module. The person who is assigned the job of producing a module takes the requirements specified by the first activity level and produces a detailed design to be programmed.

3.8 Design Tools

Design tools are used by the software engineer to help with the design. Here are some qualities of a good design tool.

1. It should be easy to use and it should allow design modifications to be made easily.
2. It should support structured programming.
3. It should allow us to see easily the design at many levels.
4. It should have good documentation facilities.

Structured Programming

In the mid-1960s, people writing software for large systems were appalled at the cost of these systems and at the amount of time needed to develop them. A landmark paper, by Bohm and Jacopini[3] in 1966, stated that any proper program was equivalent to a program that contains only three structures. That is, we can construct *any program* with only three basic structures, none of which is a GOTO statement. No one paid much attention until two years later when Dijkstra wrote a provocative paper that stated the GOTO statement in a

---

[3] C. Bohm and G. Jacopini, "Flow Diagrams, Turing Machines and Languages with Only Two Formation Rules," *CACM* **9**(5): 266–371 (1966).

Any program can be written with just *sequence*, *decision*, and *repetition* structures.

program is harmful.[4] The three structures that Bohm and Jacopini suggested (and a few more that software designers could not resist adding) form the basis of structured programming and the structured languages we know today.

The three basic structures are (1) a *sequence*, (2) a *decision*, and (3) a *repetition*. Beyond these, several general principles of structured programming can be enumerated:

1. Use these simple structures to aid in minimizing the number of interactions and interconnections between elements of the program.
2. Keep program segments small to keep them manageable.
3. Organize the problem solution hierarchically.
4. Organize each program segment so there is one input and one output (in terms of the program flow). This is not a data restriction but a restriction on the program flow. We would like to draw a box around a program segment and see that we enter that segment in only one place and leave it at only one place.

Structured programming really is not a design tool. It is a way of writing programs. However, because the principles of structured programming fit so well with the top-down design procedure, the elements of structured programming have been adapted for use as a design tool. With that in mind, let us look at the sequence, decision, and repetition structures used in structured programming along with the pseudocode design tool. We will then see how pseudocode can be used in a top-down design exercise.

Pseudocode

A frequently used design tool is the pseudocode technique. This is popular because it is easy to modify, does not require special graphical tools, and fits well with the documentation required for all design. Furthermore, the design text can be included in the software code as comments.

Many texts give a complete treatment of the pseudocode design tool. Here is an abbreviated approach that shows how to pseudocode the three simple design structures: sequence, decision, and repetition.

Sequence: A sequence structure is a sequence of functions or operations that the program is to perform. A sequence usually does not show any logic. It should show the function provided by a process block and must have a beginning and end. These are explicitly stated to show the single input–single output form we would like to achieve in the design. Thus, a sequence of A, B, C would be as shown in Table 3-2.

TABLE 3-2 Sequence Pseudocode

```
BEGIN A
 .  .  .
END A
BEGIN B
 .  .  .
END B
BEGIN C
 .  .  .
END C
```

---

[4] E. Dijkstra, "GOTO Statement Considered Harmful," *CACM* **11**(3): 147–148, (1968). Some programmers suggest that Dijkstra claimed that GOTO is a four-letter word.

TABLE 3-3 Decision Pseudocode

```
IF X
THEN
   BEGIN A
   . . .
   END A
ELSE
   BEGIN B
   . . .
   END B
ENDIF X
```

The ellipsis points (. . .) represent the elements of the design provided in the A, B, and C blocks.

Decision: The decision structure is called an *IF-THEN-ELSE* and Table 3-3 shows how to write the structure in pseudocode. The decision structure shows us that we do one of the two elements in the program. *IF* X is true, *THEN* the process A is executed, otherwise (*ELSE*) B is executed.

Another view of the IF-THEN-ELSE structure is shown in Figure 3-5. This is a structured flow chart symbol, and while flow charts are not generally used as design tools these days, they are useful to help visualize the proper structure of a good design. It is easier to see that only one process block is executed in Figure 3-5 than in the corresponding pseudocode of Table 3-3.

Inspection of Table 3-3 and Figure 3-5 shows that a Boolean or logic decision is made and must be either true or false. It doesn't have to be a simple decision; you may use any of the Boolean logic learned in your logic design course. For example, the following is a Boolean function:

F is TRUE if (A is TRUE AND B is FALSE) OR C is TRUE OR D is FALSE.

The decision structure may be single-sided. That is, there might not be an ELSE part to the decision. Table 3-4 and Figure 3-6 show the pseudocode and the corresponding structure.

Note that the IF-THEN-ELSE pseudocode block ends with an ENDIF X statement. The ENDIF, of course, signifies the end of the block. Repeating the conditional at this point is a

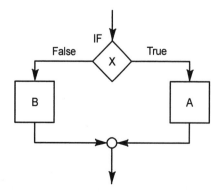

Figure 3-5 IF-THEN-ELSE decision element.

TABLE 3-4 Single-Sided Decision Pseudocode

```
IF X
THEN
  BEGIN A
  . . .
  END A
ENDIF X
```

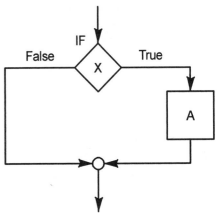

Figure 3-6 Single-sided decision structure.

good technique to help you remember what the decision was all about. This is especially useful when looking at the design later.

Repetition: The pseudocode for a repetition structure is shown in Table 3-5. This structure is called a *WHILE-DO*, and as you can see in Figure 3-7, *WHILE* the Boolean X is true, the process elements S1, S2, and S3 are *DONE*.

There are some other variations of the repetition structure. One particularly useful in assembly language programming is the *DO-WHILE* shown in Table 3-6 and Figure 3-8. Here the processing blocks, S1, S2 and S3, are done before the Boolean decision block. Thus the code in the DO block is executed at least once.

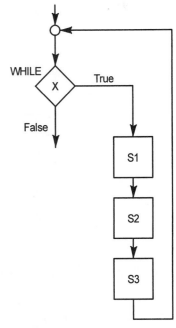

TABLE 3-5 WHILE-DO Pseudocode

```
WHILE X
DO
  BEGIN S1
  . . .
  END S1
  BEGIN S2
  . . .
  END S2
  BEGIN S3
  . . .
  END S3
ENDDO
```

Figure 3-7 WHILE-DO structure.

TABLE 3-6 DO-WHILE Pseudocode

```
DO
  BEGIN S1
    . . .
  END S1
  BEGIN S2
    . . .
  END S2
  BEGIN S3
    . . .
  END S3
ENDDO
WHILE X
```

Indentation: Indentation is often used in pseudocode. The code statements (or design requirements) for each block (bracketed by BEGIN and END) are indented to help show the structure of the design.

Single Input, Single Output: A principle of structured programming is to keep things simple without many interconnections between different parts of the program. A way to do this is to write the program so that elements of it (sequences, if-then-elses, and repetitions) have single entry and exit points.

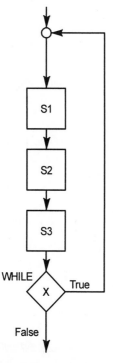

Figure 3-8 DO-WHILE structure.

Using Pseudocode Structured Elements as a Design Tool

A top-down design can be done in several levels of pseudocode. For example, when you first start the design, you might know only that A and B have to be done. The level-one design is shown in Table 3-7.

As we start to know more about the sequence block A, we can begin to fill in its details. The level-two design becomes that shown in Table 3-8.

The design goes on to level three, Table 3-9, where the C and D sequence blocks can be expanded.

In each of the design levels shown here, elements have been enclosed in boxes. This is to emphasize the single-input, single-output nature of the program flow. In Table 3-7 we can see that the A and B blocks are quite separate. By the time we get to Table 3-9, the separate blocks for A and B are still apparent even though A has been expanded.

TABLE 3-7 Level-One Design

TABLE 3-8 Level-Two Design

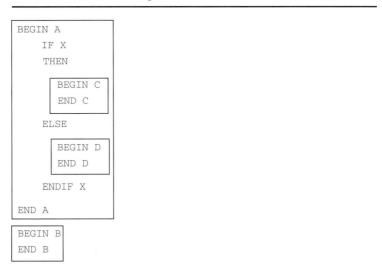

TABLE 3-9 Level-Three Design

```
BEGIN A

  IF X

  THEN

      ┌──────────────────────────────────┐
      │ BEGIN C                          │
      │                                  │
      │     IF Y                         │
      │                                  │
      │     THEN                         │
      │         ┌──────────────────┐     │
      │         │ BEGIN E          │     │
      │         │ END E            │     │
      │         └──────────────────┘     │
      │     ELSE                         │
      │         ┌──────────────────┐     │
      │         │ BEGIN F          │     │
      │         │ END F            │     │
      │         └──────────────────┘     │
      │     ENDIF Y                      │
      │                                  │
      │ END C                            │
      └──────────────────────────────────┘

  ELSE

      ┌──────────────────────────────────┐
      │ BEGIN D                          │
      │                                  │
      │     IF Z                         │
      │                                  │
      │     THEN                         │
      │                                  │
      │         BEGIN G                  │
      │                                  │
      │         END G                    │
      │                                  │
      │     ELSE                         │
      │                                  │
      │         BEGIN F                  │
      │                                  │
      │         END F                    │
      │                                  │
      │     ENDIF Z                      │
      │                                  │
      │ END D                            │
      └──────────────────────────────────┘

    ENDIF X

END A
```

```
┌──────────────────┐
│ BEGIN B          │
│ END B            │
└──────────────────┘
```

3.9 Top-Down Debugging and Testing

In the discussion of the bottom-up design technique in Section 3.5, we showed that extra code is required to test the modules as we develop them. However, if we develop the code in a top-down fashion by writing the higher level modules first and postponing the details of lower level modules, we have a program structure that tests itself. The program development might progress as illustrated in Tables 3-7 through 3–9. The upper level of the program is coded as calls to the modules that provide the functions A and B. As we start to work on function A, delaying our work on the others, we must have something to substitute for function B so the top-level program will run. Thus any of the lower level modules that have not yet been coded are temporarily coded as a *stub*. A stub is just a return with no processing done. Table 3-10 shows how the program might look for the level-two design. Here, functions B, C, and D are coded as stubs while we work on A. When the function must process some data and return a value, a dummy or test value can be returned by the stub to be

> *Stubs* are dummy programs for functions or subroutines that have not been written yet.

TABLE 3-10 Level-Two Design Using Stubs for Unfinished Modules

```
BEGIN PROGRAM
    Call Sub_A
    Call Sub_B
END PROGRAM
```
The main program consists of calls to subroutines only.

```
BEGIN Sub_A
    IF X
    THEN
       Call Sub_C
    ELSE
       Call Sub_D
    ENDIF X
END Sub_A
```
This is the module we are working on and it calls two more subroutines.

```
BEGIN Sub_B
    (Return)
END Sub_B
```
We have not started working on this module yet, so it is just a stub. It returns without doing any processing.

```
BEGIN Sub_C
    (Return)
END Sub_C
```
This is a stub too, but we will probably start to work on it next.

```
BEGIN Sub_D
    (Return)
END Sub_D
```
Another stub awaiting attention.

used by the calling program. In this way, you delay the actual programming of the lower level functions but can still develop and test the upper levels. The whole design becomes the test jig for itself. It allows you to design and test program logic and to see how data are passed back and forth at higher levels before coding the lower levels.

3.10 Structured Programming in Assembly Language

Structured programming and structured languages were invented to increase our efficiency as programmers and to make it easier to produce software without bugs and problems. What if the program has to be written in assembly language? Although assembly languages are vastly different from high-level languages and do not have built-in structured language elements, it is still possible to write structured code in assembly language. You must remember the principles of structured programming, particularly the idea of a single-input, single-output for a process block. Make your process blocks small, with jumps that do not span over great chunks of code, and ensure that there are no jumps into the middle of a block of code. This is what a compiler does for programs written in a high-level language. Chapter 8 shows how to code the sequence, decision, and repetition structures in HCS12 assembly language.

3.11 Software Documentation

Each of the software development phases—design, coding, and testing—has associated documentation.

Software Requirements Specification (SRS)

The SRS is a document or series of documents that define what is required of the software. At the upper levels of the best-designed systems, the SRS should completely define what the user of the system is to see (i.e., the user interface). This document can form the basis of the user operator's manual. It must be written first and agreed upon by the customer and the software developer. As you continue with the lower levels of the design, where one starts to think about how things are going to be done, the SRS documentation begins to define the functions required by modules in the system. You should be able to give an SRS document to a colleague to code the function and return a working module to be included in the system.

Software Design Document (SDD)

The SDD is the document produced for the detailed design of a module. It defines the logic required to produce a particular function. You start with the SRS for the module and use a design tool such as pseudocode described in Section 3.8

Software Code

The coding phase has an element of software documentation. This means including comments in the software. We would like the code to be written clearly enough that extra comments are not necessary. High-level languages allow us to do some of this but rarely should we write a program without any extra comments describing what is going on. In assembly language programs, comments are mandatory because the language is not as design oriented as high-level languages. It is particularly effective to use the pseudocode produced for the SDD for the comments in an assembly language program.

Software Verification Plan (SVP)

The SVP is a document that describes how we are going to test and verify that a particular module or system meets its specifications. The SVP should give the details of limiting values to be tested and the expected results. There may be levels of SVPs associated with the various levels of our design.

User Manuals

The four types of documents described earlier are often considered design documents to be used within the company and not delivered to the customer. Beyond these, there must be manuals for the customer's use. These include instructions on how to install the software (if appropriate) and instructions on using the software.

3.12 A Top Down Design Example

As a final exercise, let us tackle a design problem using the top down design approach. As a review, the principal steps of top down design are

- Understand the problem completely.
- Design in levels.
- Ensure correctness at each level.
- Postpone details.
- Successively refine your design.
- Design without using a programming language.

Seat Belt Alarm—Problem Statement

In many cars the seat belt alarm buzzer is also used to warn against leaving the key in the ignition or leaving the lights on. The following statement describes how such a system might operate:

The alarm is to sound if the key is in the ignition when the door is open and the motor is not running, or if the lights are on when the key is not in the ignition, or if the driver belt is not fastened when the motor is running, or if the passenger seat is occupied and the passenger belt is not fastened when the motor is running.

The Top Down Design

Understand the Problem: It is often useful to restate the problem to understand it better. Often a tabular form, as shown in Table 3-11, can help clarify the logic needed.

TABLE 3-11 Problem Solution Logic

Alarm Sounds	
When the key is in the ignition	
and the motor is not running	
and the door is open	Yes
When the key is in the ignition	
and the motor is running	
and the driver belt is not fastened	Yes
When the key is in the ignition	
and the passenger seat is occupied	
and passenger belt is not fastened	Yes
When the key is not in the ignition	
and the lights are on	Yes

First Level Design: By reading the problem statement and perhaps restating it we begin to understand the problem better, but we need a place to start the design. Table 3-11 shows there are situations when the alarm is to sound if the key is in the ignition and other conditions for sounding the alarm when the key is not in the ignition. Our first cut at the design, using the pseudocode tool and postponing details looks like the following:

```
IF the key is in the ignition
THEN
   DO the alarms if the key is in the ignition
   ENDDO the alarms if the key is in the ignition
ELSE (the key is not in the ignition)
   DO the alarms if the key is not in the ignition
   ENDDO the alarms if the key is not in the ignition
ENDIF (the key is in the ignition)
```

Second Level Design: We have obviously left out all the details that will sound the alarm, but we do have a starting structure to which we can now add details. First, though, we should look back at the design to make sure it is correct. Notice that we have put comments after the ELSE and ENDIF statements. These will help us keep track of where we are in the logic as we add details. There is not much logic in the design at this stage so we continue with the second level and start to fill in some of the details. Details that are added in each of the following levels are shown in bold type.

```
IF the key is in the ignition
THEN
   DO the alarms if the key is in the ignition
      IF the motor is not running
      THEN
         Do the alarms if the motor is not running
         ENDDO the alarms if the motor is not running
      ELSE (the motor is running)
       DO the alarms if the motor is running
       ENDDO the alarms if the motor is running
```

```
   ENDIF (the motor is not running)
  ENDDO the alarms if the key is in the ignition
 ELSE (the key is not in the ignition)
  DO the alarms if the key is not in the ignition
  ENDDO the alarms if the key is not in the ignition
 ENDIF (the key is in the ignition)
```

Third Level Design: Check back to ensure the second level design is correct and continue adding details.

```
  IF the key is in the ignition
 THEN
   DO the alarms if the key is in the ignition
    IF the motor is not running
    THEN
      Do the alarms if the motor is not running
       IF the door is open
       THEN
         Sound the alarm
       ENDIF (the door is open)
     ENDDO the alarms if the motor is not running
    ELSE (the motor is running)
     DO the alarms if the motor is running
     ENDDO the alarms if the motor is running
    ENDIF (the motor is not running)
   ENDDO the alarms if the key is in the ignition
  ELSE (the key is not in the ignition)
    DO the alarms if the key is not in the ignition
    ENDDO the alarms if the key is not in the ignition
  ENDIF (the key is in the ignition)
```

Fourth Level Design: Check back to ensure the third level design is correct and continue adding details. You do not have to continue with the "key is in the ignition" logic if it makes sense to do something else. Let us add some details in the ELSE (the key is not in the ignition) part of the logic.

```
  IF the key is in the ignition
 THEN
   DO the alarms if the key is in the ignition
    IF the motor is not running
    THEN
      Do the alarms if the motor is not running
       IF the door is open
       THEN
         Sound the alarm
       ENDIF (the door is open)
     ENDDO the alarms if the motor is not running
    ELSE (the motor is running)
     DO the alarms if the motor is running
     ENDDO the alarms if the motor is running
```

```
            ENDIF (the motor is not running)
         ENDDO the alarms if the key is in the ignition
      ELSE (the key is not in the ignition)
        DO the alarms if the key is not in the ignition
          IF the lights are on
            Sound the alarm
          ENDIF (the lights are on)
        ENDDO the alarms if the key is not in the ignition
      ENDIF (the key is in the ignition)
```

Successive Levels Design: We continue this design process by refining the design and adding the details needed to implement the solution. You may take several more design steps to complete the design.

```
IF the key is in the ignition
THEN
  DO the alarms if the key is in the ignition
    IF the motor is not running
    THEN
      Do the alarms if the motor is not running
        IF the door is open
        THEN
          Sound the alarm
        ENDIF (the door is open)
      ENDDO the alarms if the motor is not running
    ELSE (the motor is running)
      DO the alarms if the motor is running
        IF the driver's belt is not fastened
        THEN
          Sound the alarm
        ENDIF (the driver's belt is not fastened)
        IF the passenger seat is occupied
        THEN
          IF the passenger belt is not fastened
          THEN
            Sound the alarm
          ENDIF (the passenger belt is not fastened)
        ENDIF (the passenger seat is occupied)
      ENDDO the alarms if the motor is running
    ENDIF (the motor is not running)
  ENDDO the alarms if the key is in the ignition
ELSE (the key is not in the ignition)
  DO the alarms if the key is not in the ignition
    IF the lights are on
      Sound the alarm
    ENDIF (the lights are on)
  ENDDO the alarms if the key is not in the ignition
ENDIF (the key is in the ignition)
```

TABLE 3-12 Problem Solution Logic Final Check

Alarm Sounds	OK?
When the key is in the ignition	
and the motor is not running	
and the door is open	Yes
When the key is in the ignition	
and the motor is running	
and the driver belt is not fastened	Yes
When the key is in the ignition	
and the passenger seat is occupied	
and passenger belt is not fastened	Yes
When the key is not in the ignition	
and the lights are on	Yes

Final Check: Table 3-11 can be used to help check the final solution for correctness. Trace through your program logic for each of the cases that sound the alarm shown in Table 3-11. See Table 3-12.

3.13 Conclusion and Chapter Summary Points

It is vital that your software solutions be designed before written. Many problems (bugs) can be avoided by designing before writing. You must adopt a design practice such as the top-down methodology shown in this chapter.

In this chapter we learned the following points.

- The top-down design method is our choice of design approaches.
- The top-down design steps are as follows:
    Understand the problem completely before writing code.
    Design in levels.
    Ensure correctness at each level.
    Postpone details.
    Successively refine your design.
    Design without using a programming language.
- By doing bottom-up design and coding, decisions at lower levels may adversely affect the upper levels of the design.
- In the real world, we try to follow the principles of top-down design but we pragmatically use functions that have already been designed, coded, and tested.
- The elements of structured programming can be listed:
    Use three simple structures—sequence, decision, and repetition—to write all programs.
    Keep program segments small to keep them manageable.
    Organize the problem solution hierarchically (use top-down design).
    Use single-input, single-output program flow.

- The pseudocode technique is an effective design tool for all levels of top-down design.
- The top-down design method can lead to a top-down debugging and testing strategy, where the structure of the design tests itself.
- Software documentation is a vital part of all stages of software development and consists of the following:

   Software requirements specification (SRS)

   Software design documentation (SDD)

   Software code with comments

   Software verification plan (SVP)

   Users' manuals

3.14 Problems

Basic

3.1 List at least five principles of top-down design. **[a, c]**

3.2 What are the three basic elements of structured programming? **[a]**

3.3 Write the pseudocode and draw the flow chart symbol to represent the decision IF A is TRUE THEN B ELSE C. **[a, c]**

3.4 Write the pseudocode and draw the flow chart symbol to represent the decision IF A is TRUE THEN B. **[a, c]**

3.5 Write the pseudocode and draw the flow chart symbol to represent the repetition WHILE A is TRUE DO B. **[a, c]**

3.6 Write the pseudocode and draw the flow chart symbol to represent the repetition DO B WHILE A is TRUE. **[a, c]**

Intermediate

3.7 Write a design using structured flow charts or pseudocode to implement the following problem description: **[c]**

   Prompt for and input a character from a user at the keyboard.

   If the character is alphabetic and is uppercase, change it to lowercase and output it to the screen.

   If the character is alphabetic and is lowercase, change it to uppercase and output it to the screen.

   If the character is numeric, output it with no change.

   If it is any other character, beep the bell.

   Repeat this process until an ESC character is typed by the user.

Advanced

3.8 Design a program that initializes an 8-bit data storage accumulator to 0 and then inputs 10 successive 8-bit values from an input device located at address $70, adding each of them to the 8-bit data storage accumulator. If during this process an unsigned binary overflow occurs, print an error message and repeat from the beginning. Otherwise, after the 10 values have been input and added, output the result to an output device at location $71. Run the process forever. Your design must be a structured design and must show *repetition*, *decision*, and *sequence*. **[c]**

3.9 Give a design using structured pseudocode to accomplish the following. **[c]**

   A user is to input a character to select one of three processes. Valid characters are A, B, and C. A, B, and C select processes A, B, or C, respectively. Process A requires a byte of information to be input from an A/D converter, which it then converts to an integer decimal number in the range of 0 to 5, and displayed on the screen. Processes B and C are not defined at this stage. Prompts and error messages are to be displayed. You do not have to give details of the decimal conversion required in process A.

3.10 Give a design using structured pseudocode to accomplish the following: **[c]**

   A byte of data is to be input from an analog-to-digital converter and a critical value is to be input from a set of switches. If the A/D value is greater than the critical value, the microcontroller is to sound an alarm. Otherwise the alarm is to be turned off. This process is to continue forever.

3.11 Design a traffic light controller: **[c]**

Imagine an intersection with North/South and East/West streets. There are to be six traffic light signals:

RedE_W, YellowE_W, GreenE_W
RedN_S, YellowN_S, GreenN_S

Assume the time elements in the table shown are 10 seconds and that a timer delay is available as a function or subroutine. Give the pseudocode structured design for the light controller.

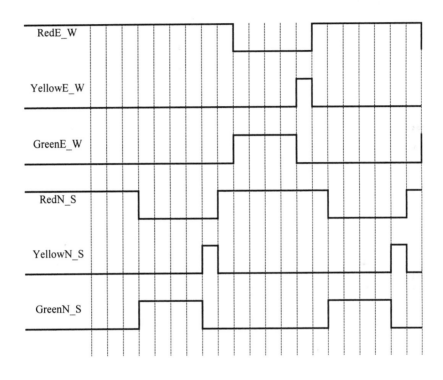

Introduction to the HCS12 Hardware

OBJECTIVES

This chapter describes the register resources and the memory addressing modes of the HCS12[1] microcontroller family of processors. Our goal is to have you understand enough about the system to be able to start programming exercises. We will tackle more advanced features of the CPU after you have started programming.

4.1 Introduction

The stored program computer shown in Figure 1-1 serves as a model for the basic operation and architecture of the Motorola HCS12 microcontroller. The HCS12 is a *h*igh-density *c*omplementary *m*etal-*o*xide *s*emiconductor (HCMOS) integrated circuit,[2] which contains the CPU with its registers and ALU, memory (RAM, EEPROM and Flash), a powerful timer section, and a variety of input and output features.

Figure 4-1 shows the block diagram of the MC9S12C family single-chip microcontroller. When you start to learn about a new microcontroller, look at its block diagram. The first thing to look for is the complement of I/O ports. Figure 4-1 shows a variety of I/O ports. We will have to learn more about the details of these, but at this point it is important to see, in general, what features may be available. Starting in the upper right corner, we see an A/D converter module (ATD). There are eight input channels, accessed by port PTAD. We will see in Chapter 17 that these can be digital as well as analog inputs, and we will learn that it is a 10-bit A/D. The next block, labeled *timer module*, is the timer I/O port. It has eight channels, each of which can be selected to provide the *input capture* or *output compare* functions. These pins can also be general purpose, bidirectional, digital I/O. Chapter 14 will cover the timer functions. Next is the *pulse-width modulation* (*PWM*) module. Up to six independent pulse-width-modulated waveforms can be generated. In some IC package versions of the microcontroller, some of the port P pins are not connected. For these cases, up to five PWM channels may be multiplexed onto port T pins.

---

[1] In this and the following chapters we will refer to the characteristics of the HCS12 family of microcontrollers. Much of this text is oriented toward a specific family, which we will refer to as the MC9S12 family or a specific member of the family—the MC9S12C32.

[2] This is the HC in the part designator HCS12.

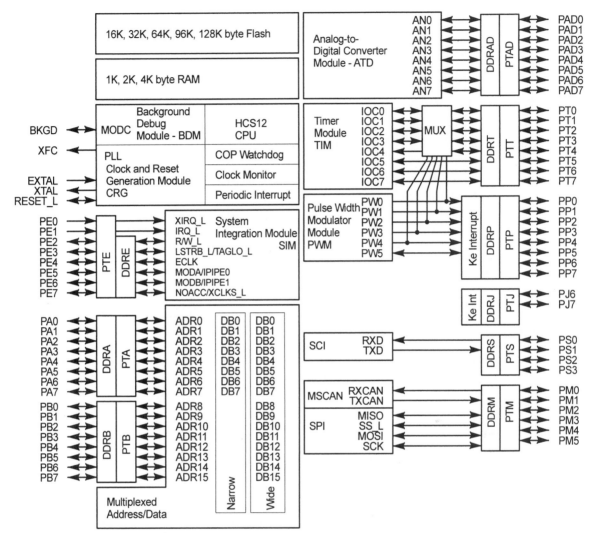

Figure 4-1 MC9S12C family block diagram.

The *serial communications interface* (*SCI*) is next. There is one, asynchronous serial port. In the MC9S12C family there is a *controller area network–MSCAN* interface. This interface implements the bit level protocols of a controller area network.

The *serial peripheral interface* (*SPI*) is another serial device. The SPI is a synchronous interface because it has a clock signal (SCK) to synchronize data bit transfers. When not in use as serial I/O functions, any of these port bits can be digital input or output pins. Chapter 15 discusses the serial I/O (SCI and SPI) of the MC9S12C, and Chapter 16 covers the MSCAN interface.

The top left side of Figure 4-1 shows us that the MC9S12C has Flash EEPROM and static RAM. Table 4-1 shows the amount of Flash and RAM in each member of this family.

The microcontroller also has a single-wire background debug module, which will prove to be extremely useful for debugging products using the MC9S12C.

TABLE 4-1 MCS9S12C Family Memory Complement

Family Member	Fixed Flash EEPROM	Paged Flash EEPROM	RAM
MC9S12C128	48K bytes	8 × 16K bytes	4K bytes
MC9S12C96	48K bytes	6 × 16K bytes	4K bytes
MC9S12C64	48K bytes	4 × 16K bytes	4K bytes
MC9S12C32	16K bytes	2 × 16K bytes	2K bytes
MC9S12GC16	16K bytes	—	2K bytes

Port E bits 0 and 1 are used for external interrupt requests and bits 2–6 may be used for bidirectional general purpose I/O. Port E also provides control signals when the microcontroller is in expanded mode with external address and data buses.

Figure 4-1 shows two bidirectional ports, *port A* and *port B*. These may be general purpose input or output but are used for multiplexed external address and data buses in expanded modes.

Some of the I/O ports can generate interrupts and are useful for interrupt driven I/O. The MC9S12C interrupt capabilities are discussed in Chapter 12.

4.2 The CPU, Registers, and Condition Code Bits

The HCS12 is a 16-bit microcontroller. It has 16-bit data and address buses allowing a memory space of 64K bytes. A built-in memory management system, discussed in Chapter 13, allows the family members to address up to 1M bytes of memory.

The Programmer's CPU Model

> The *programmer's model* includes two 8-bit accumulators, two 16-bit index registers, a 16-bit stack pointer register, and a condition code register.

The programmer's model of the CPU, that is, the set of registers that may be manipulated using the instruction set, is shown in Figure 4-2.

Accumulators A, B, and D: There are two 8-bit accumulators, *A* and *B*. Each may be a source or destination operand for 8-bit instructions. Some instructions have 16-bit operands and treat the two 8-bit accumulators as a single, 16-bit accumulator, with A being the most significant byte. When used in these instructions, the concatenation of A and B is called accumulator *D*. D is not a register in addition to A and B. Instructions that modify D also modify A and B.

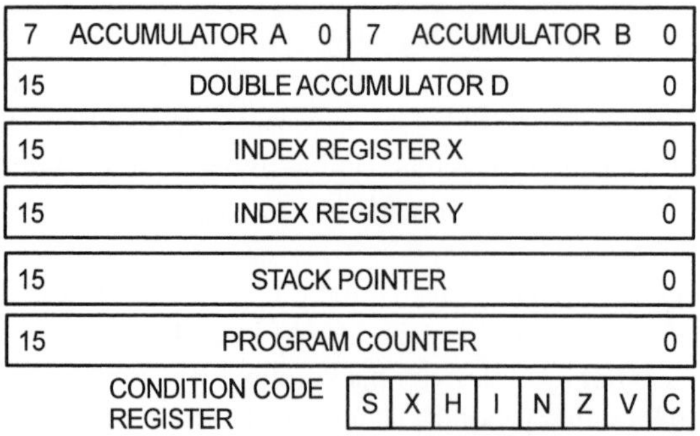

7 ACCUMULATOR A 0	7 ACCUMULATOR B 0
15 DOUBLE ACCUMULATOR D 0	
15 INDEX REGISTER X 0	
15 INDEX REGISTER Y 0	
15 STACK POINTER 0	
15 PROGRAM COUNTER 0	

CONDITION CODE REGISTER | S | X | H | I | N | Z | V | C |

Figure 4-2 Programmer's model.

Index Registers X and Y: The two 16-bit index registers are used primarily for indexed addressing, although there are some arithmetic instructions involving the index registers.

Stack Pointer: The stack pointer maintains a program stack in RAM and you must initialize it to point to RAM before it is used. The stack pointer always points to the last used memory location for a push operation. It is automatically decremented when pushing data onto the stack and incremented when data are removed or pulled.

Program Counter: Although the program counter is usually shown in the programmer's model, the programmer does not have direct control over it like the other registers. The number of bits in the program counter shows the amount of memory that can be directly addressed. In the HCS12, the program counter can be used as the base register for certain indexed addressing modes.

Condition Code Register: The HCS12 has four bits that are set or reset during arithmetic or other operations. These are the *carry* (*C*), *two's complement overflow* (*V*), *zero* (*Z*), and sign or

> The HCS12 has *carry, two's complement overflow, zero,* and *sign* condition code register bits.

negative (*N*) bits. A fifth bit, the half-carry (*H*), is set if there is a carry out of bit 3 in an arithmetic operation. There are no conditional branching instructions that test this bit, but it is used by the Decimal Adjust for Addition (DAA) instruction. Figure 4-2 and Table 4-2 show other bits to control the HCS12. The I bit (Interrupt Request Mask) may be used to globally mask and unmask the interrupt features of the processor. Bit 6, the

X bit, is a mask bit for the XIRQ_L interrupt input. These bits are described in more detail in Chapter 12. Finally, bit 7, the S or Stop disable bit, allows or disallows the STOP instruction. The STOP instruction is important in applications where low power consumption is a design goal. We will talk more about this instruction in Chapter 21.

Control Registers

Another important part of the programmer's responsibility is the set of 1024 memory locations (initially located at $0000–$03FF) called the control registers. These registers contain

TABLE 4-2 HCS12 Condition Code Register Bits

Bit	Flag	Conditions for Setting
Bits Modified by Various Instructions		
0	C	If a carry or borrow occurs
1	V	If a two's-complement overflow occurs
2	Z	If the result is zero
3	N	If the most significant bit of the result is set
5	H	This is the half-carry bit and is set if a carry or borrow out of bit 3 of the result occurs

Bit	Flag	Use
Bits Associated with HCS12 Control		
4	I	Interrupt mask
6	X	X interrupt mask
7	S	Stop disable

bits to control various aspects of the microcontroller in addition to being used to input and output data. We will cover the specific details of these registers and give examples showing their use in other chapters.

4.3 Operating Modes

When the microcontroller is reset, the status of three signals determines the operating mode of the HCS12. Tables 4-3 and 4-4 show the modes. The signals listed there are latched into the bits shown by the rising edge of the reset signal and they have the following meanings:

BKGD/MODC: The state of the BKGD/MODC pin is latched on the rising edge of reset and MODC determines, with MODB and MODA, the operating mode. After reset, this pin is used for the background debugger. See Chapter 20.

PE6/MODB and PE5/MODA: The state of these pins is latched into MODB and MODA and then the pins become available for use as a general purpose I/O pin.

PP6/ROMCTL: The state of this pin is latched into the RMON bit in the MISC register on the rising edge of reset. RMON controls the visibility of the internal Flash memory in the memory map. If RMON = 1, the memory is available in the memory map. See Chapter 13 for more details.

Normal Single-Chip Mode

In *single-chip mode*, all I/O and memory are contained within the microcontroller.

In the single-chip mode, the microcontroller is totally self-contained, except for an external clock source (an internal one is also available) and a reset circuit. The single-chip mode, with only a few parts, is ideally suited for many systems. All input/output and memory reside on the microcontroller chip, and only specialized I/O circuitry needed for the particular application must be designed.

TABLE 4-3 HCS12 Operating Mode Selection

BKGD/ MODC	PE6/ MODB	PE5/ MODA	PP6/ ROMCTL	RMON Bit	Mode Description
0	0	0	X	1	Special single-chip. Background debugger mode (BDM) is allowed and active. BDM is allowed in the other modes listed later but a serial command is required to make BDM active.
0	0	1	0	1	Emulation expanded narrow. BDM allowed.
			1	0	
0	1	0	X	0	Special test (expanded wide). BDM allowed.
0	1	1	0	1	Emulation expanded wide. BDM allowed.
			1	0	
1	0	0	X	1	Normal single-chip. BDM allowed.
1	0	1	0	0	Normal expanded narrow. BDM allowed.
			1	1	
1	1	0	X	1	Peripheral. BDM allowed but bus operations would cause bus conflicts (must not be used).
1	1	1	0	0	Normal expanded wide. BDM allowed.
			1	1	

TABLE 4-4 HCS12 Operating Modes

Operating Mode	Mode Description
Normal single-chip	Many embedded systems use this mode. The processor operates entirely within its internal memory and I/O resources because there is no expansion bus in this mode. Ports A, B, K (in some version) and most of port E are available as general purpose I/O.
Normal expanded wide	Ports A and B are used as a 16-bit multiplexed address and data bus. Port E provides bus controls and status signals. In this mode, 16-bit external memory and I/O devices can be interfaced to the processor.
Normal expanded narrow	Ports A and B are configured as a 16-bit address bus and port A is used as a multiplexed, 8-bit data bus. Port E provides bus control and status signals.
Special single-chip	This mode may be used for debugging single-chip operation, bootstrapping, or security-related operations. The active background mode controls the CPU and BDM firmware is waiting for additional serial commands through the BKGD pin. There is no expansion bus in this mode.
Emulation wide and emulation narrow	Developers use these two modes for emulating systems in which the target application is normal expanded wide or narrow modes.
Special test	Ports A and B are used as a 16-bit multiplexed address and data bus and port E provides bus control and status signals. This mode is used for factory testing.
Special peripheral	This mode is used for factory testing. The CPU is inactive and an external bus master drives the address and data buses and the control pins.

The MC9S12C32 is most suited for single-chip systems. It contains 32K bytes of Flash memory, enough program memory space for many applications. It also has 1K bytes of RAM for variable data storage.

Normal Expanded Mode

In *expanded mode*, the microcontroller gives up the normal use of I/O ports A and B to create multiplexed address and data buses. Parts of port E are used for bus control signals. All other I/O features remain in the microcontroller.

An MC9S12C32 operating in single-chip mode may not have enough resources in some applications. This is particularly true when more memory, especially RAM, is needed. The expanded mode provides address, data, and control buses at the expense of ports A and B and some pins in port E. Ports A and B give full 16-bit expansion multiplexed address and data buses. In the expanded modes, port E, bit 2 provides the R/W control signal.

The expanded wide mode implements a full 16-bit multiplexed address bus and data bus while the expanded narrow mode used a multiplexed 16-bit address bus and 8-bit data bus.

4.4 Background Debug Mode

This is a special mode used for system debugging and development. As you develop systems for special applications, there may not be a standard user interface, such as a keyboard and display, which you find in your computer laboratories. The background debug mode uses special hardware and firmware built into the HCS12. You can interact with (debug) the HCS12 system

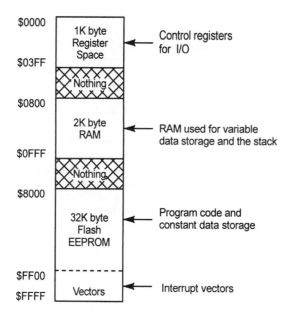

Figure 4-3 Memory map for MC9S12C32 microcontroller.

by connecting a debugging POD to the *BKGD* pin. Chapter 20 will discuss this important feature in more detail.

4.5 Memory Map

Each derivative of the HCS12 contains RAM, EEPROM, and Flash EEPROM internally. The memory map, which shows what kind of memory is at what address, depends on the operation mode and on the family member. Figure 4-3 shows the memory map for the MC9S12C32.

Registers to control input and output and other microcontroller functions occupy the first 1024 bytes of memory space. Even though these are not "memory" they are accessed like memory with memory addresses. From memory address $03FF to $0800 there is no memory. This space could be used in other versions of the MC9S12. The next memory space is 2048 bytes of RAM used for variable data storage and the stack. This is followed by a blank space and then 32K bytes of Flash EEPROM.

4.6 Addressing Modes

There are seven addressing modes for memory and I/O locations. These are (1) inherent, (2) immediate, (3) direct, (4) extended, (5) indexed, (6) indexed-indirect, and (7) relative. We use the notation shown in Table 4-5 to describe addressing modes and other operations in the next several chapters.

Inherent Addressing

Inherent addressing means that all data for the instruction are within the CPU. See Example 4-1.

TABLE 4-5 Notation

Notation	Meaning
Register Name	Indicates a register and its contents. **Example:** A refers to accumulator A and its contents.
→	Right arrow indicates a data transfer operation. **Example:** A → B indicates the contents of A are copied to B.
(...)	Contents of a memory location. **Example:** ($1234) → B indicates the contents of memory location $1234 are transferred to B.
((...))	This is for indirect addressing modes. The inner parentheses specify a memory address whose contents are the address of the data. **Example:** A → (($1234)) indicates that the contents of A are transferred to a memory location whose address is in $1234:1235.

Example 4-1 Inherent Addressing

```
85 1806                aba               ; A + B -> A
86 08                  inx               ; X+1 -> X
87 B781                exg    a,b         ; A <-> B
```

Immediate Addressing

Immediate addressing is used when an operand is a *constant* known at the time the program is written. If this is the case, the data can *immediately* follow the instruction in the memory. A memory map of the immediate addressing mode is shown in Figure 4-4, where the data may be 8 or 16 bits.

> *Immediate addressing* **requires** that the data (a number) be prefixed with the number sign (#).

You use immediate addressing to initialize registers with constants known at the time you write the program. Several examples of immediate addressing are given in Example 4-2.

Notice in Example 4-2 that a # sign appears before each of the numerical operands. **It is <u>VERY</u> important to remember to include this when using immediate addressing. It is also <u>VERY</u> easy to forget it.** The # (number sign) is a symbol that tells the assembler to use immediate addressing and not another addressing mode. It means that what is coming is a number (the data), not the *address* of the data. For example, if you write *LDAA $64*, the assembler will generate an instruction that loads A from *memory location $64*, not with the *value $64*. **Beware of this problem in your programs.** We will discuss more of the assembler's syntax in Chapter 5.

The data immediately follows the opcode.

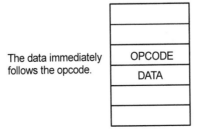

Figure 4-4 Immediate addressing.

```
Example 4-2  Immediate Addressing Examples
21                    ; Immediate Addressing
22
23  8640              ldaa  #64      ; Decimal 64 -> A
24  8664              ldaa  #$64     ; Hexadecimal 64 -> A
25  CE12 34           ldx   #$1234   ; Hexadecimal 1234 -> X
26  180B 4012 34      movb  #64,$1234; Decimal 64 -> $1234
```

Direct and Extended Addressing

Although these are listed as two separate modes by Freescale, both are commonly called direct memory addressing.

Direct addressing in the HCS12 can address an operand in the first 256 bytes of memory.

Direct addressing in the HCS12 can address an operand in the first 256 bytes of memory. Direct addressing, in the terminology used by Freescale, is also known as base page, reduced direct, or zero-page addressing. In direct addressing, the instruction contains an 8-bit memory address from or to which data are read or written. The instruction supplies the least significant byte of the address and the CPU sets the high byte equal to $00. Thus direct addressing can access the first 256 bytes of memory (addresses $0000–$00FF).

Extended addressing uses a 16-bit address to specify a location in the entire 64K-byte address space. These are 3-byte instructions. Figure 4-5 shows a memory map for an extended addressing instruction.

Extended addressing can address the full 64K-byte address space.

Examples 4-3 and 4-4 show the direct and extended addressing modes. The # that was used in immediate addressing is not used for these addressing modes. An address can be specified in either decimal or hexadecimal, although when specifying addresses, it is much better to use a label on a memory location.

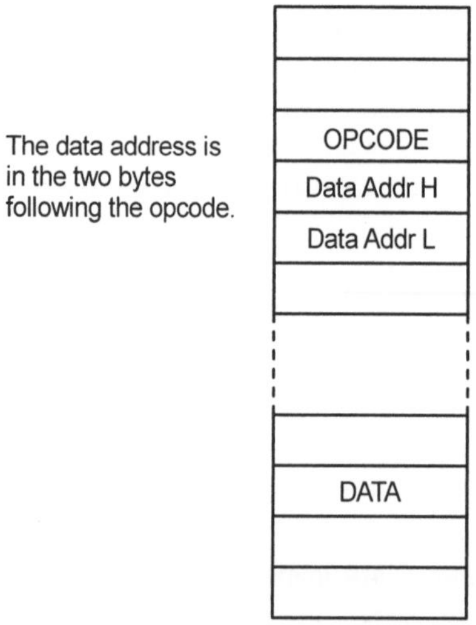

The data address is in the two bytes following the opcode.

Figure 4-5 Extended memory addressing.

```
Example 4-3  Direct Addressing
29 9664                  ldaa    $64     ; ($0064) -> A
30 5BFF                  stab    255     ; B -> ($00FF)
31 DE0A                  ldx     10      ; ($000A:000B) -> X
```

```
Example 4-4  Extended Addressing
36 B612 34               ldaa    $1234   ; ($1234) -> A
37 FC12 34               ldd     $1234   ; ($1234:1235) -> D
38 7E08 00               stx     $0800   ; (X -> ($0800:0801)
```

Indexed Addressing

The *effective address* in *indexed addressing* is the sum of a 5-, 9-, or 16-bit signed *constant* and the contents of the X, Y, SP, or PC register.

The HCS12 includes a variety of indexed addressing modes. In addition to indexed addressing using an offset, the HCS12 indexed addressing modes offer pre- and post-, incrementing and decrementing the index register. The A, B, or D registers provide the offset, and there are two indexed-indirect addressing modes.

Indexed addressing may be used with the X, Y, stack pointer register (SP), and, for some instructions, the program counter (PC). Table 4-6 shows a summary of the

TABLE 4-6 Summary of HCS12 Indexed Operations

Operand	Syntax	Comments
ldaa	, r	**5-, 9-, or 16-bit signed, constant offset**
ldaa	n, r	$n = -16$ to $+15$ for 5-bit offset
		$n = -256$ to $+255$ for 9-bit offset
		$n = -32,768$ to $32,767$ for 16-bit offset
		r can be X, Y, SP, or PC; *r is not changed* by the instruction
ldaa	n, −r	**Automatic pre-decrement**
		$n = 1$ to 8 and is subtracted from the contents of register r *before* the data value is fetched; r can be X, Y, or SP (not PC); *r is modified* by the instruction
ldaa	n, +r	**Automatic pre-increment**
		$n = 1$ to 8 and is added to the contents of register r *before* the data value is fetched; r can be X, Y, or SP (not PC); *r is modified* by the instruction
ldaa	n, r−	**Automatic post-decrement**
		$n = 1$ to 8 and is subtracted from the contents of register r *after* the data value is fetched; r can be X, Y, or SP (not PC); *r is modified* by the instruction
ldaa	n, r+	**Automatic post-increment**
		$n = 1$ to 8 and is added to the contents of register r *after* the data value is fetched; r can be X, Y, or SP (not PC); *r is modified* by the instruction
ldaa	A, r	**Accumulator offset**
ldaa	B, r	The contents of A, B, or D are used as a 16-bit, *unsigned* offset
ldaa	D, r	r can be X, Y, SP, or PC; *r is not changed* by the instruction
ldaa	[n, r]	**16-bit offset indexed-indirect**
		r can be X, Y, SP, or PC; *r is not changed* by the instruction
ldaa	[D, r]	**Accumulator D offset indexed-indirect**
		r can be X, Y, SP, or PC; *r is not changed* by the instruction

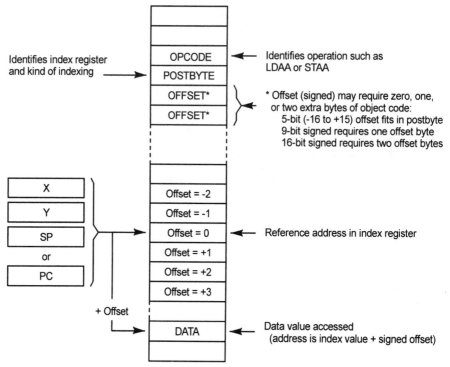

Figure 4-6 HCS12 indexed addressing.

HCS12 indexed instruction operations. The format of indexed addressing instructions is

$$\text{operation} \qquad \text{offset,index\_register}$$

where *operation* is a memory reference instruction, such as ldaa, *index_register* is *X*, *Y*, *SP*, or *PC*, and *offset* is a *signed 5- 9-*, or *16-bit value* added to the contents of the index register. This addition specifies the effective address but does not change the contents of the index register. The addition is modulo 65,536.[3]

Figure 4-6 shows how indexed addressing works in the HCS12. This addressing mode is called *based addressing* in some other systems.

INDEXED ADDRESSING USING A CONSTANT OFFSET

The first two instructions in Example 4-5 use an offset value of zero. This effectively provides a *register indirect addressing* mode. The constant may range from $-32,768$ to $+32,767$. The assembler will select the most efficient instruction (5, 9, or 16 bits) depending on the magnitude of the offset.

INDEXED ADDRESSING WITH AUTOMATIC INCREMENTING AND DECREMENTING

The HCS12 includes automatic incrementing and decrementing of the index register similar to the $++$ and $--$ operator in C. You may post- and pre- increment and decrement the

---

[3] Modulo N addition means that if the sum, say, M, is greater than N, the result returned is $M - N$. For example, for indexed addressing where the content of the X register is \$FFFE and the offset is \$10, the effective address is \$000E.

```
Example 4-5  Indexed Addressing, Constant Offset
42 A600            ldaa   ,x        ; (X+0) -> A (5-bit offset)

43 A600            ldaa   0,x       ; Same as ldaa ,x

44 A6E0 40         ldaa   64,x      ; (X+64) -> A (9-bit offset)

45 A6E9 C0         ldaa   -64,y     ; (Y-64) -> A (9-bit offset)

46 6A9F            staa   -1,SP     ;  A -> (SP-1) (5-bit offset)

47 A6FA 1388       ldaa   5000,PC   ; (PC+5000) -> A (16-bit offset)
```

register. The instruction format is

```
operation          value,-index_register        to pre-decrement
operation          value,index_register-        to post-decrement
operation          value,+index_register        to pre-increment
operation          value,index_register+        to post-increment
```

where *value* is an integer from 1 to 8 and *index_register* is *X*, *Y*, or *SP*; *value* is added or subtracted from the register before (*pre*) or after (*post*) the data value is transferred. See Example 4-5.

Warning: Unlike the constant offset-type instructions, the contents of the index register *are changed* by the instruction. Also, *value* is *not* an offset. It is a constant added or subtracted from the register contents.

ACCUMULATOR OFFSET INDEXED ADDRESSING

A limitation of the indexed addressing modes described previously is that only a constant off-set may be used to find the effective address. In the HCS12, a register, whose contents obviously can be a variable, can be used as the offset. The instruction format is

```
operation                register,index_register
```

where *register* is *A*, *B*, or *D* and *index_register* is *X*, *Y*, *SP*, or *PC*. The instruction calculates the effective address by adding the 8- or 16-bit value to the index_register. Like

```
Example 4-6  Indexed Addressing with Post- and Pre- Incrementing and Decrementing
52            ; Pre-decrement

53 A629             ldaa  7,-x    ; X-7 -> X, then (X) -> A

54            ; Post-decrement

55 A63E             ldaa  2,x-    ; (X) -> A, then X-2 -> X

56            ; Pre-increment

57 A620             ldaa  1,+x    ; X+1 -> X, then (X) -> A

58            ; Post-increment

59 A630             ldaa  1,x+    ; (X) -> A, then X+1 -> X
```

Example 4-7 Accumulator Offset Indexed Addressing

```
64  A6E5              ldaa  B,x      ;  (X+B)  -> A
65  E6EC              ldab  A,y      ;  (Y+A)  -> B
66  EDE6              ldy   D,x      ;  (X+D:X+D+1)  -> Y
```

---

the other constant offset instructions, this *does not change* the actual contents of the index register. Unlike the other constant offset instructions, the offset is *unsigned*.

INDEXED-INDIRECT ADDRESSING

The two indirect addressing modes in the HCS12 are shown in Figure 4-7. The address of the data to be manipulated is held in a memory location. The address of this memory

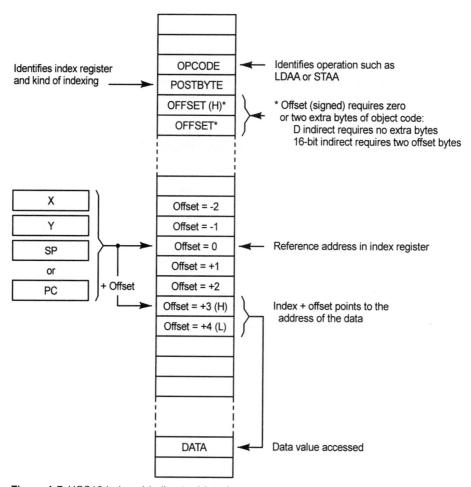

Figure 4-7 HCS12 indexed-indirect addressing.

Example 4-8 16-bit Constant Indexed-Indirect Addressing

```
71 CE50 00              ldx    #$5000  ; $5000 -> X, initialize X
72 A6E3 0064            ldaa   [$64,x] ; (($5064)) -> A
73 6AE3 FFFF            staa   [-1,x]  ; A -> (($4FFF))
```

Example 4-9 Indexed-Indirect Addressing Using Accumulator D

```
77 CE50 00              ldx    #$5000  ; $5000 -> X, initialize X
78 CC00 64              ldd    #$0064  ; $0064 -> D, initialize D
79 A6E7                 ldaa   [D,X]   ; (($5064)) -> A
```

---

location, then, is generated by the instruction. This mode is called *indexed-indirect* because indexed addressing is used first to find the address of the data. That address is then used to find the data.

The instruction format for indexed-indirect addressing is

```
operation               [offset,index_register] for constant, 16-bit offset or
operation               [D,index_register] to use the D register for the offset
```

The value `offset` is a 16-bit value, `D` is accumulator D, and `index_register` is *X*, *Y*, *SP,* or *PC*. The square brackets distinguish this addressing mode from the other constant offset addressing modes. See Examples 4-8 and 4-9.

Relative Addressing

Branch instructions often do not jump very far from the current program location. If this is the case, and the new address is within an 8-bit displacement from the program counter, relative addressing may be used. These are called *short branches*. If the address to which to program is branching is more than +127 or −128 bytes from the current PC location, a *long branch* with a 16-bit offset can be used. In the HCS12 instruction set, branch instructions use relative addressing while jump instructions use extended or indexed addressing. For relative instructions the assembler correctly calculates the branch instruction offset based on the label to which you are branching. Both short (8-bit, two's complement with an allowable range of −128 to +127) and long (16-bit, −32,768 to +32,767) relative branches may be used in the HCS12 (see Figure 4-8). Loop primitive instructions, including DBEQ, DBNE, IBEQ, IBNE, TBEQ, and TBNE, use 9-bit offsets allowing a range of −256 to +255.

Relative addressing is used for branch instructions.

Shortly after reset, the CPU finds the address of the first instruction to be executed in the *reset vector location $FFFE:FFFF*.

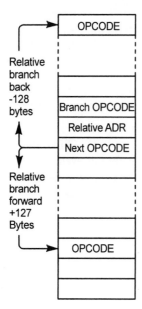

Figure 4-8 HCS12 relative addressing.

4.7 Reset

The HCS12 reset is an active low signal applied to the RESET_L pin. When RESET_L is asserted, some internal registers and control bits are forced to an initial state, but the stack pointer and other CPU registers are indeterminate. They must be initialized before they are used. The condition code register I and X bits are set to mask interrupts (interrupts cannot occur until you unmask them) and the S bit is set to disable the STOP mode. After the reset initialization, and assuming a system clock is present, the CPU fetches a *vector* from memory locations $FFFE:FFFF. This vector is the address of the first instruction to be executed. Thus, to turn on and run the HCS12, these memory locations must contain the location of the program to be executed after reset.

There are other consequences of the RESET_L signal. The HCS12 can configure its internal memory space. On reset, for example, the MC9S12C32 allocates the first 1K bytes to the Control Register Block and assigns 2K bytes of RAM to locations $0800–$0FFF. RAM and the I/O control registers can be relocated to any 2K byte boundary in the memory map if certain instructions are executed after the reset. In some versions of the HCS12, the EEPROM can be relocated as well. Chapter 13 shows us how to change these memory map locations.

The parallel I/O system is also affected by RESET_L. Details are given in Chapter 11, but for now, it is sufficient to know that bidirectional I/O lines are configured as high-impedance inputs. This is a configuration that is safe; that is, two outputs will not be connected.

Also associated with the I/O system and RESET_L is the operating mode. During the reset process, the BKGD, MODA, MODB, and MODC pins are read to select the operating mode for the CPU.

The timer system is also reset. Some registers are set to initial values and some are indeterminate. Details of the timer section are covered in Chapter 14.

All interrupt flags are cleared and the interrupt system is disabled because the interrupt handling capabilities must be programmed before they can be used. HCS12 interrupts are covered in detail in Chapter 12.

The serial I/O and the analog-to-digital converter capabilities are disabled on reset also. Serial I/O is discussed in detail in Chapter 15. The analog-to-digital converter is described in Chapter 17.

Causes of Reset

Other on-chip systems can sense failures and reset other parts of the system. If the clock oscillator stops or is running too slowly, the CPU is reset as described in the previous section with the exception that the reset vector address is at $FFFC:FFFD. If this failure occurs, the processor can execute code written especially for this event.

A *watchdog timer* generates a reset if your program runs away or goes into some error condition.

Another interesting reset is from the *COP watchdog timer*. *COP* stands for *CPU operating properly* and a watchdog timer is a device that generates a reset if the program does not keep the timer from timing out. This allows us to regain control of the processor if something happens to the program. Memory locations $FFFA:FFFB contain the vector to the code to be executed if the watchdog timer times out.

Reset Summary

As we start to program the HCS12, it is sufficient to know that upon reset all registers in the programmer's model are indeterminate, except the I, X, and S bits in the condition code register are all set. The address of the first instruction to be executed is fetched from $FFFE:FFFF.

4.8 Conclusion and Chapter Summary Points

These are enough hardware details to understand before learning the instruction set. There is, of course, an enormous amount of information and details to be learned before becoming proficient at programming and applying the microcontroller in a variety of applications. After you have learned the instruction set and how to write simple assembly language programs, we will return to more hardware topics in Chapters 11–20.

In this chapter we learned the following points:

- The HCS12 is a 16-bit microcontroller with a 16-bit address bus.
- There are two 8-bit accumulators, A and B, two 16-bit index registers, X and Y, and a 16-bit stack pointer register.
- The A and B accumulators can be concatenated to become a 16-bit accumulator D.
- The condition code register contains the carry, two's-complement overflow, zero, and sign bits used by the conditional branching instructions.
- There are 1024 I/O control registers starting at memory location $0000. The control registers can be relocated to any 2K-byte boundary.
- Internal RAM can be relocated to any 2K-byte boundary.
- The HCS12 supports the immediate, direct, extended, indexed, indexed-indirect, inherent, and relative addressing modes.
- The A, B, and D registers can be used to provide a variable offset for indexed and indexed-indirect, addressing.

- A # must be used with the operand in immediate addressing instructions.
- Direct addressing is limited to the first 256 bytes of memory.
- Extended addressing can address the entire 64K-byte address space.
- The effective address in indexed addressing is the sum of a 5-, 9-, or 16-bit signed offset and the contents of the X, Y, SP, or PC registers.
- When the HCS12 receives a power-on reset signal, it fetches the address of the first instruction to be executed from the vector location $FFFE:FFFF.

4.9 Bibliography and Further Reading

Cady, F. M., *Microcontrollers and Microprocessors*, Oxford University Press, New York, 1997.

MC9S12C Family Device User Guide V01.05, 9S12C128DGV1/D, Freescale Semiconductor, Inc., 2004.

HCS12 Microcontrollers MC9S12C Family, Rev 1.15, Freescale Semiconductor, Inc., July 2005.

CPU12 Reference Manual, M68HC12 & HCS12 Microcontrollers, CPU12RM/AD, Rev. 3, Freescale Semiconductor, Inc., April 2002.

4.10 Problems

Basic

4.1 Which of the HCS12 ports is used for the A/D converter inputs? **[a]**

4.2 Which of the HCS12 ports is used with serial I/O? **[a]**

4.3 Draw the programmer's model for the HCS12. **[a]**

4.4 Which bits in the HCS12 condition code register may be tested with conditional branching instructions? **[a]**

4.5 Complete the following sentences to describe the operation of the bits in the condition code register: **[a]**

a. The N bit is set when

b. The Z bit is set when

c. The V bit is set when

d. The C bit is set when

4.6 Describe the following HCS12 addressing modes: immediate, direct, extended, indexed, indexed-indirect, inherent, relative. **[g]**

Intermediate

4.7 Calculate the effective address for each of the following examples of indexed addressing. **[a]**

a. X = $0800

```
ldaa        0,X        EA =
```
b. Y = $0800
```
staa        $10,Y      EA =
```
c. X = $080D
```
ldaa        $25,X      EA =
```

4.8 Discuss the relative advantages and disadvantages of direct and extended addressing. **[g]**

4.9 Discuss the relative advantages and disadvantages of extended and indexed addressing. **[g]**

4.10 What is in the following CPU registers after a system reset? **[a]**

A, B, CCR, stack pointer.

Advanced

4.11 What HCS12 addressing mode is best to use when you want to access a number of sequential elements in a data array? **[a]**

a. Immediate

b. Direct

c. Extended

d. Indexed

e. None of these

4.12 What HCS12 addressing mode is best to use when you want to compare what is in accumulator A with a constant? **[a]**

a. Immediate
b. Direct
c. Extended
d. Indexed
e. None of these

4.13 Discuss how the CPU fetches the first operation code of the first instruction to be executed following a system reset. **[g]**

4.14 The HCS12 uses an instruction-fetch pipeline (or cache) that is filled by the CPU fetching bytes from memory while it (the CPU) is doing some other operations. For example, at the same time the CPU is adding the A to the B registers, which is an internal operation, it can be fetching another byte from memory to put into the instruction pipeline. **[g]**

a. Briefly explain how this strategy can improve the performance of the CPU.
b. Briefly explain under what conditions this strategy might fail to improve CPU performance.

5 An Assembler Program

OBJECTIVES

This chapter discusses the operation of a typical assembler. We will learn the assembler syntax now to be able to more easily understand examples showing the instruction set in Chapter 6.

5.1 Assembly Language Example

In this chapter we learn about some of the features of a typical assembler, at least enough to be able to understand the syntax of the examples in Chapter 6. Although this assembler is a component of the CodeWarrior® development system for the Freescale HCS12 microcontroller, we will not be giving detailed information on the user interface for operating in that development environment.

We briefly discuss the operation of an *absolute* assembler, the simplest to learn, at least for beginning assembly language programmers. We then transition quickly to the more advanced features of the CodeWarrior development software. We will be able to develop more complex and longer assembly language programs by learning about the *relocatable* assemblers and *linkers*.

An assembler converts source files to machine code, but before we look at how our assembler operates, let us consider a short example. At this stage you probably will not know what the instructions mean or what they do, nor will you understand all that the assembler does. Our goal is to give an overview of the process before we show the component parts of an assembly language program and how the assembler works.

Probably the most famous of all beginning programs, at least for C language programmers, is one that prints "Hello World!." Example 5-1 is a simple program doing just that.

The sample program in Example 5-1 has several parts. The listing you see is called the *.LST* file and is produced by the assembler to use when debugging and documenting your work. It shows, from left to right in columns, (1) *Abs.*, the absolute line number for the source code including all include files and macro lines, (2) *Rel.,* the relative line number showing each line in the current source file, (3) *Loc*, the address in memory in which the assembled code is found, (4) *Obj. code*, the assembled code bytes, and finally (5) *Source line*, the source code with label, opcode, operand, and comment fields. In this program *lines 1–4* are comments that introduce the program. *Line 7* uses an *assembler directive*, called an *equate* or *EQU*, to define the value used for the *symbol EOS*. This defines the code byte that signifies the end of the message to be printed on the terminal (such as the null byte at the end of a string in a C program). *Lines*

Example 5-1 Hello World! Example Program

```
Metrowerks HC12-Assembler

(c) COPYRIGHT METROWERKS 1987-2003
 Abs. Rel.   Loc    Obj. code    Source line
 ---- ----      ------ ---------    -----------
   1   1                           ; Example program to print "Hello
                                     World"
   2   2                           ; Source File:  hello.asm
   3   3                           ; Author: F. M. Cady
   4   4                           ; Created: 9/02
   5   5                           ; *****************************
   6   6                           ; Constant equates
   7   7               0000 0000   EOS:  EQU   0     ; End of string
   8   8                           ; Linker symbols
   9   9                           ; export 'Entry' symbol
  10  10                                   XDEF   Entry
  11  11                           ; Subroutine to printf
  12  12                                   XREF   printf
  13  13                           ; Stack pointer
  14  14                                   XREF   __SEG_END_SSTACK
  15  15                           ; *****************************
  16  16                           Entry:
  17  17                           ; Initialize stack pointer
  18  18   000000 CFxx xx                  lds #__SEG_END_SSTACK
  19  19                           ; Print Hello World! string
  20  20   000003 CCxx xx                  ldd #HELLO  ; Pass adr of
                                                       string.
  21  21   000006 16xx xx                  jsr printf  ; Jump to
                                                        subroutine.
  22  22                           ; Spin forever
  23  23                           spin:
  24  24   000009 20FE                     bra spin
  25  25                           ; Define the string to print
  26  26   00000B 4865 6C6C        HELLO: DC.B  'Hello World!',EOS
           00000F 6F20 576F
           000013 726C 6421
           000017 00
```

8–14 define symbols to be used by the linker including the *Entry* point and a subroutine, *printf*, which is to do the actual printing. The actual program code appears in *lines 17–24*. The stack pointer register is initialized (*line 18*), the message is printed (*lines 20* and *21*), and the program terminates by entering a spin loop where it stays until the microcontroller is reset. At the bottom of the program (*line 36*), an assembler directive, DC.B, defines the message *Hello World!*

This is a complete assembly language program for the HCS12, and you will be producing programs that look very much like this. In the following sections we will examine each component part of a program and describe the operation of the HC12 assembler.

5.2 CodeWarrior HC12 Assembler

The CodeWarrior HCS12 assembler converts HCS12 assembly language source files into *S Record* files that can be loaded into the microcontroller's memory.

The CodeWarrior HCS12 assembler runs on an IBM compatible personal computer (and other platforms) and *cross-assembles* code for the HCS12 microcontroller. It can produce a variety of output files and we will start using what is known as an *S-Record* file that is loaded into the HCS12 memory.

The assembler converts the program's operation mnemonics to opcodes and its operands to operand codes. Your primary task at this time is to learn the syntax of this assembler to be able to specify operations and operands. You will also learn about assembler pseudo-operations and directives that help the assembler do its job and make programs easier for us to read.

The HCS12 assembler can operate as an *absolute* or a *relocatable* assembler.

When you are using an *absolute assembler*, all source code for the program must be in one file or group of files assembled together. This is the easiest way to operate and we will use it to begin your programming chores. As your programs grow, however, we will change to relocatable assembler mode.

5.3 Assembler Source Code Fields

Each source code line has four fields—label, operation mnemonic or opcode field, one or more operands, and comments, as shown in Table 5-1. The fields are separated by a white space (usually one or more spaces or tab characters, shown as <ws> so the fields line up) and there are specific rules for each field.

Label Field

The *label field* starts in the first column of the source code line.

The *label field* is the first field of a source statement. A label is a symbol followed (optionally) by a colon. The label is optional but when used it can provide a symbolic memory reference, such as a branch instruction address, or a symbol for a constant. A valid label is the following:

- Alphanumeric characters, which may be uppercase or lowercase letters a–z, digits 0–9, underscore (_), or period (.)

TABLE 5-1 Source Code Fields

Label Field		Opcode Field		Operand(s) Field		Comment Field
Example:	<ws>	1daa	<ws>	#64	<ws>	; Initialize word counter

Example 5-2 Labels

```
; Labels Examples
TEST:                   ; Legal label
Test:                   ; Different label than TEST
_TEST:                  ; Legal label too
Test:                   ; Illegal - Duplicate label
TestData:               ; Legal
Test_Data:              ; Legal, more readable
Label                   ; Legal, a label doesn't need a colon
                        ; if it starts in the first column
    Label2              ; Illegal. It does need the colon here.
```

- A label must start with an alphabetic character.
- The label must start in the first column of the source code line unless it ends with a colon.
- Uppercase and lowercase characters are distinct by default but case sensitivity may be turned off with the assembler –*Ci* switch case sensitivity option. When this is active, the assembler treats lowercase the same as uppercase.
- A label may end with a colon (:).
- A label may appear on a line by itself.

A *whitespace* is a space or <tab> character.

A whitespace character (blank or <tab>) must be in the first character position in the line when there is no label, and there must be a white space between the label and the following opcode. See Example 5-2 for different kinds of labels.

A label may not occur more than once in your program. If it does, the assembler will give an error message noting that the symbol has been redefined. Remember also that a label must start with an alphabetic character and must start in the first column unless it ends in a colon.

Opcode or Operation Field

The *opcode field* begins after the first whitespace character.

The *opcode* field contains either a *mnemonic* for the operation, an *assembler directive* or *pseudo-operation*, or a *macro name*. It must be preceded by at least one white space. The assembler is insensitive to the case of the mnemonic; all uppercase letters are converted to lowercase. See Example 5-3.

Operand Field

The *operand field* follows the opcode with at least one whitespace character between.

The assembler uses the *operand* field to produce the binary code for the operand, and the interpretation of this field depends to a certain extent on the opcode. The operand must follow the opcode and be preceded by at least one whitespace. Operands can be the *symbols*, *constants*, or *expressions* that are evaluated by the assembler. The operand field also specifies the addressing mode for the instruction as shown in Table 5-2.

Example 5-3 Operation Field

```
Metrowerks HC12-Assembler
(c) COPYRIGHT METROWERKS 1987-2003
Rel. Loc    Obj. code Source line
---- ---     --------- -----------
  1 000000 87          CLRA     ; Legal mnemonic
  2 000001 87          clra     ; Mnemonics are not case
                                  sensitive
  3                    clra     ; Not legal, there has to be
  4                             ; at least one white space in
  5                             ; front of the mnemonic
  6 000002 08          DC.B  $08 ; Legal assembler directive
```

Symbols: A symbol represents an 8- or 16-bit integer value that *replaces* the symbol during the assembler's evaluation of the operand. For example, if the symbol CRLF is defined as $0D0A, the assembler replaces each occurrence of CRLF in your program with $0D0A. Special symbols are the asterisk (*) and dollar sign ($) that represent the current 16-bit value of the location (program) counter.

A symbol may be defined in another program module. In that case it is called an *external* symbol and must be "declared" in the current program as being external by using the *XREF* directive. We will show how this is done later in this chapter when we cover the operation of the relocatable assembler.

Constants: Constants are numerical values that do not change during the program. Constants may be specified in one of four formats—decimal, hexadecimal, binary, or ASCII characters

TABLE 5-2 Operand Formats and Addressing Modes

Operand Format	Addressing Mode/Instruction Type
No operand	Inherent
Expression	Direct, Extended, or Relative
#Expression	Immediate[b]
Expression,R[a]	Indexed offset with X, Y, SP, or PC
Expression,−R	Indexed auto pre-decrement
Expression,+R	Indexed auto pre-increment
Expression,R−	Indexed auto post-decrement
Expression,R+	Indexed auto post-increment
Accumulator,R	Indexed accumulator
[Expression,R]	Indexed-indirect
[D,R]	Indexed-indirect D accumulator
M,Expression	Bit set or clear
M,Expression,Expression	Bit test and branch

[a] R = register, M = memory location.
[b] It is excruciatingly important that you remember to include the # when you want the immediate addressing mode.

TABLE 5-3 Base Designators for Constants

Base	Prefix
Binary (2)	%
Decimal (10)	None (default)
Hexadecimal (16)	$
Octal (8)	@

The default base for numbers is *decimal*.

or strings. The format indicators shown in Table 5-3 can be given as a prefix to indicate the base of the number.[1]

The default base is normally decimal and is chosen if no other format specifier is given. The default base for the assembler can be changed by the *BASE* directive.

Decimal Constants: The decimal constant is specified no suffix. In *lines 3* and *4* in Example 5-4, the data value to be loaded into the register is a decimal value. See Example 5-15 to see how to use the *define constant byte (DC.B)* directive to define decimal constants that can be stored in memory.

Hexadecimal Constants: Hexadecimal numbers are identified with the prefix $. Hexadecimal values are a string of digits from the hexadecimal symbol set (0–9, A–F). See Example 5-4. Hexadecimal constants are used more frequently in assembly language programs than decimal constants, particularly when specifying addresses. However, if it makes sense to write a decimal constant, do not convert the decimal value to hexadecimal. Write it as a decimal constant

```
Example 5-4  Decimal, Hexadecimal, and Binary Constants

Metrowerks HC12-Assembler
(c) COPYRIGHT METROWERKS 1987-2003

 Rel. Loc    Obj. code Source line
 ---- ------ --------- -----------
   1                       ; Decimal Constants
   2                       ;        Op     Operand      Result
   3   000000 8664              ldaa   #100        ; A = 100
   4   000002 CE04 D2           ldx    #1234       ; X = 1234
   5                       ; Hexadecimal Constants
   6   000005 8664              ldaa   #$64        ; A = $64 = 100
   7   000007 869C              ldaa   #$9c        ; A = $9c = -100
   8   000009 CE12 34           ldx    #$1234      ; X = $1234 =
                                                        4660
   9                       ; Binary Constants
  10   00000C 8664              ldaa   #%01100100 ; A = $64 = 100
  11   00000E 86F0              ldaa   #%11110000 ; Most sig mask
```

[1] The assembler can be configured to be compatible with the MCUAsm and Avocet number base designators. Please refer to the full CodeWarrior HCS12 assembler manual.

Example 5-5

Show four ways to specify the code for the ASCII code for the character C and choose the best way to load that code into the accumulator A in a program.

Solution:

> ASCII—'C' or "C"
> Hexadecimal—$43
> Decimal—67
> Binary—%01000011

The best way to load accumulator A with the ASCII code for the character C is

```
ldaa    #'C'
```

and let the assembler convert it. Inspect the *Obj. code* in Example 5-4 to see that a constant, such as 100_{10} can be specified as either a decimal, a hexadecimal, or a binary value. You should choose the format that makes the most sense to you when you read the program.

Binary Constants: Binary constants are specified by the percent sign (%) prefix and are comprised of 1's and 0's. See Example 5-4. Use binary constants to make programs more readable. Suppose you wanted to define a mask for the four least significant bits of a byte; using %00001111 is more readable than using hexadecimal $0F and is far better than decimal 15.

ASCII Constants: Single *ASCII constants* or *strings* of one or more ASCII characters are enclosed in single (' ') or double quotation marks (" "). Single quotes are allowed only in strings delimited by double quotes. Double quotes can only appear in single-quote delimited strings. The assembler can assign the ASCII code for any printable character. Use this feature to specify ASCII characters instead of writing the hexadecimal code. The assembler will always make the conversion from the character to the code correctly. It is better to specify 'A' than $41, although they are equivalent. See Examples 5-5 and 5-16.

Expressions make programs more readable and easier to use in other applications.

Expressions: An expression is a combination of symbols, constants, and algebraic operators. The assembler evaluates the expression to produce a value for the operand. The HCS12 assembler algebraic operators are shown in Table 5-4.

Expressions are evaluated with normal algebraic operator precedence that can be altered by using parentheses. Because expressions are evaluated by the assembler,

TABLE 5-4 Assembler Expressions

+	Addition	−	Subtraction
*	Multiplication		
/	Division produces truncated result	%	Modulo division
>>	Shift right	<<	Shift left
&	Bitwise AND	|	Bitwise OR
∧	Bitwise Exclusive OR (XOR)	~	1's complement
!	Logical NOT		
!= or <>	Not equal	= or = =	Equal
<=	Less than or equal	<	Less than
>=	Greater than or equal	>	Greater than
HIGH	High byte of an address	LOW	Low byte of an address
PAGE	Page byte of an address		

they may be used for constants only. Nevertheless, the use of expressions is very powerful and can make a program more readable. It can also make it more portable and useful in other applications. See Examples 5-6 and 5-7.

The assembler allows *binary relational operators*. These compare two operands and return '1' if the condition is true or '0' if the condition is false. These are used most often for conditional assembly. See Example 5-8.

Example 5-6 Expressions

```
Metrowerks HC12-Assembler
(c) COPYRIGHT METROWERKS 1987-2003


Rel. Loc    Obj. code Source line
---- ----   --------- -----------
   1                  ; Test of all expression operators
   2        0000 0001 One:  EQU  1
   3        0000 0002 Two:  EQU  2
   4        0000 00FF Small:EQU  $ff
   5        0000 1234 Adr:  EQU  $1234
   6        0002 3456 Adr1: EQU  $23456
   7                  ;
   8 000000 03        Add:  DC.B  One + Two    ; Addition
   9 000001 01        Sub:  DC.B  Two - One    ; Subtraction
  10 000002 FF        SUB:  DC.B  One - Two    ; Subtraction
  11 000003 04        Mul:  DC.B  Two * Two    ; Multiplication
  12 000004 02        Div:  DC.B  Two / One    ; Division
  13 000005 00        DIV:  DC.B  One / Two    ; Division
  14 000006 01        DiV:  DC.B  One % Two    ; Modulo
                                                 division
  15 000007 7F        Shr:  DC.B  Small >> 1  ; Shift right
                                                 1 bit
  16 000008 0F        SHR:  DC.B  Small >> 4  ; Shift right
                                                 4 bits
  17 000009 08        Shl:  DC.B  Two << 2    ; Shift left 2 bits
  18 00000A 02        And:  DC.B  Small & Two ; Bitwise AND
  19 00000B 03        Or:   DC.B  Two | One   ; Bitwise OR
  20 00000C FD        Xor:  DC.B  Small ^ Two ; Bitwise XOR
  21 00000D FD        Cmpl: DC.B  ~ Two       ; 1's Complement
  22 00000E 12        Upper:DC.B  HIGH (Adr)  ; High byte of adr
  23 00000F 34        Lower:DC.B  LOW  (Adr)  ; Low byte of adr
  24 000010 02        Pg:   DC.B  PAGE (Adr1) ; Page byte of adr1
```

Example 5-7 Using an Assembler Expression

Assume an assembler program with two data buffers with the start of each signified by the labels Data_1 and Data_2. The two buffers are sequential in memory and the amount of data in each buffer changes in programs for different applications. Assume that somewhere in your program you want to load accumulator B with the number of bytes in the Data_1 buffer. Use an expression to do that.

Solution:

```
Metrowerks HC12-Assembler
(c) COPYRIGHT METROWERKS 1987-2003


Rel. Loc    Obj. code Source line
---- ------ --------- -----------
   1                     ;
   2 000000 C664             ldab   #(Data_2 - Data_1)
   3                     ; Immediate addressing
   4                     ; The assembler computes the difference
                         between
   5                     ; the address of Data_2 and Data_1
   6                     ;
   7 000002              Data_1: DS.B  100   ; Allocate 100 bytes
   8 000066              Data_2: DS.B  2*100 ; Allocate 200 bytes
```

Example 5-8 Relational Operators

```
Metrowerks HC12-Assembler
(c) COPYRIGHT METROWERKS 1987-2003


Rel. Loc    Obj. code Source line
---- ------ --------- -----------
   1                     ; Relational expressions
   2          0000 0001 One:    EQU   1
   3          0000 0002 Two:    EQU   2
   4                     ;
   5 000000 00           Equal:  DC.B  One = Two  ; Equal
   6 000001 01           EQUAL:  DC.B  One = One  ; Equal
   7 000002 01           NotEq:  DC.B  One != Two ; Not equal
   8 000003 01           Less:   DC.B  One < Two  ; Less than
   9 000004 00           Greater:DC.B  One > Two  ; Greater than
  10 000005 01           LTEQ:   DC.B  One <= Two ; Less or equal
```

```
11 000006 00          GTEQ:    DC.B  One >= Two ; Greater or equal
12 000007 01          Not:     DC.B  !0          ; Logical not
13 000008 00          NOT:     DC.B  !Two
```

Solution: The `Obj. code` bytes in the preceding listing show the results of the relational expressions. Each relational operator is evaluated true ($01) or false ($00).

Comment Field

The last field in the source statement is the *comment*. Comments start with a semicolon (;) and comments can be a complete line. Any line starting with a (;) or an (*) in column 1 is a comment line. The source program may have blank lines also.

5.4 Assembler Control

The assembler is controlled by directives (sometimes called *pseudo-operations*) that you place in your program. These are an important and vital part of an assembler program. Assembler directives are like opcode mnemonics because they appear in the opcode field, but they are not part of the microcontroller's instruction set. Directives can *define* the program's *location* in memory so all memory addresses are correct. They allow *symbols* and the *contents of memory* locations to be defined. Assembler directives also allocate memory locations for variable data storage. In short, directives help the assembler generate code for the program. Assembler directives also control how the assembler creates its output files, especially the list file. Table 5-5 shows directives available in the HCS12 assembler.

> Assembler *directives* instruct the assembler how to do its job.

TABLE 5-5 CodeWarrior HCS12 Assembler Directives

Section Definition	
ORG	Set program counter to the origin of the program in an absolute assembler mode.
SECTION	Define a relocatable section.
OFFSET	Define an offset section.
Constant Definition	
EQU	Equate symbol to an expression (cannot be redefined).
SET	Assign a symbol to an expression (can be redefined).
Reserving or Allocating Memory. Locations	
DS	Define Storage
Defining Constants in Memory	
DC.B	Define byte constant.
DC.W	Define word constant.
DCB	Define a constant block.
RAD50	RAD50 encoded string constants.

Export or Import Global
  Symbols

ABSENTRY	Specify the entry point in an absolute assembly file.
XDEF	Make a symbol public (visible to some other file).
XREF	Import reference to an external symbol.
XREFB	Import reference to an external symbol located on the direct page.

Assembly Control

ALIGN	Define alignment constraint.
BASE	Specify default base for constants.
END	End of assembly unit.
EVEN	Define two-byte alignment constraint.
FAIL	Generate user-defined error or warning messages.
INCLUDE	Include text from another file.
LONGEVEN	Define four-byte alignment constraint.

Repetitive Assembly Control

FOR	Repeat assembly blocks.
ENDFOR	End of FOR block.

Listing Control

CLIST	Include conditional assembly block.
LIST	Specify that all following assembly lines are in the list file.
LLEN	Define line length.
MLIST	Include macro expansions.
NOLIST	Specify that all following assembly lines are not in the list file.
NOPAGE	Disable pagination in the list file.
PAGE	Insert page break.
PLEN	Define page length.
SPC	Insert empty or blank line.
TABS	Define number of characters to insert for the <tab>.
TITLE	User-defined title.

Macro Definition

ENDM	End of user-defined macro.
MACRO	Start of user-defined macro.
MEXIT	Exit from macro expansion.

Conditional Assembly

ELSE	Alternate block, code included if IF statement not true.
ENDIF	End of conditional block.
IF	Start of conditional block.
IFC	Test if two string expressions are equal.
IFDEF	Test if a symbol is defined.
IFEQ	Test if an expression is null.
IFGE	Test if an expression is greater than or equal to 0.
IFGT	Test if an expression is greater than 0.
IFLE	Test if an expression is less than or equal to 0.
IFLT	Test if an expression is less than 0.
IFNC	Test if two string expressions are different.
IFNDEF	Test if a symbol is undefined.
IFNE	Test if an expression is not null.

In the syntax discussion for each of the assembler directives, the following notation is used:

[] Parentheses denote an optional element.

<> Angle brackets enclose a syntactic variable to be replaced by a user-entered value.

5.5 Assembler Directives

Section Definition

> *ORG* is used to locate sections of the program in the correct type of memory.

ORG (Set Program Counter to Origin): The ORG directive changes the assembler's location counter to the value in the expression. An ORG defines where your program is to be located in the various sections of ROM and RAM and is used in *absolute* assembly programs. In *relocatable* assembly programs the *linker* program provides this location function. See Example 5-9 for an absolute code example. See Chapter 6 for a discussion of the linker and Chapter 8 for programming examples.

$$ORG \quad <Expression> \quad [; Comment]$$

SECTION (Declare Relocatable Section): This directive establishes a relocatable section and initializes the assembler's location counter to zero for the first SECTION directive. Any

---

Example 5-9 ORG—Set Program Counter to Origin for Absolute Assembly

```
Metrowerks HC12-Assembler
Metrowerks HC12-Assembler
(c) COPYRIGHT METROWERKS 1987-2003

Rel. Loc   Obj. code Source line
---- -----  --------- -----------
  1                    ;
  2         0000 C000 ROM:    EQU   $c000   ; Location of ROM
  3         0000 0800 RAM:    EQU   $0800   ; Location of RAM
  4         0000 0A00 STACK:  EQU   $0a00   ; Location of stack
  5                    ;
  6                            ORG   ROM     ; Set program counter
  7                                          ; to ROM for code
  8                    ; The following code is located at memory
  9                    ; address ROM
 10 00C000 CF0A 00             lds   #STACK  ; Initialize SP
 11 00C003 B608 00             ldaa  Data_1  ; Load from memory
 12                                          ; address RAM
 13                    ;        -   -   -
 14                            ORG   RAM     ; Set program counter
 15                                          ; to RAM for the data
 16 000800             Data_1: DS.B  $20     ; Set aside $20 bytes
```

subsequent SECTION directive for that section restores the location counter to the value that follows the address of the last code in the section.

```
<name>:        SECTION      [SHORT]      [<number>]
```

SECTION is used to define sections of the program that will be located *later* by the *linker* program.

<name> is the name assigned to the section. You may have multiple SECTION directives with the same name to refer to the same section.

<number> is optional and is only specified for compatibility with the MCUASM assembler.

SHORT is an optional qualifier so that you can specify a short section small enough to be located in the base page for using direct addressing.

Sections may be *code*, *constant*, or *data* sections. A code section is one that contains at least one assembly instruction. Constant sections contain only Define Constant (DC) or Define Constant Block (DCB) directives and data sections contain at least a Define Storage (DS) directive. In your embedded application, code and constant sections must be *located* in read only memory (ROM) and data sections in read/write memory (RAM). See Example 5-10.

An *OFFSET* section may be used to simulate a data structure.

OFFSET (Create Absolute Symbols): OFFSET declares a section and initializes the location counter to the value in <expression>. An OFFSET section allows you to create and access data elements in a structure, for example. See Example 5-11.

```
OFFSET        <expression>
```

Example 5-10 SECTION

```
Metrowerks HC12-Assembler
(c) COPYRIGHT METROWERKS 1987-2003


Rel. Loc    Obj. code Source line
---- ------ --------- -----------
   1                  ;
   2                  RAM:  SECTION ; Define the section that
   3                              ; will go into RAM for
                                  variable
   4                              ; storage
   5 000000           Buf_1:DS  10
   6
   7                  ROM:  SECTION ; Define the section that goes
   8                              ; into ROM for the code
   9                  entry:
  10 000000 B6xx xx       ldaa  Buf_1
  11 000003 7Axx xx       staa  Buf_1+1
```

Example 5-11 Using OFFSET to Define a Structure

```
Metrowerks HC12-Assembler
(c) COPYRIGHT METROWERKS 1987-2003

Rel. Loc    Obj. code Source line
----  ------ --------- -----------
  1                      ; Example code showing OFFSET directive
  2                      ; to create and access data structures.
  3                      ;
  4       0000 C000 ROM:    EQU   $c000   ; ROM location
  5       0000 0800 RAM:    EQU   $0800   ; RAM location
  6       0000 0A00 STACK: EQU   $0a00   ; Stack pointer
                                            location
  7                    ; Define the structures
  8                            OFFSET 0
  9                    ; Structure 1
 10 000000            Count1: DS.B  1     ; An 8-bit counter
 11 000001            Value1: DS.W  1     ; The 16-bit value
 12       0000 0003 Size1:  EQU   *     ; This defines the size
 13                                      ; of the first structure
 14                            OFFSET 0
 15                    ; Structure 2
 16 000000            Count2: DS.W  1     ; A 16-bit counter
 17 000002            Value2: DS.W  1     ; 16-bit value
 18       0000 0004 Size2:  EQU   *     ; Size of second
                                            structure
 19                            ORG   ROM
 20 00C000 CF0A 00           lds   #STACK ; Initialize stack
                                            pointer
 21                    ;      - - -
 22 00C003 CE08 03           ldx   #Struct2; Point to second
                                             structure
 23 00C006 CD00 00           ldy   #0
 24 00C009 6D00              sty   Count2,x; Initialize counter
 25 00C00B ED02              ldy   Value2,x; Get the current
                                            value
 26 00C00D 02                iny         ; Increment it
 27 00C00E 6D02              sty   Value2,x; Save it again
 28 00C010 6200              inc   Count2,x; Increment the
                                            counter
 29                    ;      - - -
```

```
30                          ORG    RAM
31 000800              Struct1:DS.B  Size1      ; Define the first
                                                  structure
32 000803              Struct2:DS.B  Size2      ; Define the second
                                                  structure
33 000807              Data_1: DS.B  6          ; A data buffer
34                     end:
```

Constant Definition

An EQU may be the most used assembler directive.

EQU (Equate a Symbol to a Value): EQU is probably used more than any other directive in assembly language programming because it is good programming practice to use symbols where constants are required. Then, if the constant needs to be changed, only the equate is changed. When the program is reassembled, all occurrences of constants are changed.

```
<label:>     EQU     <expression>     [; Comment]
```

Any constant value can be defined for the assembler using the EQU. The EQU directive must have a label and an expression. It is a useful documentation technique to have a comment with each EQU to tell another programmer reading your program what the symbol is to be used for. An EQU does not generate any code that is placed into memory. See Example 5-12. The default base used to evaluate expressions and other data is decimal, but you can change the default with the BASE directive.

Example 5-12 EQU—Equate Symbol

```
Metrowerks HC12-Assembler
(c) COPYRIGHT METROWERKS 1987-2003


Rel. Loc    Obj. code Source   line
---- ------ --------- -----------
 1                     ;
 2          0000 0D0A CRLF:   EQU    $0D0A
 3                     ; For each occurrence of
 4                     ; CRLF, the assembler will
 5                     ; substitute the value $0D0A
 6          0000 0006 COUNT:   EQU    6
 7                     ; Loop counters often need to be initialized
 8          0000 001E COUNT1: EQU    COUNT*5
 9                     ; The assembler can evaluate an expression to
10                     ; provide a value of 30 for COUNT1
```

```
11          0000  000F  LS_MASK:EQU    $0F
12                            ; A mask that picks off the least significant
13                            ; nibble in a byte
14          0000  000F  ls_mask:EQU    %00001111
15                            ; A binary mask equate is more readable and
16                            ; informative than one given in hexadecimal
17                            ; Here are some code examples using the EQUs
18  000000 0D0A                    DC.W   CRLF
19  000002                         DS.B   COUNT
20  000008 861E                    ldaa   #COUNT1
21  00000A 840F                    anda   #ls_mask
22  00000C C40F                    andb   #LS_MASK
```

SET (Set a Symbol to a Value): SET is similar to the EQU directive except the value set is temporary in that it can be redefined later in the program. Any symbol defined by an EQU cannot be redefined.

```
<label:>    SET    <Expression>    [; Comment]
```

When defining symbols in include files it is useful to use the SET directive. If you do this, the assembler will not generate an error message if the symbol is SET more than once.

Reserving or Allocating Memory Locations

> The *Define Storage* directive is used to allocate memory for variable data storage.

DS (Define Storage): The DS sets aside memory locations by incrementing the assembler's location counter by the number of bytes specified in the expression. The block of memory reserved is *not* initialized with any value.

```
[label:]    DS[.<size>]    <n>    [; Comment]
```

Use this directive to allocate storage for variable data areas in RAM and then initialize the variables, if required, in the program at run time. `<size>` may be B or W for byte or word allocation. If `<size>` is not given, the default if B. See Examples 5-13 and 5-14.

Defining Constants in Memory

The following pseudo-operations define constants for ROM. We recommend highly that you do not use them to initialize variable data areas in RAM. RAM data areas should be *allocated* with the DS and then *initialized* at run time as shown in Example 5-14.

> Strings of ASCII characters may be defined with the *DC.B* directive.

DC (Define Constant): The DC directive allocates memory locations and assigns (initializes) values to each.

```
[<label>:]    DC[.<size>]    <expression>[,<expression>, . . .]
```

Example 5-13 DS—Define Storage

```
Metrowerks HC12-Assembler
(c) COPYRIGHT METROWERKS 1987-2003


Rel. Loc     Obj. code Source  line
---- ---     --------- ------------
  1                        ;
  2           0000 0010 COUNT_3: EQU   $10
  3  000000             BUFFER:  DS    COUNT_3   ; Allocates $10 bytes
  4  000010             BUFFER1: DS.B  2*COUNT_3 ; Allocates $20 bytes
  5  000030             BUFFER2: DS.W  COUNT_3   ; Allocates $10 words
```

Example 5-14

Show how to use the DS directive to reserve 10 bytes for data. Initialize each byte to zero in a small program segment.

Solution:

```
Metrowerks HC12-Assembler
(c) COPYRIGHT METROWERKS 1987-2003


Rel. Loc     Obj. code Source line
---- ------  --------- -----------
  1                        ;
  2           0000  000A NUMBER: EQU 10       ; Number of bytes
                                                allocated
  3           0000  0800 PROG:   EQU $0800    ; Program location
  4           0000  0900 RAM:    EQU $0900    ;  Location of RAM
  5                          ORG PROG
  6                        ;  - - -
  7  000800 C60A              ldab #NUMBER ; Initialize B with a
                                             loop
  8                                        ; counter
  9  000802 CE09 00           ldx   #BUF   ;  X points to the
                                             start of the
 10                                        ; buffer
 11  000805 6930         loop: clr  1,x1   ; Clear each location
                                             and
 12                                        ; point to the next
                                             location
```

```
13   000807 0431   FB      dbne b,loop          ; Decrement the loop
                                                   counter
14                                              ; and branch if the
                                                   loop
15                                              ; counter is not zero
16                            ;  - - -
17                            ORG  RAM           ; Locate the data area
18   000900    BUF:   DS  NUMBER                ; Allocate the data
                                                   area
```

You can define constants of different sizes, including bytes, words (two bytes), and long words (four bytes). These size definitions are given by the following:

DC.B: One byte is allocated for each expression. If the expression is an ASCII string, one byte is allocated per character.

DC.W: Two bytes are allocated for numeric expressions. ASCII strings are right aligned on a two-byte boundary.

DC.L: Four bytes are allocated for numeric expressions and ASCII strings are right aligned on a four-byte boundary.

If `<size>` is not given, the default is for a byte to be allocated and defined. See Examples 5-15 and 5-16.

Example 5-15 Define Constant

```
Metrowerks HC12-Assembler
(c) COPYRIGHT METROWERKS 1987-2003

Rel. Loc    Obj. code Source line
 ---- -----    --------- -----------
   1                     ;
   2                     ; Define decimal constants
   3 000000 64                     DC.B   100     ; Define a byte
   4 000001 0064                   DC.W   100     ; Define a word
   5                     ; Hexadecimal constants
   6 000003 23                     DC.B   $23
   7 000004 1234                   DC.W   $1234
   8 000006 BEEF                   DC.W   $beef
   9 000008 BEEF                   DC.W   $BEEF ;  Uppercase   same   as
                                                     lowercase
  10                     ; Binary constants
```

```
11 00000A 05                    DC    %0101 ; Valid (.B is default
                                                size)
12 00000B 65                    DC.B  0101  ; Invalid, missing %.
13                                          ; Assembler thinks
                                              this is
14                                          ; decimal 101.
15                    ; Initialize four memory locations with the
                      data
16                    ; 01, 02, $10, and $ff
17   0000 00FF  MAX: EQU   255
18   FFFF FF80  MID: EQU   -128
19 00000C 0102 10FF  Data: DC.B  01, 02, $10, MAX, MID
   000010 80
20                    ; Initialize four locations with $ff using
21                    ; the define constant block directive
22 000011 FFFF FFFF      DCB.B 4,MAX
23                    ; Initialize four words with -128
24 000015 FF80 FF80      DCB.W 4,MID
   000019 FF80 FF80
```

Example 5-16 ASCII Strings

```
Metrowerks HC12-Assembler
(c) COPYRIGHT METROWERKS 1987-2003

Rel. Loc    Obj. code Source line
---- -----  --------- -----------
   1                    ;
   2 000000 4161 62        DC.B  "A",'a',"b"
   3 000003 5468 6973      DC.B  "This is a string."
     000007 2069 7320
     00000B 6120 7374
     00000F 7269 6E67
     000013 2E
   4        0000 000D CR:   EQU   $0d ; ASCII code for carriage
                                        return
   5        0000 000A LF:   EQU   $0a ; ASCII code for line feed
   6 000014 4865 7265 Msg:  DC.B  "Here is a string with",CR,LF
     000018 2069 7320
```

```
       00001C 6120 7374
       000020 7269 6E67
       000024 2077 6974
       000028 680D 0A
     7 00002B 6361 7272                 DC.B 'carriage return and line feed'
       00002F 6961 6765
       000033 2072 6574
       000037 7572 6E20
       00003B 616E 6420
       00003F 6C69 6E65
       000043 2066 6565
       000047 6420
     8 000049 6368 6172                 DC.B  "characters."
       00004D 6163 7465
       000051 7273 2E
     9 000054 00                        DC.B  0 ; Null terminator for printf
```

A block of memory can be initialized using the *DCB*.

DCB (Define Constant Block): You can allocate and initialize a block of memory with
the DCB directive.

```
[<label>:]    DCB[.<size>]    <count>,<value>
```

Define Constant `<size>` may be B, W, or L with B the default. The `<count>` defines
the number of elements to be defined and may range from 1 to 4096. The value stored in each
location is the sign-extended expression `<value>`. See Example 5-15.

RAD50 (Rad50 Encoded String Constants): This directive places strings encoded with the
RAD50 encoding scheme into constants. This encoding places three string characters of a
reduced character set into two bytes, thereby saving memory. Only 40 different character values are supported and strings have to be decoded before they can be used.

Export or Import Global Symbols

ABSENTRY (Application Entry Point): ABSENTRY is used when the absolute assembly
option is being used. It creates an entry in the absolute code file (.abs) that is used by the
debugger.

```
ABSENTRY    <label>
```

The following directives, XDEF, XREF, and XREFB, are used when the relocatable assembler
features of the CodeWarrior HCS12 assembler are being used.

In a *relocatable* assembler, a symbol may be defined in module other than the one currently being assembled by using *XDEF* and *XREF*.

XDEF (External Symbol Definition): XDEF allows you to define a label in the current module to be made visible (public, or global) in other modules.

```
XDEF[.<size>]       <label>[,<label>] . . .
```

The default `<size>` is W.

XREF (External Symbol Reference): XREF is the "other hand" of XDEF. It declares, for the current module, that a variable used in this module is defined in another module. If you use an XREF in a module, there must be an accompanying XDEF in some other module.

```
XREF[.<size>]       <label>[,<label>] . . .
```

Again, the default `<size>` is W.

XREFB (External Reference for Symbols Located on the Direct Page): XREFB is similar to XREF except that symbols enumerated here may be located on the base page in another module. This allows direct addressing to be used.

```
XREFB      <symbol>[,<symbol>] . . .
```

Assembly Control

The assembly control directives give you control over how the assembler operates.

ALIGN (Align Location Counter): The ALIGN directive forces the next instruction to a boundary that is a multiple of `<n>`.

```
ALIGN     <n>
```

The value of `<n>` must be between 1 and 32767. Any bytes that are needed to fill the alignment block are initialized with 0.

The default base for numbers is decimal but this may be changed by the *BASE* directive.

BASE (Set Number Base): The default base when specifying constants is decimal. This can be changed with the BASE directive.

```
BASE     <n>
```

The base is set by `<n>`, which may be 2, 8, 10, or 16.

END (End Assembly): This directive causes the assembler to terminate assembling code and any subsequent statements are ignored.

```
END
```

EVEN (Force Word Alignment): The even directive forces the next instruction to the next even address relative to the start of the section.

```
EVEN
```

FAIL (Generate Error Message):

```
FAIL      <arg> | <string>
```

Parts of your source file may be
kept in another file called an
include file.

INCLUDE (Include Text from Another File): Another file can be inserted in the source
input stream.

```
INCLUDE    <file specification>
```

The assembler attempts to open the file in the current working directory. If it is not found
there, it searches for the file in every path specified in the environment variable GENPATH.
The file specification must be enclosed in quotation marks. Include files are often used to
define register addresses and other constants.

Example 5-17 INCLUDE

The source file is:

```
; include.asm
;
        INCLUDE "boilerplate.inc"
; Other code follows
;        - - -
```

and the assembler list file shows the file with the include file.

```
Metrowerks HC12-Assembler
(c) COPYRIGHT METROWERKS 1987-2003


 Abs. Rel. Loc     Obj. code Source line
 ---- ---- ----    --------- -----------
    1    1                   ; include.asm
    2    2                   ;
    3    3                           INCLUDE "boilerplate.inc"
    4   1i                   ; An include file
    5   2i                   ; boilerplate.inc
    6   3i                   ; This file contains a variety of commonly
    7   4i                   ; used equates.
    8   5i     0000 000D     CR: EQU $0d ; ASCII for carriage return
    9   6i     0000 000A     LF: EQU $0a ; Line feed
   10   7i     0000 0000     NULL: EQU   0 ; Null terminator
   11   8i                   ; End of include file
   12    4                   ; Other code follows
   13    5                   ;        - - -
```

LONGEVEN (Forcing Long-Word Alignment): The next instruction will be forced to the next long-word address relative to the start of the section.

```
                        LONGEVEN
```

Repetitive Assembly

The CodeWarrior HCS12 assembler can generate multiple lines of code from one line of input code. This is useful when defining a block of memory constants that have some algorithmic relationship that can be defined at assembly time. Repetitive assembly can only be done if the assembler option −Compat=b is used. By default this is turned off.

FOR (Repeat Assembly Block): See Example 5-18.

ENDFOR (End of FOR Block): See Example 5-18.

Example 5-18 FOR Assembly Block

```
;
; Define a lookup table for a sawtooth
; wave whose output ranges from 0 to 24 in steps
; of 4, i.e. 0, 4, 8, . . .
; The assembly option -Compat=b must be used
        FOR label = 0 TO 6
          DC.B  label*4
        ENDFOR
```

The following assembly code is generated:

```
Metrowerks HC12-Assembler
(c) COPYRIGHT METROWERKS 1987-2003


Rel. Loc  Obj. code Source line
---- ---  --------- -----------


     1                   ;
     2                   ; Define a lookup table for a sawtooth
     3                   ; wave whose output ranges from 0 to 24 in steps
     4                   ; of 4, i.e. 0, 4, 8, . . .
     5                   ; The assembly option -Compat=b must be used
     6                           FOR label = 0 TO 6
     7                             DC.B  label*4
     8                           ENDFOR
```

```
7 000000 00                    DC.B   label*4
8                              ENDFOR
7 000001 04                    DC.B   label*4
8                              ENDFOR
7 000002 08                    DC.B   label*4
8                              ENDFOR
7 000003 0C                    DC.B   label*4
8                              ENDFOR
7 000004 10                    DC.B   label*4
8                              ENDFOR
7 000005 14                    DC.B   label*4
8                              ENDFOR
7 000006 18                    DC.B   label*4
8                              ENDFOR
```

Listing Control

The following directives let you control how your list file looks.

CLIST (List Conditional Assembly): The CLIST directive controls the listing of conditional assembly blocks.

```
CLIST  [ON | OFF]
```

When CLIST is on, the list file includes all directives and instructions in the conditional assembly block. Otherwise, only directives and instructions that generate code are listed.

LIST (Enable Listing) and NOLIST (Disable Listing): LIST and NOLIST allow you to control portions of your code to be listing. You could use this to turn off listing of sections of the code that you are not interested in looking at, say, for debugging. See Example 5-19.

```
LIST
NOLIST
```

LLEN (Set Line Length): Set the number of characters from the source file that are included on the listing line to <n>, which may be in the range of 0 to 132. Lines are truncated to this length.

```
LLEN  <n>
```

MLIST (List Macro Expansions): MLIST is similar to CLIST in that it controls the listing of macro expansions in the list file. When ON (the default), macro expansions are included in the list file.

```
MSLIT  [ON | OFF]
```

Example 5-19   LIST and NOLIST

```
; Illustration of the LIST and NOLIST directives.
;
; This "code" shows in the listing.
      NOLIST
; While this line will not show.
      LIST
; Now listing is turned back on.
```

This source generates this assembly listing:

```
Metrowerks HC12-Assembler
(c) COPYRIGHT METROWERKS 1987-2003


Rel. Loc  Obj. code Source line
---- ---- --------- -----------
   1                ; Illustration of the LIST and NOLIST directives.
   2                ;
   3                ; This "code" shows in the listing.
   7                ; Now listing is turned back on.
```

NOPAGE (Disable Paging): Program lines are listed continuously without headings or top or bottom margins.

```
                          NOPAGE
```

PAGE (Insert Page Break): Insert page break in the list file.

```
                          PAGE
```

PLEN (Set Page Length): Set the list page length to <n> lines, where <n> may range from 10 to 10000. The default page length is 65 lines.

```
                        PLEN   <n>
```

SPC (Insert Blank Lines): Insert <count> blank lines in the list file, where <count> may range from 0 to 65.

```
                        SPC    <count>
```

TABS (Set Tab Length): Sets the tab length to <n> spaces where <n> may be 0 to 128. The default is eight.

```
                        TABS   <n>
```

TITLE (Provide Listing Title): The title given will be printed at the head of every page in the listing. This directive must be the first source code line and consists of a string of characters enclosed in quotes (").

```
TITLE    "This is a title"
```

Macros

The following directives are associated with the macro assembler features.

MACRO (Begin Macro Definition): The `<label>` is the name by which the macro is *called* to invoke it.

```
<label>    MACRO
```

ENDM (End Macro Definition):

```
ENDM
```

MEXIT (Terminate Macro Expansion): The MEXIT directive allows a macro expansion to be terminated before the ENDM directive. This is usually done with conditional assembly within the macro.

```
MEXIT
```

MACRO ASSEMBLER OPERATION

A macro assembler is one in which frequently used assembly instructions can be collected into a single statement. It makes the assembler more like a high-level language. For example, the problem might require a short code sequence to divide A by four

```
asra       ; Divide A by 2
asra       ; Divide by 2 again
```

in many parts of your program. This code is too short to be written as a subroutine so a macro is appropriate. Macros are often used for short segments of code, say, fewer than ten assembler statements. There are three stages of using a macro. The first is the *macro definition*, and a typical definition is shown in Example 5-20. The assembler directives *MACRO* and *ENDM* encapsulate the code to be substituted when the macro is invoked. The macro definition may include any code or directive except for another macro definition. It may include a previously defined macro as well. The label *Delay_100* is used in the second stage—the *macro invocation*. It is written in the source program where the lines of code would normally be placed. The third stage, *macro expansion*, occurs when the assembler encounters the macro name in the source code. The macro name is expanded into the full code that was defined in the definition stage.

The syntax for macro invocation in the source file where you want to use it is

```
[<label>:]    <macro_name>    [<argument_list>]
```

Example 5-20 Macro Definition

```
Metrowerks HC12-Assembler
(c) COPYRIGHT METROWERKS 1987-2003

Rel. Loc    Obj. code Source line
---- ---     --------- -----------
   1                   ;
   2                   ; Here is the macro definition
   3                   ;
   4     0000 00C7 MU100:  EQU 199        ; Delay loop counter
   5                   Delay_100 MACRO
   6                   ; Macro to delay approximately 100
                       microseconds
   7                        psha           ; Save the A reg
   8                        ldaa   #MU100
   9                   \@loop: deca        ; Label gets automatic
                                             number
  10                        bne    \@loop ; Delay until A=0
  11                        pula           ; Restore A reg
  12                        ENDM
  13                   ;
  14                   ; Now Invoke the macro
  15                        Delay_100
   6m                  ; Macro to delay approximately 100
                       microseconds
   7m000000 36              psha           ; Save the A reg
   8m000001 86C7            ldaa   #MU100
   9m000003 43        _00001loop: deca     ; Label gets automatic
                                             number
  10m000004 26FD           bne    _00001loop; Delay until A=0
  11m000006 32             pula           ; Restore A reg
  16                   ;
  17                   ; Do it again to illustrate the label change
  18                        Delay_100
   6m                  ; Macro to delay approximately 100
                       microseconds
   7m000007 36              psha           ; Save the A reg
   8m000008 86C7            ldaa   #MU100
   9m00000A 43 _00002loop: deca            ; Label gets automatic
                                             number
  10m00000B 26FD           bne    _00002loop; Delay until A=0
  11m00000D 32             pula           ; Restore A reg
```

This is *calling* a macro, much like calling a subroutine, except for each instance of the macro call the code is expanded and included in the assembly code.

Example 5-20 shows the macro definition, invocation, and expansion. Relative (*Rel*) *lines 6–13* show the macro definition bounded by the *MACRO* and *ENDM* directives. The first macro invocation occurs at *Rel line 16* and the second at *19*. The macro expansions are shown in *Abs lines 17–22* and *26–31*. Note that the relative line numbers show the line numbers from the macro definition and are denoted by the "*m*."

LABELS IN A MACRO

Multiple occurrences of a label in a program are normally not allowed, and for this reason the programmer can direct the assembler to create unique labels for each macro invocation. These assembler-generated labels are in the form _nnnnn, where nnnnn is a five digit number. The programmer specifies this to occur by specifying \@ in the label field within the macro body. You can see how this works in Example 5-20 where *lines 15:9m* and *18:9m* have unique labels generated by the assembler.

MACRO PARAMETERS

Substitutable parameters can be used in the source statement when the macro is called. These allow you to use a macro in different applications. Up to 36 parameters may be specified in the macro definition by a backslash character (\n), where n is the nth parameter and may be the digits 0–9 or an uppercase letter A–Z. When the macro is called, arguments from the argument list are substituted into the body of the macro as literal string substitutions. Example 5-21 shows a macro definition and call using a parameter to control how many times the A register is shifted left. The parameter, Num_shift, appears in the macro definition to remind you how it is used and to be helpful in documentation. Multiple parameters are separated by commas and a null argument may be passed by two commas (,,).

Example 5-21 Macro Parameters

```
Metrowerks HC12-Assembler
(c) COPYRIGHT METROWERKS 1987-2003


Rel. Loc   Obj. code Source line
---- ---    --------- -----------

   1                  ;
   2                  ; Macro definition for a variable arithmetic
   3                  ; shift left of accumulator A
   4                  alsa_n  MACRO Num_shift
   5                  ; Shift accumulator A left Num_shift bits
   6                  ; where Num_shift is a parameter in
   7                  ; the macro call.
   8                  ; Save B to set up a loop counter
```

```
 9                              pshb        ; Save B on the stack.
10                              ldab  #\1   ; Get Num_shift
11                    \@loop: asla          ; Shift A
12                          dbne  b,\@loop
13                          pulb        ; Restore the B
14                          ENDM
15
16                      ; The macro call is with a parameter
17                              alsa_n  4   ; Shift A 4 bits left
 5m                     ; Shift accumulator A left Num_shift bits
 6m                     ; where Num_shift is a parameter in
 7m                     ; the macro call.
 8m                     ; Save B to set up a loop counter
 9m000000 37                    pshb        ; Save B on the stack.
10m000001 C604                  ldab  #4    ; Get Num_shift
11m000003 48        _00001loop:   asla  ; Shift A
12m000004 0431 FC               dbne  b,_00001loop
13m000007 33                    pulb        ; Restore the B
```

Parameter zero ($\backslash$0) corresponds to a size argument that may follow the macro name, separated by a period (.). Another useful feature is macro argument grouping. If you wish to pass text with commas as a macro parameter, you may group these using a special syntax. The argument group is delimited by a *[?* as a prefix and *?]* as the suffix. See Example 5-22.

MACROS AND SUBROUTINES

Macros and subroutines have similar properties.

• Both allow the programmer to reuse segments of code. However, each time a macro is invoked, the assembler expands the macro and the code appears "in line." A subroutine code is included only once. Thus macro expansions make the program larger.

• The subroutine requires a call or jump-to-subroutine and the macro does not. This means that the subroutine is a little slower to execute than the macro and the subroutine call uses stack space temporarily.

• Both macros and subroutines allow changes to be made in one place (the macro definition or the subroutine).

• Macros and subroutines make the program easier to read by hiding details of the program. Usually, when reading a program, you do not need the details of how it is doing something, just an indication of what it is doing.

Conditional Assembly

You may have seen the IF-THEN-ELSE structure in a high-level language. We know the code for the THEN part is executed if the conditional is true, and the ELSE part if the conditional

Example 5-22 Macro Parameter \0

```
Metrowerks HC12-Assembler
(c) COPYRIGHT METROWERKS 1987-2003


Rel. Loc   Obj. code Source line
---- ---    --------- -----------
    1                                   ;
    2                                   ; Illustration of macro parameter
                                        \0 and
    3                                   ; macro argument grouping
    4                                   ;
    5                                   ; Define data macro
    6                                   define:   MACRO
    7
    8                                             DC.\0 \1,\2
    9                                             ENDM
   10
   11                                   ; Use the define macro to define bytes
   12                                   define.B  $10, 'A'
   7m
   8m000000 1041                                  DC.B $10,'A'
   13                                   ; Use the define macro to define words
   14                                   define.W  $10,$0d0a
   7m
   8m000002 0010 0D0A                             DC.W $10,$0d0a
   15                                   ; Use the define macro with argument
                                        grouping define.B $10,[?'A','B',
   16                                   'C'?]
   7m
   8m000006 1041 4243                             DC.B $10,'A','B','C'
   17                                   ; You can use argument grouping
                                        with strings define.B   $10,[?'This
   18                                   is a string'?]
   7m
   8m00000A 1054 6869                             DC.B $10,'This is a string'
      00000E 7320 6973
      000012 2061 2073
      000016 7472 696E
      00001A 67
```

is false. A conditional assembly is very similar but only the appropriate segment of code is included in the assembled program. Note that this *does not* produce code that is an executable IF-THEN-ELSE structure. We will see how to write structured code in Chapter 8.

The conditional assembly feature allows you to write code that can be customized at assembly time. Example 5-23 shows how to choose a set of equates for one of two different versions of the software. Conditional assembly may be invoked with the following directives.

IF (Conditional Assembly): This is the start of conditional assembly block.

```
IF      <condition>
        [assembly language statements if true]
[ELSE]
        [assembly language statements if false]
ENDIF
```

If <condition> is true, the [assembly language statements if true] code is assembled. Assembly continues until the ELSE or ENDIF directive is encountered. Nesting of conditional blocks is allowed, limited only by the amount of memory at assembly time.

Example 5-23 Conditional Assembly
The conditional assembly code is the following:

```
;
TRUE:   EQU    1
FALSE:  EQU    0
; Define parameters for each version
Param1: EQU    $76    ; Use in version 1
Param2: EQU    $77    ; Use in version 2
; Set the version number for this software
Ver1:   EQU    TRUE

; Here is the conditional assembly:
    IF  Ver1 = TRUE
        ldaa  #Param1
    ELSE
        ldaa  #Param2
    ENDIF
;
; You can also use the following form:
    IFNE  Ver1  ; If Ver1 not equal to zero
        ldaa  #Param1
    ELSE
        ldaa  #Param2
    ENDIF
```

The assembler generates the following code:

```
Metrowerks HC12-Assembler
(c) COPYRIGHT METROWERKS 1987-2003


Rel. Loc    Obj. code Source line
---- ---    --------- -----------
    1                   ;
    2       0000 0001 TRUE:   EQU   1
    3       0000 0000 FALSE:  EQU   0
    4                   ; Define parameters for each version
    5       0000 0076 Param1: EQU   $76   ; Use in version 1
    6       0000 0077 Param2: EQU   $77   ; Use in version 2
    7                   ; Set the version number for this software
    8       0000 0001 Ver1:   EQU   TRUE
    9
   10                   ; Here is the conditional assembly:
   11      0000 0001     IF  Ver1 = TRUE
   12 000000 8676            ldaa  #Param1
   13                        ELSE
   15                        ENDIF
   16                   ;
   17                   ; You can also use the following form:
   18      0000 0001     IFNE  Ver1  ; If Ver1 not equal to zero
   19 000002 8676            ldaa  #Param1
   20                        ELSE
   22                        ENDIF
```

The <condition> is a Boolean and its syntax is

```
<condition> := <expression> <relation> <expression>
<relation> := "=" | "!=" | ">=" | ">" | "<=" | "<" | "<>"
```

The <expression> must be able to be evaluated at assembly time.

IFcc (Conditional Assembly): An alternative conditional assembly syntax may be used.

```
IFcc    <condition>
        [assembly language statements if true]
[ELSE]
        [assembly language statements if false]
ENDIF
```

Table 5-6 shows the available conditional types.

TABLE 5-6 Conditional Assembly Types

IFcc	Condition	Meaning
IFeq	`<expression>`	If `<expression>` = = 0
IFne	`<expression>`	If `<expression>` ! = 0
IFlt	`<expression>`	If `<expression>` < 0
IFle	`<expression>`	If `<expression>` < = 0
IFgt	`<expression>`	If `<expression>` > 0
IFge	`<expression>`	If `<expression>` > = 0
IFc	`<string1>,<string2>`	If `<string1>` = = `<string2>`
IFnc	`<string1>,<string2>`	If `<string1>` ! = `<string2>`
IFdef	`<label>`	If `<label>` was defined
IFndef	`<label>`	If `<label>` was not defined

5.6 Assembler Files

The assembler produces a variety of output files including the *assembler listing* and the *debug listing* in addition to object output files. One way or another, the object files are destined to be loaded into the microcontroller's memory while the listing files are used for documentation and debugging. Table 5-7 shows the extension used and the location for all assembler input and output files.

Assembler Listing

The assembler listing file contains information about the generated code and is generated when the –L assembler option is activated. If an assembler error occurs, no listing file is generated. The file has the format shown in Example 5-1 and contains the following fields of information. The assembler list file has the extension *.lst*.

Abs: This column contains the line number for each instruction. It is *absolute* in the sense that it enumerates all included files and where all macro calls have been expanded.

Rel: The *relative* line number is the line number in the source file. For included files, the relative line number is the line number in the included file and for macro expansion it is the line

TABLE 5-7 Assembler Files

Extension	File	Where Found
.o	Binary object	Source file directory or OBJPATH
.dbg	Debug listing	Source file directory or OBJPATH
.abs	Absolute file	Source file directory or ABSPATH
.asm	Source file	Project directory or GENPATH
.inc	Include file	Project directory or GENPATH
.sx	Freescale S record	Source file directory or ABSPATH
.s1	Freescale S record if SRECORD = 1	Source file directory or ABSPATH
.s2	Freescale S record if SRECORD = 2	Source file directory or ABSPATH
.s3	Freescale S record if SRECORD = 3	Source file directory or ABSPATH
Err.txt	Error listing	Current directory

number of the instruction in the macro definition. In each of these cases, an "i" and "m" suffix is appended to the relative line number.

Loc: The *loc* column contains the address of the instruction. In absolute sections of your program an "a" precedes the address to show that this address has been fixed at assembly time. In relocatable sections, the address is the offset from the beginning of the section. The address is given in hexadecimal with up to six digits.

Obj. code: This column contains the hexadecimal code for each instruction. Parts of the code in the relocatable section that have not been resolved (to be done by the linker) are displayed as "x."

Source Line: Each line in the source file is repeated in this column.

Debug Listing

The debug listing file, created with the extension *dbg*, contains debugging information for the CodeWarrior True-Time Simulator.

Object Files

Object files always have the extension *.o* and contain the target code as well as some debugging information.

Absolute Files

Absolute files may be created when an application is encoded in a single module and all sections are absolute sections. The absolute file has the extension *.abs*.

Freescale S Files

When an application is encoded in a single module with all absolute sections, you may generate an ELF absolute file instead of an object file. A Freescale S record file is generated at the same time and can be burned into an EPROM or downloaded to a development board equipped with a debugging monitor such as D-Bug12. The S record file will have an extension *.sx* or *.s1*, *.s2*, or *.s3*.

5.7 Remaining Questions

- How do I get my program into the microcontroller? *In Chapter 6 we will show how the* Linker *program works to bring together all elements of the program to produce a file you will* download *to the microcontroller.*
- Isn't writing programs in assembly language a lot more work than writing programs in a high-level language like C? *Well, yes it is, but sometimes there are advantages to writing programs in assembly language, such as when the program needs to run fast or when intricate timing requirements must be met. We will show how to use C for producing programs in Chapter 10 so you can have the best of both worlds.*

5.8 Conclusion and Chapter Summary Points

In this chapter we learned about an assembler program to convert assembly language statements into machine code for the microcontroller's memory. Here are some of the key points of this chapter.

- The CodeWarrior assembler can operate in relocatable and absolute mode.
- Relocatable assembly code allows you to create modules and link them into the final program without reassembling everything every time a change is made.
- There must be a white space between each of the four fields in a source code line.
- Labels and symbols are case sensitive; ABCD is different from AbCd. However, case sensitivity can be turned off in the assembler.
- You must not have duplication of labels in any single source file.
- Labels may optionally end in a colon (:).
- If a symbol or label is defined in another module, the label must be XREFed in the module that references it and XDEFed in the module where it is defined.
- The opcode field contains operation mnemonics, assembler directives, or macro names.
- The opcodes and directives are not case sensitive.
- The operand field may have symbols, constants, or expressions.
- The default base for constants is decimal.
- Hexadecimal constants are signified by a $.
- Binary constants are signified by a %.
- ASCII constants are signified by enclosing the character in single (' ') or double (" ") quotes.
- Assembler directives allow you to control the assembler's operation.
- Program code and other constant data should be located in a SECTION that is located in ROM by the linker.
- Data variables should be located in a SECTION that is located in RAM by the linker.
- Memory space for data variables must be allocated by the DS.B or DS.W assembler directives.
- An EQU directive allows us to define symbols and constants.
- Byte constants to be in ROM may be defined using the DC.B directive.
- Sixteen-bit constants in ROM are defined using the DC.W directive.
- The assembler may produce a listing file as well as object files.
- A current listing file should be printed to help with debugging.

5.9 Bibliography and Further Reading

Cady, F. M., *Microcontrollers and Microprocessors*, Oxford University Press, New York, 1997.

Motorola HC12 Assembler, Metrowerks Corporation, Austin, TX, 2003.

5.10 Problems

Basic

5.1 Give the assembler fields (label, opcode, operand, Comment) for the following instructions: **[a]**

```
a.              staa  $800  ; Save the variable
b.              bra   loop
c.     loop:   ldaa  #$23  ; Initialize counter
```

5.2 Give four ways to specify each of the following constants. **[k]**

a. The ASCII character X.

b. The ASCII character x.

c. 100_{10}

d. 64_{16}

5.3 Give the symbol used when specifying a constant in the following bases: hexadecimal, decimal, binary, ASCII. **[k]**

5.4 Write a statement that allocates memory for a one-byte variable called COUNT. **[k]**

5.5 Write a statement that creates a string constant in ROM named MSG that contains the text 'This is a message.'. **[k]**

5.6 Write a statement that allocates a byte constant COUNT in ROM and initializes it with the value $26. **[k]**

5.7 Write a statement that allocates a word-size constant BIGCOUNT in ROM and initializes it with the value $1234. **[k]**

5.8 Show how to allocate storage for ten (decimal) bytes. **[k]**

Intermediate

5.9 In the CodeWarrior assembler, describe the difference between DC.B and DS.B. **[k]**

5.10 Consider the term "Allocate data storage." **[g]**

a. What is meant by this term?

b. What are the two lines in an assembly language program to allocate storage for a variable?

c. Why are variable data storage locations initialized at run time instead of at assembly time?

5.11 What is an assembler expression? **[k]**

5.12 Give an example of an assembler expression. **[k]**

Advanced

5.13 Give the result in hexadecimal for each of the following assembler expressions. **[k]**

```
MS_MASK:        EQU        %11110000
LS_MASK:        EQU        %00001111
MOST:           EQU        255
NEG:            EQU        -128
SIX:            EQU        6
TEN:            EQU        10
DATA1:          DS.B       10
DATA2:          DS.B       10
```

```
a. Q1:    DC.B      SIX+TEN
b. Q2:    DC.B      DATA2-DATA1
c. Q3:    DC.B      SIX | TEN
d. Q4:    DC.B      MOST ^ LS_MASK
e. Q5:    DC.B      ~MS_MASK
```

5.14 An assembly language programmer wants to multiply a variable data value stored in DATA1 by a constant 5. What is wrong with the following code segment? **[k]**

```
        ldaa      DATA1*5
        staa      DATA1
```

6 The Linker

OBJECTIVES

This chapter introduces the CodeWarrior® Smart Linker and shows how to link together separately assembled or compiled modules to form a complete executable program.

6.1 Introduction

A *linker program* combines *object* files to produce an *absolute* file to be loaded into the microcontroller's memory.

In this chapter, we investigate the operation of a program used to link together separately assembled and compiled source files into one file. The *linker* produces a file to be loaded into the microcontroller using the correct types of memory for each portion of the program.

6.2 Assumptions

We assume you are using the CodeWarrior tool set and that you will be running the linker from within the CodeWarrior Integrated Development Environment (IDE). Furthermore, at least until we get to Chapter 10 where we consider C programs, we assume you are developing assembly language programs. In the following sections, we describe how the linker locates each of the various parts of an assembly language program in the proper memory.

6.3 Code Development

Figure 6-1 gives a flow chart showing a typical program development.

1. *Design and document software.* Use the top-down design strategy discussed in Chapter 3 to design and document your software modules.
2. *Code the modules.* Write your software modules in either assembly language or a high-level language like C.
3. *Assemble or compile the modules.* This produces *object* code files that will be *linked* together to form the executable program.
4. *Determine the hardware memory map.* Find out where ROM and RAM are located in the target hardware so you can link everything together and place the code in ROM and the data in RAM. See Figure 6-2.

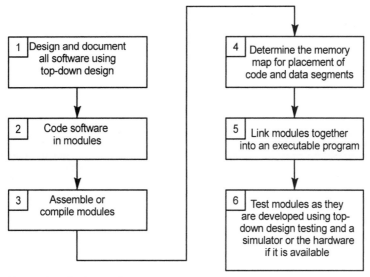

Figure 6-1 Code development sequence.

5. *Link the modules together.* The linker program described in this chapter ties all modules together and locates code and data in the proper memory type.

6. *Test the modules.* Chapter 3 shows a top-down testing strategy that is very effective. In some cases, for part of your testing, you may be able to use a simulator such as the CodeWarrior system. However, simulators are difficult to use to test hardware components, so at some point you will have to use the hardware for final testing. Chapters 9 and 20 discuss debugging techniques useful for both simulation and hardware testing.

6.4 Why Use a Relocatable Assembler/Compiler and a Linker?

A relocatable assembler and compiler allow efficient and effective program development using modular programming methods like you use for C, C++, or other high-level language programs. You may develop functional modules separately, and independently, from one another and then *link* them to *build* the final application to be loaded into the microcontroller's memory. Thus portions of a large project can be allocated to different members of the development team. The linker also allows you to have local labels and to expose only a smaller subset of labels to the global program and other modules. Imagine the problems you could have with duplicate labels if you took many modules built by separate groups and tried to assemble them in one giant assembly language file having thousands of lines.

6.5 Memory Types, Sections, and Section Types

In Chapter 2 we described the operation of the relocatable assembler and compiler. We also described the different types of memory, RAM and ROM, and showed that each type is used for different purposes in our embedded system programs. When using a relocatable assembler, various parts of our programs must be placed in different kinds of memory. This is called *locating*. That is, we must locate the different parts of the program in the proper memory.

Another task to be done is linking together the separate modules. The need for this arises when code in one module calls or refers to code or data elements in another module. This process is also achieved by the linker.

Every program, whether written in assembly language or C, will have various *segments* or *sections*. A section refers to global objects declared in a source file, while segment refers to a type of memory that may not be contiguous. We will generally use the term section, although you will see segment used in some of the CodeWarrior files. The sections that we define are the *code section*, *constant data section*, *variable data section*, and *stack section*.

Code Section

All parts of the program that must remain when power is turned off, including the *code* and *constant data*, are placed in the code section, which is then located in a ROM segment.

The code section contains all of the program code. The program *must* be located in read only memory in an embedded system. In a development environment, where you may have a *development system* or an *evaluation board* (*EVB*), you may be placing your program and constants into RAM. This makes it easier to change and modify as you are developing and debugging your software. Ultimately, however, you will change the physical location of the code section to ROM for the embedded application.

Constant Data Section

Constants are to be located in ROM too.

Constant data elements are also located in ROM. Examples of constant data include variable data initialization values and messages to be printed. In the CodeWarrior system a separate, constant data section does not have to be defined. It is useful to do so, however, to give you additional flexibility in locating all constant data from various modules together.

Variable Data Section

The *variable data* section is located in a RAM segment.

The variable data section contains the storage locations for all variable data used in the program. There are several points to make about variable data storage used in a program:

- All variable data storage locations must be allocated in the program by the assembler or compiler using a define storage (DS) directive.
- All variable data storage must be located in RAM.
- The program code must initialize all variable data, if needed, when the processor is running. You *must not* assume any data value will be in a RAM data storage location when the system power is turned on.

Stack Section

The *stack* section is also located in RAM.

As we will see in later chapters, the stack is an allocation of RAM used for temporary information storage. This could be data saved temporarily or return addresses for subroutines. The stack pointer register must be initialized pointing to RAM before the stack can be used. In the CodeWarrior system, there does not need to be a specific stack section.

6.6 Linker Operation

Locating in Proper Memory

Before we show how to set up the linker to create your program to be loaded into the microcontroller memory, let us look at what it is supposed to do.

Examples 6-1 and 6-2 show the *main* and *subroutine* programs for two separately assembled modules. This program does not do anything sensible so we will not worry particularly about the assembly language code.

MAIN MODULE

In the main program, Example 6-1, there are four distinct uses of memory. The order in which they appear is not important, although we will see some programming style suggestions in Chapter 8. The first is a variable data section shown between *lines 13 and 15*. A memory allocation for a single byte of data storage is made at *line 15* by the assembler directive

```
main_data:     DS.B 1
```

The label `main_data` marks a single byte of storage and will be assigned an address during the linking process. Any other *variable* data storage you might need must be defined in this section or other named sections that can be located in RAM. This section continues until another `SECTION` directive or something other than the define storage (`DS`) directive is found. Everything in this section must be located in RAM.

The next memory use starts at *line 17*. This is the *code section* and it begins with the directive

```
MainCode:     SECTION
```

Everything in this section, from *line 18* to *line 29* is to be located in ROM or EEPROM. You should choose the label `MainCode` to be something meaningful so you can see where it is located when you view the linker map file.

A *constant data section*,

```
MainConst:     SECTION
```

is defined at *line 31* and a constant byte is defined by

```
main_const:     DC.B $44
```

at *line 33*. Because constants go into the same type of memory as the code, you could simply define the bytes following the program code. However, it is better to define a constant section to allow all constants to be grouped together in the linking process or to locate them in a different part of the ROM than the code.

A fourth memory use is implied in *line 11*, where the label __SEG_END_SSTACK[1] is defined to be external. This means that although the label is used in this module, it is defined somewhere else. The value of this label will be defined by the linker and is used with the stack

---

[1] This label starts with two underbar (_) characters.

Example 6-1 Relocatable Main Program

```
Metrowerks HC12-Assembler
(c) COPYRIGHT METROWERKS 1987-2003

Rel. Loc    Obj. code Source line

---- ----     --------- -----------
   1                    ; The main module in a relocatable example.
   2                    ; This program does nothing that is sensible.
   3                    ; *******************************************
   4                    ; Define the entry point for the main program.
   5                            XDEF   Entry, main
   6                    ; Define symbols used in the subroutine.
   7                            XREF   sub_1
   8                    ; *******************************************
   9                    ; Define a symbol to be used to initialize
  10                    ; the stack pointer register
  11                            XREF   __SEG_END_SSTACK
  12                    ; *******************************************
  13                    ; Variable Data Section
  14                    MainData: SECTION
  15 000000            main_data:  DS.B 1
  16                    ; *******************************************
  17                    ; Code section
  18                    MainCode:  SECTION
  19                    Entry:
  20                    main:
  21                    ; Initialize the stack pointer
  22 000000   CFxx xx           lds    #__SEG_END_SSTACK
  23                    ; Initialize the variable main_data.
  24 000003   180C xxxx         movb  main_const,main_data
     000007 xxxx
  25 000009   B6xx xx loop:   ldaa  main_data ; Get the data.
  26 00000C   16xx xx          jsr   sub_1     ; Pass to sub
  27                    ; Increment data for the next loop.
  28 00000F   72xx xx          inc   main_data
  29 000012   06xx xx          jmp   loop    ; Continue forever.
  30                    ; *******************************************
  31                    MainConst:  SECTION
  32                    ; Define constants in the code section
  33 000000   44       main_const: DC.B  $44
```

Example 6-2 Relocatable Subroutine

```
Metrowerks HC12-Assembler
(c) COPYRIGHT METROWERKS 1987-2003

 Rel. Loc   Obj. code Source line

 ---  ----   --------- -----------

    1                     ; Relocatable subroutine example with
    2                     ; variable and constant data.
    3                     ; Input Registers:
    4                     ;   A = some input data
    5                     ; Output Registers:
    6                     ;   A = some output data
    7                     ; Registers modified:
    8                     ;   A, CCR
    9                     ; ****************************************
   10                     ; Define symbols used in the subroutine.
   11                             XDEF   sub_1
   12                     ; ****************************************
   13                     ; Code Section
   14                     SubCode:  SECTION
   15                     sub_1:
   16                     ; On entry, the data is in Accumulator A.
   17                     ; Store the data in the subroutine.
   18 000000 7Axx xx         staa   sub_data
   19                     ; Get some data to return to the main.
   20 000003 B6xx xx         ldaa   sub_const
   21 000006 3D             rts     ; Return to main program.
   22                     ; ****************************************
   23                     ; Define constant data.
   24                     SubConst: SECTION
   25 000000 33          sub_const:DC.B  $33
   26                     ; ****************************************
   27                     ; Variable Data Section.
   28                     SubData:  SECTION
   29 000000             sub_data: DS.B  1
```

pointer initialization instruction at *line 22* in the main program. As we will learn later, the stack pointer register must be initialized before calling any subroutines. In addition, the stack itself must be located in RAM.

TABLE 6-1 Program Memory Requirements

Module		ROM			RAM	
Main	Code	21 bytes		Data	1 byte	
	Constant Data	1 byte		Stack	2 bytes	
Subroutine	Code	7 bytes		Data	1 byte	
	Constant Data	1 byte				

SUBROUTINE MODULE

Let us now look at how memory is used in the subroutine shown in Example 6-2. There are three memory uses in this example. There is a section of code starting at *line 14* and continuing through *line 21*. We see a constant data section at *line 24* and a variable data section at *line 28*.

CODEWARRIOR NOTE

Using SECTIONs as shown in Examples 6-1 and 6-2 is optional in the CodeWarrior system. The assembler and linker will automatically locate code and constants in ROM and variables in RAM based on operation mnemonics, define constant byte (DC.B), and define storage byte (DS.B) directives. However, defining SECTIONs is good documentation and improves readability.

MEMORY NEEDS

Our analysis of memory requirements for the main and subroutine is summarized in Table 6-1. Notice that we have included an estimate of the size of the stack needed for this program. We will defer how we find this to a later chapter.

THE HARDWARE MEMORY MAP

Software engineers must know the system memory map to be able to load their code into the microcontroller's memory.

As programmers, now we know how much ROM and RAM our program will require. As embedded system engineers, we must look at the hardware upon which our program is to be installed. We do that by looking at the *memory map* of the system. In a single-chip microcontroller, such as the one we described in Chapter 2, the manufacturer determines the memory map when the chip is designed and manufactured. Figure 6-2 shows a MC9S12C32 microcontroller operating in single-chip mode in an embedded system with no external RAM.

Linking the Code

Before we show how to instruct the linker to locate our code correctly, we need to discuss the linking part of this job. Linking refers to establishing the final address for certain instructions, such as a *jsr—jump-to-subroutine*, when the starting address of the subroutine is in another module. This address resolution must be performed also when one module refers to a data location defined in another module. *Line 26* in Example 6-1 has a `jsr sub_1` instruction, where `sub_1` is in the second, separately assembled module. The assembler has three directives that organize the information needed to accomplish this linking task.

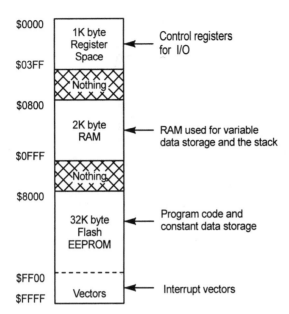

Figure 6-2 Memory map for MC9S12C32 microcontroller.

XREF, XDEF, AND XREFB

> Every XREF must have a companion XDEF in some other module.
> XREF means the symbol is *referenced* here but defined someplace else.
> XDEF means the symbol is *defined* here and referenced someplace else.

When a symbol is used in one module but is defined in another, it is said to be an *external* symbol. In this case, the *XREF—External Reference* directive is used. In the module where the symbol is defined, the *XDEF—External Definition* directive is used. In Example 6-1 the symbol sub_1 is the address for the jump-to-subroutine instruction on *line 26*. Because that symbol is not defined in the main module, it is declared external by XREF sub_1 on *line 7*. Similarly, the sub_1 symbol must be made visible in the subroutine module. This is done by the XDEF sub_1 on *line 11*. For every XREF in our program modules, there must be a partner XDEF in some other module.[2]

Example 6-1 shows XDEF Entry at *line 5*. This makes the code entry point (where the Entry label is at *line 19*) visible to the linker.

There is another directive, *XREFB—External Reference for Symbols on the Direct Page*, included in the assembler directives. This is used when an external symbol is on the direct addressing page and thus fewer addressing bits are required.

6.7 The Linker Parameter File

> The linker parameter file allows us to specify easily where code and data are located.

The linker is controlled by a *linker parameter file (*.prm)*. You put the memory map information shown in Figure 6-2 in the linker parameter file so the linker can locate each section properly. Example 6-3 shows a linker parameter file used to link (and locate) the two modules in Examples 6-1 and 6-2.

---

[2] If you forget to declare a symbol as external with XREF, the assembler will give you an error message. If you forget the XDEF, the linker will generate an error message.

Example 6-3 Linker Parameter File

```
NAMES
END

SECTIONS
    RAM = READ_WRITE 0x0800 TO 0x0FFF;
/* unbanked FLASH ROM */
    ROM_C000 = READ_ONLY  0xC000 TO 0xFEFF;
END

PLACEMENT
/* Place the code and constant sections in ROM */
    .init, MainCode, SubCode,MainConst, SubConst,
    DEFAULT_ROM                                 INTO   ROM_C000;
/* Place the Variable Data */
    MyData, MySubData,DEFAULT_RAM               INTO   RAM;
END
/* Define the stack size needed */
STACKSIZE 0x100
/* Specify the initial reset vector entry point */
VECTOR 0 Entry  /* reset vector */
/* INIT is needed for assembly applications */
INIT Entry
```

Parameter File Commands

Although the CodeWarrior linker is very powerful with many features, we will describe only those few necessary to link assembly and C program modules. Keywords and commands in the parameter file must be capitalized. Each section in the parameter file starts with the command name in capitals and ends with the END. All commands and qualifiers are case sensitive.

Names: The *NAMES* command lists the files building the application. For an application linked in the CodeWarrior tool set, the files that have been "added" to the application are automatically linked. Other modules may be linked by including their names in the NAMES section.

Sections:

```
SECTIONS
    RAM = READ_WRITE 0x0800 TO 0x0FFF;
/* unbanked FLASH ROM */
    ROM_C000 = READ_ONLY  0xC000 TO 0xFEFF;
END
```

TABLE 6-2 Memory Type Qualifiers

Qualifier	Initialized Variables	Noninitialized Variables	Constants	Code
READ_ONLY	Not applicable.[a]	Not applicable.[a]	Content written to target address.	Content written to target address.
READ_WRITE	Content written into copydown area with information to allow it to be copied at startup.[b]	Area contained in zero out information.[c]	Content written into copydown area.[b,c]	To allocate code in a RAM area for system development, you should declare this area as READ_ONLY. In an embedded system, this is not applicable.[a]
NO_INIT	Not applicable.[a]	Handled as allocated.	Not applicable.[a]	Not applicable.[a]
PAGED	Not applicable.[a]	Handled as allocated.	Not applicable.[a]	Not applicable.[a]

[a] These cases are not intended but the linker does allow them. When used, the qualifier controls what is written into the application.

[b] Initialized objects and constants in READ_WRITE must be initialized at the start of the program at run time. In a C program, this is done by the startup code but in an assembly language program, you must ensure this is done. The copydown area contains information to initialize the constants in the startup code.

[c] The zero out information specifies areas that must be initialized with 0 at startup.

The *SECTIONS*[3] part defines the memory map for your application. It allows you to assign meaningful names to address ranges. These can be used in the subsequent placement section to increase the readability of your parameter file. The syntax of a SECTIONS entry is the following:

```
<your_name> = MEMORY_TYPE  address TO address;
```

where `<your_name>` is the name you wish to use for the particular address range, and `MEMORY_TYPE` defines the type of memory as shown by the qualifiers in Table 6-2.

Placement:

```
PLACEMENT
/* Place the code and constant sections in ROM */
    .init, MainCode, SubCode, MainConst, SubConst,
    DEFAULT_ROM                                     INTO   ROM_C000;
/* Place the Variable Data */
    MyData, MySubData, DEFAULT_RAM                  INTO   RAM;
END
```

The *PLACEMENT* section places your various sections into the proper type of memory as defined in the SECTIONS section. You must include `DEFAULT_RAM` and `DEFAULT_ROM` placements. The linker, by default, will place your sections into the proper placement. For example, the linker knows to place any memory allocation (DS) parts of the program into DEFAULT_RAM and any code and constants (DC) into DEFAULT_ROM. You may, for the sake of documentation readability or to control where sections are placed, "place" the named sections from your program as shown in Example 6-3. This is optional but is a good documentation practice.

---

[3] It can also be called SEGMENTS.

Stacksize:

```
/* Define the stack size needed */
STACKSIZE 0x100
```

An important part of your code development is the allocation of storage and placement of the stack. As we will learn in Chapter 7, the stack is an area of RAM used for temporary variable storage and for return addresses for subroutines. Because both the stack and the variable data storage sections are in READ_WRITE memory, it is important that neither one interfere with or overwrite the other. You must estimate the amount of stack storage that is needed and specify at least this amount. Because stack overflow (using more stack memory than is allocated) can cause many problems in your running program, most embedded system engineers are very conservative and allocate plenty of stack space. In this example, STACKSIZE allocates $0100 bytes, places the stack into RAM, and defines the variable __SEG_END_SSTACK to be used in your program to initialize the stack pointer register (as shown in *line 22* in Example 6-1).

Vector:

```
/* Specify the initial reset vector entry point */
VECTOR 0 Entry  /* reset vector */
```

The vector section allows you to initialize vectors. This example shows how to initialize the reset vector to allow the microcomputer to be *vectored* to the start of the program when power is turned on. We will see other vectors and their uses for interrupts in Chapter 12.

Init:

```
/* INIT is needed for assembly applications */
INIT Entry
```

When linking an assembly application, an *entry point* is required. The *INIT* command with its operand, in this case *Entry*, provides this information.

Comments: Comments are entered into the parameter file using standard C or C++ syntax for comments.

6.8 The Linker Output Files

Two output files created by the linker are important to us after we have located and linked our object files. These are the linker map file, which shows us where the linker has placed our code, and the absolute code file to be loaded into the microcontroller's memory.

The Linker Map File

The *linker map file* shows us where all elements of our program are located.

The *linker map file* shows us where our code, constant, and variable data are located, along with other information. After linking Examples 6-1 and 6-2 using the linker parameter file in Example 6-3, the linker map file shows the following sections.

Target Section:

```
TARGET SECTION
------------------------------------------------------------------
Processor    : Motorola HC12
Memory Model : SMALL
File Format  : ELF\Dwarf 2.0
Linker       : SmartLinker V-5.0.22 Build 4047, Feb 17 2004
```

The target section shows the processor, the memory model (more important when programming in C), the file format, and the linker version.

File Section:

```
FILE SECTION
------------------------------------------------------------------
rel_ex_1a_main.asm.o        Model: SMALL,        Lang: Assembler
rel_ex_1a_sub_1.asm.o       Model: SMALL,        Lang: Assembler
```

This lists all files that have been linked.

Startup Section:

```
STARTUP SECTION
------------------------------------------------------------------
Entry point: 0xC000 (Entry)
```

The startup section shows the entry point, that is, where the program will start.

Section-Allocation Section:

```
SECTION-ALLOCATION SECTION
Section Name          Size  Type    From     To       Segment
------------------------------------------------------------------
.init                  21    R      0xC000   0xC014   ROM_C000
SubCode                 7    R      0xC015   0xC01B   ROM_C000
MainConst               1    R      0xC01C   0xC01C   ROM_C000
SubConst                1    R      0xC01D   0xC01D   ROM_C000
MainData                1    R/W    0x800    0x800    RAM
SubData                 1    R/W    0x801    0x801    RAM
.stack                256    R/W    0x802    0x901    RAM
.vectSeg0_vect          2    R      0xFFFE   0xFFFF   .vectSeg0

Summary of section sizes per section type:
READ_ONLY (R):         20  (dec:       32)
READ_WRITE (R/W):     102  (dec:      258)
```

Each of the software components that have been allocated into the various sections is described. Note that the code in the module that is first executed, that is, the main module, is placed into a section named *.init*. In a C program, the startup code will be placed in .init. All other named sections, such as SubCode (in the subroutine), MainConst, MainData, and Subdata, are listed with their size, type of memory, the range of address, and the segment assignment.

Vector-Allocation Section:

```
VECTOR-ALLOCATION SECTION
    Address      InitValue    InitFunction
------------------------------------------------------------------
    0xFFFE         0xC000     Entry
```

This section shows how vectors have been initialized. We will see more of this when we discuss interrupts in Chapter 12.

Object-Allocation Section:

```
OBJECT-ALLOCATION SECTION
    Name           Module        Addr  hSize  dSize  Ref Section RLIB
------------------------------------------------------------------
MODULE:        -- rel_ex_1a_main.asm.o --
- PROCEDURES:
    Entry                        C000    9      9     0 .init
    loop                         C009    C     12     1 .init
- VARIABLES:
    main_data                    800     1      1     3 MainData
    main_const                   C01C    1      1     1 MainConst
- LABELS:
    __SEG_END_SSTACK             902     0      0     1
MODULE:     -- rel_ex_1a_sub_1.asm.o --
- PROCEDURES:
    sub_1                        C015    7      7     1 SubCode
- VARIABLES:
    sub_const                    C01D    1      1     1 SubConst
    sub_data                     801     1      1     1 SubData
```

One of the most important sections is the object-allocation section. It shows, for each of the modules (in this case the main and subroutine), where the procedures, labels, and data elements are located. We will need this information when it comes time to debug our program. The hSize and dSize give the number of bytes in hexadecimal and decimal.

Module Statistic:

```
MODULE STATISTIC
    Name                                   Data    Code   Const
------------------------------------------------------------------
    rel_ex_1a_main.asm.o                     2      42       2
    rel_ex_1a_sub_1.asm.o                    2      14       2
    other                                  256       2       0
```

This shows for each module the number of code, data, and constants bytes that have been allocated. In this listing, *other* refers to the space allocated to the stack.

Section Use in Object-Allocation Section:

```
SECTION USE IN OBJECT-ALLOCATION SECTION
SECTION: ".init"
  Entry loop
SECTION: "SubCode"
  sub_1
SECTION: "MainConst"
  main_const
SECTION: "SubConst"
  sub_const
SECTION: "MainData"
  main_data
SECTION: "SubData"
  sub_data
```

This section shows which elements (or objects) of our program use which memory section.

Object List Sorted by Address:

```
OBJECT LIST SORTED BY ADDRESS
    Name              Addr    hSize    dSize    Ref Section RLIB
-----------------------------------------------------------------
    main_data         800       1        1       3 MainData
    sub_data          801       1        1       1 SubData
    Entry             C000      9        9       0 .init
    loop              C009      C       12       1 .init
    sub_1             C015      7        7       1 SubCode
    main_const        C01C      1        1       1 MainConst
    sub_const         C01D      1        1       1 SubConst
```

The address of each object (denoted by a label) is given here.

Unused Object Section: This lists any objects found in the object files that were not linked, such as a label not referenced.

Copydown Section: This section lists all blocks that are copied from ROM to RAM at program startup. This is more relevant in C programming.

Object-Dependencies Section:

```
OBJECT-DEPENDENCIES SECTION
-----------------------------------------------------------------
Entry                   USES __SEG_END_SSTACK main_const main_data
loop                    USES main_data sub_1 loop
sub_1                   USES sub_data sub_const
```

This section lists the names of global objects that every function and variable specifies.

Dependency Tree: An object dependency tree shows in a tree format all detected dependencies between functions.

Statistic Section:

```
STATISTIC SECTION
-----------------------------------------------------------------------
ExeFile:
--------
Number of blocks to be downloaded: 5
Total size of all blocks to be downloaded: 32
```

This section gives the number of bytes of code in the application.

Absolute Files

The linker can produce two kinds of output files to be loaded in the microcontroller memory. These are absolute (*.abs*) and S files.

Absolute files, which have the extension .abs, contain the program code plus some debugging information. S files, which have extensions .s1, .s2, .s3, or .s19, contain all code from the READ_ONLY sections of the application. These can be downloaded to the microcontroller or programmed into the ROM.

6.9 Remaining Questions

In this chapter, we have described enough of the features of the linker to get started in assembly language programming. We still have a number of issues to cover and we will do that in succeeding chapters. Some of these details yet to be explained include the following topics:

- What is startup code? *Startup code is the code that is executed when the processor is first powered up or the program is first run. In C programs a standard startup code sequence is used. We will discuss this issue further in Chapter 9.*
- Do I have to run the linker if I have only one module with all the code I need? *Yes, you do. In this case no linking will be done but you still need to locate the code in ROM and the variable data in RAM. You could, however, use the absolute assembler features and assign all locations in the program. It is better to use the relocatable assembler.*

6.10 Conclusion and Chapter Summary Points

In this chapter, we discussed the operation of the linker. We included enough details to be able to link assembly language programs using separately assembled relocatable modules.

- There are two kinds of memory in a microcontroller system—RAM and ROM.
- There can be two types of ROM—EEPROM and Flash EEPROM.
- You must know what the memory map is to be able to locate the various parts of the program in the proper memory types.

- Variable data storage and the stack use RAM.
- The program code and constants use ROM.
- The linker is controlled by a linker parameter file.
- It is very easy to change the parameter file to relocate the code for different hardware configurations.

6.11 Bibliography and Further Reading

Smart Linker, Metrowerks Corporation, Austin, TX, 2003.

6.12 Problems

Intermediate

6.1 Answer the following questions based on the given linker parameter file. **[k]**

    a. How much RAM is in this microcontroller?

    b. Where is the RAM located?

    c. Where is the program memory located?

    d. How many bytes have been allocated for the stack?

```
SEGMENTS
      RAM = READ_WRITE 0x0800 TO 0x0FFF;
      /* unbanked FLASH ROM */
      ROM_4000 = READ_ONLY   0x4000 TO 0x7FFF;
      ROM_C000 = READ_ONLY   0xC000 TO 0xFEFF;
      /* banked FLASH ROM */
END

PLACEMENT
      DEFAULT_ROM                 INTO   ROM_C000;
      DEFAULT_RAM                 INTO   RAM;
END

STACKSIZE 0x100
VECTOR 0 Entry /* reset vector: this is the default
entry point for an Assembly application. */
VECTOR ADDRESS 0xFFDE tof_pa0_isr
```

6.2 Answer the following questions based on the given linker parameter file. **[k]**

    a. How much RAM is in this microcontroller?

    b. Where is the RAM located?

    c. Where will MySubData be located?

    d. Where will MyData be located?

    e. Where will MainConst be located?

```
SECTIONS
      RAM = READ_WRITE 0x0800 TO 0x1FFF;
/* unbanked FLASH ROM */
      ROM_C000 = READ_ONLY  0xC000 TO 0xFEFF;
END

PLACEMENT
/* Place the code and constant sections in ROM
*/
      .init, MainCode, SubCode, MainConst,
      SubConst,
      DEFAULT_ROM              INTO  ROM_C000;
/* Place the Variable Data */
      MyData, MySubData,DEFAULT_RAM  INTO  RAM;
END
/* Define the stack size needed */
STACKSIZE 0x100
/* Specify the initial reset vector entry point
*/
VECTOR 0 Entry  /* reset vector */
/* INIT is needed for assembly applications */
INIT Entry
```

Advanced

6.3 Inspect the given (partial) map file and answer the following questions. **[k]**

    a. What source files are used in this program?

    b. What is the program entry point?

    c. How many bytes of memory are allocated in the RAM?

    d. To what memory location does the label "loop" refer?

e. In what memory location will you find the main_data variable?

f. In what memory location will you find the sub_data variable?

g. What is the total number of ROM bytes needed for the program and constant data?

h. If you wished to set a breakpoint at the label sub_1, what address would you specify?

```
*****************************************************************************************
FILE SECTION
-----------------------------------------------------------------------------------------
rel_ex_1a_main.asm.o                           Model: SMALL,        Lang: Assembler
rel_ex_1a_sub_1.asm.o                          Model: SMALL,        Lang: Assembler
*****************************************************************************************
STARTUP SECTION
-----------------------------------------------------------------------------------------
Entry point: 0xC000 (Entry)
*****************************************************************************************
SECTION-ALLOCATION SECTION
Section Name               Size        Type        From          To          Segment
-----------------------------------------------------------------------------------------
.init                       21          R         0xC000       0xC014       ROM_C000
SubCode                      7          R         0xC015       0xC01B       ROM_C000
MainConst                    1          R         0xC01C       0xC01C       ROM_C000
SubConst                     1          R         0xC01D       0xC01D       ROM_C000
MainData                     1         R/W        0x800        0x800        RAM
SubData                      1         R/W        0x801        0x801        RAM
.stack                     256         R/W        0x802        0x901        RAM
.vectSeg0_vect               2          R         0xFFFE       0xFFFF       .vectSeg0

Summary of section sizes per section type:
READ_ONLY (R):         20 (dec:       32)
READ_WRITE (R/W):     102 (dec:      258)
*****************************************************************************************
VECTOR-ALLOCATION SECTION
     Address      InitValue    InitFunction
-----------------------------------------------------------------------------------------
     0xFFFE        0xC000      Entry
*****************************************************************************************
OBJECT-ALLOCATION SECTION
     Name                    Module      Addr       hSize       dSize       Ref
-----------------------------------------------------------------------------------------
MODULE:                     -- rel_ex_1a_main.asm.o --
PROCEDURES:
     Entry                               C000        9           9          0  .init
     loop                                C009        C          12          1  .init
VARIABLES:
     main_data                           800         1           1          3  MainData
     main_const                          C01C        1           1          1  MainConst
LABELS:
     __SEG_END_SSTACK                    902         0           0          1
MODULE:                     -- rel_ex_1a_sub_1.asm.o --
PROCEDURES:
     sub_1                               C015        7           7          1  SubCode
VARIABLES:
     sub_const                           C01D        1           1          1  SubConst
     sub_data                            801         1           1          1  SubData
*****************************************************************************************
```

```
OBJECT LIST SORTED BY ADDRESS
     Name                              Addr      hSize      dSize      Ref      Section
RLIB
-----------------------------------------------------------------------------------------
     main_data                         800       1          1          3        MainData
     sub_data                          801       1          1          1        SubData
     Entry                             C000      9          9          0
     loop                              C009      C          12         1
     sub_1                             C015      7          7          1        SubCode
     main_const                        C01C      1          1          1        MainConst
     sub_const                         C01D      1          1          1        SubConst
*****************************************************************************************
```

7 HCS12 Instruction Set

OBJECTIVES

This chapter describes the HCS12 instruction set and gives examples showing how to use many of the instructions. The instructions are grouped into categories, each with its own functional types of instructions. When you learn these categories, you will be able to find a particular instruction to give you the function you need in your program.

7.1 Introduction

Learning a new instruction set is easier if you *first learn the categories of instructions* to be found and then *learn what instructions are in each category.*

You are about to start on what seems to be a difficult and frustrating task—learning the instruction set of a computer. The HCS12 has over 1000 instructions. Remembering all of these is a daunting task. If one counts the number of different operations, there are 188, still a considerable number. However, there are only five different *categories* of instructions into which 17 different types of operations can be grouped (see Table 7-1). Our strategy for learning the instruction set is to learn first the different instruction categories, which are based on the function or service supplied by the instruction, and then to see the operation types in each category. Programming then becomes much simpler. We *know* what has to be done, for example, temporarily saving a variable for later use; we then *look* in the *instruction category* for the correct operation and *choose* an *addressing mode* to complete the instruction. Of course, it is not quite as simple as this. Simultaneously, we have to manage the resources in the programmer's model and plan what will be happening to those resources a few instructions later.

The following sections describe the operation of instructions in the various instruction categories, with numerous examples. A Freescale Semiconductor, Inc. publication, *CPU Reference Manual*, CPU12RM/AD, contains complete descriptions, including cycle-by-cycle details, of all HCS12 instructions.

7.2 HCS12 Instruction Set

In this chapter we cover all HCS12 instructions in five categories. For example, if data in a register are to be stored in memory, we want to *move* data and Table 7-1 shows that the store register instructions could be used. At each step in your program, you will pick the operation and the appropriate addressing mode. A summary of all instructions is given in Table 7-2. Keep a copy of this list in your programming notebook because it will allow you to easily look up the correct mnemonic for an instruction when you have determined the category and instruction type.

TABLE 7-1 HCS12 Instruction Categories

Instruction Category	Instruction Type	Reference Table	Examples
Move Data	Load Register	Table 7-5	Section 7.5
	Store Register	Table 7-6	Section 7.5
	Transfer Register	Table 7-7	Section 7.6
	Move Memory	Table 7-8	Section 7.7
Modify Data	Decrement/Increment	Table 7-9	Section 7.8
	Clear/Set	Table 7-10	Section 7.9
	Shift/Rotate	Table 7-11	Section 7.10
	Arithmetic	Table 7-12	Section 7.11
	Logic	Table 7-13	Section 7.12
	Condition Code	Table 7-14	Section 7.17
Decision Making	Data Test	Table 7-15	Section 7.13
	Conditional Branch	Tables 7-14 and 7-17	Section 7.14
	Loop Primitives	Table 7-18	Section 7.15
	Branch Bit Set or Clear	Table 7-17	Section 7.14
Flow Control	Jump/Branch	Table 7-19	Section 7.16
	Interrupt	Table 7-20	Section 7.18
Other	Fuzzy Logic	Table 7-21	Section 7.19
	Miscellaneous	Table 7-22	Section 7.20

TABLE 7-2 HCS12 Instruction Set[a]

Mnemonic	Operation	Mnemonic	Operation
Load Registers (see Section 7.5)			
LDAA	$(M) \rightarrow A$	LDAB	$(M) \rightarrow B$
LDD	$(M:M + 1) \rightarrow D$	LDS	$(M:M + 1) \rightarrow SP$
LDX	$(M:M + 1) \rightarrow X$	LDY	$(M:M + 1) \rightarrow Y$
LEAS	$EA \rightarrow SP$	LEAX	$EA \rightarrow X$
LEAY	$EA \rightarrow Y$		
PULA	$(SP) \rightarrow A$	PULB	$(SP) \rightarrow B$
PULD	$(SP:SP + 1) \rightarrow D$	PULC	$(SP) \rightarrow CCR$
PULX	$(SP:SP + 1) \rightarrow X$	PULY	$(SP:SP + 1) \rightarrow Y$
Store Registers (see Section 7.5)			
STAA	$A \rightarrow (M)$	STAB	$B \rightarrow (M)$
STD	$D \rightarrow (M:M + 1)$	STS	$SP \rightarrow (M:M + 1)$
STX	$X \rightarrow (M:M + 1)$	STY	$Y \rightarrow (M:M + 1)$
PSHA	$A \rightarrow (SP)$	PSHB	$B \rightarrow (SP)$
PSHD	$D \rightarrow (SP:SP + 1)$	PSHC	$CCR \rightarrow (SP)$
PSHY	$Y \rightarrow (SP:SP + 1)$	PSHX	$X \rightarrow (SP:SP + 1)$
Transfer/Exchange Registers (see Section 7.6)			
TFR	Any Reg $\rightarrow$ Any Reg	EXG	Any Reg $\leftarrow \rightarrow$ Any Reg
Move Memory Contents (see Section 7.7)			
MOVB	$(M1) \rightarrow (M2)$	MOVW	$(M1:M1 + 1) \rightarrow (M2:M2 + 1)$

TABLE 7-2 Continued

Mnemonic	Operation	Mnemonic	Operation
Decrement/Increment (see Section 7.8)			
DEC	$(M) - 1 \rightarrow (M)$	DECA	$A - 1 \rightarrow A$
DECB	$B - 1 \rightarrow B$	DES	$SP - 1 \rightarrow SP$
DEX	$X - 1 \rightarrow X$	DEY	$Y - 1 \rightarrow Y$
INC	$(M) + 1 \rightarrow (M)$	INCA	$A + 1 \rightarrow A$
INCB	$B + 1 \rightarrow B$	INS	$SP + 1 \rightarrow SP$
INX	$X + 1 \rightarrow X$	INY	$Y + 1 \rightarrow Y$
Clear/Set (see Section 7.9)			
CLR	$0 \rightarrow (M)$	CLRA	$0 \rightarrow A$
CLRB	$0 \rightarrow B$		
BCLR	$0 \rightarrow (M \text{ bits})$	BSET	$1 \rightarrow (M \text{ bits})$
Arithmetic (see Section 7.11)			
ABA	$A + B \rightarrow A$	ABX	$B + X \rightarrow X$ (see LEAX)
ABY	$B + Y \rightarrow Y$ (see LEAY)	ADDA	$A + (M) \rightarrow A$
ADDB	$B + (M) \rightarrow B$	ADDD	$D + (M{:}M + 1) \rightarrow D$
ADCA	$A + (M) + C \rightarrow A$	ADCB	$B + (M) + C \rightarrow B$
DAA	Decimal adjust		
SUBA	$A - (M) \rightarrow A$	SBA	$A - B \rightarrow A$
SUBD	$D - (M{:}M + 1) \rightarrow D$	SUBB	$B - (M) \rightarrow B$
SBCB	$B - (M) - C \rightarrow B$	SBCA	$A - (M) - C \rightarrow A$
NEG	Two's complement (M)	NEGA	Two's complement $\rightarrow A$
NEGB	Two's complement B	SEX	Sign extend A,B,CCR
MUL	Unsigned $A * B \rightarrow D$	EMUL	Unsigned $D*Y \rightarrow Y{:}D$
EMULS	Signed $D * Y \rightarrow Y : D$		
IDIV	Unsigned $D/X \rightarrow X,D$	EDIV	Unsigned $Y{:}D/X \rightarrow Y,D$
EDIVS	Signed $Y : D/X \rightarrow Y,D$	IDIVS	Signed $D/X \rightarrow X,D$
FDIV	Fractional $D/X \rightarrow X,D$		
Logic (see Section 7.12)			
ANDA	$A \bullet (M) \rightarrow A$	ANDB	$B \bullet (M) \rightarrow B$
ANDCC	$CCR \bullet (M) \rightarrow CCR$		
EORB	$B \text{ EOR } (M) \rightarrow B$	EORA	$A \text{ EOR } (M) \rightarrow A$
ORAB	$B \text{ OR } (M) \rightarrow B$	ORAA	$A \text{ OR } (M) \rightarrow A$
ORCC	$CCR \text{ OR } (M) \rightarrow CCR$		
COM	Ones' complement (M)	COMA	Ones' complement A
COMB	Ones' complement B		
Rotates and Shifts (see Section 7.10)			
ROL	Rotate Left (M)	ROLA	Rotate Left A
ROLB	Rotate Left B	ROR	Rotate Right (M)
RORA	Rotate Right A	RORB	Rotate Right B
ASL	Arith Shift Left (M)	ASLA	Arith Shift Left A
ASLB	Arith Shift Left B	ASLD	Arith Shift Left D
ASR	Arith Shift Right (M)	ASRA	Arith Shift Right A
ASRB	Arith Shift Right B		
LSLA	Logic Shift Left A	LSL	Logic Shift Left (M)
LSLD	Logic Shift Left D	LSLB	Logic Shift Left B
LSRA	Logic Shift Right A	LSR	Logic Shift Right (M)
LSRD	Logic Shift Right D	LSRB	Logic Shift Right B
Data Test (see Section 7.13)			
BITA	Test bits in A	BITB	Test bits in B
CBA	$A - B$	CMPA	$A - (M)$
CMPB	$B - (M)$	CPD	$D - (M : M + 1)$

TABLE 7-2 Continued

Mnemonic	Operation	Mnemonic	Operation
CPX	$X - (M : M + 1)$	CPY	$Y - (M : M + 1)$
CPS	$SP - (M : M + 1)$		
TST	Test (M) = 0 or negative	TSTA	Test A = 0 or negative
TSTB	Test B = 0 or negative		

Fuzzy Logic and Specialized Math (see Chapter 19)

Mnemonic	Operation	Mnemonic	Operation
MEM	Membership function	REV	MIN-MAX Rule Evaluation
REVW	Weighted rule evaluation	WAV	Weighted average
EMINM	$MIN(D, (M : M + 1)) \rightarrow (M : M + 1)$	EMIND	$MIN(D, (M : M + 1)) \rightarrow D$
MINM	$MIN(A, (M)) \rightarrow (M)$	MINA	$MIN (A, (M)) \rightarrow A$
EMAXM	$MAX(D, (M : M + 1)) \rightarrow (M : M + 1)$	EMAXD	$MAX(D, (M : M + 1)) \rightarrow D$
MAXM	$MAX(A, (M)) \rightarrow (M)$	MAXA	$MAX(A, (M)) \rightarrow A$
ETBL	16-bit table interpolate	EMACS	Multiply and accumulate
TBL	8-bit table interpolate		

Conditional Branch (see Section 7.14)

Mnemonic	Operation	Mnemonic	Operation
BMI	Short branch minus	LBMI	Long branch minus
BPL	Short branch plus	LBPL	Long branch plus
BVS	Short branch two's complement overflow set	LBVS	Long branch two's-complement overflow set
BVC	Short branch two's-complement overflow clear	LBVC	Long branch two's-complement overflow clear
BLT	Short branch two's-complement less than	LBLT	Long branch two's complement less than
BGE	Short branch two's complement greater than or equal	LBGE	Long branch two's complement greater than or equal
BLE	Short branch two's complement less than or equal	LBLE	Long branch two's complement less than or equal
BGT	Short branch two's complement greater than	LBGT	Long branch two's complement greater than
BEQ	Short branch equal	LBEQ	Long branch equal
BNE	Short branch not equal	LBNE	Long branch not equal
BHI	Short branch higher	LBHI	Long branch higher
BLS	Short branch lower or same	LBLS	Long branch lower or same
BHS	Short branch higher or same	LBHS	Long branch higher or same
BLO	Short branch lower	LBLO	Long branch lower
BCC	Short branch carry clear	LBCC	Long branch carry clear
BCS	Short branch carry set	LBCS	Long branch carry set

Loop Primitive (see Section 7.15)

Mnemonic	Operation	Mnemonic	Operation
DBEQ	Decrement and branch = 0	DBNE	Decrement and branch <> 0
IBEQ	Increment and branch = 0	IBNE	Increment and branch <> 0
TBEQ	Test and branch = 0	TBNE	Test and branch <> 0

Jump and Branch (see Section 7.16)

Mnemonic	Operation	Mnemonic	Operation
JMP	Jump to address		
JSR	Jump to subroutine	Call	Call subroutine
RTS	Return to subroutine	RTC	Return from CALL
BSR	Branch to subroutine		
BRN	Short branch never	LBRN	Long branch never
BRA	Short branch always	LBRA	Long branch always
BRSET	Branch bits set		
BRCLR	Branch bits clear		

TABLE 7-2 Continued

Mnemonic	Operation	Mnemonic	Operation
Condition Code (see Section 7.17)			
ANDCC	Clear CCR Bits	ORCC	Set CCR Bits
Interrupt (see Chapter 12)			
CLI	Clear interrupt mask	SEI	Set interrupt mask
SWI	S/W interrupt	RTI	Return from interrupt
WAI	Wait for interrupt	TRAP	S/W interrupt
Miscellaneous (see Section 7.20)			
NOP	No operation	STOP	Stop clocks
BGND	Background debug mode		

^{a} (M) indicates the instruction addresses memory using immediate, direct, extended, or index addressing. Register Name (A, B, D, X, Y, SP, PC) indicates the contents of that register. (SP) means on the stack. C denotes the contents of carry flag. CCR denotes the contents of the condition code register. EA means Effective Address.

7.3 HCS12 Instruction and Operand Syntax

As we learned in Chapter 5, each assembly language program step is an *operation*, signified by the instruction *mnemonic*, followed by one or more *operands*. Table 7-3 shows a short-hand notation used to describe the operands. Operands may be data or references to memory locations. In this section we describe the syntax used to form the instructions, and starting in Section 7.5 we will describe fully all of the HCS12 instructions using tables and examples.

Addressing Modes

IMMEDIATE ADDRESSING

Look at Table 7-2, which shows the instructions in the load register category, and Table 7-3, which shows the operand syntax for these instructions. For example, to load the A register with 8-bit data using immediate addressing the instruction syntax is

```
ldaa    #data8i
```

which means "load accumulator A with 8-bit immediate data." A typical 8-bit load instruction would be

```
ldaa    #$28
```

and a 16-bit instruction would be

```
ldx     #$1234
```

DIRECT AND EXTENDED ADDRESSING

The instructions that read data from or write data to memory locations are called *memory reference instructions*. The direct addressing modes—direct and extended—are useful when

TABLE 7-3 HCS12 Operand Syntax

data8i	8-bit immediate data
data16i	16-bit immediate data
adr8a	8-bit direct memory address
adr16a	16-bit extended memory address
constx	Indexed addressing constant: 0 or any 5-, 9-, or 16-bit value
xysp	Either of the X, Y, SP, or PC registers
xys	Either of the X, Y, or SP registers
abd	Either of the A, B, or D registers
const16	16-bit indexed addressing constant
data3	3-bit, pre- or post-, increment or decrement value in the range 1–8.
−xys+	Pre- or post-, increment or decrement of register X, Y, or SP
rel9	Label of a branch destination −512 to +511 locations

accessing a specific memory location each time the instruction is executed. A direct (8-bit) or extended (16-bit) address follows the operation mnemonic. We would rarely specify the address using a hexadecimal or decimal number. It is far preferable to refer to a memory location symbolically.

```
ldaa     Data1
 jmp     Loop
```

INDEXED ADDRESSING

There are three forms for indexed addressing:

```
ldaa     const,xysp
ldx      register,xysp
ldaa     data3,-xys+
```

The first means "load accumulator A with an 8-bit value using indexed addressing." The offset is given by *const* that may be 0 or any 5-, 9-, or 16-bit value. The index register may be X, Y, the stack pointer, or the program counter (*xysp*). A typical indexed addressing instruction is

```
ldaa     3,x
```

The second form means "load index register x with a 16-bit value using indexed addressing with a register (A, B, or D) providing the offset." Unlike the first form, where the offset is a constant, the offset here is in a register and can be a variable.

The third form means "load accumulator A with an 8-bit value using automatic pre- or post-, decrement or increment, indexed addressing." The increment or decrement value is given by *data3*, which must be in the range of 1 to 8, and the index register that may be used is X, Y, or the stack pointer (*xys*). A typical instruction of this type is

```
ldaa     1,x+
```

When you need to access different memory locations with an instruction in a loop, say, when reading data from or writing to a table, indexed addressing must be used. To refresh your memory, review the sections in Chapter 4 that describe indexed addressing.

TABLE 7-4 Indexed Addressing Summary

Offset	Registers Used	Instruction Format	
5-, 9-, and 16-bit signed integer	X, Y, SP, or PC	opcode	offset,register
A, B, or D register	X, Y, SP, or PC	opcode	A, register
		opcode	B,register
		opcode	D,register
+1 to +8 automatic increment or decrement	X, Y, SP	opcode	decrement,−register
		opcode	decrement,register−
		opcode	increment,+register
		opcode	increment,register+
16-bit or D register indexed-indirect	S, Y, SP, or PC	opcode	[offset,register]
		opcode	[D,register]

Table 7-4 gives a summary of indexed addressing modes and shows registers that can be used and the instruction format for the different addressing offsets.

7.4 Instruction Categories

In this section we summarize all instructions, giving their opcodes, symbolic operation, addressing modes, and effect on the condition code register. You will find examples showing how to use these instructions in the next sections.

Move Data Instructions

See Tables 7-5 to 7-8.

TABLE 7-5 Load Register Instructions

			Addressing Modes[a]						Condition Codes[b]			
Function	Opcode	Symbolic Operation	IMM	DIR	EXT	IDX	IDR	INH	N	Z	V	C
Load Accumulator A	LDAA	$(M) \to A$	x	x	x	x	x		$\updownarrow$	$\updownarrow$	0	—
Load Accumulator B	LDAB	$(M) \to B$	x	x	x	x	x		$\updownarrow$	$\updownarrow$	0	—
Load Accumulator D	LDD	$(M : M + 1) \to D$	x	x	x	x	x		$\updownarrow$	$\updownarrow$	0	—
Load Stack Pointer	LDS	$(M : M + 1) \to SP$	x	x	x	x	x		$\updownarrow$	$\updownarrow$	0	—
Load Index Register X	LDX	$(M : M + 1) \to X$	x	x	x	x	x		$\updownarrow$	$\updownarrow$	0	—
Load Index Register Y	LDY	$(M : M + 1) \to Y$	x	x	x	x	x		$\updownarrow$	$\updownarrow$	0	—
Load SP Effective Address	LEAS	$EA \to SP$				x			—	—	—	—
Load X Effective Address	LEAX	$EA \to X$				x			—	—	—	—
Load Y Effective Address	LEAY	$EA \to Y$				x			—	—	—	—
Pull A from Stack	PULA	$(SP) \to A$						x	—	—	—	—
Pull B from Stack	PULB	$(SP) \to B$						x	—	—	—	—
Pull CCR from Stack	PULC	$(SP) \to CR$						x	$\updownarrow$	$\updownarrow$	$\updownarrow$	$\updownarrow$
Pull D from Stack	PULD	$(SP : SP + 1) \to D$						x	—	—	—	—

TABLE 7-5 Continued

			Addressing Modes[a]						Condition Codes[b]			
Function	Opcode	Symbolic Operation	I M M	D I R	E X T	I D X	I D R	I N H	N	Z	V	C
Pull X from Stack	PULX	$(SP:SP + 1) \rightarrow X$						x	–	–	–	–
Pull Y from Stack	PULY	$(SP:SP + 1) \rightarrow Y$						x	–	–	–	–

[a]IDX, indexed addressing; IDR, indexed-indirect addressing.
[b]Only the condition code register bits of interest to the programmer are shown in these tables. The S, X, and I bits are related to the STOP and interrupt group instructions, and, although the H bit is modified by many instructions, it is of no concern to us. The following notation is used to show changes in the condition code register:

- No change
0 Reset to zero
1 Set to one
↕ Changed to one or zero
↑ May be set or remain cleared
↓ May be cleared or remain set
? May be changed but final state unknown

TABLE 7-6 Store Register Instructions

			Addressing Modes						Condition Codes			
Function	Opcode	Symbolic Operation	I M M	D I R	E X T	I D X	I D R	I N H	N	Z	V	C
Store Accumulator A	STAA	$A \rightarrow (M)$		x	x	x	x		↕	↕	0	–
Store Accumulator B	STAB	$B \rightarrow (M)$		x	x	x	x		↕	↕	0	–
Store Accumulator D	STD	$D \rightarrow (M : M + 1)$		x	x	x	x		↕	↕	0	–
Store Stack Pointer	STS	$SP \rightarrow (M : M + 1)$		x	x	x	x		↕	↕	0	–
Store Index Register X	STX	$X \rightarrow (M : M + 1)$		x	x	x	x		↕	↕	0	–
Store Index Register Y	STY	$Y \rightarrow (M : M + 1)$		x	x	x	x		↕	↕	0	–
Push A to Stack	PSHA	$A \rightarrow (SP)$						x	–	–	–	–
Push B to Stack	PSHB	$B \rightarrow (SP)$						x	–	–	–	–
Push CCR to Stack	PSHC	$CCR \rightarrow (SP)$						x	–	–	–	–
Push D to Stack	PSHD	$D \rightarrow (SP : SP + 1)$						x	–	–	–	–
Push X to Stack	PSHX	$X \rightarrow (SP : SP + 1)$						x	–	–	–	–
Push Y to Stack	PSHY	$Y \rightarrow (SP + SP + 1)$						x	–	–	–	–

TABLE 7-7 Transfer Register Instructions

			Addressing Modes						Condition Codes			
Function	Opcode	Symbolic Operation	I M M	D I R	E X T	I D X	I D R	I N H	N	Z	V	C
Transfer A to B	TAB	$A \rightarrow B$						x	↕	↕	0	–
Transfer A to CCR	TAP	$A \rightarrow CCR$						x	↕	↕	↕	↕
Transfer B to A	TBA	$B \rightarrow A$						x	↕	↕	0	–
Transfer Registers	TFR Reg,Reg	$Reg \rightarrow Reg$						x	–	–	–	–
Transfer Reg to CCR	TFR Reg,CCR	$Reg \rightarrow CCR$						x	↕	↕	↕	↕
Transfer CCR to A	TPA	$CCR \rightarrow A$						x	–	–	–	–
Exchange Registers	EXG Reg,Reg	$Reg \leftrightarrow Reg$						x	–	–	–	–

TABLE 7-8 Move Instructions

			Addressing Modes						Condition Codes			
Function	Opcode	Symbolic Operation	IMM	DIR	EXT	IDX	IDR	INH	N	Z	V	C
Move Byte	MOVB	(M1) → (M2)	X		X	X			–	–	–	–
Move Word	MOVW	(M1 : M1 + 1) → (M2 : M2 + 1)	X		X	X			–	–	–	–

Modify Data Instructions

See Tables 7-9 to 7-14.

TABLE 7-9 Decrement and Increment Instructions

			Addressing Modes						Condition Codes			
Function	Opcode	Symbolic Operation	IMM	DIR	EXT	IDX	IDR	INH	N	Z	V	C
Decrement Memory	DEC	(M) − 1 → (M)			x	x	x		↕	↕	↕	–
Decrement A	DECA	A − 1 → A						x	↕	↕	↕	–
Decrement B	DECB	B − 1 → B						x	↕	↕	↕	–
Decrement X	DEX	X − 1 → X						x	–	↕	–	–
Decrement Y	DEY	Y − 1 → Y						x	–	↕	–	–
Decrement SP	DES[a]	S − 1 → S						x	–	–	–	–
Increment Memory	INC	(M) + 1 → (M)			x	x	x		↕	↕	↕	–
Increment A	INCA	A + 1 → A						x	↕	↕	↕	–
Increment B	INCB	B + 1 → B						x	↕	↕	↕	–
Increment X	INX	X + 1 → X						x	–	↕	–	–
Increment Y	INY	Y + 1 → Y						x	–	↕	–	–
Increment SP	INS[a]	S + 1 → S						x	–	–	–	–

[a] DES and INS are equivalent to LEAS −1,S and LEAS 1,S.

TABLE 7-10 Clear and Set Instructions

			Addressing Modes						Condition Codes			
Function	Opcode	Symbolic Operation	IMM	DIR	EXT	IDX	IDR	INH	N	Z	V	C
Clear Memory	CLR	0 → (M)			x	x	x		0	1	0	0
Clear A	CLRA	0 → A						x	0	1	0	0
Clear B	CLRB	0 → B						x	0	1	0	0
Clear Bits in Memory	BCLR				x	x	x		↕	↕	0	–
Set Bits in Memory	BSET				x	x	x		↕	↕	0	–

TABLE 7-11 Shift and Rotate Instructions

Function	Opcode	Symbolic Operation	IMM	DIR	EXT	IDX	IDR	INH	N	Z	V	C
Arithmetic Shift Left Memory	ASL	Figure 7-5			x	x	x		↕	↕	↕	↕
Arithmetic Shift Left A	ASLA							x	↕	↕	↕	↕
Arithmetic Shift Left B	ASLB							x	↕	↕	↕	↕
Arithmetic Shift Left D (16-bit)	ASLD							x	↕	↕	↕	↕
Arithmetic Shift Right Memory	ASR	Figure 7-6			x	x	x		↕	↕	↕	↕
Arithmetic Shift Right A	ASRA							x	↕	↕	↕	↕
Arithmetic Shift Right B	ASRB							x	↕	↕	↕	↕
Arithmetic Shift Right D (16-bit)	ASRD							x	↕	↕	↕	↕
Logical Shift Left Memory	LSL	Figure 7-7			x	x	x		↕	↕	↕	↕
Logical Shift Left A	LSLA							x	↕	↕	↕	↕
Logical Shift Left B	LSLB							x	↕	↕	↕	↕
Logical Shift Left D (16-bit)	LSLD							x	↕	↕	↕	↕
Logical Shift Right Memory	LSR	Figure 7-8			x	x	x		0	↕	↕	↕
Logical Shift Right A	LSRA							x	0	↕	↕	↕
Logical Shift Right B	LSRB							x	0	↕	↕	↕
Logical Shift Right D (16-bit)	LSRD							x	0	↕	↕	↕
Rotate Left Memory	ROL	Figure 7-9			x	x	x		↕	↕	↕	↕
Rotate Left A	ROLA							x	↕	↕	↕	↕
Rotate Left B	ROLB							x	↕	↕	↕	↕
Rotate Right Memory	ROR	Figure 7-10			x	x	x		↕	↕	↕	↕
Rotate Right A	RORA							x	↕	↕	↕	↕
Rotate Right B	RORB							x	↕	↕	↕	↕

TABLE 7-12 Arithmetic Instructions

Function	Opcode	Symbolic Operation	IMM	DIR	EXT	IDX	IDR	INH	N	Z	V	C
Add B to A	ABA	$A + B \to A$						x	↕	↕	↕	↕
Add Memory to A	ADDA	$A + (M) \to A$	x	x	x	x	x		↕	↕	↕	↕
Add Memory to B	ADDB	$B + (M) \to B$	x	x	x	x	x		↕	↕	↕	↕
Add Memory to D (16-bit)	ADDD	$D + (M : M +1) \to D$	x	x	x	x	x		↕	↕	↕	↕
Add with Carry to A	ADCA	$A + (M) + C \to A$	x	x	x	x	x		↕	↕	↕	↕
Add with Carry to B	ADCB	$B + (M) + C \to B$	x	x	x	x	x		↕	↕	↕	↕
Decimal Adjust	DAA							x	↕	↕	↕	↕
Subtract B from A	SBA	$A - B \to A$						x	↕	↕	↕	↕
Subtract Memory from A	SUBA	$A - (M) \to A$	x	x	x	x	x		↕	↕	↕	↕
Subtract Memory from B	SUBB	$B - (M) \to B$	x	x	x	x	x		↕	↕	↕	↕
Subtract with Carry from A	SBCA	$A - (M) - C \to A$	x	x	x	x	x		↕	↕	↕	↕
Subtract with Carry from B	SBCB	$B - (M) - C \to B$	x	x	x	x	x		↕	↕	↕	↕
Subtract Memory from D (16-bit)	SUBD	$D - (M : M +1) \to D$	x	x	x	x	x		↕	↕	↕	↕
Sign Extend A, B, CCR	SEX	See Example 7-30						x	—	—	—	—
Two's-Complement Memory	NEG	$-(M) \to (M)$			x	x	x		↕	↕	↕	↕

TABLE 7-12 Continued

Function	Opcode	Symbolic Operation	IMM	DIR	EXT	IDX	IDR	INH	N	Z	V	C
Two's Complement A	NEGA	−A → A						x	↕	↕	↕	↕
Two's Complement B	NEGB	−B → B						x	↕	↕	↕	↕
Unsigned 8-bit Multiply A * B	MUL	A * B → D						x	—	—	—	↕
Unsigned 16-bit Multiply	EMUL	D * Y → Y : D						x	↕	↕	—	↕
Signed 16-bit Multiply	EMULS	D * Y → Y : D						x	↕	↕	—	↕
Unsigned 32/16-bit Division	EDIV	Y : D/S → Y,D						x	↕	↕	↕	↕
Signed 32/16-bit Division	EDIVS	Y : D/X → Y,D						x	↕	↕	↕	↕
Unsigned 16/16-bit Division	IDIV	D/X → X,D						x	—	↕	0	↕
Signed 16/16-bit Division	IDIVS	D/X → X,D						x	↕	↕	↕	↕
Fractional Division	FDIV	D/X → X,D						x	—	↕	↕	↕

TABLE 7-13 Logic Instructions

Function	Opcode	Symbolic Operation	IMM	DIR	EXT	IDX	IDR	INH	N	Z	V	C
AND A with Memory	ANDA	A AND (M) → A	x	x	x	x	x		↕	↕	0	—
AND B with Memory	ANDB	B AND (M) → B	x	x	x	x	x		↕	↕	0	—
AND CCR with Memory	ANDCC	CCR AND #data → CCR	x						↓	↓	↓	↓
Exclusive OR A with Memory	EORA	A EOR (M) → A	x	x	x	x	x		↕	↕	0	—
Exclusive OR B with Memory	EORB	B EOR (M) → B	x	x	x	x	x		↕	↕	0	—
Inclusive OR A with Memory	ORAA	A OR (M) → A	x	x	x	x	x		↕	↕	0	—
Inclusive OR B with Memory	ORAB	B OR (M) → B	x	x	x	x	x		↕	↕	0	—
Inclusive OR CCR with Constant	ORCC	CCR OR #data → CCR	x						↑	↑	↑	↑
One's-Complement Memory	COM	(M)* → (M)			x	x	x		↕	↕	0	1
One's Complement A	COMA	A* → A						x	↕	↕	0	1
One's Complement B	COMB	B* → B						x	↕	↕	0	1

TABLE 7-14 Condition Code Register Instructions

Function	Opcode	Symbolic Operation	IMM	DIR	EXT	IDX	IDR	INH	N	Z	V	C
AND CCR with Constant	ANDCC	CCR AND #data → CCR	x						↓	↓	↓	↓
Inclusive OR CCR with Constant	ORCC	CCR OR #data → CCR	x						↑	↑	↑	↑
Set Carry Bit	SEC	ORCC #$01	x						—	—	—	↑
Clear Carry Bit	CLC	ANDCC #$FE	x						—	—	—	↓
Set Overflow Bit	SEV	ORCC #$02	x						—	—	↑	—
Clear Overflow Bit	CLV	ANDCC #$FD	x						—	—	↓	—

Decision Making Instructions

See Tables 7-15 to 7-18.

TABLE 7-15 Data Test Instructions

			Addressing Modes						Condition Codes			
Function	Opcode	Symbolic Operation	I M M	D I R	E X T	I D X	I D R	I N H	N	Z	V	C
Test Bits in A	BITA	A AND (M)	x	x	x	x	x		$\updownarrow$	$\updownarrow$	0	—
Test Bits in B	BITB	B AND (M)	x	x	x	x	x		$\updownarrow$	$\updownarrow$	0	—
Compare A to B	CBA	A − B						x	$\updownarrow$	$\updownarrow$	$\updownarrow$	$\updownarrow$
Compare A to Memory	CMPA	A − (M)	x	x	x	x	x		$\updownarrow$	$\updownarrow$	$\updownarrow$	$\updownarrow$
Compare B to Memory	CMPB	B − (M)	x	x	x	x	x		$\updownarrow$	$\updownarrow$	$\updownarrow$	$\updownarrow$
Compare D to Memory (16-bit)	CPD	D − (M : M + 1)	x	x	x	x	x		$\updownarrow$	$\updownarrow$	$\updownarrow$	$\updownarrow$
Compare X to Memory (16-bit)	CPX	X − (M : M + 1)	x	x	x	x	x		$\updownarrow$	$\updownarrow$	$\updownarrow$	$\updownarrow$
Compare Y to Memory (16-bit)	CPY	Y − (M : M + 1)	x	x	x	x	x		$\updownarrow$	$\updownarrow$	$\updownarrow$	$\updownarrow$
Compare S to Memory (16-bit)	CPS	S − (M : M + 1)	x	x	x	x	x		$\updownarrow$	$\updownarrow$	$\updownarrow$	$\updownarrow$
Test Memory for Zero or Negative	TST	(M) − 0			x	x	x		$\updownarrow$	$\updownarrow$	0	0
Test A for Zero or Negative	TSTA	A − 0						x	$\updownarrow$	$\updownarrow$	0	0
Test B for Zero or Negative	TSTB	B − 0						x	$\updownarrow$	$\updownarrow$	0	0

TABLE 7-16 Conditional Branch Instructions

			Addressing Modes	Condition Codes			
Function	Opcode	Symbolic Operation	R E L	N	Z	V	C
Branch if Minus	BMI, LBMI	Branch if N = 1	x	—	—	—	—
Branch if Plus	BPL, LBPL	Branch if N = 0	x	—	—	—	—
Branch if Overflow Set	BVS, LBVS	Branch if V = 1	x	—	—	—	—
Branch if Overflow Clear	BVC, LBVC	Branch if V = 0	x	—	—	—	—
Branch Less Than	BLT, LBLT	Branch if N EOR V = 1	x	—	—	—	—
Branch Greater Than or Equal	BGE, LBGE	Branch if N EOR V = 0	x	—	—	—	—
Branch Less Than or Equal	BLE, LBLE	Branch if Z OR [N EOR V] = 1	x	—	—	—	—
Branch Greater Than	BGT, LBGT	Branch if Z OR [N EOR V] = 0	x	—	—	—	—
Branch Equal	BEQ, LBEQ	Branch if Z = 1	x	—	—	—	—
Branch Not Equal	BNE, LBNE	Branch if Z = 0	x	—	—	—	—
Branch Higher	BHI, LBHI	Branch if C OR Z = 0	x	—	—	—	—
Branch Lower or the Same	BLS, LBLS	Branch if C OR Z = 1	x	—	—	—	—
Branch Higher or the Same	BHS, LBHS	Branch if C = 0	x	—	—	—	—
Branch Lower	BLO, LBLO	Branch if C = 1	x	—	—	—	—
Branch Carry Clear	BCC, LBCC	Branch if C = 0	x	—	—	—	—
Branch Carry Set	BCS, LBCS	Branch if C = 1	x	—	—	—	—

TABLE 7-17 Conditional Branch if Bits Set or Clear

Function	Opcode	Symbolic Operation	Addressing Modes						Condition Codes			
			REL	DIR	EXT	IDX	IDR	INH	N	Z	V	C
Branch if Bit Set	BRSET	Branch if bits in mask are set		X	X	X			—	—	—	—
Branch if Bit Clear	BRCLR	Branch if bits in mask are clear		X	X	X			—	—	—	—

TABLE 7-18 Loop Primitive Instructions

Function	Opcode	Symbolic Operation	Addressing Modes	Condition Codes			
			REL	N	Z	V	C
Decr and Branch = 0	DBEQ	Reg − 1 → Reg, Branch if Zero	X	—	—	—	—
Decr and Branch <> 0	DBNE	Reg − 1 → Reg, Branch if not Zero	X	—	—	—	—
Incr and Branch = 0	IBEQ	Reg + 1 → Reg, Branch if Zero	X	—	—	—	—
Incr and Branch <> 0	IBNE	Reg + 1 → Reg, Branch if not Zero	X	—	—	—	—
Test and Branch = 0	TBEQ	Test Reg, Branch if Zero	X	—	—	—	—
Test and Branch <> 0	TBNE	Test Reg, Branch if not Zero	X	—	—	—	—

Flow Control Instructions

See Tables 7-19 and 7-20.

TABLE 7-19 Unconditional Jump and Branch Instructions

Function	Opcode	Symbolic Operation	Addressing Modes						Condition Codes			
			REL	DIR	EXT	IDX	IDR	INH	N	Z	V	C
Jump to Address	JMP	EA → PC			X	X	X		—	—	—	—
Jump to Subroutine	JSR	EA → PC		X	X	X	X		—	—	—	—
Branch to Subroutine	BSR	EA → PC	X						—	—	—	—
Call Subroutine	CALL				X	X	X		—	—	—	—
Return from Subroutine	RTS	EA → PC						X	—	—	—	—
Return from Call	RTC							X	—	—	—	—
Branch Always	BRA		X						—	—	—	—
	LBRA		X						—	—	—	—
Branch Never	BRN		X						—	—	—	—
	LBRN		X						—	—	—	—

TABLE 7-20 Interrupt Instructions

| Function | Opcode | Symbolic Operation | Addressing Modes | | | | | | Condition Codes | | | |
			IMM	DIR	EXT	IDX	IDR	INH	N	Z	V	C
Clear Interrupt Mask	CLI	$0 \to I$						x	—	—	—	—
Set Interrupt Mask	SEI	$1 \to I$						x	—	—	—	—
Return from Interrupt	RTI	$(SP) \to PC$						x	↕	↕	↕	↕
S/W Interrupt	SWI	$Vector \to PC$						x	—	—	—	—
Wait for Interrupt	WAI							x	—	—	—	—
Opcode Trap	TRAP							x	—	—	—	—

Other Instructions

See Tables 7-21 and 7-22.

TABLE 7-21 Fuzzy Logic and Maximum/Minimum Instructions

| Function | Opcode | Symbolic Operation | Addressing Modes | | | | | | Condition Codes | | | |
			IMM	DIR	EXT	IDX	IDR	INH	N	Z	V	C
Membership Function	MEM								?	?	?	?
Rule Evaluation	REV								?	?	↕	?
Weighted Rule Evaluation	REVW								?	?	↕	↕
Weighted Average	WAV								?	1	?	?
Minimum $\to$ D	EMIND	$MIN(D,(M:M+1)) \to D$					x	x	↕	↕	↕	↕
Minimum $\to$ (M)	EMINM	$MIN(D,(M:M+1)) \to M$					x	x	↕	↕	↕	↕
Minimum $\to$ A	MINA	$MIN(A,(M)) \to A$					x	x	↕	↕	↕	↕
Minimum $\to$ (M)	MINM	$MIN(A,(M)) \to (M)$					x	x	↕	↕	↕	↕
Maximum $\to$ D	EMAXD	$MAX(D,(M:M+1)) \to D$					x	x	↕	↕	↕	↕
Maximum $\to$ (M)	EMAXM	$MAX(D,(M:M+1)) \to M$					x	x	↕	↕	↕	↕
Maximum $\to$ A	MAXA	$MAX(A,(M)) \to A$					x	x	↕	↕	↕	↕
Maximum $\to$ (M)	MAXM	$MAX(A,(M)) \to (M)$					x	x	↕	↕	↕	↕
Ext Mult and Accum	EMACS								↕	↕	↕	↕
Table Lookup	ETBL						x		↕	↕	—	↕
Table Lookup	TBL						x		↕	↕	—	↕

TABLE 7-22 Miscellaneous Instructions

| Function | Opcode | Symbolic Operation | Addressing Modes | | | | | | Condition Codes | | | |
			IMM	DIR	EXT	IDX	IDR	INH	N	Z	V	C
Background Debug	BGND							x	—	—	—	—
No Operation	NOP							x	—	—	—	—
Stop Clock	STOP							x	—	—	—	—

7.5 Load and Store Register Instructions

Eight-Bit Load and Store Instructions

The load and store instructions are shown in Tables 7-5 and 7-6. The main choice you make when choosing the load and store instructions is the type of addressing. Examples 7-1 to 7-4 show how to use the 8-bit load and store instructions.

LOAD AND STORE INSTRUCTIONS MODIFY THE CONDITION CODE REGISTER BITS

> Most load and store instructions modify the condition code register.

Notice in Tables 7-5 and 7-6 that load and store instructions modify the condition code register. This could adversely affect your program as shown in Example 7-3, in which the bne loop instruction is supposed to branch until the decb instruction decrements the B accumulator to zero, setting the Z bit to one. By putting the ldaa #$64 instruction between the decb and the bne loop, the Z bit will never be set and the program will stay in the loop forever. Example 7-4 shows how to fix the problem.

Example 7-1 Load and Store Register Instructions

```
Metrowerks HC12-Assembler
(c) COPYRIGHT METROWERKS 1987-2003

Rel. Loc    Obj. code Source line
---- ------ --------- -----------

   5                     ; Immediate addressing
   6 000000 8640             ldaa   #64      ; Decimal 64 -> A
   7 000002 CF09 02          lds    #$0902   ; $0902 -> SP
   8                     ; Direct addressing
   9 000005 D664             ldab   $64      ; ($0064) -> B
  10 000007 5A65             staa   $65      ; A -> ($0065)
  11                     ; Extended addressing
  12 000009 B612 34          ldaa   $1234    ; ($1234) -> A
  13 00000C FE12 34          ldx    $1234    ; ($1234:1235) -> X
  14 00000F 7E08 00          stx    $0800    ; X -> ($0800)
  15                     ; Indexed addressing
  16 000012 E6E0 17          ldab   23,x     ; (X+23) -> B
  17 000015 6B40             stab   ,y       ; B -> (Y)
  18 000017 A63E             ldaa   2,x-     ; (X) -> A
  19                                         ; X-2 -> X (Auto
                                             ;   decrement)
  20                     ; Indexed-indirect addressing
  21 000019 EEEF             ldx    [d,y]    ; ((D+Y)) -> X
  22 00001B A6E3 0100        ldaa   [$100,x]; ((X+$0100)) -> A
```

Example 7-2

A. What is in A and the N, Z, V, and C CCR bits after an `ldaa #$70` instruction is executed?

Solution: A = $70, NZVC = 000−. The − means that the carry bit is not changed.

B. What does the instruction `ldaa #$94` do?

Solution: It loads A with the value $94 and sets the NZVC bits to 100−.

C. What does the instruction `ldaa #64` do?

Solution: It loads A with the value 64_{10} and sets the NZVC bits to 000−.

D. What does the instruction `ldaa $64` do?

Solution: It loads A from memory location $0064 and sets the N and Z bits according to the data value, clears V, and does not change C.

E. The X register contains $0804. What does the instruction `staa $10,x` do?

Solution: This stores the contents of A into memory location $0814 and sets the N and Z bits according to the data value, clears V, and does not change C.

F. What does the instruction `stab DATA` do?

Solution: It stores B into the memory location at the label `DATA`. It also modifies the N and Z bits according to the data and resets the V bit to 0. The carry bit is unchanged.

Example 7-3 Load and Store Instructions Modify the CCR

```
Metrowerks HC12-Assembler
(c) COPYRIGHT METROWERKS 1987-2003

Rel. Loc    Obj. code Source line
---- ------ --------- -----------
   4
   5        0000 0008 COUNT:  EQU   8   ; Loop counter
   6                          ; . . .
   7 000000 C608                 ldab  #COUNT ; Initialize loop
                                                counter
   8                          ; . . .
   9                          loop:
  10                          ; Here is the code for whatever has to be
  11                          ; done in a loop. At the end of the loop, we
  12                          ; decrement the loop counter and branch back
```

```
13                              ; if it hasn't been decremented to zero.
14                              ; . . .
15 000002 53             decb      ; Decrement B register
16                                  ; and branch back if not
                                    zero
17 000003 C664           ldaa  #$64 ; But first load A with
                                     some data
18 000005 26FB           bne   loop
```

Example 7-4

How could you reorganize the code in Example 7-3 to ensure the branch is taken properly?

Solution:

```
Metrowerks HC12-Assembler

(c) COPYRIGHT METROWERKS 1987-2003

Rel. Loc    Obj. code Source line
---- ------ --------- -----------
  1                             XDEF Entry, main
  2                   Entry:
  3                   main:
  4
  5      0000 0008 COUNT:  EQU   8    ; Loop counter
  6                   ; . . .
  7 000000 C608             ldab  #COUNT ; Initialize loop
                                      counter
  8                   ; . . .
  9                   loop:
 10                   ; Here is the code for whatever has to be
 11                   ; done in a loop. At the end of the loop, we
 12                   ; decrement the loop counter and branch back
 13                   ; if it hasn't been decremented to zero.
 14                   ; . . .
 15 000002 8664             ldaa  #$64 ; Load A with some data
 16                                  ; BEFORE the decrement
 17 000004 53               decb      ; Decrement B register
 18                                  ; and branch back if not
                                      zero
 19 000005 26FB             bne   loop
```

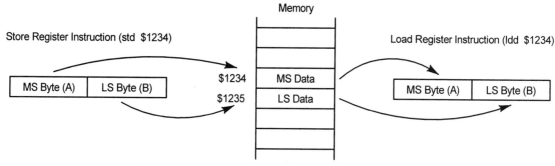

Figure 7-1 Sixteen-bit load and store instructions.

Sixteen-Bit Load and Store Instructions

The 16-bit load and store instructions move data to and from memory. In Freescale microcontrollers, the most significant byte is stored at the effective memory address and the least significant byte in the next location. Figure 7-1 shows how the load and store instructions access memory.

Most programs use a variety of load and store instructions. Example 7-5 contains examples of 16-bit load and store operations, and Example 7-6 shows a program that uses the HCS12 A/D converter to sense a temperature from a nonlinear transducer, such as the thermocouple shown in Figure 7-2. A *lookup table* is used to convert the A/D value to its temperature value. Indexed addressing using the B register is used to access the table values based on the input A/D value.

Example 7-5

A. What does the instruction `ldx #DATA` do?

Solution: It load the X register with the 16-bit value of `DATA`. `DATA` may be a label on a memory location or may be defined by an assembler EQU directive. The N and Z bits are modified, the V bit is reset to 0, and the C bit is unchanged.

B. What does the instruction `ldx #$1234` do?

Solution: It loads the X register with the number $1234 and sets the NZVC bits to 000–.

C. What does the instruction `ldx $1234` do?

Solution: It loads the X register from memory locations $1234:1235 and modifies the NZVC bits according to the data.

D. What does the instruction `ldx DATA` do?

Solution: It loads the X register from the memory locations starting at the label `DATA`. Two locations are used, `DATA` and `DATA+1`. The NZVC bits are modified according to the data.

Example 7-6 Indexed Addressing Instruction Example

Figure 7-2 shows the transfer function of a nonlinear temperature transducer being read by an analog-to-digital converter. Instead of calculating the temperature using a nonlinear equation, a lookup table can be used. Each position (address) in the table will contain the value of the temperature for each reading. The A/D value is used to provide an *index* into the table to retrieve the temperature. Here is a code example to do this.

```
Metrowerks HC12-Assembler
(c) COPYRIGHT METROWERKS 1987-2003

Rel. Loc    Obj. code Source line
---- ------ --------- -----------

  1                               XDEF Entry, main
  2                    Entry:
  3                    main:
  4
  5        0000 0091 ADR1L:  EQU   $0091   ; A/D result register
  6                    ; . . .
  7                    ; Now get an A/D value and linearize it
  8 000000 D691                ldab  ADR1L  ; Get the A/D value
  9 000002 CDxx xx              ldy   #TBLE  ; Init Y to start of
                                               table
 10 000005 A6ED                 ldaa  b,y    ; A now has
                                               linearized value
 11                    ; . . .
 12 000007 6E6D 6A64 TBLE:   DC.B  110,109,106,100,95,89
    00000B 5F59
 13                    ; (Note: All 256 values for an 8-bit A/D
                          input are
 14                    ; not shown here. When the table is
                          completely
 15                    ; populated each location in the table
                          has the
 16                    ; temperature value corresponding to the
 17                    ; thermistor value read by the A/D.)
```

Stack Instructions

The stack is an area of RAM used to *temporarily* save data. The 16-bit *stack pointer register* (*SP*) points to the last byte placed on the stack. The stack pointer must be initialized, with a `ld sp, #address` instruction, where `address` is a valid RAM address, or at least no more than one higher than the highest address of the RAM to be used for the stack.

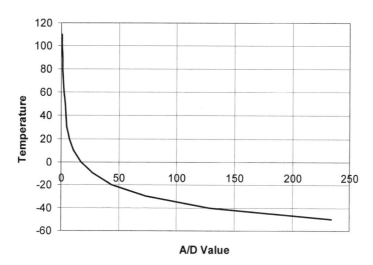

Figure 7-2 Nonlinear thermistor temperature transducer transfer characteristic.

There are three main uses for the stack:

- Temporarily saving a few bytes of data from registers; this is commonly done when entering a subroutine in which you do not want to disturb the contents of a register.
- Saving the return address when the program jumps to or calls a subroutine. This is done automatically by the jump-to-subroutine instruction.
- Allocating data space for temporary variables such as "automatic" variables in C.

The push (PSH) and pull (PUL) instructions store data to and load data from the stack. Figure 7-3 shows push and pull operations. Figure 7-3(a) shows the stack pointer pointing to the last data byte pushed onto the stack. The result of pushing 2 bytes is shown in (b) and (c). New Data 1 and New Data 2 are now in memory and the stack pointer has been decremented twice. The result of a pull is shown in Figure 7-3(d) and a subsequent push in Figure 7-3(e).

Normally, stack operations must be balanced; that is, when retrieving data from the stack, say, when restoring the contents of registers at the end of a subroutine, there must be the same number of PULs as PSHes. In addition, PULs must be in the reverse order of the PSHes. See Examples 7-7 to 7-10.

The jump and call instructions for transferring to a subroutine also use the stack. When a jsr sub instruction is executed, the address of the instruction immediately following the JSR is called the return address. This is the location where the program must return after the subroutine is over. It is pushed onto the stack. Thus a JSR is like a push of the program counter followed by a jump to the subroutine. The return from subroutine instruction, RTS, pulls the return address from the stack.

High level languages such as C use the stack to temporarily store data. The HCS12 supports a clean, practical way to allocate and deallocate stack space. The *load effective address* instructions covered in the next section provide this capability.

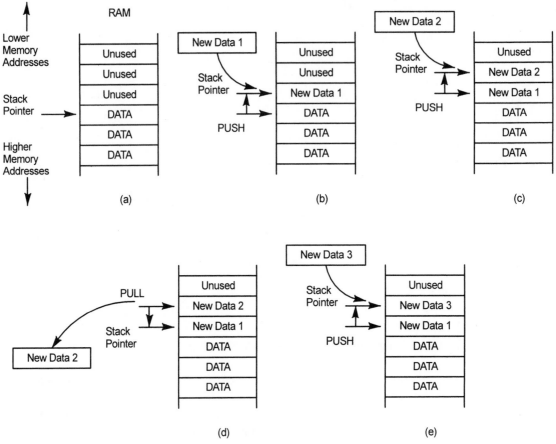

Figure 7-3 Stack operations. (a) Stack pointer before stack operations. (b) Stack pointer after a push. (c) Stack pointer after a second push. (d) Stack pointer after a pull. (e) Stack pointer after a third push.

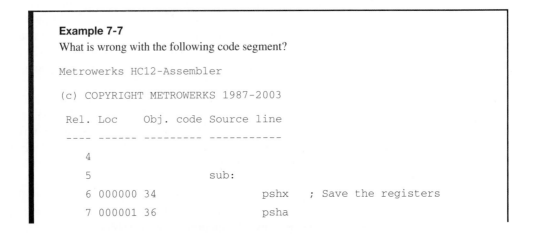

Example 7-7

What is wrong with the following code segment?

```
Metrowerks HC12-Assembler

(c) COPYRIGHT METROWERKS 1987-2003

Rel. Loc    Obj. code Source line
----  ------ --------- -----------
   4
   5                   sub:
   6 000000 34              pshx   ; Save the registers
   7 000001 36              psha
```

```
 8                     ; . . .
 9 000002 37                   pshb    ; Temp save some data
10                     ; . . .
11                     ; . . .
12 000003 32                   pula    ; Restore the registers
13 000004 30                   pulx
14 000005 3D                   rts
```

Solution: The stack operations are unbalanced. There is one more push than pull. The subroutine will not return to the proper place in the calling program.

Example 7-8

What is wrong with the following code segment?

```
Metrowerks HC12-Assembler
(c) COPYRIGHT METROWERKS 1987-2003

Rel. Loc     Obj. code Source line
---- ------  --------- -----------
   4          0000 0008 COUNT:  EQU    8    ; Loop counter
   5                    ; . . .
   6 000000 C608                ldab   #COUNT ; Initialize loop
                                              counter
   7                    ; . . .
   8                    loop:
   9                    ; . . .
  10 000002 36                  psha          ; Temp save A reg
  11                    ; . . .
  12 000003 0431 FC             dbne   b,loop ; Decrement B
                                              register
  13                                          ; and branch if zero
  14 000006 32                  pula
  15                    ; . . .
```

Solution: There is also unbalanced stack usage. There is a push with no corresponding pull inside the loop.

Example 7-9

What is wrong with the following code in a subroutine?

```
Metrowerks HC12-Assembler

(c) COPYRIGHT METROWERKS 1987-2003

 Rel. Loc    Obj. code Source line

 ---- ------ --------- ----------

    4
    5                       sub:
    6 000000 34                     pshx   ; Save the registers
    7 000001 36                     psha
    8 000002 37                     pshb
    9                         ; . . .
   10                         ; . . .
   11 000003 30                     pulx   ; Restore the registers
   12 000004 32                     pula
   13 000005 33                     pulb
   14 000006 3D                     rts
```

Solution: The pull operations must be in the reverse order to properly restore the registers.

Example 7-10

What does the following program sequence do?

```
                        pshx
                        puly
```

Solution: It copies the contents of the X register into the Y register. The condition code register is not modified. An equivalent (and probably better) instruction is TFR X,Y.

Load Effective Address Instructions

An *effective address* (*EA*) is the memory address from or to which data are transferred. Indexed addressing instructions form the effective address by combining an *offset* with a *register*. For example, the instruction

```
        ldaa    $10,x
```

loads the A register from memory location ($10 + the contents of the X register). Thus, the effective address is ($10 + the contents of the X register).

Example 7-11 Load Effective Address Instructions
Assume X = $1234, Y = $1000, and SP = $0A00. Give the contents of each affected register after the following instructions are executed:

Instruction	Result
`leax  10,X`	$X = X + 10_{10} = \$1234 + \$000A = \$123E$
`leax  $10,Y`	$X = Y + \$10 = \$1000 + \$0010 = \$1010$
`leas  -10,SP`	$SP = SP - 10_{10} = \$0A00 - \$000A = \$09F6$

The HCS12 has three instructions that allow the programmer to load the effective address into the stack pointer (LEAS), the X register (LEAX), or the Y register (LEAY). The instruction

```
leay    $10,x
```

will load this effective address into the Y register. The load effective address instructions allow us to calculate memory addresses at run time without modifying the contents of the condition code register.

7.6 Transfer Register Instructions

Table 7-7 shows the transfer register instructions, which transfer data only within the CPU. When saving data temporarily, use one of these instructions if the destination register is not otherwise in use. Transfer instructions copy the source data to the destination register and exchange instructions swap the contents of the two. The TAB, TBA, TFR <reg>,C, and EXG <reg>,C instructions affect the condition code register but the rest of the transfer and exchange instructions do not. The rules for transferring or exchanging between 8- and 16-bit registers are shown in Table 7-23. See Example 7-12.

7.7 Move Instructions

The move instructions are shown in Table 7-8. Move instructions allow you to transfer data from one memory location to another without using any CPU registers. This is an important feature in a processor with only a few registers. For these instructions, indexed addressing

TABLE 7-23 The 8-bit and 16-bit Transfer and Exchange Instructions

Transfer	8-bit → 16-bit	The 8-bit source is transferred to the low byte of the destination; the sign of the source is extended into the high byte of the destination; see the SEX instruction.
	16-bit → 8-bit	The low byte of the 16-bit source is transferred to the 8-bit destination.
Exchange	8-bit ↔ 16-bit	The low bytes of the registers are exchanged and the high byte of the 16-bit register is set to $00.

Example 7-12 Transfer Register Instructions

```
Metrowerks HC12-Assembler
(c) COPYRIGHT METROWERKS 1987-2003

Rel. Loc    Obj. code Source line
---- ------ --------- -----------

     4
     5 000000 180E              tab         ; A -> B
     6 000002 180F              tba         ; B -> A
     7 000004 B701              tfr    a,b  ; A -> B
     8 000006 B710              tfr    b,a  ; B -> A
     9 000008 B781              exg    a,b  ; A <-> B
    10 00000A B750              tfr    x,a  ; Low byte X -> A
    11 00000C B705              tfr    a,x  ; A sign extended -> X
    12 00000E B796              exg    b,y  ; B <-> Low Byte Y, High
                                              Byte Y = $00
```

---

with 9- and 16-bit constant offsets and indexed-indirect addressing are not allowed. Five-bit constant, accumulator, and automatic increment and decrement indexing are allowed. See Examples 7-13 to 7-15.

Example 7-13 MOVB and MOVW Instructions

```
Metrowerks HC12-Assembler
(c) COPYRIGHT METROWERKS 1987-2003

Rel. Loc    Obj. code Source line
---- ------ --------- -----------

     4
     5                    ; Initialize 8-bit memory
     6 000000 180B 64xx          movb   #$64,Data1
       000004 xx
     7                    ; Initialize 16-bit memory
     8 000005 1803 1234          movw   #$1234,Data2
       000009 xxxx
     9 00000B 1804 xxxx          movw   Data2,Data3
       00000F xxxx
    10 000011             Data1: ds.b   1
    11 000012             Data2: ds.w   1
    12 000014             Data3: ds.w   1
```

Example 7-14 Reversing Order of Data

Write a program segment to reverse the order of data in a 100-byte table.

Solution:

```
Metrowerks HC12-Assembler

(c) COPYRIGHT METROWERKS 1987-2003

Rel. Loc    Obj. code Source line
---- ---     --------- -----------
  1                                  XDEF Entry, main
  2                  Entry:
  3                  main:
  4
  5   0000 0064 COUNT:  EQU    100        ; Length of table
  6                  ;  . . .
  7                  ; Reverse the data in the table
  8 000000 CExx xx      ldx   #Table        ; Point to start of
                                              table
  9 000003 CDxx xx      ldy   #Table+COUNT-1 ; Point to the end
                                              byte
 10 000006 C632         ldab  #COUNT/2       ; Initialize counter
 11              loop:
 12 000008 A600         ldaa  0,x            ; Get a byte out of
                                              the way
 13 00000A 180A 4030    movb  0,y,1,x+       ; Get from bottom,
                                              put in top
 14 00000E 6A7F         staa  1,y-           ; Put in bottom and
                                              decr ptr
 15 000010 0431 F5      dbne  b,loop         ; Decr count and
                                              branch <> 0
 16              ;  . . .
 17 000013      Table: DS.B  COUNT
```

Example 7-15 Transferring Data from One Buffer to Another

```
Metrowerks HC12-Assembler

(c) COPYRIGHT METROWERKS 1987-2003

Rel. Loc    Obj. code Source line
---- ---     --------- -----------
  1                                  XDEF Entry, main
  2                  Entry:
```

```
 3                     main:
 4
 5    0000 0064 COUNT:   EQU    100          ; Length of table
 6                  ; . . .
 7                  ; Move from Table1 to Table2
 8 000000 CExx xx         ldx    #Table1    ; Point to Table 1
 9 000003 CDxx xx         ldy    #Table2    ; Point to Table 2
10 000006 C664            ldab   #COUNT     ; Initialize counter
11                  ; . . .
12                  ; Move data from Table1 to Table2 and
                    increment the
13                  ; pointers using auto-increment and indexed
                    addressing
14                  loop:
15 000008 180A 3070  movb  1,x+,1,y+
16 00000C 0431 F9    dbne  b,loop  ; Decr count and branch <> 0
17                  ; . . .
18 00000F           Table1: DS.B    COUNT
19 000073           Table2: DS.B    COUNT
```

7.8 Decrement and Increment Instructions

Decrement and increment instructions (see Table 7-9) are used in many assembler language programs. All of these, except the DES and INS instructions, modify one or more CCR bits. Examples 7-3 and 7-15 show how to decrement a counter in a register. Examples 7-16 and 7-18 show how to decrement 8- and 16-bit counters in memory.

Example 7-16 Eight-Bit Memory Counter
Counters are often used in assembly language programs. If there is no free register, you can use a memory location. Show a segment of code that will create and use a counter in memory.

Solution:

```
(c) COPYRIGHT METROWERKS 1987-2003

Rel. Loc    Obj. code Source line
---- ---    --------- -----------
    4
    5          0000 0064 COUNT:   EQU    100
```

```
 6                          ; Initialize the counter in memory
 7 000000 180B 64xx         movb   #COUNT,Counter
   000004 xx
 8                       ; . . .
 9                       loop:
10                       ; . . .
11 000005 73xx xx            dec    Counter
12 000008 26FB               bne    loop
13                       ; . . .
14 00000A                 Counter: DS.B 1    ; 8-bit counter
```

Example 7-17
Sometimes a counter is needed that is bigger than 255, or 8-bits. A 16-bit counter can be kept in memory like the 8-bit counter shown in Example 7-16. What is wrong with the code sequence shown below?

```
Metrowerks HC12-Assembler

(c) COPYRIGHT METROWERKS 1987-2003

Rel. Loc    Obj. code Source line
---- ------ --------- -----------
    4
    5        0000 03E8 COUNT:  EQU    1000    ; 16-bit value
    6                       ; Initialize the counter in memory
    7 000000 1803 03E8        movw   #COUNT,Counter
      000004 xxxx
    8                       ; . . .
    9                       loop:
   10                       ; . . .
   11 000006 73xx xx            dec    Counter
   12 000009 26FB               bne    loop
   13                       ; . . .
   14 00000B                 Counter: DS.W 1    ; 16-bit counter
```

Solution: The DEC instruction decrements an 8-bit operand. Thus dec Counter will decrement only the most significant byte of the 2-byte counter. In this case, the loop code will be executed only three times instead of 1000 because Counter:Counter + 1 = $03:E8.

Example 7-18 Sixteen-Bit Memory Counter

Show how to fix the problem illustrated in Example 7-17.

Solution: Instead of decrementing the memory location with the dec Counter instruction, you have to load the counter into a 16-bit register, decrement that, and then save the counter.

```
(c) COPYRIGHT METROWERKS 1987-2003

Rel. Loc    Obj. code Source line
---- ---    --------- -----------
  4
  5         0000 03E8 COUNT:  EQU   1000    ; 16-bit value
  6                   ; Initialize the counter in memory
  7 000000 1803 03E8         movw  #COUNT,Counter
    000004 xxxx
  8                   ; . . .
  9                   loop:
 10                   ; . . .
 11                   ; Decrement the 16-bit counter
 12 000006 34                 pshx        ; Save X register
 13 000007 FExx xx            ldx   Counter ; Get the counter and
 14 00000A 09                 dex         ; decrement it
 15 00000B 7Exx xx            stx   Counter
 16 00000E 30                 pulx
 17 00000F 26F5               bne   loop
 18                   ; . . .
 19 000011           Counter: DS.W 1     ; 16-bit counter
```

Example 7-19

What affect do the stx Counter and pulx instructions have on the condition code register and the bne loop instruction in Example 7-18.

Solution: The pulx does not modify the condition code register but the stx instruction does. However, the state of the Z bit, which was set when the dex instruction was executed will not be changed by the stx because the data in X will still be either zero or not zero so the bne loop instruction will work properly.

7.9 Clear and Set Instructions

Clear instructions can clear all 8 bits in an operand. Bit Set and Bit Clear can set and clear individual bits in an 8-bit operand.

See Table 7-10 for a summary of the clear and set instructions. These instructions clear and set bits; CLR and CLRA and CLRB clear memory and accumulators A and B, respectively. BSET and BCLR are *bit addressing* instructions. Any bit in an operand can be modified without affecting the others. The bit clear (BCLR) and bit set (BSET) instructions clear or set individual bits in the selected memory location. The format of these instructions is

```
bclr        Operand,Mask
bset        Operand,Mask
```

Operand is a memory location specified using direct addressing in the first 256 memory locations and extended or indexed addressing for the rest of memory. The bits to be cleared or set are specified by 1's in the *Mask* byte. Any bits in the mask that are zero indicate bits in the operand that are *not* affected by the instruction.

BCLR and BSET are useful when controlling external devices one bit at a time. Figure 7-4 shows eight LEDs connected to port A of the MC9S12C32. As we will see in Chapter 11, data may be output to port A by writing to memory location $0000. Writing a 1 to a bit in port A turns the LED off and writing a 0 turns it on. Example 7-20 shows a program segment to flash the LEDs one at a time in a loop.

In Example 7-20, *lines 10–18*, we see instructions that seem to be the opposite of what we might expect. *Line 11* is an instruction to clear bit 0 and yet the operand mask has a one in the bit-0 position. This is correct. The BSET and BCLR instructions have 1's in the bit positions that are to be set or cleared.

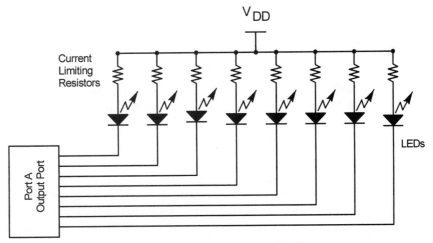

Figure 7-4 BCLR and BSET used for turning on and off LEDs.

Example 7-20 LED Program Using BCLR and BSET

```
Metrowerks HC12-Assembler
(c) COPYRIGHT METROWERKS 1987-2003
Rel. Loc    Obj. code  Source line
---- ------ ---------- -----------

  4
  5           0000 0002 DDRA:  EQU   $0002  ; Data direction
                                              register
  6           0000 0000 PTA:   EQU   $0000  ; Port A register
  7                     ; . . .
  8  000000 4C02 FF            bset  DDRA,%11111111 ; Make all lines
                                              output
  9                     loop:
 10  000003 4C00 FF            bset  PTA,%11111111  ; Set all bits,
                                              LEDs off
 11  000006 4D00 01            bclr  PTA,%00000001  ; Clear bit 0,
                                              LED on
 12  000009 4D00 02            bclr  PTA,%00000010  ; Clear bit 1
 13  00000C 4D00 04            bclr  PTA,%00000100  ; Clear bit 2
 14  00000F 4D00 08            bclr  PTA,%00001000  ; Clear bit 3
 15  000012 4D00 10            bclr  PTA,%00010000  ; Clear bit 4
 16  000015 4D00 20            bclr  PTA,%00100000  ; Clear bit 5
 17  000018 4D00 40            bclr  PTA,%01000000  ; Clear bit 6
 18  00001B 4D00 80            bclr  PTA,%10000000  ; Clear bit 7
 19  00001E 20E3              bra   loop           ; Do it forever
```

Example 7-21

What do you expect to see on the LEDs as the program in Example 7-20 runs?

Solution: We expect to see the LEDs go out and then come on one at a time, starting from the least significant bit, until all are on and then to repeat forever.

What do we actually see?

Solution: All lights will appear to be on because the bclr and bset instructions take less than a microsecond to execute, much too fast for our eyes to respond. If you traced the program one step at a time, you would see the expected behavior. If we wanted to see the LEDs flash on one at a time, we would have to put a delay between each of the instructions in lines 10–18. (The number of clock cycles, and thus the time, required to execute an instruction can be found in the *CPU12 Reference Guide*, CPU12RG/AD available from Freescale Semiconductor, Inc.

7.10 Shift and Rotate Instructions

The arithmetic shift right instruction *preserves the sign* of the number.

There are 34 shift and rotate instructions (see Table 7-11). Each one shifts or rotates the operand only one bit position each time the instruction is executed. Let us look at the shift instructions first. There are two kinds of shifts—arithmetic and logical—and each shifts left and right. Inspect Figures 7-5 to 7-8.

The logical shift instructions shift a zero into either the least significant or most significant bit position and are used for nonnumerical data, such as shifting an LED display pattern. The arithmetic shift instructions are used for numbers. As shown in Figures 7-5 and 7-7, the logical and arithmetic shift left instructions are identical because a zero is shifted into the least significant bit position. Shifting numerical data to the left one bit is equivalent to multiplying the number by 2. The arithmetic and logical shift right instructions are different, as can be seen by comparing Figures 7-6 and 7-8. Shifting a number right is equivalent to dividing by 2 and shifting the most significant bit to the right *preserves the sign* in the arithmetic shift right. The two rotate instructions can be seen in Figures 7-9 and 7-10. These rotate bits around in the operand, including the carry bit, instead of just shifting them.

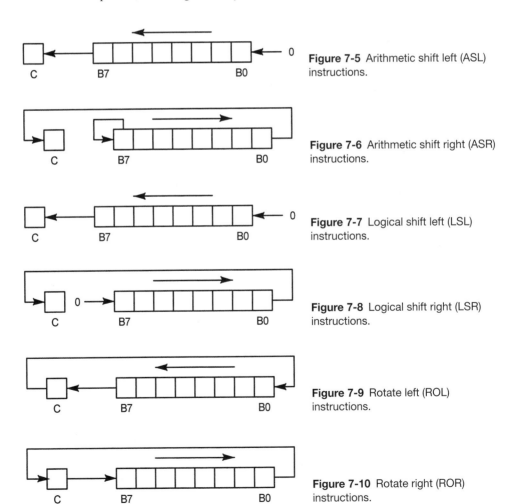

Figure 7-5 Arithmetic shift left (ASL) instructions.

Figure 7-6 Arithmetic shift right (ASR) instructions.

Figure 7-7 Logical shift left (LSL) instructions.

Figure 7-8 Logical shift right (LSR) instructions.

Figure 7-9 Rotate left (ROL) instructions.

Figure 7-10 Rotate right (ROR) instructions.

Shifts and rotates maneuver bits around in an operand. Arithmetic shifts can multiply and divide by powers of 2 sometimes faster than multiply and divide instructions. Example 7-22 shows how to multiply a number by 10 using shift instructions. This demonstrates a use of the ASLD instruction as an alternative to the EMUL multiply instruction. Examples 7-23 and 7-24 show other uses of the shift and rotate instructions.

Example 7-22 Multiply a 16-bit Number by 10

Write HCS12 code to multiply a 16-bit number in the D register by 10 using arithmetic left shift instructions instead of the `emul` instruction.

Solution:

```
Metrowerks HC12-Assembler
(c) COPYRIGHT METROWERKS 1987-2003
 Rel. Loc    Obj. code Source line
 ---- ------ --------- -----------
    4 000000 7Cxx xx         std    Temp  ; Save in location Temp
    5 000003 59              asld         ; Now it is x2
    6 000004 59              asld         ; Now it is x4
    7 000005 F3xx xx         addd   Temp  ; Add the original.
                                            Now x5
    8 000008 59              asld         ; x2 again makes x10
    9                    ;
   10                    Data:   SECTION
   11 000000            Temp:   ds.w  1
```

Example 7-23

Assume the A value is $A8. What is the result of each of the following instructions: ASLA, ASRA, LSLA, LSRA, ROLA, RORA?

Solution: The easiest way to look at these instructions is to show the values in binary. Before each instruction is executed, A contains %1010 1000. After each instruction, then, we find the following:

	Before			After	
C	**A Reg**	**Instruction**	**C**	**A Reg**	**Comments**
x	1010 1000	ASLA	1	0101 0000	Zero shifted into bit 0.
x	1010 1000	ASRA	0	1101 0100	Sign bit is preserved.
x	1010 1000	LSLA	1	0101 0000	Same result as ASLA.
x	1010 1000	LSRA	0	0101 0100	Different from the ASRA.
C	1010 1000	ROLA	1	0101 000C	Carry bit is rotated into bit 0.
C	1010 1000	RORA	0	C101 0100	Carry bit is rotated into bit 7.

Example 7-24

Show how to load the ASCII code for the number 4 into accumulator B and then shift the most significant nibble into the least significant.

Solution:

```
Metrowerks HC12-Assembler
(c) COPYRIGHT METROWERKS 1987-2003
 Rel. Loc    Obj. code Source line
 ---- ------ --------- -----------
    4 000000 C634                  ldab   #'4'  ; ASCII code for 4
    5 000002 54                    lsrb
    6 000003 54                    lsrb
    7 000004 54                    lsrb
    8 000005 54                    lsrb
```

What is in B before and after the four `lsrb` instructions are executed?

Solution: Before B = $34; after B = $03.

7.11 Arithmetic Instructions

Add and Subtract

The HCS12 *arithmetic instructions* can add, subtract, decimal adjust, negate, multiply, and divide.

Table 7-12 shows the arithmetic instructions. All, except SEX, modify bits in the condition code register. The ADCA, ADCB, SBCA, and SBCB instructions add or subtract the carry bit to the operands. This allows multiple byte arithmetic because the carry bit acts as a *link* between one 8-bit addition (or subtraction) and another. See Example 7-25.

The ADDD adds 16-bit data from memory to the D accumulator. Remember that the D accumulator is the concatenation of the A and B accumulators. See Example 7-26.

The load effective address instructions, LEAX, LEAY, and LEAS, discussed in Section 7.5, are also 16-bit arithmetic instructions because they add 16-bit values to the X, Y, and SP registers.

Decimal Arithmetic

Packed BCD addition must be corrected with the *DAA* instruction.

The *Decimal Adjust A* (DAA) instruction is useful when performing arithmetic on packed binary coded decimal numbers that contain the codes for two decimal digits in one byte. When packed BCD numbers are added, the result can be

---

Example 7-25 Multibyte Addition

Add two 16-bit numbers stored in `Data1:Data1+1` and `Data2:Data2+1` using the 8-bit addition instructions. The result is to be stored in `Data3:Data3+1`. Do this using add-with-carry to demonstrate the algorithm used for multiple-byte arithmetic.

Solution:

```
Metrowerks HC12-Assembler
(c) COPYRIGHT METROWERKS 1987-2003
 Rel. Loc    Obj. code Source line
 ---- ------ --------- -----------
    1                          XDEF Entry, main
    2                  Entry:
    3                  main:
    4                  ; Add the least significant bytes first
    5                  ;   Get least sig byte of 16-bit Data1
    6 000000 B6xx xx          ldaa  Data1+1
    7                  ;   Add in the least sig byte of Data2
    8 000003 BBxx xx          adda  Data2+1
    9                  ; The sum is in A and the carry bit now
                       has the
   10                  ; carry out that must be added into
                       the most
   11                  ; significant byte addition.
   12                  ;   Get most sig byte; ldab does not
                       change the carry
   13 000006 F6xx xx          ldab  Data1
   14                  ;   Add most significant plus carry
   15 000009 F9xx xx          adcb  Data2
   16 00000C 7Cxx xx          std   Data3    ; Stores 16 bits
   17                  ; . . .
   18                  Data:   SECTION
   19 000000          Data1:  ds.w  1
   20 000002          Data2:  ds.w  1
   21 000004          Data3:  ds.w  1
```

---

incorrect. For example, the addition of 34_{10} and 29_{10} gives an incorrect result if BCD codes are being used.

Decimal	BCD Code
34	0 0 1 1 0 1 0 0
<u>29</u>	<u>0 0 1 0 1 0 0 1</u>
63	0 1 0 1 1 1 0 1

Example 7-26 Multibyte Addition with D Register

Add two 16-bit numbers stored in Data1:Data1+1 and Data2:Data2+1 using a 16-bit add instruction. The result is to be stored in Data3:Data3+1.

Solution:

```
Metrowerks HC12-Assembler
(c) COPYRIGHT METROWERKS 1987-2003

 Rel. Loc    Obj. code Source line
 ---- ------ --------- -----------
    4                  ; Load 16 bits from Data1:Data1+1
    5 000000 FCxx xx        ldd   Data1
    6                  ; Add 16 bits from Data2:Data2+1
    7 000003 F3xx xx        addd  Data2
    8                  ; Save the result
    9 000006 7Cxx xx        std   Data3   ; Stores 16 bits
   10                  ;  . . .
   11                  Data:   SECTION
   12 000000           Data1:  ds.w  1
   13 000002           Data2:  ds.w  1
   14 000004           Data3:  ds.w  1
```

The binary result, 0101 1101, is incorrect, both as a binary code and as a BCD code. The DAA instruction, if executed immediately after the addition of the two BCD numbers, adds a correction factor to the binary result. In this case, $06 is added.

Decimal	BCD Code	
34	0 0 1 1 0 1 0 0	
29	0 0 1 0 1 0 0 1	
	0 1 0 1 1 1 0 1	
	0 0 0 0 0 1 1 0	Correction added by DAA
63	0 1 1 0 0 0 1 1	Correct BCD result

The DAA instruction automatically determines the correction factor to add, depending, in part, on the half-carry bit (H) in the condition code register. See Table 7-24.

Negating and Sign Extension

Instructions to negate (two's complement) numeric information are included in the arithmetic group. See Example 7-27.

TABLE 7-24 DAA Instruction Correction Values

Initial C Bit	Value of A[7:4]	Initial H Bit	Value of A[3:0]	Correction Value	Corrected C Bit
0	0-9	0	0-9	00	0
0	0-8	0	A-F	06	0
0	0-9	1	0-3	06	0
0	A-F	0	0-9	60	1
0	9-F	0	A-F	66	1
0	A-F	1	0-3	66	1
1	0-2	0	0-9	60	1
1	0-2	0	A-F	66	1
1	0-2	1	0-3	66	1

The sign extend instruction, SEX, is another mnemonic for any 8-bit to 16-bit register transfer, such as `tfr b,d`. In this case, the sign bit of the B register is *extended* into the most significant byte of the D register (i.e., the A register). SEX may have the A, B, or CCR registers as the source and D, X, Y, or SP as the destination. The instruction would be useful if, when doing 8-bit signed arithmetic, you wanted to multiply two 8-bit signed numbers using the EMULS instruction (the 8-bit MUL instruction is for unsigned data). See Example 7-30.

Example 7-27 Negating Data

Assume A contains the following data before the HCS12 executes the NEGA instruction. What are the results in A and the N, Z, V, and C bits after the negation of each byte?
A = $00, $7F, $FF, $80.

Solution:

		After				
Before	**A Reg**	**N**	**Z**	**V**	**C**	**Comments**
$00	$00	0	0	0	0	Negating zero gives us zero
$7F	$81	1	0	0	0	Negating +127 give us −127
$01	$FF	1	0	0	0	Negating +1 gives us −1
$FF	$01	0	0	0	0	Negating −1 gives us +1
$80	$80	1	0	1	1	Negating −128 gives us −128 and overflow and carry

Explanation: The logic for the condition code register bits for all NEG instructions is as follows:

N: The N bit is set if the most significant bit of the result is set; it is cleared otherwise.

Z: Z is set if the result is $00; it is cleared otherwise.

V: V is set if there is a two's-complement overflow from the implied subtraction from zero; it is cleared otherwise. Two's-complement overflow occurs only if the data being negated is $80.

C: C is set if there is a borrow in the implied subtraction from zero and thus is set in all cases except when the data being negated is $100.

Multiplication

Signed and unsigned numbers can be multiplied or divided.

Multiply instructions for 16- and 32-bit signed numbers have been included in the HCS12 instruction set. Table 7-12 gives a summary of the multiply instructions and Examples 7-28 and 7-30 show short code examples.

The MUL instruction multiplies *8-bit, unsigned numbers* in the A and B accumulators. The 16-bit product resides in the D accumulator.

Sixteen-bit signed and unsigned multiply instructions are EMULS and EMUL, respectively. The algorithms, and thus the hardware required, are different for signed and unsigned multiplication; thus different instructions are needed. See Example 7-30.

Example 7-28 An 8-bit × 8-bit Unsigned Multiply

Write a small program to multiply the 8-bit contents of Data1 and Data2. Store the 16-bit result in Data3:Data3+1.

Solution:

```
Metrowerks HC12-Assembler
(c) COPYRIGHT METROWERKS 1987-2003
Rel. Loc    Obj. code Source line
---- ------ --------- -----------
   4 000000 B6xx xx            ldaa    Data1 ; Get the multiplier
   5 000003 F6xx xx            ldab    Data2 ; Get the multiplicand
   6 000006 12                 mul           ; The product is in D
   7 000007 7Cxx xx            std     Data3
   8                   Data:   SECTION
   9 000000           Data1:   ds.b    1     ; 8-bit multiplier
  10 000001           Data2:   ds.b    1     ; 8-bit multiplicand
  11 000002           Data3:   ds.w    2     ; 16-bit product
```

Example 7-29

Two 8-bit unsigned numbers, 51_{10} and 144_{10}, are to be multiplied. Give the instruction to be used and the registers used for the input and the result. Give the input operands and the result in hexadecimal after the multiply is accomplished.

Solution: The mul instruction is to be used. Accumulator A and B hold the input operands, $33 and $90, and the result, 7344_{10} = $1CB0, is in accumulator D.

Example 7-30 An 8-bit × 8-bit Signed Multiply

```
Metrowerks HC12-Assembler
(c) COPYRIGHT METROWERKS 1987-2003
 Rel. Loc    Obj. code Source line
 ---- ------ --------- -----------
    4                        ; 8-bit * 8-bit signed multiply
    5 000000 B6xx xx         ldaa    Data1   ; Get the
                                             multiplicand
    6 000003 B706            sex     a,y     ; Sign extend into Y
    7 000005 B6xx xx         ldaa    Data2   ; Get the multiplier
    8 000008 B704            tfr     a,d     ; Same as sex a,d
    9 00000A 1813            emuls           ; Extended signed
                                             multiply
   10                     ; The 32-bit product is in Y:D
   11                     ; The 16-bits we need are in D
   12 00000C 7Cxx xx         std     Data3
   13                     Data:   SECTION
   14 000000             Data1:  ds.b  1
   15 000001             Data2:  ds.b  1
   16 000002             Data3:  ds.w  1
```

Example 7-31

Assume Data1 and Data2 in Example 7-30 are $+10_{10}$ and -10_{10}, respectively. For each of the instructions in Example 7-30 give the results in the A, B, Y, and D accumulators.

Solution:

Instruction	A	B	Y	D
ldaa Data1	$0A	xx	$xxxx	$0Axx
sex a,y	$0A	xx	$000A	$0Axx
ldaa Data2	$F6	xx	$000A	$0Axx
tfr a,d	$FF	$F6	$000A	$FFF6
emuls	$FF	$9C	$FFFF	$FF9C
std Data3	$FF	$9C	$FFFF	$FF9C

Fractional Number Arithmetic

The add, subtract, and multiply arithmetic instructions can be used also for fractional numbers. A fractional number is one in which the binary point is somewhere other than to the right of the least significant bit. Example 7-32 shows unsigned fractional arithmetic.

Example 7-32 Unsigned Fractional Arithmetic

```
 0.50 =   .1000      0.75 =   .1100
+0.25 = +.0100     -0.25 = -.0100
 0.75 =   .1100      0.50 =   .1000

 0.50 =       .1000           0.375/0.5 = 0.75
x0.25 =      x.0100                              .1100
  250          0000          .0100  =  1000 /0100.000
  100          0000          .1000
             1000
             0000
 0.125 = .00100000
```

The 4-bit by 4-bit multiplication in Example 7-32 gives an 8-bit result. When multiplying in the HCS12 using the MUL instruction, an 8-bit fractional multiply gives a 16-bit fractional result. At times, it may be convenient to discard the least significant eight bits and round-up the result to the most significant 8 bits. The MUL instruction provides a convenient way to do this by automatically setting the carry if bit 7 in accumulator B is one. See Example 7-33. The EMUL instructions operate the same for 16-bit fractional multiplications.

Example 7-33 Fractional Multiplication with Rounding

```
Metrowerks HC12-Assembler
(c) COPYRIGHT METROWERKS 1987-2003
 Rel. Loc    Obj. code Source line
 ---- ------ --------- -----------
    4                         ; Multiply fractional numbers
    5 000000 B6xx xx            ldaa    Data1   ; 8-bit fraction
    6 000003 F6xx xx            ldab    Data2   ; 8-bit fraction
    7 000006 12                 mul             ; 16-bit fraction
                                                result
    8 000007 8900               adca    #0      ; Incr A if B is
                                                0.5 or
                                                ; greater
    9 000009 7Axx xx            staa    Data3
   10                   Data:   SECTION
   11 000000           Data1:  ds.b  1
   12 000001           Data2:  ds.b  1
   13 000002           Data3:  ds.w  1
```

TABLE 7-25 HCS12 Division Instructions

Opcode	Data Type	Operation	Comment
IDIV	16-bit, integer, unsigned	D/X	Quotient → X; Remainder → D The radix point is the same for both numerator and denominator and is to the right of the LSB. If X = 0 (divide by zero) $FFFF → X and C is set.
IDIVS	16-bit, integer, signed	D/X	Quotient → X; Remainder → D If divide by zero, D and X are unchanged and C is set; if overflow occurs, V is set.
FDIV	16-bit, fractional, unsigned	D/X	Quotient → X; Remainder → D Numerator assumed < denominator; radix point for each operand assumed to be the same; overflow (V) bit is set if denominator <= numerator. If divide by zero, $FFFF → X and C is set.
EDIV	32-bit, integer, unsigned	Y:D/X	Quotient → Y; Remainder → D If divide by zero, C is set; if result is > $FFFF, V is set.
EDIVS	32-bit, integer, signed	Y:D/X	Quotient → Y; Remainder → D If divide by zero, C is set.

Division

Division instructions for 16- and 32-bit signed numbers have been included in the HCS12 instruction set. IDIVS and EDIVS are 16- and 32-bit signed division instructions. Tables 7-12 and 7-25 give summaries of the division instructions and Examples 7-34 to 7-36 show small sample programs.

IDIV, FDIV, and EDIV are for unsigned operands only. IDIV divides the 16-bit integer in the D accumulator by the 16-bit integer in the X register. The 16-bit quotient is placed in the X register and a 16-bit remainder in the D accumulator. If the denominator is zero (divide by 0) the carry bit is set and the quotient X is set to $FFFF. The radix point is the same for both numerator and denominator and is to the right of the least significant bit (bit 0) of the quotient. See Example 7-34.

Example 7-34 Integer Division

Assume D contains 176_{10} and X = 10_{10}. What is in A, B, and X before and after an `idiv` instruction?

Solution:

Before the `idiv` instruction:

$$D = 176_{10} = \$00B0, \text{ therefore A} = \$00 \text{ and B} = \$B0.$$

$$X = 10_{10} = \$000A$$

After the `idiv` instruction:

$$176_{10}/10_{10} = 17_{10} \text{ with a remainder of 6, therefore}$$

$$X = 17_{10} = \$0011, D = 6_{10} = \$0006, A = \$00, B = \$06$$

Example 7-35 Fractional Division

Assume D contains 100_{10} and $X = 400_{10}$. What is in A, B, and X before and after an `fdiv` instruction?

Solution:

Before the `fdiv` instruction:

$$D = 100_{10} = \$0064, \text{ therefore A} = \$00, \text{ B} = \$64$$
$$X = 400_{10} = \$0190$$

After the `fdiv` instruction:

$$100_{10}/400_{10} = 0.25_{10} \text{ with a remainder of 0, therefore}$$
$$X = 0.25_{10} = \$4000, \text{ D} = 0 = \$0000, \text{ A} = \$00, \text{ B} = \$00$$

Example 7-36 Extended, Unsigned Multiply and Divide Instructions

Write a small program segment to produce `Data1*Data2/100`, where `Data1` and `Data2` are both 16-bit unsigned numbers.

Solution:

```
Metrowerks HC12-Assembler
(c) COPYRIGHT METROWERKS 1987-2003

Rel. Loc    Obj. code Source line
---- ------ --------- -----------
 4                        ; Data1 * Data2/100
 5 000000 FCxx xx            ldd     Data1 ; 16-bit integer
 6 000003 FDxx xx            ldy     Data2 ; 16-bit integer
 7 000006 13                 emul          ; 32-bit product in Y:D
 8 000007 CE00 64            ldx     #100
 9 00000A 11                 ediv          ; Do the division
10 00000B 7Dxx xx            sty     Data3 ; Save the quotient
11 00000E 7Cxx xx            std     Data4 ; Save the remainder
12                  Data:    SECTION
13 000000           Data1:   ds.w  1
14 000002           Data2:   ds.w  1
15 000004           Data3:   ds.w  1
16 000006           Data4:   ds.w  2
```

Example 7-37

Assume that in Example 7-36 Data1 and Data 2 are 4096_{10} and 16_{10}, respectively. For each of the instructions in Example 7-36 give the contents of all registers.

Solution:

Instruction	A	B	D	X	Y
ldd Data1	$10	$00	$1000	$xxxx	$xxx
ldy Data2	$10	$00	$1000	$xxxx	$0010
emul	$00	$00	$0000	$0000	$0001
ldx #100	$00	$00	$0000	$0064	$0001
ediv	$00	$24	$0024	$0064	$028F

Explanation: Data1 $= 4096_{10} = \$1000$, Data2 $= 16_{10} = \$0010$; when these are multiplied by the emul the result is $65536_{10} = \$0001{:}0000$ in Y:X. Dividing 65536_{10} by $100_{10} = 655$ with a remainder of 36. The quotient, $655 = \$028F$, is in Y and the remainder, $36 = \$24$, is in D.

FDIV performs a 16-bit unsigned fractional divide. The numerator is in the D accumulator and is assumed to be less than the denominator. The denominator is in the X register and the radix point for each of the operands is assumed to be the same; the result is a binary-weighted fraction. As in the IDIV instruction, the carry bit is set to indicate a divide by zero. The overflow bit (V) is set if the denominator is less than or equal to the numerator. The binary fraction quotient is in X and the remainder in D. See Example 7-35.

EDIV and EDIVS are 32-bit divide-by-16-bit unsigned and signed division instructions. The 32-bit dividend in each case is the Y:D register combination and the 16-bit divisor is the X register. A 16-bit quotient is produced in the Y register with a 16-bit remainder in the D. See Example 7-36.

7.12 Logic Instructions

The *logic instructions* can be AND, OR, Exclusive-OR, and complement.

The logic instructions perform bit-wise logic operations with the two operands (Table 7-13). The one's-complement instructions are included here because one's complementing is typically a logic rather than numeric operation. Example 7-38 gives some sample logic operations.

ANDing is sometimes called *masking*. You can mask off certain bits in an operand to use them in some way. Example 7-39 shows a masking operation used to convert packed BCD numbers to ASCII for printing.

Example 7-38 Logic Operations

If memory location $0010 contains $B3 and A contains $64, what is the result of the following instructions?

```
anda $10; anda #$10; oraa $10; oraa #$10; eroa $10; coma; com $10
```

Treat each instruction as an independent separate instruction, not a sequence of instructions.

Solution:

```
anda  $10              A = 0110 0100
                  ($0010) = 1011 0011
                            0010 0000
anda  #$10             A = 0110 0100
                      $10 = 0001 0000
                            0000 0000
oraa  $10              A = 0110 0100
                  ($0010) = 1011 0011
                            1111 0111
oraa #$10              A = 0110 0100
                      $10 = 0001 0000
                            0111 0100
eora $10               A = 0110 0100
                  ($0010) = 1011 0011
                            1101 0111
coma                   A = 0110 0100
                            1001 1011
com   $10         ($0010) = 1011 0011
                            0100 1100
```

Example 7-39 Making Operations

Assume the A accumulator has a packed BCD number to be converted to ASCII and printed (using a subroutine called `print_digit`). Write a small segment of code using logic instructions to do this.

Solution:

```
Metrowerks HC12-Assembler
(c) COPYRIGHT METROWERKS 1987-2003
 Rel. Loc    Obj. code Source line
 ---- ------ --------- -----------
    5        0000 000F LS_MASK:EQU  %00001111  ; Least sig nibble
                                                 mask
    6                      ;
    7 000000 B701          tfr   a,b            ; Save the BCD
                                                  number in B
    8                      ; Need to print the most significant
                           nibble first
    9 000002 44            lsra                 ; Shift 4 bits to
                                                  right
```

```
10 000003 44            lsra
11 000004 44            lsra
12 000005 44            lsra
13                  ; Convert to ASCII
14 000006 8A30        oraa   #$30
15 000008 16xx xx     jsr    print_digit ; Go print it
16 00000B B710        tfr    b,a         ; Get the orginal back
17 00000D 840F        anda   #LS_MASK    ; Set most sig bits = 0
18 00000F 8A30        oraa   #$30        ; Convert to ASCII
19 000011 16xx xx     jsr    print_digit ; Print it
```

7.13 Data Test Instructions

The previous sections have shown that many instructions modify data and the condition code register bits. All of the test and compare instructions given in Table 7-15 modify only the condition code register bits. The operands themselves are not changed.

The BITA and BITB instructions AND the contents of the specified accumulator with the contents of the addressed memory location. These instructions are useful for determining if a particular bit in the memory operand is one or zero. Consider an example in which port T, bits 4–7, is an input port attached to a set of switches as shown in Figure 7-11. Our program is to

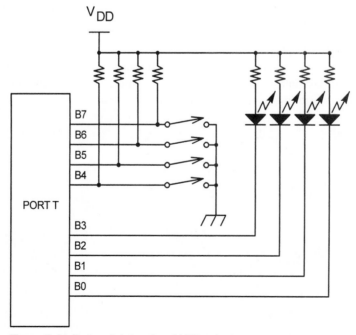

Figure 7-11 Port switch input and LED output.

Example 7-40 Using the BITA Instruction

```
Metrowerks HC12-Assembler
(c) COPYRIGHT METROWERKS 1987-2003

 Rel. Loc    Obj. code Source line
 ---- ------ --------- -----------
    4          0000 0001 BIT_7: EQU   %00000001  ; Mask for Bit-7
    5          0000 0240 PORTT: EQU   $240       ; Address for Port T
    6                    ; . . .
    7                    ; IF BIT_7 on Port T is zero
    8 000000 8601               ldaa  #BIT_7
    9 000002 B502 40            bita  PORTT      ; Test Bit-7 on
                                                 Port T
   10 000005 2602               bne   do_if_one ; Do the one part
   11                    ; THEN This is the code to do if the bit
                         is zero
   12                    ; . . .
   13 000007 2000               bra   end_if
   14                    do_if_one:
   15                    ; ELSE This is the code if the bit is one
   16                    ; . . .
   17                    end_if:
```

test the state of switch 7 (attached to bit 7 on port T) and to do one thing if the switch is zero and another if it is one. Example 7-40 shows a code segment that checks bit 7 of Port T.

The compare instructions subtract one operand from the other, without modifying either, and set the condition code bits. TST, TSTA, and TSTB simply test if the operand is zero or negative and modify the Z and N bits accordingly.

7.14 Conditional Branch Instructions

The *conditional branch* instructions allow us to test the condition code registers bits.

We have seen that many instructions modify the condition code register, and the conditional branch instructions (Table 7-16) test these bits for us. Conditional branch instructions make our machine a computer instead of just a calculator because programs can make decisions based on data that are available when the computer is running. Be careful when using a conditional branch to test for something, say, a loop counter equal to zero, that another instruction that modifies the CCR is not inserted before the conditional branch is executed. We saw this problem in Example 7-3.

The short conditional branch instructions use 8-bit relative addressing. This means that the branch can be at most $+127$ or -128 bytes from the instruction following the branch instruction. If you try to branch to a location outside these limits, the assembler program will give an error message "parameter out of Range." If this occurs, use a long branch instruction.

TABLE 7-26 Conditional Branch Instructions for Signed and Unsigned Data

Signed Data Tests		Unsigned Data Tests		Universal Tests	
BMI	Minus				
BPL	Plus				
BVS	Two's-complement overflow	BCS	Carry set = unsigned overflow	BCS	Carry set
BVC	No two's-complement overflow	BCC	Carry clear = no overflow	BCC	Carry clear
BLT	Less than	BLO	Lower than		
BGE	Greater than or equal	BHS	Higher or the same		
BLE	Less than or equal	BLS	Lower or the same		
BGT	Greater than	BHI	Higher than		
BEQ	Equal	BEQ	Equal	BEQ	Equal
BNE	Not equal	BNE	Not equal	BNE	Not Equal

Signed and Unsigned Conditional Branches

> There are different conditional branch instructions for signed and unsigned data.

Another concern is how to handle signed and unsigned numbers. For example, is $FF bigger than $00? It depends on the code. If the code is for an unsigned number, $FF is larger than $00. If a two's-complement signed number system is in use, $FF is smaller than $00. There are *different* conditional branch instructions for *signed* and *unsigned* data. Table 7-26 shows a tabulation of various instructions used for signed and unsigned data. Note the terminology used. "Greater than" and "less than" imply numerical data and thus are to be used for signed numbers. "Higher than" and "lower than" imply unsigned data.

The mnemonics used for the branch instructions make the most sense if we think of them being executed after a compare instruction. For example, if we cmpa $1234 and then use the branch greater-than-or-equal (BGE), we are comparing signed numbers. We would expect the branch to be taken if the value in A is greater than or equal to the contents of memory location $1234. The BNE and BEQ instructions test the Z bit. You may wish to think of the BNE as meaning *branch if not-equal-to-zero* and BEQ as *branch if equal-to-zero*. Make sure you understand why branches are taken or not taken as shown in Example 7-41.

Example 7-41 Conditional Branches
Assume the A = $FF and memory location DATA = $00. A cmpa data instruction in executed followed by a conditional branch. For each of the conditional branch instructions in the table, indicate by yes or no if you expect the branch to be taken and why.

BGE BLE BGT BLT BEQ BNE

BHS BLS BHI BLO

Solution:

Branch	Taken?	Reason
BGE	No	$FF = −1: Not greater than or equal to $00.
BLE	Yes	$FF = −1: Is less than or equal to $00.
BGT	No	$FF = −1: Is not greater than $00.

BLT	Yes	$FF = −1: Is less than $00.
BEQ	No	$FF is not equal to $00.
BNE	Yes	$FF is not equal to $00.
BHS	Yes	$FF = 255: Is higher than $00.
BLS	No	$FF = 255: Is not lower or the same as $00.
BHI	Yes	$FF = 255: Is higher than $00.
BLO	No	$FF = 255: Is not lower than $00.

Bit Test Conditional Branching

We frequently test bits in an operand, say, switches connected to a port or a flag in a register, to control program flow with a conditional branch. Example 7-40 shows how to use a BITA instruction with a conditional branch instruction. Table 7-17 shows two instructions that make this task easier.

The *branch if bit set* (BRSET) and the *branch if bit clear* (BRCLR) instructions check individual bits in an operand and branches if all the bits tested are set or clear. The syntax of these instructions is

                    brset   Operand,mask,target

and

                    brclr   Operand,mask,target

where `Operand` is an 8-bit memory operand such as a port or register address, `mask` is a mask byte where 1's indicate bits to check in the operand, and `target` is the relative address for the branch. See Example 7-42.

Example 7-42 BRSET and BRCLR Instructions

```
Metrowerks HC12-Assembler
(c) COPYRIGHT METROWERKS 1987-2003
Rel. Loc    Obj. code Source line
---- ------ --------- -----------
   4        0000 0080 BIT_7:  EQU   %10000000  ; Mask for Bit-7
   5        0000 0040 BIT_6:  EQU   %01000000  ; Mask for Bit-6
   6        0000 0240 PORTT:  EQU   $240       ; Address for
                                                  Port T
   7                          ; . . .
   8                          ; WHILE Bit-7 is zero, wait for it to
                              become one
   9                          wait_for_7:
  10 000000 1F02 4080              brclr PORTT,BIT_7,wait_for_7
     000004 FB
  11                          ; . . .
```

```
    12                              ; WHILE Bit-6 is one, wait for it to become
                                    zero
    13                              wait_for_6:
    14 000005 1E02 4040              brset PORTT,BIT_6,wait_for_6
       000009 FB
```

7.15 Loop Primitive Instructions

The *loop primitive* instructions (Table 7-18) provide, in one instruction, an operation that is very common in assembly language programs. We often use counters, decrement or increment them, and then branch if the counter equals (or does not equal) zero. The counter may be any of the registers A, B, D, X, Y, or SP. Note that, in contrast to the usual decrement followed by a branch instruction combination, the condition code register is not changed by these instructions. Another advantage of these instructions is that a 16-bit counter can be implemented in a register. Compare Example 7-43 with Example 7-18. The relative branching done by these instructions uses a 9-bit offset with -256 to $+255$ locations. See Example 7-43.

7.16 Unconditional Jump and Branch Instructions

Unconditional jump and branch instructions always take the branch. The *jump* (JMP) instruction, and *jump-to-subroutine* (JSR), use extended, indexed, and indexed-indirect addressing and can thus jump to any address in memory. The *branch always* (BRA) is an unconditional relative addressing branch. Use the branch instructions when you can because they save memory (2 bytes instead of 3 or more).

Branches to Subroutines

There are five instructions in this group dealing with subroutines. These are the *jump to subroutine* (JSR), *branch to subroutine* (BSR), *call subroutine in expansion memory* (CALL), *return from subroutine* (RTS), and *return from call* (RTC). As you might expect, the BSR uses relative addressing while the JSR uses all other addressing modes. In any event, the instruction contains the effective address of the subroutine to which the program must branch. The address of the instruction immediately following the jump or branch is called the return address. Before starting the subroutine, the return address is pushed onto the stack.

A *return from subroutine*, RTS, is used at the end of the subroutine. When the microcontroller executes the RTS, the return address is pulled from the stack and placed in the program counter.

Some versions of the MC9S12C family allow up to 1 megabyte of addressable program space, and two HCS12 instructions allow subroutines to be located in this *expansion* memory. This is done with a paged architecture. The CALL and the RTC instructions allow subroutines located in expansion memory to be executed and returned from.

There are three (at least) important rules to obey when using subroutines:

1. *Always* initialize the stack pointer register to point to an area of RAM *before* using BSR or JSR. If you do not, the program may or may not run. If the program is acting strangely, check to see if the stack pointer is initialized correctly.

Example 7-43 Loop Primitive Instruction

```
Metrowerks HC12-Assembler
(c) COPYRIGHT METROWERKS 1987-2003

 Rel. Loc    Obj. code Source line
 ----  ------ --------- -----------
    4                      ; Comparing the loop primitive and "normal"
    5                      ; decrement and branch instructions.
    6          0000 03E8   COUNT1: EQU   1000    ; Counter value
    7          0000 00FF   COUNT2: EQU   255     ; Max 8-bit counter
    8                      ; . . .
    9 000000 CE03 E8               ldx   #COUNT1 ; Initialize counter
   10                      ; . . .
   11                      loop:
   12                      ; Here is the repetitive code
   13                      ; . . .
   14 000003 0435 FD               dbne  x,loop  ; Using the dbne
                                                   instruction
   15                      ; Demonstrate 8-bit dec and branch
                             instruction
   16 000006 C6FF                  ldab  #COUNT2 ; Initialize counter
   17                      loop2:
   18                      ; Here is the repetitive code
   19                      ; . . .
   20 000008 53                    decb
   21 000009 26FD                  bne   loop2
```

2. *Never, ever* JMP to a subroutine. *Always* use the JSR, BSR, or CALL when transferring to a subroutine.

3. *Never, ever* JMP out from subroutine. *Always* use the RTS or RTC when returning to the calling program or the stack will be unbalanced.

7.17 Condition Code Register Instructions

The condition code register instructions (Table 7-14) use OR and AND operations to set and clear the carry and overflow bits. These instructions are useful when transferring Boolean information from a subroutine back to the calling program. Example 7-44 shows how to use the carry bit to inform the calling program if a variable is out of range or not. You may also use the V bit and *branch if overflow clear* (BVC) or *branch if overflow set* (BVS) instructions. The Z and N bits may be used as well.

Example 7-44 Using the Carry Bit for Boolean Information Transfer

This example shows that a condition code register bit can be used to transfer a Boolean result back to the calling program.

```
Metrowerks HC12-Assembler
(c) COPYRIGHT METROWERKS 1987-2003
Rel. Loc    Obj. code Source line
---- ------ --------- -----------
    1                         XDEF Entry, main
    2                 Entry:
    3                 main:
    4                 ; ...
    5                 ; Branch to a subroutine that checks if a
                      variable
    6                 ; is within a set range. If it is, it returns
                      with
    7                 ; the carry = 0, otherwise it returns
                      carry = 1.
    8 000000 0702         bsr    check_range
    9                 ; After the return, c=0 if OK, c=1 if not
   10                 ; IF range is not OK
   11                 ; THEN branch to error handling code
   12 000002 2500         bcs    out_of_range
   13                 ; ELSE Continue with the process
   14                 ; . . .
   15                 ; . . .
   16                 out_of_range:
   17                 ; Print an error message or something
   18                 ; . . .
   19
   20                 ;*********************************************
   21                 ; Subroutine to check if a variable is in range
   22                 ; If it is, clear the carry bit and return,
                      otherwise
   23                 ; set the carry bit and return.
   24                 check_range:
   25                 ; . . . Imagine the code to check the variable
                      is here
   26                 ;         and it branches to OK if it is in
                                range and
   27                 ;         NOT_OK if it is not.
   28 000004 10FE     OK:    clc  ; Clear the carry bit
   29 000006 2002            bra    done
   30 000008 1401     NOT_OK: sec   ; Set the carry bit
   31                 done:
   32 00000A 3D              rts  ; Return with carry set or clear
```

7.18 Interrupt Instructions

The interrupt instructions shown in Table 7-20 allow us to control whether or not external events can interrupt the normal program flow. Interrupts will be covered in more detail in Chapter 12.

7.19 Fuzzy Logic Instructions

The HCS12 has a number of instructions that support writing programs for fuzzy logic systems. We will investigate these features in Chapter 19.

7.20 Miscellaneous Instructions

The miscellaneous instructions are given in Table 7-22. The BGND instruction enables a special debugging feature in the HCS12. See Chapter 20 for a complete discussion.

The NOP instruction can be useful when making a short software delay because each time the NOP is executed, one clock cycle is expended.

The STOP instruction stops all microcontroller clocks and puts the microcontroller in a power saving mode. It can be "awakened" from this mode by using an interrupt. When the microcontroller is reset, the stop instruction is disabled. To use STOP, the S bit in the condition code register must be reset by an `andcc #$7F` instruction.

7.21 Conclusion and Chapter Summary Points

Learning the instruction set of a microcontroller is a difficult and sometimes frustrating job, particularly in a processor with over 1000 instructions. To make your job easier, we presented the instruction set in various categories related to what you want to accomplish in your program.

Here are some of the chapter's key points.

- Learning the instruction set is easier if you first learn the categories of instructions available.
- Basic instruction categories shown in Table 7-1 are *move data*, *modify data*, *decision making, flow control*, and *other*.
- Table 7-1 is useful for quickly finding the type of operation in the category you want.
- After finding the operation, you must specify the operand and its addressing mode.
- Load and store instructions modify the condition code register.
- Sixteen-bit load and store instructions require two memory locations for the data.
- Push, pull, and other stack operations must be balanced.
- BSET and BCLR instructions set and clear one or more bits in an operand.
- LEAX and LEAY instructions are useful for calculating effective addresses.
- The DAA instruction produces the correct result when adding packed BCD numbers.
- MUL, FDIV, IDIV, EMUL, and EDIV can be used only with unsigned numbers.
- IDIVS, EMULS, and EDIVS can be used with signed (two's-complement) numbers.

- Logical instructions perform a bit-wise logical operation between two operands.
- Data test instructions set the condition code register bits used in conditional branching.
- A compare instruction subtracts one operand from another but does not change either.
- When comparing signed numbers, the conditional branch instructions with the words "greater" and "less" are to be used.
- When comparing unsigned numbers, the conditional branch instructions with the words "higher" and "lower" are to be used.
- The branch instructions all use relative addressing.
- Always use the JSR, BSR, or CALL instruction to transfer to a subroutine.
- Always use the RTS or RTC instruction to transfer back to the calling program.
- Condition code register bits may be used to transfer binary information between parts of a program.

7.22 Bibliography and Further Reading

CPU12 Reference Manual, M68HC12 & HCS12 Microcontrollers, CPU12RM/AD, Rev. 3, Freescale Semiconductor, Inc., April 2002.

7.23 Problems

Basic

7.1 Assume the memory display of the HCS12 shows 16 bytes starting at $0800:

```
$0800:B0 53 05 2B 36 89 00 FF FE 80 91 3E
77 AB 8F 7F
```

Give the results after each of the following instructions are executed. [k]

a. ldaa $0800   A = ?, NZVC = ?

b. Assume X = $0800:

```
ldaa        0,X        A = ?, NZVC = ?
```

c. Assume X = $0800:

```
ldaa        6,X        A = ?, NZVC = ?
```

7.2 Use the contents of memory shown in Problem 7.1 and give the results of the following instructions. [k]

a. ldx $0800   X = ?

b. ldy $0802   Y = ?

c. ldx $0803

```
   pshx
   puld                  X = ?, A = ?, B = ?
```

d. ldd $0800

```
   ldx        $0802
   exg        D,X        D = ?, X = ?
```

e. Assume X = $0800:

```
   ldd        $0A,X      D = ?
```

7.3 Use the contents of memory shown in Problem 7.1 and give the results of the following instructions. [k]

a. SP = $0805:

```
   pula       A = ?,     SP = ?
```

b. SP = $0805:

```
   pula       A = ?
   pulb       B = ?
```

c. SP = $0805:

```
   psha
   pshb       SP = ?
```

d. SP = $080A:

```
   pula
   pshb       A = ?,     SP = ?
```

7.4 Use the contents of memory shown in Problem 7.1 and give the results of the following instructions. **[k]**

a. Assume X = $0800:

```
bset       0,X,$0F      ($0800) = ?
```
b. Assume X = $0800:

```
bset       6,X,$AA      EA = ?,(EA) = ?
```
c. Assume X = $0807:

```
bclr       0,X,$AA      ($0807) = ?
```
d. Assume Y = $0800:

```
bclr       0,Y,$FF      ($0800) = ?
```

7.5 Use the contents of memory shown in Problem 7.1 and give the results of the following instructions. **[k]**

a.
```
ldaa       $0803
ldab       $0804
aba                     A = ?,B = ?,NZVC = ?
```
b.
```
ldx        $0800
ldab       $0807
leax       B,X          X = ?, NZVC = ?
```
c.
```
ldab       $0809
addb       $080A        B = ?, NZVC = ?
```
d. Assume X = $0800:
```
ldaa       9,x
adda       $0A,X        A = ?, NZVC = ?
```
e. Assume X = $0800:
```
ldaa       9,x
suba       $0A,X        A = ?, NZVC = ?
```
f.
```
ldaa       $0800
ldab       $0801
aba
adca       $0802        A = ?, NZVC = ?
```

7.6 Use the contents of memory shown in Problem 7.1 and give the results of the following instructions. **[k]**

a.
```
ldaa       $0806
nega                    A = ?, NZVC = ?
```
b.
```
ldaa       $0807
nega                    A = ?, NZVC = ?
```
c.
```
neg        $0809        ($0809)= ?,NZVC = ?
```
d.
```
ldaa       $0806
coma                    A = ?, NZVC = ?
```
e.
```
ldaa       $0807
coma                    A = ?, NZVC = ?
```
f.
```
com        $0809        ($0809)=?,NZVC = ?
                        ($0809) = $7F,NZVC
                        = 0001
```

7.7 Use the contents of memory shown in Problem 7.1 and give the results of the following instructions. **[k]**

a.
```
ldaa       $0802
oraa       $0803        A = ?, NZVC = ?
```
b.
```
ldaa       $0802
eora       $0803        A = ?, NZVC = ?
```
c.
```
ldaa       $080d
anda       $080E        A = ?, NZVC = ?
```
d.
```
ldab       $0802
comb                    B = ?, NZVC = ?
```

7.8 Use the contents of memory shown in Problem 7.1 and give the results of the following instructions. **[k]**
```
ldaa       $0802
adda       $0804        A = ?
daa                     A = ?
```

7.9 Use the contents of memory shown in Problem 7.1 and give the results of the following instructions. **[k]**

a.
```
ldaa       $0800
cmpa       $0801        A = ?, NZVC = ?
```
b.
```
tst        $0806        NZVC = ?
```
c.
```
tst        $0807        NZVC = ?
```

7.10 Describe what each of the following operations does. **[a, k]**

a. `orab`

b. `rora`

c. `adcb`

d. `sty`

7.11 Assume the following information in the HCS12 registers for each of the following instructions:

A = $00 B = $01 X = $6000 Y = $6010

The data in memory locations $6000–$600F is

```
6000  11 23 42 00  60 00 65 02  11 22 48
65  6c 70 4d 65
```

Give the addressing mode used, the effective address for the source data, and the result for each of the following instructions. **[a, k]**

Instruction	Addressing Mode	Effective Address	Result
ldab 0,x			
ldab $600e			
ldab 8,x			
ldab -1,y			
ldab 1,x+			

7.12 After an addition, the carry bit in a status register indicates that a two's complement overflow has occurred—true or false? **[a]**

7.13 The condition code register is modified by every instruction—true of false? **[a]**

7.14 The memory display shows

```
4000:   08   29   3F   7F 86   99   A0   64 . . .
```

and the current value in the X register is $4000. Give the results of the following instructions. **[a, k]**

  a. `ldaa`   `$4000`   A = ?

  b. `ldab`   `$4,x`   B = ?

  c. `ldy`   `$4001`   Y = ?

Now, with the value in the B register as computed in part b, the following instructions are executed:

```
cmpb   #$50
bgt    Someplace
```

  d. Is the branch taken?

7.15 Briefly describe what each of the following instructions does. These are separate instructions, not a program. **[g]**

```
coma
cba
cmpb $10,X
tstb
brn
bita #$80
bcc LOOP
exg d,x
lsr $0800
negb
```

7.16 What is a stack? **[a]**

7.17 Where does the stack pointer register point? **[a]**

7.18 Is the stack in the HCS12 a first-in, first-out (FIFO) or last-in, first-out (LIFO)? **[a]**

Intermediate

7.19 Give the addressing mode and the effective address for each of the following instructions. **[k]**

  a. `ldaa`   `#5`

  b. `ldaa`   `$5`

  c. `ldaa`   `$5,x`

  d. `staa`   `$081a`

7.20 For each of the following instructions, assume A = $C9 and the NZVC bits are 1001. Give the result in

A and the NZVC bits for each of the following instructions. **[k]**

  a. `lsla`

  b. `lsra`

  c. `asla`

  d. `asra`

  e. `rola`

  f. `rora`

7.21 The ASLx instructions have the same operation codes as the LSLx instructions. Why? **[k]**

7.22 Why do store instructions not use the immediate addressing mode? **[a]**

7.23 For each of the following HCS12 instructions, explain where the data comes from (the source) and where it is going (the destination). **[a, k]**

Instruction	Comes from	Going to
ldab #$40		
staa 6,x		
ldx $1234		
ldaa $06		

7.24 The following unsigned binary addition was done in the HCS12. What is the binary result and what are the N, Z, V, and C flags? **[a, k]**

$$\begin{array}{r} 01010111 \\ \underline{01100110} \end{array}$$

7.25 Assume the following HCS12 code is executed in sequence. Give the hexadecimal result in each of the registers after each instruction is executed. **[a, k]**

		A	B	N Z V C
a.	ldaa	#$4A	_____	XXXX _____
b.	ldab	#$D3	_____	_____ _____
c.	aba		_____	_____ _____
d.	adca	#$70	_____	_____ _____

7.26 Assuming that in the HCS12, A = $83, B = $29, and the ABA instruction is executed:

  a. What is in the condition code register bits? **[a]**

  b. If the result represents information encoded in two's-complement code, what is the decimal equivalent of the result? **[a]**

  c. If the result represents information encoded in unsigned binary code, what is the decimal equivalent of the result? **[a]**

7.27 The registers in the HCS12 show the following: **[a, b, k]**

```
D = $2245, X = $1234, Y = 5678, SP =
$0900
```

Assume the following sequence of code is executed and then specify what is in the stack and what is in the registers:

```
psha
pshx
aba
psha
puly
```

a. What are the register contents after the program sequence is executed?

b. Show the contents of the stack after the program sequence is executed (give your answer in hexadecimal).

Memory Address	Memory Contents (hex)
08FA	
08FB	
08FC	
08FD	
08FE	
08FF	
0900	
0901	
0902	
0903	
0904	
0905	

7.28 Assume A = $00 and memory location `data` = $B0. A `cmpa data` instruction is executed followed by a conditional branch. For each of the conditional branch instructions listed, indicate by yes or no if you expect the branch to be taken. **[a, k]**

| bge | ble | bgt | blt | beq | bne |
| bhs | bls | bhi | blo | | |

7.29 Assume A = $05 and memory location `data` = $22. A `cmpa data` instruction is executed followed by a conditional branch. For each of the conditional branch instructions listed, indicate by yes or no if you expect the branch to be taken. **[a, k]**

| bge | ble | bgt | blt | beq | bne |
| bhs | bls | bhi | blo | | |

7.30 Assume A = $56 and memory location `data` = $22. A `cmpa data` instruction is executed followed by a

conditional branch. For each of the conditional branch instructions listed, indicate by yes or no if you expect the branch to be taken. **[a, k]**

| bge | ble | bgt | blt | beq | bne |
| bhs | bls | bhi | blo | | |

7.31 Assume A = $22 and memory location `data` = $22. A `cmpa data` instruction is executed followed by a conditional branch. For each of the conditional branch instructions listed, indicate by yes or no if you expect the branch to be taken. **[a, k]**

| bge | ble | bgt | blt | beq | bne |
| bhs | bls | bhi | blo | | |

7.32 What conditional branch instruction would you use in each of the following program scenarios? **[a, k]**

a. Branch to a part of the program if a two's-complement overflow occurred.

b. Branch to a part of the program if an unsigned binary overflow occurred.

c. Branch to a part of the program if an 8-bit unsigned binary number input from the switches is less than one-half the maximum value.

d. Branch back to the start of a loop a number of times based on a counter in a register.

e. Unconditional branch to a part of the program.

7.33 For each of the following questions, choose the correct HCS12 instruction from the following list. An instruction may be used more than once. The correct instruction will be found in the list. Assume the program is executed in the order that is given; that is, the pointer register is initialized pointing to data memory and then accumulator A is loaded from that data memory, and so on. **[c, k]**

bge LOOP	ldaa 0,X	jmp DONE
bgt LOOP	cmpa #$82	jsr SUB
bhi LOOP	cmpa #DATA1	bne LOOP
bhs LOOP	cmpa DATA1	bgt LOOP
ldab #DATA1	ldx #DATA1	bhi LOOP
ldab 1,+X	ldx DATA1	bhs LOOP
ldab DATA1	ldy DATA1	
ldaa #DATA1	ldy DATA1+1	

a. ; Initialize a pointer register pointing to data memory DATA1.

b. ; Load accumulator A from data memory location DATA1.

c. ; Load accumulator B from data memory location DATA1+1.

d. ; Load index register Y from DATA1:DATA1+1.

e. ; Compare what is A with what is in data memory location DATA1.

f. ; Branch to LOOP if the data in A is bigger than the data in DATA1.

(Assume the information in A and DATA1 are encoded using 8-bit unsigned binary code.)

7.34 What instruction or instructions would you use and what branch instruction would follow if you wanted to branch someplace if the contents of memory location $0800 is negative? **[a]**

7.35 What instruction or instructions would you use and what branch instruction would follow if you wanted to branch someplace if the contents of memory location $0800 is positive? **[a]**

7.36 What instruction or instructions would you use and what branch instruction would follow if you wanted to branch someplace if the contents of memory location $0800 is an odd number? **[a]**

7.37 What instruction or instructions would you use and what branch instruction would follow if you wanted to branch someplace if the contents of memory location $0800 is zero? **[a]**

7.38 What branch instruction would you use if you wanted to branch someplace if a just-completed addition resulted in a two's-complement arithmetic overflow? **[a]**

7.39 What branch instruction would you use if you wanted to branch someplace if a just-completed addition resulted in an unsigned arithmetic overflow? **[a]**

7.40 What instruction or instructions would you use and what branch instruction would follow if you wanted to branch someplace if the unsigned number in register A is bigger than $64? **[a]**

7.41 What instruction or instructions would you use and what branch instruction would follow if you wanted

to branch someplace if the two's-complement number in register A is bigger than $64? **[a]**

Advanced

7.42 Write a small HCS12 program segment to do the following: **[c]**

```
; Initialize the stack pointer to
$0900
; Initialize the Y register to point
to $0800
; Initialize memory locations $0800
- $080f to the alternating
; pattern $55, $AA, $55, $AA, . . .
```

7.43 Two eight-bit unsigned numbers, 86_{10} and 94_{10}, are to be multiplied. Give the instruction to be used and the registers used for the input and the result. Give the input operands and the result in hexadecimal after the multiply is accomplished. **[c]**

7.44 Assume Data1 and Data2 in the table that follows are $+100_{10}$ and -100_{10}, respectively. For each of the instructions give the results in the A, B, Y, and D accumulators. **[a, b]**

Instruction	A	B	Y	D
ldaa Data1				
sex a,y				
ldaa Data2				
tfr a,d				
emuls				
std Data3				

7.45 Assume D contains 4096_{10} and X = 10_{10}. What is in A, B, and X before and after an IDIV instruction?

Assembly Language Programs for the HCS12

OBJECTIVES

This chapter shows programming techniques and suggests an assembly language programming style. An example using the CodeWarrior relocatable assembler is given and explained. We also show how to write structured assembly language programs that meet the goals of top-down software design.

8.1 Assembly Language Programming Style

We discussed the syntax requirements of the CodeWarrior assembler in Chapter 5 and, like most assemblers, each program line must have its fields separated by white spaces. In addition to the syntactical requirements of each line, a standard format or style should be adopted when writing programs. This will make programs more readable for colleagues who may have to modify your code or collaborate on a software engineering project.

Source Code Style

A consistent style can make your programs easier to read.

Any program is a sequence of program elements, from the top to the bottom, and these elements should be organized in a readable and consistent style. Adopt a standard format and use it for all assembly language programs. Table 8-1 shows a format that can serve as an outline for your programs.

PROGRAM HEADER

After reading the header, you should know what the program does, not in any great detail, but at least in general. The author's name should be here so praise (or blame) can be apportioned correctly. The date of original code release and modification record is good information too.

TABLE 8-1 Assembly Language Program Elements

Program Element	Purpose
Program Header	Briefly describes the purpose of the program.
External Symbol Definitions	XREFs for symbols defined in some other source file.
Internal Symbol Definitions	XDEFs for symbols defined in this source file.
Assembler Equates	Definition of constants used in the program.
Code Section Start	Defines the following bytes to be in the code segment or section in ROM.
Program Initialization	Initializes the stack pointer, I/O devices, and other variables.
Main Program Body	This contains the main program.
Program End	Start the main program again or terminate in some way.
Program Subroutines	Subroutines and functions used in the main program.
Constant Data Section Start	Defines the following bytes to be in the constant data segment in ROM.
Constant Data Definitions	Definitions of constants in ROM.
Variable Data Section Start	Defines the following bytes to be variable data elements in RAM.
Variable Data Allocation	Allocation of space for variable data elements.

The modification record should give what has been done to the original code, when it was done, and by whom.

Program Element	Program Example
| Program Header | ```
;   HCS12 Assembler Example
;
; This program is to demonstrate a
; readable programming style.
; It initializes the A/D converter
; and a bank of LEDs. It then reads the
; value on the A/D, displays it, and delays
; about 0.5 second. It then displays the
; last value it converted for about 0.5
; second and repeats.
; Source File:  M6812EX1_REL.ASM
; Author:  F. M. Cady
; Created: 7/26/2004
; Modifications:  None
``` |

EXTERNAL SYMBOL DEFINITIONS

When you are using a relocatable assembler, source files may reference a symbol or label that is defined in some other source file. It is the job of the linker to evaluate and to provide the value for the symbol. The XREF directive tells the assembler to leave the resolution of the symbol for the linker.

Program Element	Program Example
| External Symbol Definitions | ```
;**********************************************
; External symbol definitions
        XREF   get_AD, init_AD
        XREF   enable_LED, put_LED
        XREF   delay_X_ms
        XREF   __SEG_END_SSTACK
;**********************************************
``` |

INTERNAL SYMBOL DEFINITIONS

Whenever there is an external symbol definition (XREF) in a source file, there must be an accompanying definition of the symbol (XDEF) in some other source file that is part of the project. This section of the program provides that.

Program Element	Program Example
Internal Symbol Definitions	``` ;** ; Internal symbol definitions XDEF Entry, main ;** ```

ASSEMBLER EQUATES

Equates are often found at the beginning of the program.

Some programmers put equates at the bottom of the program and some argue that it is more useful to put a constant definition right where it is used. We suggest that all equates be in one area in the program and before they are used. The assembler always assumes a referenced label refers to a 16-bit address unless it already knows it is in direct address space (the first 256 memory locations). It is always better to let the assembler know the values of labels before they are used.

Program Element	Program Example
Constant Equates	``` ;** ; Constant Equates DELAY: EQU 500 ; Used for delay subroutine ;** ```

CODE SECTION START

Each section in a relocatable assembler should have a name. This allows you to easily locate the sections with the linker parameter file and to identify them in the linker map file (see Chapter 6).

Program Element	Program Example
Code Section Start	``` ;** ; Code Section Start MyCode: SECTION ```

PROGRAM INITIALIZATION

The *stack pointer* must be initialized before it is used for subroutine calls, interrupts, and data storage. Do it as the first instruction in the program. *Variables* must be initialized at run time. Put the section of code to do this here.

Program Element	Program Example
| | ```
Entry:
main:
``` |
| | ```
; Initialization section
``` |
| Stack Pointer | ```
; Initialize stack pointer
``` |
|   Initialization | ```
        lds   #__SEG_END_SSTACK
``` |
| I/O Devices Initialization | ```
; Initialize all I/O devices
``` |
| | ```
        jsr   init_AD      ; Init the A/D
        jsr   enable_LED  ; Enable LED port
``` |
| Variable Data | ```
; Initialize the last data value
``` |
|   Initialization | ```
        clr   Last_Val
``` |

MAIN PROGRAM BODY

The main program starts here. Typically, it will be short and consist of several subroutine calls.

Program Element	Program Example
| Main Program Body | ```
;*********************************************
; Main process loop starts here:
loop:
; Get value from A/D
        jsr   get_AD
        pshb              ; Save it
; Display on LEDs
        jsr   put_LED
; Delay about 0.5 second
        ldx   #DELAY
        jsr   delay_X_ms
; Now display the last value
        ldab  Last_Val
        jsr   put_LED
        pulb              ; Get the value back
        stab  Last_Val; Save it for next time
; Delay 100 milliseconds
        ldx   #Delay1
        jsr   delay_X_ms  ; Delays # ms in X reg
``` |

PROGRAM END

When developing software on an evaluation system such as a manufacturer's evaluation board, you must return control to the monitor at the end of your program. This is often done with a *Software Interrupt* (SWI) instruction. In programs that run continuously with no need to return to the debugging monitor, a branch or jump back to the beginning of the process loop is made.

Program Element	Program Example
| Return to the beginning | ```
; Do forever
``` |
|   of main loop. | ```
        bra   loop
;*********************************************
``` |

PROGRAM SUBROUTINES

It is good programming practice to make the main program a sequence of calls to subroutines. You may place subroutines anywhere in the source program or they may be in other source files when using a relocatable assembler. In this program example we choose to do the latter. We will see some of the subroutines for initializing the analog-to-digital converter in Chapter 17.
```

Program Element	Program Example
| Subroutines and Functions | ```
;***********************************
; Subroutines and functions
;   (This relocatable assembler program does
;    not place any subroutines in the main
;    module. If you do, however, this is the
;    place to put them.)
;***********************************
``` |

CONSTANT DATA SECTION START

Constants are located in the ROM. You do not have to create a constant data section but it is good programming practice to do so.

Program Element	Program Example
| Constant Data Section Start | ```
;***********************************
; Constant data area in ROM
MyConst:SECTION
``` |

CONSTANT DATA DEFINITIONS

Constants are located in ROM. Usually it is best to have constants at the end of all code sections to decrease the danger of executing data. However, some programmers group constants with the section of code that uses them, that is, constants used in a subroutine.

Program Element	Program Example
Main Program Constants and Strings	`Delay1: DC.W   DELAY`

VARIABLE DATA SECTION START

Variable data are located in RAM. The name you give allows you to locate the variable data with the linker.

Program Element	Program Example
| Variable Data Section Start | ```
;Variable data area in RAM
MyData: SECTION
``` |

VARIABLE DATA STORAGE ALLOCATION

Use the DS to allocate storage for all variable data elements.

Program Element	Program Example
Allocation of Data Areas	`Last_Val: DS.B  1`

THE COMPLETED PROGRAM

Example 8-1 shows this program as a complete assembler list file. We have not shown the subroutines that initialize the LED display or the A/D converter or the subroutines for getting data from the A/D and displaying it on the LEDs.
```

Example 8-1 The Completed Program

```
Metrowerks HC12-Assembler
(c) COPYRIGHT METROWERKS 1987-2003

 Rel. Loc     Obj. code Source line
 ---- ----    --------- -----------
     1                    ;  HCS12 Assembler Example
     2                    ;
     3                    ; This program is to demonstrate a
     4                    ; readable programming style.
     5                    ; It initializes the A/D converter
     6                    ; and a bank of LEDs. It then reads the
     7                    ; value on the A/D, displays it, and delays
     8                    ; about 0.5 second. It then displays the
     9                    ; last value it converted for about 0.5
    10                    ; second and repeats.
    11                    ; Source File:  M6812EX1_REL.ASM
    12                    ; Author:  F. M. Cady
    13                    ; Created: 7/26/2004
    14                    ; Modifications:  None
    15                    ;
    16                    ;*****************************************
    17                    ; External symbol definitions
    18                            XREF   get_AD, init_AD
    19                            XREF   enable_LED, put_LED
    20                            XREF   delay_X_ms
    21                            XREF   __SEG_END_SSTACK
    22                    ;*****************************************
    23                    ; Internal symbol definitions
    24                            XDEF   Entry, main
    25                    ;*****************************************
    26                    ;    Constant Equates
    27       0000 01F4 DELAY:  EQU    500   ; Used for delay sub
    28                    ;*****************************************
    29                    ; Code Section Start
    30                    MyCode: SECTION
    31                    Entry:
    32                    main:
    33                    ; Initialization section
    34                    ; Initialize stack pointer
```

```
35 000000 CFxx xx              lds    #__SEG_END_SSTACK
36                     ; Initialize all I/O devices
37 000003 16xx xx              jsr    init_AD      ; Init the A/D
38 000006 16xx xx              jsr    enable_LED  ; Enable LED port
39                     ; Initialize the last data value
40 000009 79xx xx              clr    Last_Val
41                     ;*****************************************
42                     ; Main process loop starts here:
43                     loop:
44                     ; Get value from A/D
45 00000C 16xx xx              jsr    get_AD
46 00000F 37                   pshb            ; Save it
47                     ; Display on LEDs
48 000010 16xx xx              jsr    put_LED
49                     ; Delay about 0.5 second
50 000013 CE01 F4              ldx    #DELAY
51 000016 16xx xx              jsr    delay_X_ms
52                     ; Now display the last value
53 000019 F6xx xx              ldab   Last_Val
54 00001C 16xx xx              jsr    put_LED
55 00001F 33                   pulb            ; Get the value back
56 000020 7Bxx xx              stab   Last_Val; Save it for next time
57                     ; Delay about 0.5 second
58 000023 FExx xx              ldx    Delay1
59 000026 16xx xx              jsr    delay_X_ms ; Delays # ms in X
60                     ; Do forever
61 000029 20E1                 bra    loop
62                     ;*****************************************
63                     ; Subroutines and functions
64                     ; (This relocatable assembler program does
65                     ;  not have any subroutines in the main
66                     ;  module. If you do, however, this is the
67                     ;  place to put them.)
68                     ;*****************************************
69                     ; Constant data area in ROM
70                     MyConst:SECTION
71 000000 01F4         Delay1: DC.W   DELAY
72                     ;*****************************************
73                     ; Variable data area in RAM
74                     MyData: SECTION
75 000000              Last_Val: DS.B  1
```

To Indent or Not to Indent

Indentation is not used very often in assembly language programming.

In high-level languages, indentation shows lower levels of the design and makes the code more readable. Indentation is not generally used in assembly language programming. Historically, assemblers were used long before high-level language compilers that allowed indentation were developed. Also, an assembler's syntax is generally fixed. Often, labels must start in the first space on the line and there must be white space between labels, mnemonics, operands, and comments. Assembly language programmers are used to seeing the program with the fields all nicely lined up because it is easier to see the operations and operands. However, you may want to try a few programs with indented code to see how you like it.

Uppercase and Lowercase

Using *uppercase* and *lowercase* letters can make your programs more readable.

Uppercase and lowercase letters can make your code more readable. The goal is to be able to look at a name or label and tell what it is without looking further. For example, uppercase labels can be used for constants and lowercase for variables. Mixed case used for multiple word labels can make the label easier to read. Table 8-2 shows uppercase, lowercase, and mixed case examples. Some assemblers are not case sensitive and some are.

Use Equates, Not Magic Numbers

Equates make programs easier to read and easier to change in the future.

A number that just appears in the code is called a magic number. For example, if the program statement

```
Ldab #8
```

TABLE 8-2 Examples of Uppercase, Lowercase, and Mixed Case

Case	Examples
Uppercase	
Constants defined by EQU	`NULL:    EQU   $0` `PORT_H:  EQU   $24`
Constants defined by DC.B, DC.W	`STRING:  DC.B  "This is a string."` `CRLF:    DC.W  $0D0A`
Assembler directives	`ORG, EQU, DC`
Lowercase	
Instruction mnemonics	`ldaa, jsr, bne`
Labels	`loop:   . . .` `          bne    loop`
Variables	`data:   DS    10`
Mixed Case	

Multiword variables and labels	`PrintData:`
	`NumChars:`
	`InputDataBuffer:`

Multiword subroutine names	`     Jsr    PrintData`
Comments	`; Write complete sentences for comments.`

appeared in the program, you would have to ask, "What significance does the 8 have in the program?" Is it used as a counter or as an output value? You do not know. However, the code

```
COUNTER: EQU 8
         ldab #COUNTER
```

is much clearer. Furthermore, if the counter is used in several places in the program and needs to be changed, it is easier to change the equate than to search for and change all places it is used. Always use the EQU or SET directive to define constants in your program.

Using Include Files

An *include file* can contain frequently used symbols and definitions.

Assembly language programs often use the same equates in each program. More powerful assemblers such as CodeWarrior allow include files to be used as shown in Example 8-2. This technique is similar to the use of #include in C programs.

Example 8-2 Include Example

This is the source file:

```
; include_ex.asm
; Demonstration of the use of include files
        INCLUDE    "include1.inc"
        INCLUDE    "include2.inc"
; The rest of the source file follows:
```

This is the list file from the assembler:

```
Metrowerks HC12-Assembler
(c) COPYRIGHT METROWERKS 1987-2003

Rel. Loc    Obj. code Source line
---- ---     --------- -----------

   1                      ; Demonstration of the use of include files
   2                            INCLUDE    "include1.inc"
   1i                     ; This is the include1.inc file.
   2i                     ; A useful include file is one which defines
   3i                     ; commonly used constants.
   4i         0000 0004 EOT:    SET    $04   ; End of transmission
   5i         0000 000D CR:     SET    $0d   ; Carriage-return
   6i         0000 000A LF:     SET    $0a   ; Line-feed
   3                            INCLUDE    "include2.inc"
   1i                     ; include2.inc
   2i                     ; This is the second include file
   3i                     ; You can have source code and macro
```

```
4i                          ; definitions.
5i                          ; Here is a macro definition.
6i          0000 00C7 MU100:  EQU    199     ; Delay loop counter
7i                          Delay_100 MACRO
8i                          ; Macro to delay approximately 100
9i                          ; microseconds
10i                                psha         ; Save the A reg
11i                                ldaa   #MU100
12i                    \@loop: deca             ; Automatic number
13i                                bne \@loop   ; Loop until A=0
14i                                pula         ; Restore A reg
15i                                ENDM
4                           ; The rest of the source file follows:
5
```

Commenting Style

There are various commenting styles. Some programmers would have a comment on each program line. Another style is to place comments in blocks that explain what the following section of code is to do, that is, on the design or function of each block. Then, within the block of code, place comments on lines in which further explanations may be required. Using high-level, pseudocode design statements as comments in the program is very effective also. Table 8-3 shows useful information that can be included as comments in each subroutine's header.

8.2 Structured Assembly Language Programming

Programs are often designed using pseudocode as a design tool. After we have completed our design, we must write the assembly language code for it. There are two parts of the assembly language code to do structured programming. The first is a comment. This normally can be taken from the pseudocode design document. The second part is the code that implements the comment. Let us look at the three structured programming elements as they might appear in assembly language.

TABLE 8-3 Subroutine or Function Header

```
;       Subroutine calling sequence or invocation.
;       Subroutine name.
;       Purpose of subroutine.
;       Name of file containing the source.
;       Author.
;       Date of creation or release.
;       Input and output variables.
;       Registers modified.
;       Global data elements modified.
;       Local data elements modified.
;       Brief description of the algorithm.
;       Functions or subroutines called.
```

Example 8-3 Assembly Language for Sequence Block

```
;   * * * * * * * * * * * * * * * * * * * * * * * * * * * * * * * * * * * * * * * * * * * * * * * * * * * * * *
;   DO_A
;   Comments describing the function of this sequence block
. . . .    (Assembly language code to do the function)
;   ENDO_A
;   * * * * * * * * * * * * * * * * * * * * * * * * * * * * * * * * * * * * * * * * * * * * * * * * * * * * *
```

Sequence

The sequence is straightforward. There should be a block of comments describing what the next section of assembly code is to do. Remember that the flow of the program is in at the top and out at the bottom. We must not enter or exit the code between DO_A and ENDO_A except to call and return from a subroutine. Do not jump into or out of the middle of a sequence block. See Example 8-3.

IF-THEN-ELSE Decision

A pseudocode design using the decision element and the associated assembly language code is shown in Example 8-4. See also Examples 8-5 and 8-6.

The IF-THEN-ELSE code always has the same form. The **bold** lines in Example 8-4 will appear in every decision structure. Notice that in the assembly code the indentation we are familiar with in high-level languages is not used, although we may retain the indentation of the pseudocode comments.

Example 8-4 Decision Element Assembly Language Program

Pseudocode Design

```
; Get the temperature
; IF Temperature > Allowed Maximum
;    THEN Turn the water valve off
;    ELSE Turn the water valve on
; END IF temperature > Allowed Maximum
```

Structured Assembly Code

```
Metrowerks HC12-Assembler
(c) COPYRIGHT METROWERKS 1987-2003

Rel. Loc    Obj. code Source line
---- ---    --------- -----------
   1                      ; 68HCS12 Structured assembly code
   2                      ; IF-THEN-ELSE example.
   3                      ; Equates define constants needed by the code
```

```
 4      0000 0091 AD_PORT:    EQU   $91   ; A/D Data Port
 5      0000 0080 MAX_TEMP:   EQU   128   ; Maximum temperature
 6      0000 0000 VALVE_OFF:  EQU   0     ; Bits for valve off
 7      0000 0001 VALVE_ON:   EQU   1     ; Bits for valve on
 8      0000 0258 VALVE_PORT: EQU   $258  ; Port P for the valve
 9                             ;
10                             ;
11                             ;          . . .
12                             ; Get the temperature
13 000000 9691                      ldaa  AD_PORT
14                             ; IF Temperature > Allowed Maximum
15 000002 8180                      cmpa  #MAX_TEMP
16 000004 2307                      bls   ELSE_PART
17                             ;   THEN Turn the water valve off
18 000006 9600                      ldaa  VALVE_OFF
19 000008 7A02 58                   staa  VALVE_PORT
20 00000B 2005                      bra   END_IF
21                             ;   ELSE Turn the water valve on
22                             ELSE_PART:
23 00000D 9601                      ldaa  VALVE_ON
24 00000F 7A02 58                   staa  VALVE_PORT
25                             END_IF:
26                             ; END IF Temperature > Allowed Maximum
27
```

EXPLANATION OF EXAMPLE 8-4

Lines 12, 14, 17, 21, and 26: These lines contain the pseudocode design as comments in the source code.

Line 15: Following the IF statement is code to set the condition code register for the conditional branch in *line 16* to the ELSE part.

Line 16: There will always be a conditional branch to the ELSE part, as shown here, or the THEN part. When branching to the ELSE part, the conditional branch instruction is the complement of the logic in the IF statement. In this example, the ELSE part is to be executed if the temperature is lower or the same as the allowed maximum because the THEN part is done when the temperature is higher.

Lines 18 and 19: This is the code for the THEN part.

Line 20: The THEN part *always* ends with a branch-always or jump to the END-IF label. This branches around the ELSE part code.

Line 22: The label for the ELSE part conditional branch is always here.

Lines 23 and 24: This is the code for the ELSE part.

Line 25: The IF-THEN-ELSE always ends with and END_IF label.

Example 8-5

For each of the logic statements, give the appropriate HCS12 code to set the condition code register and to branch to the ELSE part of an IF-THEN-ELSE. Assume P and Q are 8-bit, signed numbers in memory locations P and Q.

A. IF P >= Q.
B. IF Q > P.
C. IF P = Q.

Solution:

```
A.  ; IF  P >= Q
            ldaa    P
            cmpa    Q
            blt     ELSE_PART
B.  ; IF  Q > P
            ldaa    Q
            cmpa    P
            ble     ELSE_PART
C.  ; IF  P = Q
            ldaa    P
            cmpa    Q
            bne     ELSE_PART
```

Example 8-6

For each of the logic statements, give the appropriate HCS12 code to set the condition code register and to branch to the THEN part of an IF-THEN-ELSE. Assume P and Q are 8-bit, unsigned numbers in memory locations P and Q.

A. IF P >= Q.
B. IF Q > P.
C. IF P = Q.

Solution:

```
A.  ; IF  P >= Q
            ldaa    P
            cmpa    Q
            bhs     THEN_PART
B.  ; IF  Q > P
            ldaa    Q
            cmpa    P
            bhi     THEN_PART
C.  ; IF  P = Q
            ldaa    P
            cmpa    Q
            beq     THEN_PART
```

WHILE-DO Repetition

The WHILE-DO structure is shown in Example 8-7. The elements common to all WHILE-DOs are shown in **bold**.

Example 8-7 Assembly Code for a WHILE-DO

Pseudocode Design

```
; Get the temperature from the A/D
; WHILE the temperature > maximum allowed
;     DO
;        Flash light 0.5 second on, 0.5 second off
;         Get the temperature from the A/D
;     END_DO
; END_WHILE the temperature > maximum allowed
```

Structured Assembly Code

```
Metrowerks HC12-Assembler
(c) COPYRIGHT METROWERKS 1987-2003

  Rel. Loc    Obj. code Source line
  ---- ---    --------- -----------

    1                   ; 68HC12 Structured assembly code
    2                   ; WHILE - DO Example
    3                   ; Equates needed
    4          0000 0091 AD_PORT:    EQU $91    ; A/D Data port
    5          0000 0080 MAX_ALLOWED:EQU 128     ; Maximum Temp
    6          0000 0001 LIGHT_ON:   EQU 1
    7          0000 0000 LIGHT_OFF:  EQU 0
    8          0000 0258 LIGHT_PORT: EQU $258   ; Port P
    9                   ;   - - -
   10                   ; Get the temperature from the A/D
   11 000000 9691           ldaa   AD_PORT
   12                   ; WHILE the temperature > maximum allowed
   13                   WHILE_START:
   14 000002 9180           cmpa   MAX_ALLOWED
   15 000004 2314           bls    END_WHILE
   16                   ;     DO
   17                   ; Flash light 0.5 second on, 0.5 second off
   18 000006 9601           ldaa   LIGHT_ON
   19 000008 7A02 58        staa   LIGHT_PORT  ; Turn the light
```

```
20 00000B 16xx xx            jsr   delay      ; 0.5 second delay
21 00000E 9600               ldaa  LIGHT_OFF
22 000010 7A02 58            staa  LIGHT_PORT ; Turn the light off
23 000013 16xx xx            jsr   delay
24                    ;      End flashing the light
25                    ;      Get the temperature from the A/D
26 000016 9691              ldaa  AD_PORT
27                    ; END_DO
28 000018 20E8               bra   WHILE_START
29                   END_WHILE:
30                   ; END_WHILE the temperature > maximum allowed
31
32                           ; Dummy subroutine
33 00001A 3D         delay:  rts
```

EXPLANATION OF EXAMPLE 8-7

Lines 10, 12, 16, 17, 24, 25, 27, and 30: The pseudocode design appears as comments in the code.

Lines 14, and 15: A WHILE-DO tests the condition at the top of the code to be repeated. Thus the conditional branch in *line 15* must be preceded by code that initializes the variable to be tested. The A register is initialized with the A/D value in *line 11*.

Line 13: There must be a label at the start of the conditional test code. This is the address for the BRA in *line 28*.

Line 14: Following the WHILE statement is code to set the condition code register for the subsequent conditional branch to the end of the WHILE-DO.

Line 15: A conditional branch allows us to exit this structure.

Lines 17–26: This is the code for the DO part.

Line 26: A special requirement of the WHILE-DO structure is code that changes whatever is being tested. If this were not here, the program would never leave the loop.

Line 28: The code block always ends with a branch back to the start.

As an assembly language programmer you might be smarter than the average compiler and realize that *line 26* could be eliminated if the code to initialize the A register with the A/D value (*line 11*) is moved below the label WHILE_START.

DO-WHILE Repetition

Another useful repetition is the DO-WHILE. In this structure, the DO part is executed at least once because the test is at the bottom of the loop. An example of the DO-WHILE is shown in Example 8-8, where again the parts common to all DO-WHILEs are shown in **bold**.

Example 8-8 DO-WHILE Assembly Language Code

Pseudocode Design

```
; DO
;    Get data from the switches
;    Output the value to the LEDs
; ENDO
; WHILE Any switch is set
```

Structured Assembly Code

```
Metrowerks HC12-Assembler
(c) COPYRIGHT METROWERKS 1987-2003

Rel. Loc    Obj. code Source line
---- ---    --------- -----------
    1                   ; 68HC12 Structured assembly code
    2                   ; DO-WHILE example
    3                   ; Equates needed for this example
    4      0000 0028 SW_PORT: EQU $28 ; Switches are on Port J
    5      0000 0024 LEDS:    EQU $24 ; The LEDs are on Port H
    6                   ;         - - -
    7                   ; DO
    8                   DO_BEGIN:
    9                   ;   Get data from the switches
10 000000 9628              ldaa  SW_PORT
11                   ;   Output the data to the LEDs
12 000002 5A24              staa  LEDS
13                   ; END_DO
14                   ; WHILE Any switch is set
15 000004 F700 28          tst   SW_PORT
16 000007 26F7             bne   DO_BEGIN
17                   ; END_WHILE
```

EXPLANATION OF EXAMPLE 8-8

Lines 7, 9, 11, 13, and 14: The pseudocode appears as comments.

Line 8: The start of the DO block has a label for the conditional branch instruction in *line 16*.

Lines 9–12: These are the code lines for the DO part.

Lines 15 and 16: The DO-WHILE always ends with a test and a conditional branch back to the beginning of the DO block.

Line 17: A comment marks the end of the WHILE test code.

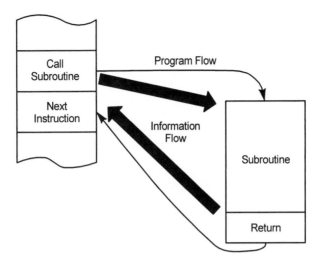

Figure 8-1 Information transfer between modules.

8.3 Interprocess Communication

Interprocess communication, also called parameter passing, refers to information that is transferred from one part of the program to another. Most information transfer in well-designed programs is between a subroutine or function and its calling function as shown in Figure 8-1. In choosing how information is transferred between modules, a goal is to reduce the chance of the subroutine accidentally changing other data. There are several methods that can be used.

Information in Registers

The most efficient and fastest way to transfer information between parts of a program when writing in assembly language is to use the registers. This method also has the advantage that the subroutine does not access any other data areas and is thus more general. Documentation must be provided to show what registers are used for what purpose. A typical subroutine header describing the registers used is shown in Table 8-4.

Using the registers is simple and straightforward. See Example 8-9.

TABLE 8-4 Subroutine Header Comments

```
;  ***********************************************************
;  *
;  * Subroutine Name:      SQRT
;  * Author:               F. M. Cady
;  * Date:                 July 19, 1993
;  * Function:             Calculate the square root of a 16
;  *                       bit integer number.
;  * Input Registers:      D = 16 bit integer number
;  * Output Registers:     B = 8 bit integer square root
;  *                       Carry flag = 1 if input number is negative
;  *                       Carry flag = 0 if input number is positive
;  * Registers modified:   B, condition code register
;  * Global data modified: none
;  * Functions called:     none
;  *
;  ***********************************************************
```

Example 8-9 Passing Information in Registers

```
11                         ;*******************************
12                         ; Parameter passing between modules
13                         ;*******************************
14                         ; Passing arguments in Registers
15                         ;*******************************
16                         ; . . .
17                         ; Get the input argument and pass to the
                           subroutine
18 000003 B6xx xx              ldaa    Input_Arg1
19 000006 16xx xx              jsr     sub1
20                         ; . . .
21                         ;*******************************
22                         ; The subroutine may be local or external
23                         ; Input: A = Input Argument
24                         ; Output: A = Output Argument
25                         ; Registers modified: A
26                         sub1:
27                         ; Push the registers used on the stack
28                         ; . . .
29                         ; Use the input argument and/or modify it
30 000009 48                   asla
31                         ; Pull the registers used from the stack
32                         ; Return with the modified data
33 00000A 3D                   rts
34                         ;*******************************
35                         MyData: SECTION
36                         ; Place variable data here
37 000000                 Input_Arg1: DS.B  1
```

Information in Global Data Areas

The main disadvantage of using registers is that most CPUs have only a few and some functions may need many bytes of data. Using global data areas is a solution with advantages and potential problems. Global data are data elements that can be reached from any part of the program. Figure 8-2 shows four modules using two global data elements.

The danger of maintaining global data is that a function may modify data that it shouldn't. For example, let's assume that Module_1 shares Data_Element_1 with Module_2 and Module_3 shares Data_Element_2 with Module_4. Now let's assume that you make a mistake (a bug) in the code that is supposed to write data into Data_Element_1 and write into Data_Element_2 instead. (This could be done by using a 16-bit store operation instead of an

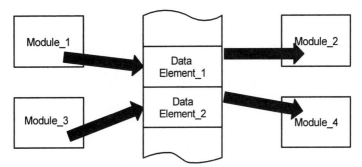

Figure 8-2 Information transfer using global data.

8-bit one, or by having an incorrectly initialized pointer register, or simply by writing the wrong label in the operand field.) Now Module_4 is working with incorrect data. This is a difficult bug to find, particularly if the code in Module_1 is executed infrequently. Experienced assembly and high-level language programmers try to avoid using global data if other methods are available. Nevertheless, global data structures are widely used in assembly and high-level language programming. See Example 8-10.

```
Example 8-10  Passing Information in Global Data Area
Metrowerks HC12-Assembler
(c) COPYRIGHT METROWERKS 1987-2003

Rel. Loc    Obj. code Source line
---- ----   -------- -----------
   1                    ;*******************************
   2                    ; Passing arguments in global data
   3                    ;*******************************
   4                    ; Define the entry point for the main program
   5                            XDEF   Entry, main
   6                            XREF   __SEG_END_SSTACK
   7                    ; Define the data names that are external in
   8                    ; a global data buffer
   9                            XREF  Data_Element_1, Data_Element_2
  10                            XREF  Data_element_3, Data_Element_4
  11              MyCode: SECTION
  12              Entry:
  13              main:
  14                    ;*******************************
  15                    ; Initialize stack pointer register
  16 000000 CFxx xx             lds    #__SEG_END_SSTACK
  17                    ;*******************************
  18                    ; Module_1 puts data into Data_Element_1
  19 000003 7Axx xx             staa  Data_Element_1
  20                    ;*******************************
```

```
21                      ; Module_2 gets data from Data_Element_1
22 000006 B6xx xx               ldaa   Data_Element_1
23                      ;********************************
24                      ; Module_3 puts data into Data_Element_2
25 000009 7Axx xx               staa   Data_Element_2
26                      ;********************************
27                      ; Module_4 gets data from Data_Element_2
28 00000C B6xx xx               ldaa   Data_Element_2
29                      ;********************************
30
```

```
Metrowerks HC12-Assembler
(c) COPYRIGHT METROWERKS 1987-2003
Rel. Loc    Obj. code Source line
--- ----   ------ --------
    1                  ;********************************
    2                  ; This is the global data definition
    3                  ; The data storage allocations are done
    4                  ; here and all data names are XDEFed to
    5                  ; make them globally available
    6                  ;********************************
    7                          XDEF   Data_Element_1, Data_Element_2
    8                          XDEF   Data_Element_3, Data_Element_4
    9                  GlobalData: SECTION
   10                  ; Place variable data here
   11 000000          Data_Element_1: DS.B  1
   12 000001          Data_Element_2: DS.B  1
   13 000002          Data_Element_3: DS.B  1
   14 000003          Data_Element_4: DS.B  1
   15
```

Information in Local Data Areas

Local data areas invoke the principle of "divide and conquer." Figure 8-3 shows modules and their common data elements that are separately assembled source files. When a relocatable assembler is used, as it must be here, any names or labels are local to that source file only, unless a special assembler directive called EXTERNAL is used. Thus the assembler will show an error if you assemble the file with Module_1, Module_2, and Data_Element_1, and accidentally refer to Data_Element_2. However, as you can see, the data elements are global within these localized structures.

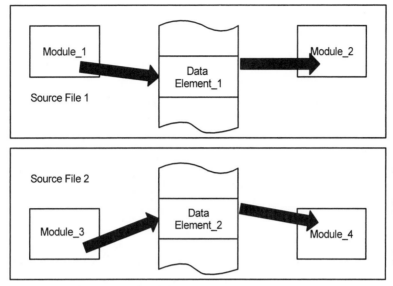

Figure 8-3 Information in local data areas.

Information on the Stack

The stack can also be used to transfer data to and from a subroutine. When this is done, the data elements are localized on the stack and the subroutine is designed to operate with them alone. This reduces the chance of global data being accidentally corrupted. You must be careful when using the stack because, in addition to the data on the stack, the return address and any bytes pushed on the stack when the subroutine is entered are there also.

Example 8-11 and Figure 8-4 show how data may be passed to and from a subroutine using the stack. Figure 8-4(a) shows the initial position of the stack pointer and the contents of the

```
Example 8-11  Passing Information on the Stack
Metrowerks HC12-Assembler
(c) COPYRIGHT METROWERKS 1987-2003
 Rel. Loc    Obj. code Source line
 ----  ----  --------  -------------
    1                   ;*******************************
    2                   ; Passing arguments on the stack
    3                   ;*******************************
    4                   ; Define the entry point for the main program
    5                            XDEF   Entry, main
    6                            XREF   __SEG_END_SSTACK
    7                   MyCode: SECTION
    8                   Entry:
    9                   main:
   10                   ;*******************************
```

```
11                        ; Initialize stack pointer register
12 000000 CFxx xx               lds    #__SEG_END_SSTACK
13                        ;*******************************
14                        ;
15 000003 CC12 34               ldd    #$1234  ; Demo data
16 000006 CE45 67               ldx    #$4567
17                        ; Put the data to be transferred on the stack
18 000009 3B                    pshd       ; Two bytes transferred
19 00000A 16xx xx               jsr    sub1
20                        ; Get the returned data and clean up the
21                        ; stack pointer
22 00000D 3A                    puld       ; Two bytes returned
23                        ; . . .
24                        ;*******************************
25                        ;*******************************
26                        ; Subroutine sub
27                        ; Input: 16-bit data on the stack
28                         ; Output: 16-bit data on the stack
29                         ; Registers modified: CCR
30                         ;*******************************
31    0000 0002 Num_B:  EQU   2  ; Number of data bytes on stack
32    0000 0004 Reg_B:  EQU   4  ; Number of register bytes on
                                       stack
33                        sub1:
34                        ; Push registers used in the subroutine onto
                          the stack
35 00000E 3B                    pshd
36 00000F 34                    pshx
37                        ; . . .
38                        ; Use indexed addressing to get the data
                          passed in
39                        ; from the stack
40 000010 EC86               ldd    Num_B+Reg_B,sp
41                        ; . . .
42                        ; Put the return data back on the stack
43 000012 CC9A BC             ldd    #$9ABC
44 000015 6C86               std    Num_B+Reg_B,sp
45                        ; . . .
46                        ; Pull the used registers from the stack
47 000017 30                    pulx
48 000018 3A                    puld
49 000019 3D                    rts
50                        ;*******************************
```

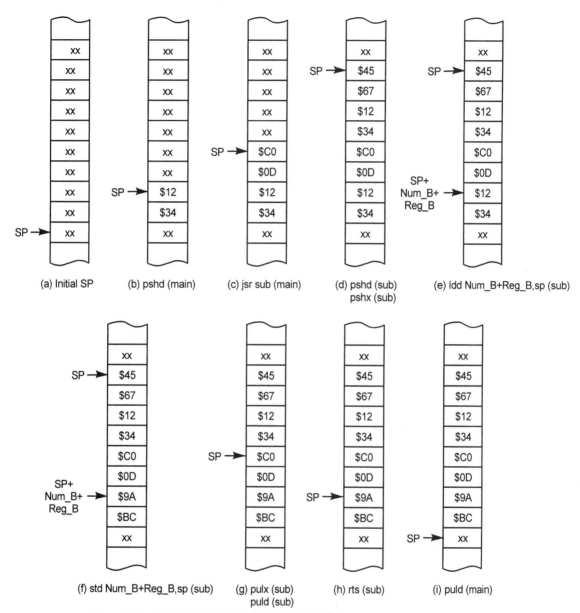

Figure 8-4 Using the stack for information transfer.

stack. *Line 18* in Example 8-11 pushes 16-bit data onto the stack (Figure 8-4(b)) and the subroutine call is made in *line 19* (Figure 8-4(c)). The D and X registers are pushed onto the stack in the subroutine (*lines 35* and *36*, Figure 8-4(d)). The subroutine retrieves data from the stack and returns data to the stack using indexed addressing and the stack pointer (*lines 40* and *44*, Figures 8-4(e) and 8-4(f)). The number of bytes between the current value of the stack pointer and the data to be pulled is given by Num_B + Reg_B. After the registers are restored (*lines*

47 and *48*, Figure 8-4(g)) and the return from subroutine (*line 49,* Figure 8-4(h)), the main program can retrieve the returned data (*line 22,* Figure 8-4(i)).

Using the stack to transfer information is very powerful and very general. Most compilers for high-level languages use this method. Programmers must be careful to make sure stack operations are balanced and good documentation must be provided.

Using Addresses Instead of Values

Using global data areas, even with potential problems and disadvantages, is common, particularly where large amounts of data are to be shared between modules. One can avoid some problems by using a register to pass the address of the data. This is useful when accessing data buffers where, perhaps, one module fills the buffer and another processes the data. The two modules can be given the starting address of the buffer and the number of data elements (to avoid running over the end of the buffer).

Passing Boolean Information

At times a Boolean, or logic, value must be returned to the calling program. For example, you might want to indicate whether a procedure was successful and then act accordingly in the calling program. A register or a memory location could be allocated for this, but if you are programming in assembly language you may use a bit in the condition code register. All processors have the capability of setting or resetting the carry flag. This can be tested with a conditional branch instruction in the calling program.

8.4 Assembly Language Tricks of the Trade

Here are some tricks of the assembly language programming trade.

Do not modify registers in a subroutine: You should ensure that a subroutine in assembly language *does not* modify the contents of any of the registers unless a register is to return a value to a calling program. In assembly language programming we use registers to hold data from one instruction to another. Unseen instructions in a subroutine should not change registers. An exception to this rule may be the condition code register.

Use register addressing when possible: Instructions that use internal registers execute faster and use less memory.

Use register indirect or indexed addressing: These modes are the next most efficient after register addressing. The address may be calculated at run time, allowing the location of the data to be a variable depending on the current state of the program. Often data must be stored in or retrieved from a buffer. An indirect addressing mode, such as register indirect or indexed addressing, is most efficient for this, especially if the register can be automatically incremented or decremented. Remember to load the register with the address before using it.

Use the stack for temporary data storage: When using the stack in subroutines, remember to pull the registers before returning to the calling program.

Do not use the assembler to initialize variable data areas: Assemblers have directives or pseudo-operations to initialize the contents of memory locations. This works well in systems where the program is downloaded each time it is run. However, in an embedded microcontroller application where the program resides in ROM, all variable data areas must be initialized at run time by the program. Use directives that allocate memory storage locations for the variables and then initialize them in the program.

Use assembler features, directives, and pseudo-operations: Study and use the assembler directives and pseudo-operations. It is mandatory to use labels for symbolic addresses. Never refer to a memory location by a direct address. Using the assembler to evaluate expressions can make programs more readable and more easily transferred to other applications. If a macro-assembler is available, use macros to make programs more readable.

8.5 Making It Look Pretty

Your programs will be judged by your peers, not only for how they run, but also for how they look. It does not take much extra effort to make your program listings look good. Here are some helpful hints.

- Adopt a consistent indentation style. Most assembly language programmers indent and align the opcode, the operand, and the comment fields.
- Use frequent comments. A very effective commenting style is to include your design statements as comments preceding the code that implements the commented design. There is no need to place a comment on every line of assembly code. Simply comment assembly lines that need some explanation.
- Place your equates in one place in the program, often at the beginning, so readers will know where to go to find them.
- Use white space—blank lines—to separate sections of code.
- Use comment lines of asterisk characters (*) to outline sections of code.
- Don't try to completely box in a section of comments with asterisk characters. If the comment lines change, you have to spend extra time cleaning up the boxes. See Example 8-12.
- Use uppercase and lowercase writing style. Don't put comments in all uppercase or all lowercase.
- Use uppercase and lowercase for constants, variables, labels, and so on. Adopt some style that looks good and be consistent with it.

Example 8-12

```
;*****************************************************
;* Even though it looks pretty, you should not      *
;* try to box in comments with asterisks (or other  *
;* characters). If the comments change, it is just  *
;* more work to change the line.                     *
;*****************************************************
```

8.6 Conclusion and Chapter Summary Points

In this chapter we showed an example of a readable program style for assembly language programs. We illustrated (and strongly urge you to adopt) a structure assembly language programming style that uses a structure pseudocode design implemented in assembly language.

8.7 Bibliography and Further Reading

Ganssle, J. G., *A Guide to Commenting*, The Ganssle Group. Baltimore, MD, February, 2006. http://www.ganssle.com/commenting.htm.

8.8 Problems

Basic

8.1 For each of the logic statements, give the appropriate HCS12 code to set the condition code register and to branch to the ELSE part of an IF-THEN-ELSE. Assume P and Q are 8-bit, unsigned numbers in memory locations P and Q. **[c, k]**

  a. IF P $>=$ Q.

  b. IF Q $>$ P.

  c. IF P $=$ Q.

8.2 For each of the logic statements, give the appropriate HCS12 code to set the condition code register and to branch to the ELSE part of an IF-THEN-ELSE. Assume P and Q are 8-bit, signed numbers in memory locations P and Q. **[c, k]**

  a. IF P $>=$ Q.

  b. IF Q $>$ P.

  c. IF P $=$ Q.

Intermediate

8.3 For each of the logic statements, give the appropriate HCS12 code to set the condition code register and to branch to the ELSE part of an IF-THEN-ELSE. Assume P, Q, and R are 8-bit, signed numbers in memory locations P, Q and R. **[c, k]**

  a. IF P $+$ Q $>=$ 1.

  b. IF Q $>$ P $-$ R.

  c. IF (P $>$ R) OR (Q $<$ R).

  d. (P $>$ R) AND (Q $<$ R).

8.4 Assume K1 and K2 are 8-bit signed (two's complement) integer variables and K3 is a 16-bit unsigned integer variable. **[c, k]**

  a. Show how to allocate storage for these variables in a relocatable assembly language program using the CodeWarrior assembler.

  b. Write structured assembly language code for the following design. (Assume K1, K2, and K3 have been initialized in some other part of the program.)

```
;  IF K1 < K2
;  THEN
;      Set K1 to the most positive number
;  ELSE
;      Set K3 to the most positive number
;      Initialize K2 to the most negative
       number
;  ENDIF K1 < K2
```

8.5 Insert code to implement the following structured design immediately after each design comment. Assume the following structured design is just a small segment of an overall program. Assume the following 8-bit two's complement variable data allocations have been made and have been initialized in some other part of the program. **[c, k]**

```
Temp1:  DS.B  1
Temp2:  DS.B  1
;  Implement the following design
;  IF Temp1 < Temp2
;  THEN Temp1 = Temp2
;  ELSE Temp2 = Temp1
;  ENDIF
endif:
```

8.6 Insert code to implement the following structured design immediately after each design comment. Assume the following structured design is just a small segment of an overall program. Assume the following 8-bit-complement unsigned variable data allocations have been made and have been initialized in some other part of the program. **[c, k]**

```
Temp3:              DS.B        1
Temp4:              DS.B        1
Temp5:              DS.B        1
; Implement the following design
; WHILE Temp3 > Temp4
; DO
; Temp4 = Temp4 + 1
; Temp5 = 2 * Temp5
; ENDWHILEDO
enddo:
```

8.7 Write a section of HCS12 code to implement the following design, where K1 and K2 are 8-bit unsigned numbers in memory locations K1 and K2. **[c, k]**

```
; IF K1<K2
; THEN K2=K1
; ELSE K1=64
; ENDIF K1<K2
```

8.8 A 16-bit number is in sequential memory positions DATA1 and DATA1+1 with the most significant byte in DATA1. Write an HCS12 code segment to store the negative of this 16-bit number in DATA2 and DATA2+1. **[a, c, k]**

8.9 Write a section of HCS12 code to implement the following design. **[c]**

```
IF Data1 > Data2
THEN Data2 = Data1
ELSE Data2 = 64₁₀
ENDIF Data1 > Data2
```

Assume Data1 and Data2 are memory locations containing 8-bit unsigned integer data. Structured code must be used and comments must be included.

Advanced

8.10 Write HCS12 assembly language code for the following pseudocode design assuming K1, K2, and K3 are 8-bit, signed or unsigned, numbers in memory locations K1, K2, and K3. Assume memory has been allocated for these data. **[c, k]**

```
; WHILE K1 does not equal $0d
while_start:
; DO
;   IF K2 = K3
;     THEN
;       K1 = K1 + 1
;       K2 = K2 - 1
;     ELSE
```

```
;       K1 = K1 - 1
;     ENDIF K2 = K3
;   ENDO
enddo:
; ENDWHILE
```

8.11 Write a structured assembly language code segment for the following pseudocode design. **[c k]**

Assume P, Q, and R are 8-bit signed integer variables in memory locations P and Q. Insert the code needed for the design in the comments below. You may add more comments if you wish. **[c, k]**

```
; IF P does not equal Q
; THEN P=R
; ENDIF P does not equal Q
; WHILE P < R
; DO
;     P = P + 1
; ENDO
; ENDWHILEDO
end_while:
```

8.12 Write a section of HCS12 code to implement the following pseudocode design, where K1, K2, and K3 are signed 8-bit integer numbers stored at memory locations K1, K2, and K3. **[c, k]**

```
; WHILE K1 < K2
; DO
;   IF K3 > K2
;     THEN K2 = K1
;     ELSE K2 = K3
;   ENDIF K3>K2
;   K1 = K1+1
; ENDO
; ENDWHILE K1<K2
```

8.13 For Problem 8.12, assume K1 = 1, K2 = 3, and K3 = −2. How many times should the code pass through the loop and what final values do you expect for K1, K2, and K3?

8.14 Write structured HCS12 code for the following pseudocode design. **[c, k]**

```
; IF A1 = B1
; THEN
;   WHILE C1 < D1
;     DO
;       Decrement D1
;       A1 = 2 * A1
;     ENDO
;   ENDWHILE C1 < D1
; ELSE
;   A1 = 2 * B1
; ENDIF A1 = B1
```

Assume that A1, B1, C1, and D1 are 16-bit unsigned binary numbers and that memory has been allocated in the program by the following code:

```
A1:   DS.B   2
B1:   DS.B   2
C1:   DS.B   2
D1:   DS.B   2
```

Assume A1, B1, C1, and D1 are initialized to some value in some other part of the program.

8.15 For Problem 8.14, assume A1 = 2, B1 = 2, C1 = 3, and D1 = 6. What final values do you expect after the code has been executed? **[b]**

8.16 Write a structured assembly language code segment for the following pseudocode design. **[c, k]**
       Assume P and Q are 8-bit unsigned integer variables in memory locations P and Q. Also assume function X is implemented in a subroutine named X. Insert the code needed for the design in the comments below.

```
; If P = $1B
; THEN
;   WHILE Q < 186
;   DO function X
;   ENDDO
;   ENDOWHILE
; ENDIF
```

8.17 Write a pseudocode design for the following program statement. **[c, k]**
       The program is to prompt for and accept a two-digit hexadecimal number from a user using a routine called getchar. If the two digits entered by the user signify a printable ASCII character, the character is to be printed with an appropriate message. Otherwise, an error message is to be printed. The program is to continue until the user types two hex numbers that are not a code for a printable character. Your design must show at least one example of a repetition and one decision.
       (Example: If the user types a 4 and then a 1, A should be printed along with an appropriate message.)

8.18 Write structured pseudocode (do not write assembly language code) using the principles of structured programming for the following problem statement. **[c]**
       The program is to prompt for and accept a two-digit hexadecimal number from a user typing characters on the keyboard. These are to be converted to an 8-bit binary number and displayed on the LEDs.

After a 1-second delay, the complement of the byte is to be displayed on the LEDs for 1 second. After this delay, the LEDs are to be turned off and the process repeated starting at the prompt. The program is to continue until the user types two zeros ("00").
       Your design should follow the principles of top-down design and you may postpone (you don't have to show the design for) details such as how to convert the two input characters to binary and the details of the prompt and how it is to be printed.

8.19 Write a program for the HCS12 to find the largest of 32, 8-bit unsigned numbers in memory locations $0800–$081F. Place the answer in $0820. **[c]**

8.20 Write a program for the HCS12 to find the smallest of 32, 8-bit unsigned numbers in memory locations $0800–$081F. Place the answer in $0820. **[c]**

8.21 Write a program for the HCS12 to find the largest (most positive) of 32, 8-bit two's-complement numbers in memory locations $0800 = $081F. Place the answer in $0820. **[c]**

8.22 Write a program for the HCS12 to find the smallest (most negative) of 32, 8-bit two's-complement numbers in memory locations $0800 = $081F. Place the answer in $0820. **[c]**

8.23 Write a program for the HCS12 to find the address of the largest of 32, 8-bit unsigned numbers in memory locations $0800–$081F. If more than one location contains the largest number, use the lowest address as the result. Place the answer in $0820:$0821. **[c]**

8.24 Write a program for the HCS12 to find the address of the largest of 32, 8-bit unsigned numbers in memory locations $0800–$081F. If more than one location contains the largest number, use the highest address as the result. Place the answer in $0820:$0821. **[c]**

8.25 There are 4 bytes of data in memory locations $0800–$0803. Write a structured assembly program to count the number of 1's in these 4 bytes. Place the result in $0804. **[c]**

8.26 Write a structured assembly program to reverse the order of $20 bytes in a buffer. Assume the buffer is in memory locations $0800–$081F. **[c]**

8.27 Write a structured assembly program to compute factorial 8. Store the result in a 2-byte memory location in RAM. **[c]**

8.28 Write a structured assembly program subroutine to search a null terminated string of characters for a specific substring and to return the address of the start of the substring. The input to the subroutine is to be the starting address of the string to be searched, the starting address of the substring to be searched for, and the number of characters in the substring. If the substring is found, return the address of the first character in the search string, otherwise return an address of $0000. **[c]**

8.29 Write an assembly program showing how to transfer 4 bytes of data from the main to a subroutine using the stack. The subroutine does not return any data to the main. Show how the main puts data onto the stack, how the subroutine retrieves the data, and how the main program restores the stack pointer after the return from the subroutine. **[c]**

8.30 An 8-bit signed/magnitude number system is in use. Write assembly subroutines for the following. **[c]**

a. Add two 8-bit signed/magnitude numbers.

b. Subtract two 8-bit signed/magnitude numbers.

c. Multiply two 8-bit signed/magnitude numbers.

d. Divide two 8-bit signed/magnitude numbers.

Debugging HCS12 Programs

OBJECTIVES

This chapter describes debugging strategies and techniques useful in helping you to find problems in your programs.

9.1 Introduction

By now you will have experienced writing and running simple assembly language programs on your laboratory equipment. You have also probably experienced the programmer's nightmare. Your program does not work. Sometimes it does not even appear to run. It is time for some program debugging.

9.2 Program Debugging

Program debugging is like solving a mystery. We start the program, fully expecting it to work perfectly, and it does not. Often, when beginning students are asked, "What is your program doing?" they respond with "Nothing!" The computer *cannot* be doing nothing. It has to be fetching opcodes, executing them, incrementing the program counter, and fetching the next opcode. Remember that *you* are responsible for the opcodes the computer is executing, and you should know what every instruction does at every step along the way. You must do some detective work to find the difference between what you expect the program to be doing and what it is actually doing. Debugging is the process of finding the clues and interpreting them to find the problem.

> Programs that are not working properly should be *analyzed* to find out what they are doing before trying to fix them.

There are two approaches to fixing bugs in programs.[1] The first is a *synthesis* approach in which you try to fix the problem by changing the code somewhere. *This is wrong!* You must first find out what the program is doing before it can be fixed. This is the second approach—*analyzing the problem*. You first find out *what* it is doing, then *why* it is doing that, and then you will probably have enough clues to be able to *fix* it.

---

[1] The first program "bug" was found on September 9, 1945 by Grace Murray Hopper, when she found a moth between two contacts of a relay used in a computer logic circuit.

What we think will happen.   What actually happens.

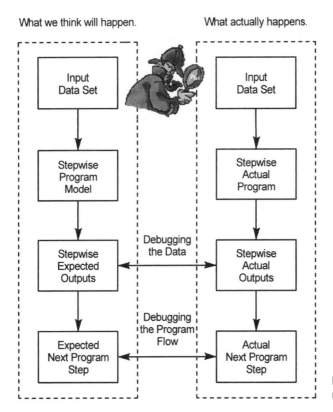

Figure 9-1 Analytical debugging model.

Programs have only two parts—the data and the logic. Data values are input to program, stored and manipulated, and output in some form. The logic determines the sequence in which program steps are executed and how the data are manipulated. Most program bugs are in the logic. We mean for the computer to do one thing and we program it to do another. Normally the data affects the program's flow, and this can help us find the debugging clues. When using the analytical debugging technique, we try to match what we think the program should do for a particular input data set with what the program actually is doing.

Figure 9-1 shows how we should do analytical debugging. We choose an input data set. We predict what the program will do with these data at each step of the program and what the program will do next. This is a model of what the program should do. Now run the program and, using the tools described next, look for data values and program steps that differ from the model. Once we find out where the program deviates from the model, we are well on our way to finding out why it is going wrong and what will be needed to fix it.

9.3 Code Walkthroughs

Probably the single most effective debugging technique is to remove bugs before they are put into the firmware. A *code walkthrough* is often used (although not used often enough, it seems) to eliminate problems before they end up in the code.

Code walkthroughs are sometimes called peer code reviews. The code developer invites other technical experts, both familiar and unfamiliar with the particular project, to review the source code. The object is to look at the code line-by-line, to develop the program model suggested in Figure 9-1, and to look for problems and for better ways to accomplish the programming task.

Code walkthroughs are very effective in increasing the quality of the code and reducing problems before they occur. The developer can learn about other ways to accomplish the tasks and, as a side benefit, reviewers can learn about techniques that may help them in their assignments. The result of a code walkthrough is better performance of the developer, the program, and the application.

9.4 The Debugging Plan

A debugging plan will help to isolate where the problem is occurring and to find out what is going wrong. Well-designed programs consist of separate, independent sections of code written to do a particular function. For example, a program may simply input data, process it, and output it as shown by the flow chart in Figure 9-2. How do you know if the program is performing correctly? You must choose test data for which you know the correct output. Using these data, you can look for the problem. Let us assume that a problem exists in the program somewhere. Your first step is simply to analyze the program output. This can sometimes give clues about where the program is going wrong. If not, plan to set breakpoints after each section and inspect the data to see where it deviates from what is expected. In this way, the problem area is isolated. You can now move your debugging strategy into the offending block of code and continue the process.

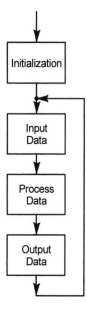

Figure 9-2 A typical program flow chart.

9.5 Debugging Tools

All debugging programs offer a variety of tools. The features in a debugger depend on the computer on which the debugging program is run. Today, personal computer-based debuggers, particularly for the high-level languages, have many features. The CodeWarrior system is an example of a software development package that contains a powerful debugger that can operate in a simulation mode or can interact directly with the microcontroller hardware. Other, older debuggers run on simple development boards, such as the Freescale and other manufacturers' single board computers, and are more limited in the features they offer. These debuggers are called monitor programs and contain rudimentary debugging tools and basic I/O functions.

Debugging Demonstration Program

Example 9-1 is a program we will use to demonstrate some of the debugging techniques described in this chapter. This program simply moves a null (zero) terminated constant data string from a block of ROM to a buffer in RAM. To debug it, we need to develop the analytical model as suggested by Figure 9-1. Indexed addressing is to be used and two pointers are initialized with the addresses of the buffers in *lines 21* and *22*. A WHILE-DO loop is entered at *line 26* by testing if the memory location pointed to by index register X is zero. If not, we enter the DO part to transfer the data. A subroutine to get the data is included to illustrate some debugging techniques. We normally would not use a subroutine for this simple task. Indexed addressing with automatic post-incrementing is used to get the data (*line 45*) and to store it (*line 30*). We expect the program to transfer each byte of data from the `String` buffer to the `Ram_Str` buffer one byte at a time.

Example 9-1 Sample Program to Demonstrate Debugging

```
Metrowerks HC12-Assembler
(c) COPYRIGHT METROWERKS 1987-2003
Rel. Loc    Obj. code Source line
---- ------ --------- -----------
   1                  ; Sample program to demonstrate
                        CodeWarrior features.
   2                  ;********************************
   3                  ; The program simply transfers a null
                        terminated
   4                  ; string from a ROM buffer to a RAM buffer.
   5                  ;********************************
   6                  ; Define the entry point for the main program
   7                          XDEF   Entry, main
   8                          XREF   __SEG_END_SSTACK ; Note double
                        underbar
   9                  ;********************************
```

```
10                      ; Code Section
11                      MyCode: SECTION
12                      Entry:
13                      main:
14                      ;********************************
15                      ; Initialize stack pointer register
16 000000 CFxx xx           lds    #__SEG_END_SSTACK
17                      ;********************************
18                      main_loop:
19                      ; Initialize pointers to source
20                      ; and destination data
21 000003 CExx xx           ldx    #String   ; Source data
22 000006 CDxx xx           ldy    #Ram_Str  ; Destination for it
23                      ; DO
24                      ; WHILE the end of the string not found
25                      while_start:
26 000009 E700             tst    0,x       ; Check for null
27 00000B 2707             beq    done      ; If found, all done
28                      ; DO move a byte from source to destination
29 00000D 16xx xx          jsr    get_byte  ; Get the data
30 000010 6B70             stab   1,y+      ; Save it and
                                              incr pointer
31 000012 20F5             bra    while_start
32                      ; ENDO
33                      ; END WHILE the end of the string not
                          found
34                      done:
35                      ; FOREVER
36 000014 20ED             bra    main_loop
37                      ;********************************
38                      ; Use a subroutine for the demonstration
39                      ; Input: X pointing to the source data
40                      ; Output: B contains the data
41                      ;        X is pointing to the next data
42                      ; Registers modified: B, X, CCR
43                      ;********************************
44                      get_byte:
45 000016 E630             ldab   1,x+        ; Get it and incr
                                                pointer
46 000018 3D               rts
47                      ;********************************
```

```
48                        MyConst:SECTION
49                        ; Place constant data here
50 000000 5468 6973       String: DC.B  "This is a string",0
   000004 2069 7320
   000008 6120 7374
   00000C 7269 6E67
   000010 00
51                        Str_End:
52                        ;********************************
53                        MyData: SECTION
54                        ; Place variable data here
55 000000                 Ram_Str:DS.B  Str_End - String
```

Debugging Program Flow and Logic

Tracing and *setting breakpoints* allow us to follow the program flow.

The first debugging task we have is to find out *where* the program is going wrong. You must follow the program flow until a deviation from the expected flow or an unexpected data modification is found. There are two ways to follow the program flow.

Program Trace: Tracing is stepping through the program a statement at a time. In the more powerful, high-level language debuggers and in CodeWarrior, data elements, including the contents of memory locations and registers, may be displayed while tracing the code. In less powerful, assembly language debuggers, the register set is shown at each step but data elements in memory must be inspected manually.

Breakpoints: The program trace is a slow way to get through a program. It is quicker to find out where problems are occurring by running the program at full speed to a breakpoint. A breakpoint is a set of conditions that interrupt the program flow and return control to the debugging program. Normally, breakpoints are set at program statements, but they also may be generated by a combination of other conditions. For example, in some debuggers, breakpoints can be generated when a particular data element becomes some specific value. In some systems, hardware breakpoint generators may create breakpoints when a condition or a set of bits on the computer's bus is detected. See Chapter 20 for the details of the in-circuit debugging tools integrated into the HCS12 microcontrollers.

Figure 9-3 shows a screen snapshot of the CodeWarrior debugger with a breakpoint in the program shown in Example 9-1. Let us say we suspect that the get_byte subroutine is not properly returning a value that we want to display. A good place to set a breakpoint is right after the subroutine returns a value in the B register. When the program hits the breakpoint, we can look at the B register to see if the value makes sense. In this case, we can see that B contains the value $54, which is correct as shown by the Data pane.

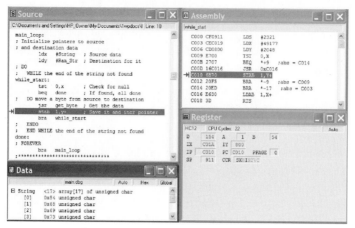

Figure 9-3 Setting breakpoints.

Debugging Data Elements

While you are following the flow of the program, you can observe data elements, both in the registers and in the memory.

Registers: The state of the registers at each step of an assembly language program should be known when you know the data input. Assembly language debuggers usually display the contents of all the registers, including the condition code register, in a register window. While tracing the program, you can watch the contents of the registers change and watch for values that are different from those expected.

Figure 9-4 shows the register display pane from the CodeWarrior debugger. All registers are shown including the condition code bits (CCR). If we were expecting the get_byte subroutine to return some value other than $54 (as shown here in the B register), we would know that there is a problem somewhere in the subroutine. It is vital that you know what the subroutine should return to be able to see if it is correct or not.

Memory: In high-level language debuggers, one can generally inspect any of the declared variables. Usually the display is formatted according to the type of declaration that has been made. In assembly language debugging monitors, the display of data elements is more crude, usually only in hexadecimal.

Figure 9-5 shows the CodeWarrior memory display. You view memory contents as a formatted data value either in the Data pane or as hexadecimal values in the Memory pane. In this display we see that the program has transferred several bytes from String to Ram_Str. A

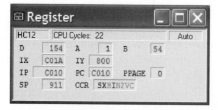

Figure 9-4 CodeWarrior register display.

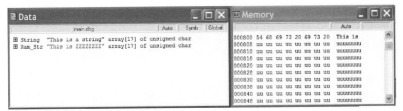

Figure 9-5 CodeWarrior memory display.

"Z" in the data pane and "uu" in the Memory pane represents a memory location that the program has not written.

The Source Code Listing

An up-to-date listing of the program is very useful. The listing should be the assembler list file, such as Example 9-1, not the source file. The list file shows the code the assembler has produced, and errors frequently can be spotted by using this listing instead of just the source file. Figure 9-3 shows the assembler source file in the Source pane and the assembled code and memory locations in the Assembly pane. When debugging C programs, a source listing, as shown in Example 9-2, showing the assembly language produced by a C compiler is often very useful too.

Example 9-2 C Program Assembly Listing

```
 4:    /*********************************************************
 5:     * Example showing store, bset and bclr operations.
 6:     *********************************************************/
 7:    #include <mc9s12c32.h>      /* derivative information */
 8:    /*********************************************************/
 9:    void main(void) {
10:       unsigned char data_A;
11:    /*********************************************************/
12:       /* Make Port A bit-0 and bit-1 outputs */
13:          DDRA_BIT0 = 1;
0000 4c0001        BSET   _DDRAB,#1
14:          DDRA_BIT1 = 1;
0003 4c0002        BSET   _DDRAB,#2
15:       /* make Port A bit-7 input (to show a bclr) */
16:          DDRA_BIT7 = 0;
0006 4d0080        BCLR   _DDRAB,#128
17:       /* The rest of the bits in DDRA will be inputs */
18:       /* Write to the output bits */
19:          PORTA = 0x83;
0009 c683          LDAB   #131
000b 5b00          STAB   _PORTAB
```

```
20:    /* Read from the port */
21:       data_A = PORTA; /* This also reads the current state of
000d 9600          LDAA   _PORTAB
22:                           * the output bits
23:                           */
24:  }
000f 3d            RTS
```

9.6 Typical Assembly Language Program Bugs

As you start your beginning programming assignments, you will commit many follies. Here are some common problems.

Stack Problems

The stack is an area of RAM used for temporarily storing data and for saving the return address for a return-from-subroutine (RTS) instruction. Here are some problems associated with using the stack.

Improper transfer to subroutines: The return address from a subroutine must be on the stack. Use a branch-to-subroutine, jump-to-subroutine, or a call instruction. Never use a branch or jump that does not put the return address on the stack.

Forgetting to initialize the stack pointer: The stack pointer must be initialized pointing to an area of RAM. Do this in the very first few lines of code in an assembly language program. It *must* be done before calling any subroutine or using the stack for data storage.

Not allocating enough memory for the stack: The data storage allocation (static data in C) grows from the bottom of memory to the top, while the stack (used for automatic variables in C) grows from the top of memory toward the bottom, as illustrated in Figure 9-6. If

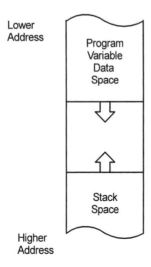

Lower Address

Program Variable Data Space

Stack Space

Higher Address

Figure 9-6 Program variable data and stack segments grow toward each other.

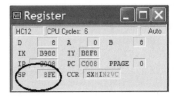 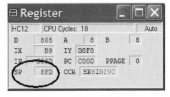

Figure 9-7 Register contents when entering and about to leave the subroutine.

the stack and data overlap, stack operations will write into data areas or vice versa with unknown, and usually dire, consequences.

Unbalanced stack operations: Make sure there are the same number of pulls as pushes. This is particularly true in subroutines where registers are temporarily saved on the stack. If the program does not return from a subroutine, it is likely there are unbalanced stack operations. Set breakpoints at the beginning and the end of the subroutine and check the stack pointer at each place. Look for errors such as unbalanced stack operations when using a stack inside a program loop. See Figure 9-7. Example 9-3 shows an unbalanced stack operation. (See Example 9-4.)

FINDING STACK PROBLEMS

Often a clue that there are problems with the stack is that the program executes properly up to a jump-to-subroutine instruction and then does not seem to return from the subroutine.

Example 9-3 Unbalanced Stack Operation

```
Metrowerks HC12-Assembler
(c) COPYRIGHT METROWERKS 1987-2003

Rel. Loc    Obj. code Source line
---- ------ --------- -----------
  1                            XDEF Entry, main
  2                   Entry:
  3                   main:
  4
  5                   sub:
  6 000000 34                pshx   ; Save the registers
  7 000001 36                psha
  8                   ; . . .
  9 000002 37                pshb   ; Temp save some data
 10                   ; . . .
 11                   ; . . .
 12 000003 32                pula   ; Restore the registers
 13 000004 30                pulx
 14 000005 3D                rts
```

Example 9-4 Unbalanced Stack Operation Analysis

Analyze the stack problem illustrated in Example 9-3.

Solution: This is an unbalanced stack operation because there is an extra `pshb` instruction in the body of the subroutine. The subroutine will not return to the calling program properly. In Figure 9-7 we see that the stack pointer register is $8FE when entering the subroutine and $8FD when we are about to execute the `rts` instruction. If the stack operations are properly balanced, these should be the same.

You can easily find this by putting breakpoints at the subroutine jump and at the next instruction following the `jsr  sub`. If you never hit the second breakpoint, you know there is a problem in the subroutine. With this simple step you have been able to isolate *where* the problem is. The next step is to look for more clues. Set a breakpoint at the start of the subroutine and at the return-from-subroutine instruction at the end, for example, *lines 6* and *14* in Example 9-3. Check the register display at each breakpoint to see if the stack pointer register is the same in each case. If it is not, you know there is an unbalanced stack operation somewhere in the subroutine and you can continue to look for clues until you find the trouble spot. See Figure 9-7.

Other stack problems can be caused by data overwriting the stack or stack overwriting the data. If you suspect these problems, and if your debugger allows it, you can write a specific data pattern, say, all $00, $FF, or some other fixed pattern into the uninitialized RAM before running the program. After problems occur, you can view the RAM to see if your data pattern is intact or if the stack or other data elements have used more locations than expected.

Register Problems

Using immediate addressing incorrectly: One of the biggest problems with the HCS12 instruction set is that the # sign is used to signify immediate addressing. Do not confuse immediate addressing with direct memory addressing. This is a particular problem for beginning assembly language programmers (and for some old-timers too). Remember that immediate addressing retrieves constant data from the memory location immediately following the operation code. Direct memory addressing retrieves data, which may be constant or variable, from some other memory location. Use the # to load a register with a number you know at the time you write the program.

A typical immediate addressing error is to forget the # when initializing a register to point to a memory address. If in Example 9-1 at *lines 21* and *22* you did not use immediate addressing with the #, the registers would not point to the correct places in memory. Figure 9-8 shows the debugger display when a breakpoint is set after the jump-to-subroutine instruction. We see that the data in B is $5A, not $54 as expected. Looking further, we see that the X register is $5469. This is unexpected because we see in the Data pane that the starting address of `String` is $C019. Thus X should be $C01A (it was incremented in the subroutine). Our debugging strategy would then lead us to find out why the pointer is not initialized correctly.[2]

---

[2] Probably the most often heard exclamation in the student programming laboratory is "Arrrg! I forgot the # sign again!"

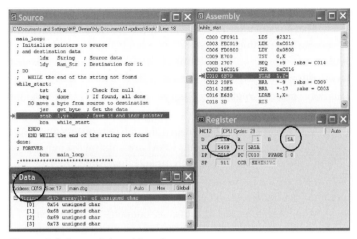

Figure 9-8 Immediate addressing error.

Using subroutines that wipe out registers: Well-designed subroutines for assembly language programs should not modify registers that may be used in the calling program. Push all registers used in the subroutine onto the stack when entering the subroutine and pull them before returning to the calling program.

Example 9-5 shows subroutine code that incorrectly restores the registers when returning to the calling program. Because the stack is a last-in, first-out operation, the registers must be pulled in the reverse order of pushing. This error can easily be found by setting a breakpoint at the jump to subroutine and at the instruction just following the `jsr`. Comparing the contents of the registers at these two points will quickly show any problems.

Example 9-5 Improper Restoration of Register Contents

```
Metrowerks HC12-Assembler
(c) COPYRIGHT METROWERKS 1987-2003


Rel. Loc    Obj. code Source line
----  ------ --------- -----------
    5                     sub:
    6 000000 34               pshx   ; Save the registers
    7 000001 36               psha
    8 000002 37               pshb
    9                     ; . . .
   10                     ; . . .
   11 000003 30               pulx   ; Restore the registers
   12 000004 32               pula
   13 000005 33               pulb
   14 000006 3D               rts
```

Transposed registers: A difficult problem to find is one in which the operands for an instruction have been transposed. For example, if data are to be moved from the B to the A register, the proper instruction is, say, `tfr b,a`. It is very easy to transpose the register operands and write `tfr a,b`. To find this error, trace the program, watch the registers, and compare them with what is expected.

Not initializing pointer registers: Register indirect and indexed addressing modes must have the register initialized with the address of the data.

Not initializing registers and data areas: Initialize registers and data areas before using them because the contents of registers and RAM are unknown when the computer's power is turned on. Do this at run time, not assembly time or load time.

Using a 16-bit counter in memory: Counters are used for many things in assembly language programming. The fastest and easiest way to implement a counter is to use a CPU register; however, when one runs out of registers, counters must be kept in memory. A typical error occurs when incrementing or decrementing a 16-bit memory counter with an 8-bit memory increment or decrement instruction. Instead of changing the full 16 bits, only 8 bits are changed. Load the 16-bit counter into a register, increment or decrement it, and store it back into memory. Make sure the store operation does not change the flags if a conditional branch based on the increment or decrement is to be done. See Example 9-6.

Example 9-6

Sometimes a counter is needed that is bigger than 255, or 8 bits. A 16-bit counter can be kept in memory like an 8-bit counter. What is wrong with the following code sequence?

```
Metrowerks HC12-Assembler
(c) COPYRIGHT METROWERKS 1987-2003


Rel. Loc    Obj. code Source line
---- ------ --------- -----------

   4

   5        0000 03E8 COUNT:  EQU   1000    ; 16-bit value

   6                          ; Initialize the counter in memory

   7 000000 1803 03E8             movw   #COUNT,Counter
     000004 xxxx

   8                          ; . . .

   9                          loop:

  10                          ; . . .

  11 000006 73xx xx             dec    Counter

  12 000009 26FB                bne    loop

  13                          ; . . .

  14 00000B              Counter: DS.W 1    ; 16-bit counter
```

Solution: The `dec` instruction decrements an 8-bit operand. Thus `dec Counter` will decrement only the most significant byte of the 2-byte counter. In this case, the loop code will be executed only three times, not 1000.

Example 9-7 Improper Counter Initialization

```
Metrowerks HC12-Assembler
(c) COPYRIGHT METROWERKS 1987-2003

    4
    5          0000 0064 COUNT:  EQU   100
    6                    ; Initialize the counter in memory
    7                    loop:
    8 000000 180B 64xx          movb  #COUNT,Counter
      000004 xx
    9                    ; . . .
   10 000005 73xx xx            dec   Counter
   11 000008 26F6              bne   loop
   12                    ; . . .
   13 00000A              Counter: DS.B 1    ; 8-bit counter
```

Modifying a counter in a loop: Another common problem with counters is reinitializing the counter inside the loop. Example 9-7 shows this problem. To fix it, move the label `loop:` below the counter initialization code at *line 8*.

FINDING REGISTER PROBLEMS

These problems are difficult to find, especially because data in registers are often variable. It is critical that you follow the analytical debugging model shown in Figure 9-1 and closely monitor what you expect the register contents to be and what the program is actually doing. When you find a deviation in these two, you have found the area of the problem and now you have to look further to find the cause.

Consider the bug shown in Example 9-6. Imagine our search process has led us to this section of code where we think there is a problem. We could trace through this section of code, but tracing through a loop that is supposed to run 1000 times is inefficient. A better approach might be to set a breakpoint at the first instruction following the `bne loop`. That way we would execute the loop at full speed and then hit the breakpoint. Inspecting registers and the variable memory location `Counter` would show that `Counter` was $00E8 and not $0000 as we expected. This would lead us to investigate the operation of the `dec Counter` instruction and we would quickly see that only the first byte of the 2-byte value was being decremented.

Condition Code Register Problems

Almost all instructions modify the condition code register, and you must know which instructions modify which bits. Problems with the condition code register lead to problems in the program flow. These problems too are hard to find, especially when variable data are used. Some of the problems you are likely to encounter are the following.

Modifying condition code contents before conditional branch instructions: Be aware of all instructions that modify the contents of the condition code register. Make sure there are

Example 9-8 Load and Store Instructions Modify the CCR

Analyze the following code for a condition code register problem.

```
Metrowerks HC12-Assembler
(c) COPYRIGHT METROWERKS 1987-2003


 Rel. Loc    Obj. code Source line
 ---- ------ --------- -----------
    4
    5        0000 0008 COUNT:  EQU   8   ; Loop counter
    6                          ; . . .
    7 000000 C608                    ldab  #COUNT ; Initialize loop
                                                    counter
    8                          ; . . .
    9                          loop:
   10                          ; Here is the code for whatever has to be
   11                          ; done in a loop. At the end of the loop, we
   12                          ; decrement the loop counter and branch back
   13                          ; if it hasn't been decremented to zero.
   14                          ; . . .
   15 000002 53                       decb       ; Decrement B register
   16                                             ; and branch back if not
                                                  zero
   17 000003 C664                     ldaa  #$64 ; But first load A with
                                                    some data
   18 000005 26FB                     bne   loop
```

Solution: The programmer follows the decb instruction with an instruction to load A with $64. The ldaa modifies the condition code register bits so that the Z bit is zero and the bne loop instruction will always be taken. The program will never exit from the loop.

no instructions that change the condition code register between the time it is set and the time the conditional branch is executed. See Example 9-8.

Using the wrong conditional branch instruction: There are different instructions for signed and unsigned numbers. Conditional branches with the words "greater" or "less" are used for signed numbers and those with the words "higher" or "lower" for unsigned numbers. See Example 9-9.

FINDING CONDITION CODE REGISTER PROBLEMS

Condition code register problems are similar to the register problems described previously and are found in a similar fashion. You must carefully watch the condition code bits as they are modified by your program while using different sets of test data.

Example 9-9 Using the Wrong Conditional Branch

A programmer intends to compare two, 8-bit unsigned data values, the first in the A register and the second in a variable memory location `Data`. The program design requires a branch to `GREATER`: if the value in A is greater than the value in `Data`. The following program segment shows the code that was written:

```
    . . .

cmpa    Data
bgt     GREATER

    . . .
```

What is wrong with this code and what data values would you suggest show that it works incorrectly?

Solution: The `bgt` instruction (branch greater than) is for signed, two's-complement data, not for unsigned data. The correct conditional branch instruction to use is the `bhi` (branch higher than) instruction. This is a particularly hard bug to find because sometimes it will work properly and sometimes not, as shown in Table 9-1.

TABLE 9-1 Data That Illustrate `bgt` Is the Incorrect Instruction to be Used

A	Data	bgt Taken?	bhi Taken?	Explanation
$55	$05	Yes	Yes	$55 is both greater than and higher than $05.
$05	$55	No	No	$05 is not greater than or higher than $55.
$7F	$80	Yes	No	$7F (+127) is greater than $80 (−128) (signed) but $7f (+127) is not higher than $80 (+128) (unsigned). `bhi` is the correct instruction.
$80	$7F	No	Yes	$80 (−128) is not greater than $7F (+127) (signed) but $80 (+128) is higher than $7f (+127) (unsigned). `bhi` is the correct instruction.

Consider the problem shown in Example 9-8. Our searching for the problem using breakpoints would allow us to find that we never exit from the loop. We could then set a breakpoint on the `decb` instruction and we would observe that the B register decrements from eight down to zero, as it should. The next pass through the loop would show that it decrements again to $FF. Tracing through the `ldaa  #$64` instruction would show us that while the Z bit is set when the B register is decremented to zero, `ldaa  #$64` clears it again because the result of that instruction is not zero.

The problem shown in Example 9-9 is very difficult to find because it depends on the data being compared. As Table 9-1 shows, for some data values the `bgt` gives the correct result and for others it does not. This is why you must carefully design your test values to test as many data cases as possible.

Test Data Strategies

Generating test data to test exhaustively your code and hardware is a complete topic itself, and we can only give some guidelines here. It is virtually impossible to test rigorously all possible combinations of inputs, outputs, and processing paths in our programs. We should try to be as thorough as we can and some strategy and judgment must be applied to make the testing program reasonable. Here are some suggestions for developing testing conditions for your programs.

- Choose data values or conditions that are representative of what you expect to test under normal operation.
- Choose data values or conditions that are at the boundaries of what you expect. For example, if your code is dealing with signed data, test the most negative and most positive numbers. For unsigned data, be sure to test zero and the most positive numbers.
- Choose conditions that are outside anything you would responsibly expect. This is very important. Your program has to work with good values and bad values too.
- If your program has user input, make sure to test all possible inputs, including more than one key pressed at a time and keys pressed rapidly.

9.7 Other Debugging Techniques

Sometimes the debugging strategies suggested previously do not lead us to a problem solution. There may be interacting problems or code that obscures the problem. A technique you can use in these cases is to start eliminating sections of code to simplify the problem. You can simply comment out sections of code to allow you to add it back in later.

When doing your hardware design, reserve an I/O bit or two for testing the software. This bit could be connected to an LED or to a pin to which you connect an oscilloscope. When you are testing the hardware and software of your system, you can output to this pin to test a variety of software events such as entering or leaving subroutines and interrupt service routines. By setting the bit when entering a section of code and resetting it when leaving, you can measure the execution time.

A common addition to many embedded systems is a "heartbeat" LED. When the code is operating normally it flashes the LED at some rate, say, 2 Hz. You can easily look at the heartbeat LED to see if the code is hung up or delayed.

9.8 Conclusion and Chapter Summary Points

In this chapter we offered some debugging strategies to help you find those annoying bugs in your programs. Key elements of those strategies are the following:

- Analyze the program; do not try to synthesize new code to fix it.
- Develop a model of how you think the program should run including data transformation, register use, and program flow.
- Compare the model of the program with what the program actually does.

- Find out where the program differs from the model.
- Use tracing and breakpoints to check the program flow.
- Use register and memory display to check the data.
- Carefully construct test cases that test the program's behavior as thoroughly as possible.
- Do not forget to test for unexpected inputs.

9.9 Bibliography and Further Reading

CPU12 Reference Manual, M68HC12 & HCS12 Microcontrollers, CPU12RM/AD, Rev. 3, Freescale Semiconductor, Inc., April 2002.

The Ganssle Group, http://www.ganssle.com.

10 Program Development Using C

In this chapter, we show some of the changes in thinking needed to program in C for an embedded system microcontroller compared to a desktop or personal computer application. We assume that you have learned to program in C in another course and now wish to use C to create programs for the HCS12 microcontroller.

10.1 Introduction

Although most of this book is dedicated to assembly language programming, many embedded applications are programmed in C because of the increased programming efficiency, portability, and documentation provided by this high-level language. There are other high-level languages, such as C++, that have been created for various microcontrollers but C is still widely used. In this chapter we show some of the changes needed in the C language to create programs for embedded microcontrollers. The following chapters will have both assembly language and C language programming examples.

10.2 Major Differences Between C for Embedded and C for Desktop Applications

C programs for embedded applications are different from desktop applications.

The C programming language gives system engineers a high-level language for developing embedded system applications. This widely used tool gives us the programming efficiency and portability we have come to expect when comparing high-level languages with assembly language programming. Embedded system developers, however, must pay closer attention to the architecture of the processor and to the interface with the real world than a developer of an application for a desktop or personal computer. Table 10-1 lists some of the differences found when comparing C programs written for these two applications.

One of the major differences is that in the embedded application the executable code is placed into read only memory (giving us the term *firmware*) instead of the copious random access memory found in desktop system. The system engineer must know where the ROM is located to be able to locate the program properly. Another difference is that often RAM in an embedded system is a scarce resource. For example, the MC9S12C32 microcontroller has

TABLE 10-1 Embedded and Desktop Applications

Program Feature	Embedded System	Desktop System
Program location	In ROM, with data and stack in RAM.	In RAM.
In-line assembly	Often used to take advantage of hardware features.	Rarely used.
Operating system support	Rarely found in embedded systems, except for some real-time operating systems.	Often used to take advantage of user-oriented I/O such as keyboard, display, and disk drives.
Use of on-chip features such as timers and A/D.	Often used.	Rarely used.
Calls to procedures written in assembly language and in-line assembly.	Often used.	Sometimes used when procedures are available for special purpose hardware.
Interrupt service routines.	Often needed and used.	If needed, often provided by operating system support.

only 2K bytes of RAM. The embedded system engineer must carefully evaluate how much RAM is needed for variable data storage and for the stack. C programs, however, may help us conserve this scarce resource because automatic variables, declared inside a function, use the stack and the storage space for these variables is released when the function exits.

Because embedded applications often control some hardware in real time, C programs can effectively use assembly language functions. To write these assembly language functions, you must know what calling convention (how the registers are used and how data are placed on the stack) is being used. A programmer can also introduce assembly language statements in-line with C statements to directly control the hardware.

Many embedded applications use hardware-generated interrupts to control the program flow. The routines written for these interrupts are called *interrupt service routines* or *interrupt handlers*. They require a special return from interrupt—*RTI*—instruction to return to the interrupted part of the program. Compilers need nonstandard ANSI C features to provide this capability.

ANSI C Versus Microcontroller Implementations

The ANSI C standard programming language is extended with features for embedded applications.

The American National Standards Institute (ANSI) established a committee in 1983 to produce "an unambiguous and machine-independent definition of the language C." The result was the ANSI standard for C. However, ANSI C was not designed for embedded controller applications. It lacks standard ways to assign pointers to specific memory locations, such as the I/O registers, a way to implement interrupt service routines, which require a return from interrupt (RTI) instruction, and extended addressing for microcontrollers with paged memory architectures. Compiler vendors tackle these problems in different ways, so you must check the documentation for your compiler's language extensions. Table 10-2 shows CodeWarrior compiler language extensions applicable for the HCS12 microcontrollers.

TABLE 10-2 CodeWarrior Compiler Language Extensions

Language Extension	Purpose
@address	Assign global variables to specific address. Useful for accessing memory mapped I/O ports.
far	Allows pointers to access the whole memory range supported by the processor.
near	The near keyword is used in cases where the default addressing is far and the near calling convention is to be used.
interrupt	Specifies a function to be an interrupt service routine.
asm	Allows assembly language instructions to be placed in the program.

PAGED MEMORY

As the semiconductor manufacturing industry has matured, far more memory can be added to the microcontroller base memory than can be addressed by the address bus. Different microcontrollers use different strategies for accessing this memory, and the HCS12 uses a *paged* memory architecture. To do this, a switching mechanism is included and special call and return instructions are implemented. The CodeWarrior compiler contains language extensions to access these memory locations. See Chapter 13 for a discussion on the paged memory architecture in the HCS12 family.

DATA TYPES

The size of data types varies from one compiler implementation to another depending on the target microcontroller. For example, in one microcontroller the size of an int variable may be 16 bits while in another it is 32 bits. As an example, Table 10-3 shows the size of data types for the CodeWarrior HCS12 compiler.

PORTABILITY

A major benefit of ANSI C is the standardization that allows a program to be "ported," or moved, to another processor. However, it is usual for the compiler vendor to extend the C language to support a specific microcontroller. Using these extensions limits the portability of the code to another, different processor. It is a good design practice to organize your code so that any processor-specific language enhancements are in easily identified modules. You would then replace these when moving to a different microcontroller. Writing the rest of the code in ANSI C will allow increased portability.

TABLE 10-3 CodeWarrior HCS12 Compiler Data Types

Data Type	Number of Bits	Data Range Minimum	Data Range Maximum
unsigned char	8	0	255
signed char	8	−128	127
unsigned short, unsigned int	16	0	65,535
signed short, enum, signed integer	16	−32,768	32,767
unsigned long, unsigned long long	32	0	4,294,967,295
signed long, signed long long	32	−2,147,483,648	2,147,483,647

10.3 Architecture of a C Program

Embedded applications have the program in ROM and use RAM for variable data and the stack.

In an embedded system, as we discovered in Chapter 2, the executable code is "burned" into read only memory (ROM), with variable data and stack segments located in read-write memory (RAM). Often these memory types are not contiguous and the software or firmware engineer must know where the various types of memory are located to be able to link and locate the final executable code.

Figure 10-1 shows the memory map of a MC9S12C32 microcontroller. We will learn more about the 1K-byte register space that contains registers and bits to control the hardware features of the microcontroller, such as the timer and analog-to-digital converter, in the following chapters. The 2K bytes of RAM are used for variable data storage and the stack, which in a C program provides the storage locations for *automatic* variables. The 32K bytes of Flash EEPROM is where our executable program resides. In comparison with a desktop personal computer system, there is not very much RAM and the entire program is located in the Flash EEPROM.

The compiler and the linker/locater for C programs written for an embedded application must allow us to position the code in the ROM and to use the RAM for the variable data storage. In addition, it must allow us to make specific reference to particular memory locations to access the control registers in the 1K-byte register space shown in Figure 10-1. The executable code consists of the code you write, starting with your main() program and all procedures linked together, plus a section of code normally provided by the C compiler called the *startup* code.

Startup Code

Startup code is automatically included by the compiler and initializes variables and hardware features before the main program executes.

The startup code is provided by the compiler vendor to initialize a variety of microcontroller hardware features plus any initialized program variables. In the CodeWarrior system, this code is called *start12.c*, and it does the following initializations:

- Defines macros to initialize and periodically reset the *Computer Operating Properly (COP)* watchdog timer. We will learn more about this important microcontroller feature in Chapter 12.

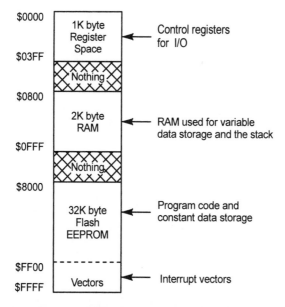

Figure 10-1 MC9S12C32 memory map.

- Defines register values for banked memory available in some members of the HCS12 family.
- Initializes registers to move (remap) RAM, EEPROM, and I/O registers from their default memory addresses to new ones if required.
- Initializes to zero any static data locations allocated in RAM.
- Initializes to their starting values any variables initialized by the program.
- Initializes the stack pointer register.
- Calls the main() program to transfer control to your embedded system.

Void Main(Void);

Your programming efforts start with the *void main(void);* program segment. The last thing the startup code does after its initialization steps is to call your main program. The first thing the compiler-generated code does in your main() program (and any other module) is to allocate storage for variables and to initialize those that have not already be taken care of in the startup code.

Variables

AUTOMATIC VARIABLES

Automatic variables are stored on the stack during the execution of the module in which they are used.

Automatic variables are those declared within a procedure; no other procedure has access to them and their lifetime (accessibility and validity) is only during the procedure's execution time. Because these variables come and go with the function, they do not retain their value from one invocation to another.

Automatic variables are placed on the stack and the compiler generates code to access these in a variety of ways. As an embedded system programmer, we must ensure that the system has enough stack memory to accommodate the automatic variables in a function. This mechanism for storing variable data is very efficient. RAM in most microcontrollers is a finite and scarce resource. By using automatic variables, succeeding functions can reuse these RAM locations.

Automatic variables may or may not be initialized to some value. If they are not initialized, the value is undefined.

STATIC VARIABLES

Static variables are allocated storage space in RAM.

Static variables may be declared either inside or outside a function. They too are located in RAM but are reserved for use throughout the program. Static variables are initialized by the startup code to either zero or whatever value is specified in the program.

VOLATILE VARIABLES

Volatile variables may be on the stack if they are automatic or in RAM if they are static.

A volatile variable is one whose value may change due to outside forces. For example, an A/D converter may be loading a register with new conversion values. A variable that is being set by reading that location should be declared as a volatile to ensure the compiler does not eliminate code that it considers redundant or unnecessary. See Example 10-1 where the volatile variable A_Val is written twice (*lines 11* and *12*) but the nonvolatile variable B_Val is written only once (*line 15*).

Example 10-1 C Volatile Variables

Source Code	Listing Showing Compiled Code
| ```
typedef unsigned char BYTE;
        BYTE PORTA;
void main(void) {
volatile static BYTE A_Val;
        static BYTE B_Val;
  /* Read from Port A */
  A_Val = PORTA;
``` | `.`<br>`.`<br>`.`<br>`.`<br>`.`<br>`.`<br><br>11:    A_Val = PORTA;<br>0000 f60000        LDAB    PORTA<br>0003 7b0000        STAB    A_Val |
`  A_Val = PORTA;`	12: A_Val = PORTA; 0006 f60000 LDAB PORTA 0009 7b0000 STAB A_Val

`  /* */`
`  B_Val = PORTA;`
`  B_Val = PORTA;`
	`.`
	14: B_Val = PORTA;
	15: B_Val = PORTA;
	000c f60000 LDAB PORTA
	000f 7b0000 STAB B_Val

`}`	`.` 0012 3d RTS

10.4 Assembly Language Interface

Compiler Produced Assembly Language Code

Compilers can produce a listing file that shows all the assembly language code that has been generated.

As embedded system designers, we must be aware of all parts of our code and be able to understand what the compiler is doing for us. In any C program, we may not be aware of code produced in support of our application program, for example, the startup code. Before the application code for any function starts to execute, the compiler generates code to initialize automatic variables that have initial values. This overhead could degrade the execution time of our program and cause unwanted effects. Fortunately, most compilers can produce a listing file showing the actual code produced. Example 10-1 shows the compiled code for declaring a volatile variable. Below each C statement we see the code assembled for that statement. In this example, we see that the two assignments of *volatile variable A_Val* in *lines 11* and *12* have code that is produced. The assignment of the *nonvolatile B_Val* at *line 14* does not produce any code because the second assignment at *line 15* would overwrite the value. The compiler has done this optimization automatically for us.

Example 10-2 shows the compiler-generated code to initialize variables in a main function. The *static int A_Val* array is initialized as part of the startup code because it is a static variable. The *automatic int B_Val* array must be initialized before any code in main is executed. As you
```

Example 10-2 C Program Overhead to Initialize Variables

Source Code	Listing Showing Compiled Code

/****************************/
 void main(void) {
   static char A_Val[] =
 {2,90,53,8};
       char B_Val[] =
 {0,7,255,34};
	.
	 .
	 .
	 .
	 9: char B_Val[] =
	 {0,7,255,34};
	 0000 69a8 CLR 8,-SP
	 0002 c607 LDAB #7
	 0004 6b81 STAB 1,SP
	 0006 86ff LDAA #255
	 0008 6a82 STAA 2,SP
	 000a c622 LDAB #34
	 000c 6b83 STAB 3,SP

char C_Val[4], i;
 /****************************/
   i = 2;
   C_Val[i] = A_Val[i]+B_Val[i];
	.
	 .
	 .
	 12: i = 2;
	 13: C_Val[i] =
	       A_Val[i]+B_Val[i];
	 000e f60000 LDAB A_Val:2
	 0011 eb82 ADDB 2,SP
	 0013 6b86 STAB 6,SP

}	. 0015 1b88 LEAS 8,SP 0017 3d RTS

can see from the listing, this is done by the code immediately following *line 8*. The *C_Val array* and the *i variable* are not initialized.

EXPLANATION OF EXAMPLE 10-2

The listing of the compiled code in Example 10-2 shows how invoking a function like main() produces code we may not see. The static A_Val has been initialized in the startup code but the automatic variable B_Val is initialized here and placed on the stack. Although this example is code produced for main(), similar code will be found for any procedure with initialized automatic variables.

Using Assembly Language in C

Embedded system programmers sometimes resort to assembly language functions to reduce the amount of program ROM needed or to solve some timing problem with modules that execute

TABLE 10-4 C Compiler Assembly Language Interface

	CodeWarrior	COSMIC C
Multiple Arguments Passed to the Function		
Fixed number of parameters	Pushed onto the stack in left-to-right order. If possible, the last argument is transferred in a register. If it cannot, it is pushed onto the stack. (Pascal convention.)	All arguments pushed onto the stack in right-to-left order. (Normal C convention.)
Variable number of parameters	Pushed onto the stack in right-to-left order up to the last argument. If possible, the last argument is transferred in a register. If it cannot, it is pushed onto the stack. (C calling convention.)	
Single Arguments Passed into the Function		
Byte argument (char, unsigned char)	Register B.	Bytes are extended to 16 bits (short) and returned in Register D.
16-Bit argument (int, unsigned int, pointer)	Register D.	Register D.
3 Bytes (far data pointer)	Register B (high byte): Register X (low word)	
32-Bit argument (double, long, float)	Register X (high word): Register D (low word)	Register X (high word): Register D (low word)
Single Arguments Returned by the Function		
Byte argument (char, unsigned char)	Register B.	Bytes are extended to 16 bits (short) and returned in Register D.
16-Bit argument (int, unsigned int, pointer)	Register D.	Register D.
3 Bytes (far data pointer)	Register B (high byte): Register X (low word).	
32-Bit argument (double, long, float)	Register X (high word): Register D (low word).	Register X (high word): Register D (low word).
Registers on return	Except for the return value in the registers defined earlier, registers and the condition code registers are undefined.	Except for the return value in Register D, registers and the condition code registers are undefined.

faster than a comparable C program. A frequent scenario is that an application program is written entirely in C and then analyzed to find the bottlenecks. This procedure is called *profiling* and software contributing to the bottlenecks can be rewritten in assembly language to improve performance. The overriding C program can call these assembly modules the same as any C function.

To create an assembly module called by a compiled C program, we must know how the compiler transfers arguments into and back from the assembled module. Table 10-4 shows methods used by two popular compilers for the HCS12 microcontroller and Example 10-3 shows the code produced by the CodeWarrior compiler. See Examples 10-4 and 10-5. If you are using another vendor's compiler, you should be sure to check its documentation to find out what calling convention is used.

Example 10-3 CodeWarrior Calling Convention

Source Code	Listing Showing Compiled Code
`/************************`	.
`* Test showing the Pascal`	.
`* calling convention used by`	.
`* CodeWarrior.`	.
`************************/`	.
`int function1(char arg1, int`	.
`arg2, float arg3);`	.
`int function2(char arg1);`	.
`void main(void) {`	.
`static  char stacy;`	.
`static  int sam, mike;`	.
`static  float susan;`	.

```
  mike = function1(stacy, sam,      13:    mike = function1(stacy,
susan );                         sam, susan );
                                    0000 f60000    LDAB   stacy
                                    0003 37        PSHB
                                    0004 fc0000    LDD    sam
                                    0007 3b        PSHD
                                    0008 fc0000    LDD    susan:2
                                    000b fe0000    LDX    susan
                                    000e 160000    JSR    function1
                                    0011 1b83      LEAS   3,SP
                                    0013 7c0000    STD    mike
  sam = function2(stacy);           14:    sam = function2(stacy);
                                    0016 f60000    LDAB   stacy
                                    0019 160000    JSR    function2
                                    001c 7c0000    STD    sam

}                                   001f 3d        RTS
```

EXPLANATION OF EXAMPLE 10-3

The Pascal calling convention is used by the CodeWarrior compiler. In Example 10-3 we see two functions being called. At line 13 *mike = function1(stacy, sam, susan)* is called. The char argument *stacy* is pushed onto the stack followed by the int argument *sam*. The float argument *susan* is transferred in the D and X registers. The function returns an int in the D register to be assigned to *mike*. In *function2*, the input char argument *stacy* is passed in the B register and the output argument *sam* is returned in the D register.

Example 10-4 Assembly Language Module with int Input Argument
Write an assembly language to C interface for an assembly module to which the C program
passes an integer value.

```
                void function_1( int arg1 );
```

Solution:

```
;************************************************
; Assembly C callable function
;   void function_1( int arg1 );
;
        XDEF   function_1
;************************************************
; Assembly language interface
;
function_1:
; The D register has the integer arg1
; This module does whatever it has to do
; with arg1. There is no other interface
; required.
; . . .
; Return to the calling program
        rts
;************************************************
```

Example 10-5 Assembly Language Module with int Input Argument
Write an assembly language to C interface for an assembly module that returns an integer
value to the C program .

```
                int function_2( void );
```

Solution:

```
;************************************************
; Assembly C callable function
;   int function_2( void );
;
        XDEF   function_2
;************************************************
; Assembly language interface
;
function_2:
; The D register is used to return an integer
; value
```

```
; . . .
        ldd    Int_val
; Return to the calling program
        rts
;*************************************************
DSEG:   SECTION
Int_val: DS.W 1
;*************************************************
```

In-line Assembly

Assembly language statements may be intermixed with C program statements.

You can insert assembly language instructions into a C program by using *in-line assembly*. All compilers have slightly different syntax for implementing this useful feature. You must be cautious when mixing assembly language instructions and C program statements. In general, the C program does not save or restore register contents in normal operation, although there are some compilers that allow a register type variable, where a variable is kept in a register and the register's contents are maintained. Typically, when you use in-line assembly instructions you cannot rely on register contents to be preserved from one assembly instruction to another when there are C statements in between. On the other hand, the C compiler will sometimes recognize when it can assign a variable in a register, thus making the contents vulnerable to being clobbered by in-line assembly code. See Example 10-6. If you have a reason to write complex in-line assembly code, you should put the code in a separate, assembly-only function.

Example 10-6 In-line Assembly Code that Destroys a Register Value

Source Code	Listing Showing Compiled Code
`    int A_Val;`	.
`void main(void) {`	.
`    int fred;`	.
`    fred = 1;`	`    10:    fred = 1;`
	`    0000 c601          LDAB   #1`
	`    0002 87            CLRA`
	`    0003 3b            PSHD`
	.
`    asm {`	`    12:        asm {`
`        ldd  #5555`	`    0004 cc15b3        LDD    #5555`
`        std  A_Val`	`    0007 7c0000        STD    A_Val`
`    }`	`    15:        }`
`    fred = 2;`	.

```
                                000a 58              ASLB
                                000b 6c80            STD    0,SP
         for (;;) {              17:        for (;;) {
                                000d 20fe            BRA    *+0
         }                       18:        }        ;abs = 000d
```

EXPLANATION OF EXAMPLE 10-6

In Example 10-6 the variable *fred* is assigned the value 1 in *line 10* and the compiler keeps it in the D register. When *fred* is assigned the value 2 in *line 16*, the compiler assumes the D register still has the original value of *fred*. We can see, because of the in-line code, that the D register has been changed and no longer has the original value of *fred*.

INSERTING IN-LINE ASSEMBLY INSTRUCTIONS

In the CodeWarrior compiler, in-line assembly statements may appear anywhere a C statement can appear and there are several forms as shown in Example 10-7. The required syntax elements are shown in bold. Example 10-8 shows the list file with the assembled in-line code shown italicized.

Example 10-7 CodeWarrior In-line Assembly

```
void main(void) {
static char static_var;  /* Static variable in RAM */
      char auto_var;         /* Automatic variable on the stack */
  /* In-Line Assembly Examples */
  asm nop            /* Comment */
  asm ldaa #55;      // Comment
  /* A block of in-line assembly */
  asm {
    nop                /* Comment */
    staa static_var; /* Comment */
    staa auto_var;
  }
  /* Multiple assembly statements on a line */
  asm nop; asm nop;  /* Comment */

  asm ( nop; /* comment */ );

  /* Another block of assembly */
  #asm
```

```
    nop;
    bset static_var,0x01 /* Comment */
    bclr auto_var,0x02


    #endasm


}
```

Example 10-8 CodeWarrior In-line Assembly Listing

```
    7:  void main(void) {
    8:  static char static_var;  /* Static variable in RAM */
    9:        char auto_var;       /* Automatic variable on the
                                      stack */
   10:    /* In-Line Assembly Examples */
   11:    asm nop            /* Comment */
0001 a7          NOP
   12:    asm ldaa #55;      // Comment
0002 8637        LDAA  #55
   13:    /* A block of in-line assembly */
   14:    asm {
   15:      nop              /* Comment */
0004 a7          NOP
   16:      staa static_var  /* Comment */
0005 7a0000      STAA  static_var
   17:      staa auto_var;
0008 6a80        STAA  0,SP
   18:    }
   19:    /* Multiple assembly statements on a line */
   20:    asm nop; asm nop;  /* Comment */
000a a7          NOP
000b a7          NOP
   21:
   22:    asm ( nop; /* comment */ );
000c a7          NOP
   23:
   24:    /* Another block of assembly */
   25:    #asm
   26:    nop
000d a7          NOP
```

```
 27:     bset static_var,0x01 /* Comment */
000e 1c000001     BSET  static_var,#1
 28:     bclr auto_var,0x02
0012 0d8002       BCLR  0,SP,#2
 29:
 30:     #endasm
 31:
 32:   }
```

10.5 Bits and Bytes—Accessing I/O Registers

Sometimes it is better to read or write a complete byte and other times it is better to access individual bits in an I/O port or memory location.

There are many control registers and control bits in our microcontrollers that must be initialized to enable and disable hardware features. Our choices for setting and resetting control register bits in assembly language are to write the whole byte using load and store instructions,

```
ldaa      #CONTROL
staa      PORT
```

or to set or reset bits using bit-set and bit-clear instructions,

```
bset      PORT,#SET_BIT_MASK
bclr      PORT,#CLEAR_BIT_MASK
```

When writing to a byte with a store instruction, all bits in the byte are written with the CONTROL byte. The bit-set and bit-clear instructions modify only those bits set to 1 in the BIT_MASK byte. In general, when enabling or disabling control bits, it is better to use the bit-set or bit-clear instruction so that you do not modify other bits in the control register that may be set in some other part of the program. An exception to this rule is when flags are being reset by the program after being set by the hardware. The flag registers are constructed so that resetting a bit requires that the program write a 1 to the bit. Perversely, writing a 0 to the bit does not reset it or alter the contents of the bit in any way! Furthermore, the bset and bclr instructions do not work as we might expect to reset flags in registers. Thus the preferred way to reset flags is to store a data byte with 1's in the bit positions to be reset knowing that 0's will not affect bit values.

In C we have a variety of ways to access specific memory locations such as I/O registers. As Figure 10-1 shows, the MC9S12C32 I/O registers are located, at least initially, in memory locations 0x0000 to 0x03FF. Each register has a specific address in this space and our C code must allow reading from and writing to these addresses.

Byte Addressing

Example 10-9 shows two ways to address a port when it is appropriate to access a byte such as when you are writing to or reading from an I/O port or when resetting flags in a flag register.

```
/*************************************************************/
/* Define a port type = unsigned char */
typedef unsigned char PORT;
/*************************************************************/
/* Define PORTA to be a volatile unsigned char at address 0x0000 */
volatile PORT PORTA @0x0000;
/*************************************************************/
/* Declare PORTB to be the contents of a memory location pointed to
 * by the volatile unsigned char pointer 0x0001 */
#define PORTB  (*(volatile PORT *) 0x0001)
/*************************************************************/

void main(void) {
unsigned char p_a_val;

   PORTA = 6;       /* Write to PORTA */
   p_a_val = PORTA; /* Read from PORTA */
   PORTB = 26;      /* Write to PORTB */
}
```

Example 10-9 Port Addressing for Byte Accesses in C

In the CodeWarrior system you can assign global variables to specific addresses with the *global address modifier*. These are called *absolute variables* where absolute refers to the address of the variable. The code

```
typedef unsigned char PORT;
volatile PORT PORTA @0x0000;
```

defines PORTA to be a volatile unsigned char at address 0x0000. Writing to and reading from the port are shown in Example 10-9.

For a compiler that does not have the global address modifier, you can accomplish a similar result by the line of code

```
#define PORTB  (*(volatile PORT *) 0x0001)
```

which declares the PORTB to be the contents of the volatile unsigned char pointer 0x0001. This line of code works as follows:

```
volatile PORT
```

declares the unsigned char to be a volatile;

$$(\text{volatile PORT } *)$$

defines a pointer to this type;

$$(\text{volatile PORT } *) \text{ 0x0001}$$

sets the pointer value to 0x0001; and

$$(*(\text{volatile PORT } *) \text{ 0x0001})$$

defines PORTB to be the contents of this memory address.

This definition would be more portable to other compilers than the global address modifier in the CodeWarrior system. Again, you can simply read and write to PORTB. See Example 10-9.

Bit Addressing

ANSI C provides a way to define bit fields that are addressed individually using a bit-field structure. This is ideal for bit-set and bit-clear operations. Example 10-10 shows an 8-bit field defined for the volatile variable PORTB located at 0x0001. When compiled, statements such as PORTB.BIT0 = 1 are conveniently treated as bit-set and bit-clear instructions.

Example 10-10 Bit Addressing

```
/***************************************************************/
/* Define a bitfield type as unsigned int */
typedef unsigned int BITFIELD;
/***************************************************************/
/* Define an eight-bit field for the PORTB
 * which is volatile at address 0x0001 */
struct {
  BITFIELD BIT0 : 1;
  BITFIELD BIT1 : 1;
  BITFIELD BIT2 : 1;
  BITFIELD BIT3 : 1;
  BITFIELD BIT4 : 1;
  BITFIELD BIT5 : 1;
  BITFIELD BIT6 : 1;
  BITFIELD BIT7 : 1;
} volatile PORTB @ 0x0001;
/***************************************************************/
```

```
/* Define a different way to access the bits */
#define PORTB_BIT1 PORTB.BIT1
/*****************************************************************/

void main(void) {
   /* These instructions generate bit-set and bit-clr
    * instructions */
   /* Strobe PORTB bit-0 */
   PORTB.BIT0 = 1;
   PORTB.BIT0 = 0;
   /* Strobe PORTB bit-1 */
   PORTB_BIT1 = 1;
   PORTB_BIT1 = 0;
}
```

Byte and Bit Addressing

It is convenient to be able to address a register as a byte in some situations and as a bit in others. For example, consider the hardware shown in Figure 10-2. Four switches are connected to bits 7 to 4 and four LEDs to bits 3 to 0. It would be convenient to read all four switches at once by reading a byte and to control individually the four LEDs.

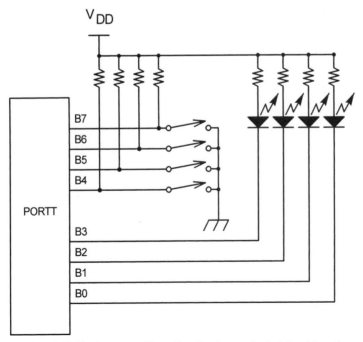

Figure 10-2 Hardware configuration to demonstrate bit addressing.

A program to accomplish this is shown in Example 10-11. A *union* of two storage classes, *unsigned char PortByte* and an 8-bit *bit-field structure PortBits*, is defined. Two registers of this type, *PORTT* and *DDRT*, are declared and located appropriately. One can access the register as a byte as in the statement

`DDRT.PortByte = 0x0F; and`

`switches = PTT;`

or as bits using

`PTT_BIT0 = 0;`

and

`PTT.PortBit.BIT0 = 0;`

Example 10-11 Byte and Bit Addressing

```
/****************************************************************/
/* Define a port type = unsigned char */
typedef unsigned char PORT;
/* Define a bitfield type as unsigned int */
typedef unsigned int BITFIELD;
/****************************************************************/
/* Define IOPort as a union of an unsigned char, PortByte, and eight
 * bit-fields, PortBits  */
typedef union {
  PORT PortByte;
  struct {
    BITFIELD BIT0  :1;
    BITFIELD BIT1  :1;
    BITFIELD BIT2  :1;
    BITFIELD BIT3  :1;
    BITFIELD BIT4  :1;
    BITFIELD BIT5  :1;
    BITFIELD BIT6  :1;
    BITFIELD BIT7  :1;
  } PortBits;
} IOPort;
/****************************************************************/
/* Locate two volatile registers with this union */
volatile IOPort PORTT @0x0240;
```

```
    volatile IOPort DDRT @0x0242;
    /*****************************************************************/
    /* Define some different ways to access the registers */
    #define PTT PORTT.PortByte
    #define PTT_BIT0 PORTT.PortBits.BIT0
    #define PTT_BIT4 PORTT.PortBits.BIT4
    void main(void) {
      volatile unsigned char switches;
    /*****************************************************************/
      /* Initialize the data direction register */
        DDRT.PortByte = 0x0F;        /* Bits 3 - 0 output */
      /* Read from the switches */
        switches = PTT;
      /* Output the switches as a byte to the LEDs */
        PORTT.PortByte = switches >> 4;
      /* Demonstrate one bit read and write */
        if (PTT_BIT4 == 1)           /* Read bit-4 switch */
          PTT_BIT0 = 0;              /* Turn LED0 on */
        else PTT_BIT0 = 1;           /* Turn LED0 off */
      /* Here is another way to access a bit field */
        PORTT.PortBits.BIT3 = 1;
    }
```

CodeWarrior Bytes and Bits

The *mc9s12c32.h* header file provides definitions for all registers and control bits.

The CodeWarrior C compiler chooses bit-set, bit-clear, or store instructions depending on the data type of the destination operand as defined in the *mc9s12c32.h header file*. A variable such as DDRA is defined as a byte at its I/O location (0x0002) allowing byte-wide (char) memory accesses with load and store operations. DDRA_BIT0 is defined as a bit in the DDRA register allowing bset and bclr operations. See Example 10-12 where the compiler-generated code is italicized. Also, Chapter 12 discusses the resetting of interrupt flags in more detail.

Caution on Compiler Implementation Dependent Bit Addressing

The order (least or most significant bit first) of the bit-field addressing described previously is implementation dependent. While for some applications, where internally defined structures are being maintained, this may not be an issue, but for accessing the defined control bits in register, the order is important. In Example 10-10, bit-0, the least significant bit, is defined first. This is the default order for the CodeWarrior compiler although there is a compiler predefined macro that can be used to reverse the bit order. With any compiler, you must check what order is being used.

Example 10-12 Store, Bit-Set, and Bit-Clear Instructions

```
 4:   /*********************************8**************
 5:    * Example showing store, bset and bclr operations.
 6:      **********************************************************/
 7:   #include <mc9s12c32.h>        /* derivative information */
 8:    /**********************************************************/
 9:   void main(void) {
10:      unsigned char data_A;
11:   /**********************************************************/
12:     /* Make Port A bit-0 and bit-1 outputs */
13:        DDRA_BIT0 = 1;
0000 4c0001          BSET   _DDRAB,#1
14:        DDRA_BIT1 = 1;
0003 4c0002          BSET   _DDRAB,#2
15:     /* make Port A bit-7 input (to show a bclr) */
16:        DDRA_BIT7 = 0;
0006 4d0080          BCLR   _DDRAB,#128
17:     /* The rest of the bits in DDRA will be inputs */
18:     /* Write to the output bits */
19:        PORTA = 0x83;
0009 c683            LDAB   #131
000b 5b00            STAB   _PORTAB
20:     /* Read from the port */
21:         data_A = PORTA; /* This also reads the current state of
000d 9600            LDAA   _PORTAB
22:                                 * the output bits
23:                                 */
24:   }
000f 3d              RTS
```

10.6 Interrupts

The Interrupt Service Routine or Interrupt Handler

An *interrupt* is an important, asynchronous event that requires immediate attention.

We will discuss interrupts in much more detail in Chapter 12, but at this point, it is sufficient to know that an interrupt is an important event signaled by some hardware mechanism. For example, the HCS12 microcontroller has a powerful timer system (see Chapter 14) that can generate periodic interrupts to help generate specific frequency waveforms. Other hardware features can generate interrupt signals or requests. Whenever one of these events happens, the software must branch to a routine called the *interrupt service routine* (*ISR*) or *interrupt handler* to process the interrupt. This is a hardware function call and it is processed like any function call except that no arguments may pass into or back from the function except for globally declared variables. In addition to this restriction, a special return from interrupt instruction, the *RTI*, is used instead of the return from subroutine (*RTS*) instruction. Furthermore, all registers are pushed onto the stack before entering and are pulled while leaving.

An interrupt service routine function uses the keyword *interrupt* and executes when the interrupt request occurs. You may use automatic variables whose lifetimes are the life of the function execution. Sometimes, however, an interrupt service routine must modify a variable each time the ISR is entered. In this case, the variable must be declared static so that it "lives" from one interrupt to another. Finally, in all interrupt service routines the interrupting source, often a flag or bit in a register, must be reset before leaving the routine.

Locating the Interrupt Service Routine

The CPU uses an *interrupt vector* to find the correct interrupt service routine when the interrupt request is received. The interrupt vector is simply the address of the start of the interrupt service routine, and each interrupting source has a specific address where its own vector is stored. It is our job to provide the linker program with the starting address so it can be placed in the correct vector location. The CodeWarrior system can do this in two ways.

We learned in Chapter 6 that the linker receives its address information from the *linker parameter file*. We will learn in Chapter 12 that the interrupt vectors are numbered 0 through 58 (see Table 12-1 in Chapter 12). One way to assign the correct address to the vector is to include the following lines in the linker parameter file:

```
VECTOR 0 Entry              /* Reset Vector */
VECTOR 8 Timer_Ch_0_ISR /* Timer Channel 0 Interrupt Handler */
```

The number after VECTOR is the interrupt number given in Table 12-1 and the label is the function name for the interrupt handler.

Another method is to include the vector number in the function definition of the interrupt handler in your C program.

```
void interrupt 8 Timer_Ch_0_ISR( void ) {
    . . .
}
```

Example 10-13 shows a complete interrupt routine. The main program initializes an LED display and enables falling edge interrupts on the IRQ_L line. Each time the interrupt occurs, the interrupt handler toggles the display.

Example 10-13 Interrupt Handler for IRQ_L

```
/*****************************************************************
 * IRQ_L Interrupt Example
 * Turns on the LED1 on the CSM-12C32 CPU module
 * and enables IRQ_L for falling edge interrupts.
 * Each time the interrupt occurs, the display LED
 * is toggled.
```

```
*****************************************************************/
#include<mc9s12c32.h>1                   /* derivative information */
/*****************************************************************/
void main(void) {
/*****************************************************************/
  /* Initialize I/O */
  /* Make Port A bit-0 output */
    DDRA_BIT0 = 1;
  /* Turn LED1 on (active low) */
    PORTA_BIT0 = 0;
  /* Initialize INTCR:IRQE for falling edge interrupts */
    INTCR_IRQE = 1;
  /* Enable the IRQ_L Interrupt */
    INTCR_IRQEN = 1;
  /* Enable Interrupts */
    EnableInterrupts;         /* Clears I bit with CLI instruction */
  /* DO Nothing */
    for(;;) {
    }
  /* FOREVER */
}
/****************************************************************
 * Interrupt handler for the IRQ_L interrupt

 ****************************************************************/
void interrupt 6 irq_handler( void ) {
/*****************************************************************/
  /* Complement the display bits */
    if (PORTA_BIT0 == 1)
      PORTA_BIT0 = 0;
    else
      PORTA_BIT0 = 1;
  /* There may need to be some debouncing done here!
   * and there is no flag to be reset in this example. */
} /* End of IRQ interrupt handler */
```

---

[1] All registers and bits in registers are defined in the mc9s12c32.h header file. The header defines the memory addresses for the control registers, such as DDRA, and assignment to these are unsigned char types. The header file defines bits within registers as, for example, DDRA_BIT0. Assignments to these are bit assignments using bset or bclr instructions. Your C program can access registers as bytes or individual bits.

10.7 Remaining Questions

- Where can I find more examples of C programs? *In each of following chapters describing features of the HCS12, we will provide programming examples showing how to access and use these features.*
- I would like to know more about interrupts. *See Chapter 12 for general information on interrupts and specific chapters, for example, Chapter 14 for the timer, for information on each feature.*
- What are some other resources where I can learn more about embedded C programs? *Please check Section 10.9 for a bibliography of useful information.*

10.8 Conclusion and Chapter Summary Points

In this chapter we discussed the use of the C programming language for programming embedded systems using microcontrollers.

- There are two kinds of memory in a microcontroller system—RAM and ROM.
- An embedded system's program is in ROM with variables in RAM.
- A desktop system's program and data are in RAM.
- The embedded system engineer must know the microcontroller's memory map to be able to locate the program and data correctly.
- The ANSI C standard does not provide some of the features needed in an embedded system such as direct memory access, interrupt control, and in-line assembly statements.
- Compilers that are ANSI C compliant also offer extensions necessary for embedded systems.
- When resetting flags in the M68HCS12, we must be sure that the compiler does not use a BSET or BCLR instruction.

10.9 Bibliography and Further Reading

Cady, F. M., *Microcontrollers and Microprocessors,* Oxford University Press, New York, 1997.

CPU12 Reference Manual, M68HC12 & HCS12 Microcontrollers, CPU12RM/AD, Rev. 3, Freescale Semiconductor, Inc., April 2002.

Doughman, G., *Using the HCS12 NVM Standard Software Drivers with the Cosmic Compiler,* Application Note AN2678/D, Freescale Semiconductor, Inc., 2004.

Freescale HC12 Compiler, Metrowerks, Austin, TX, 2003.

Robb, S., *Creating Efficient C Code for the MC68HC08,* Application Note AN2093/D, Freescale Semiconductor, Inc., 2000.

10.10 Problems

Basic

10.1 Write a program in C to reverse the order of 32 bytes in a buffer. Assume the buffer is in memory locations DATA[0]–DATA[31]. **[c]**

Intermediate

10.2 Write structured assembly language code for the following pseudocode design. Assume all variables A–F are static unsigned chars and G is static unsigned int. Then write and compile a C program and compare the assembly code produced by the compiler with your assembly language solution. How many instructions does each program have? How many bytes? **[c]**

```
; WHILE A < B
; DO
; A = C + D;
; ENDO
; ENDWHILE

; IF A = B
; THEN C = D
; ELSE E = F
; ENDIF

; G = A * B
```

10.3 Using the compiled code listing for Example 10-2 and assuming the stack pointer register is 0x0900, show a memory map giving the location and value (when known) for the automatic variables declared in line 9. **[a, b]**

Advanced

10.4 Write an assembly language module to which the C program passes an unsigned char and receives the unsigned char with the most significant mibble set to 0000. in return. **[c, k]**

```
unsigned char function( unsigned char );
```

10.5 Write an assembly language module to which the C program passes an unsigned char and receives an unsigned int equal to five times the unsigned char in return. **[c, k]**

```
unsigned int function( unsigned char );
```

10.6 Write an assembly language module to which the C program passes three integer values and receives the sum of these as an int in return. **[c, k]**

```
int function(int arg1, int arg2, int arg3 );
```

10.7 Write an assembly language module to which the C program passes two integer values and receives a long equal to the product of the two input values in return. **[c, k]**

```
long function(int arg1, int arg2 );
```

10.8 Write a program in C and then in assembly (or vice versa) for the M68HCS12 to find the largest of 32, 8-bit unsigned numbers in memory locations $0800 – $081F. Place the answer in $0820. Compare the two programs for size and the time taken to run. **[c]**

10.9 Write a program in C and then in assembly (or vice versa) for the M68HCS12 to find the largest of 32, 8-bit two's complement signed numbers in memory locations $0800–$081F. Place the answer in $0820. Compare the two programs for size and the time taken to run. **[c]**

10.10 There are 4 bytes of data in variable data array DATA[0]–DATA[1]. Write a program in C to count the number of bits that are 1's in these 4 bytes. Place the result in NUM_ONES.

10.11 Write a program in C to compute factorial 8. Store the result in a 2-byte memory location in RAM. **[c]**

10.12 Write a C function to search a null terminated string of characters for a specific substring and to return the address of the start of the substring. The input to the subroutine is to be the starting address of the string to be searched, the starting address of the substring to be searched for, and the number of characters in the substring. If the substring is found, return the address of the first character in the search string; otherwise return an address of $0000.

10.13 An 8-bit signed/magnitude number system where bit 7 is a sign bit and bits 6-0 are a 7-bit number holding the magnitude is in use. Write assembly or C subroutines or functions for the following:

a. Add two 8-bit signed/magnitude numbers.

b. Subtract two 8-bit signed/magnitude numbers.

c. Multiply two 8-bit signed/magnitude numbers.

d. Divide two 8-bit signed/magnitude numbers.

11 HCS12 Parallel I/O

OBJECTIVES

This chapter describes the parallel I/O capabilities of the HCS12. You must program or initialize almost all I/O features before using them by setting and resetting bits in control registers. Examples show how this is done.

11.1 Introduction

1024 registers contain bits to control all aspects of the HCS12 I/O.

The HCS12 has a particularly rich, fully integrated, suite of I/O capabilities, including parallel and serial I/O, analog input, and timer functions. Many of the I/O pins have shared or dual purpose functions. All I/O, and control of I/O, is done using a set of 1024 *control registers*, which are initially located in memory locations $0000–$03FF.

In the following sections we will describe the parallel I/O features of the HCS12. Except for very few cases, most of the I/O ports require initialization before use. This is done by programming bits in the control registers.

Families of HCS12 Processors

There are a number of versions of the HCS12 family of microcontrollers. Each may have different parallel I/O features. Even though this may seem very confusing at first, we will find that all parallel I/O devices are similar and thus what we will learn about the MC9S12C32 processor will apply to most other versions of the family.

11.2 The Register Base Address

The I/O registers may be relocated from the starting position at $0000 to any 2K-byte boundary.

The starting, or *Base*, address for the input and output registers and their control registers is initially located at $0000 when the microcontroller comes out of reset. In the following sections, all addresses are given as an *offset* from this starting, or base, location. The base address may be specified and the registers relocated in the application program by setting REG14–REG11 in the INITRG register.

INITRG—$0011—Initialization of Internal Register Position

	Bit 7	6	5	4	3	2	1	0
Read:	0	REG14	REG13	REG12	REG11	0	0	0
Write:								
Reset:								

░░░░ = reserved, unimplemented, or cannot be written to.

Read: Anytime. Write: Once in normal and emulation modes, anytime in special modes.

REG[14:11]

Internal Register Map Position

These bits specify the starting address of the I/O registers on a 2K boundary for addresses between $0000 and $7800.

See Also:

Topic	Register	Chapter
Initialize RAM Location	INITRM ($0010)	13
Initialize EEPROM Location	INITEE ($0012)	13
Miscellaneous	MISC ($0013)	13

11.3 The Port Integration Module (PIM)

The port integration module establishes the connections between internal I/O devices and the external world.

One of the best ways to visualize the I/O features available in any microcontroller is to inspect a block diagram as shown in Figure 11-1. The *port integration module* describes the interface between I/O features within the microcontroller, such as the timer, and the I/O pins for all ports. Internal I/O devices are connected to the external world through an *I/O port*. Most I/O ports can serve as a general purpose input or output for the CPU or as a connection between the specialized internal device and the external world.

Let us tour the port integration module shown in Figure 11-1, starting in the upper right and proceeding clockwise.

TIM—Timer Module: The timer module is described in detail in Chapter 14. The timer system includes a free-running counter and eight timer channels. These channels may be configured in any combination of timer comparison channels, called *output compare* channels, or *input capture* channels, which capture the time when an external event occurs. There is a *real-time periodic interrupt* and a counter for external events called the *pulse accumulator*. The timer uses *port T* as its connection to the outside world.

Port T Multiplexer: Figure 11-2 shows a 5-bit multiplexer (MUX) between the timer and port T. This allows up to five of the six channel pulse-width modulators (PWMs) to be output on

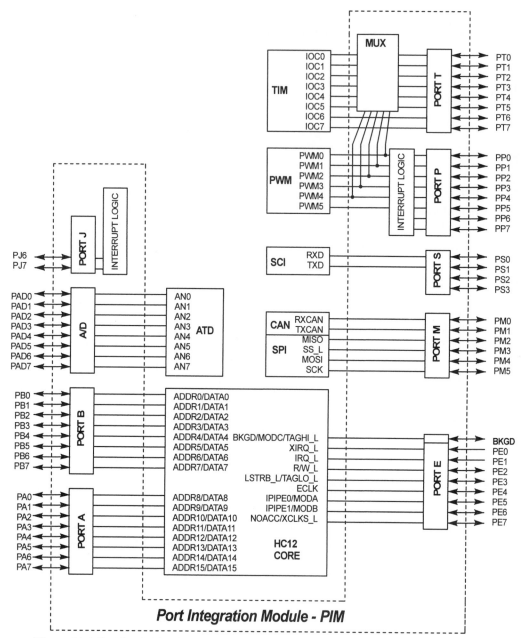

Figure 11-1 Port integration module.

port T. This would allow you to use port P for other purposes, such as interrupt request inputs. In addition, in some versions of the HCS12 family, such as the MC9S12C32, port P pins may not be available. This multiplexer gives you access to the PWM, although at the expense of losing some port T pins. A module routing register, MODDR, associated with port T controls the multiplexer.

PWM—Pulse-Width Modulator: The pulse-width modulator can generate up to six pulse-width modulated waveforms. After we initialize and enable the PWM it will automatically output

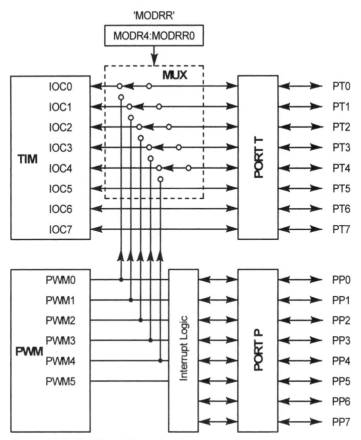

Figure 11-2 Port T multiplexer.

PWM waveforms with no further action required by the program. This is very useful in applications such as controlling stepper motors. The PWM is described with the timer in Chapter 14.

Port P is normally used by the pulse-width module but it can also be used for interrupts. When port P is configured as an input port, and interrupts are enabled, the *Interrupt Logic* block can generate interrupts. See Chapter 12 for a complete discussion of the interrupting capabilities of the MC9S12C32 microcontroller.

SCI—Serial Communications Interface: The SCI is a full duplex, asynchronous, serial interface, and it can provide a serial interface like the COM port in a personal computer. The SCI uses port S and we will describe it fully in Chapter 15. An RS232 interfacing example is given in Chapter 18.

CAN—Controller Area Network: A special serial interface, called the CAN interface, is included and uses port M. The CAN interface is described in Chapter 16.

SPI—Synchronous Peripheral Interface: The SPI is a high-speed, synchronous, serial interface used between the microcontroller and a peripheral such as a serial D/A converter or serial memory. We describe the operation of the SPI in Chapter 15 and we show an example of its use in Chapter 18.

BKGD—Background Debug: The BKGD pin is used for the background debugging capability of the microcontroller.

Port E: As we can see in Figure 11-1, port E can be used as a parallel I/O port. It is also used for special purpose input and output depending on the microcontroller's operation mode. We will discuss these modes in more detail in Chapter 21.

Port A and Port B: These two general purpose parallel I/O ports can be used also for expanded mode address and data bus lines. Expanded mode operation and the use of external memory are covered in Chapter 21.

Port AD: The microcontroller's 10-bit analog-to-digital converter uses port AD. The A/D converter is an eight-channel, multiplexed, successive approximation converter. Its operation is covered in Chapter 17. When the A/D is not in use, the port AD lines can be used as normal, digital I/O.

Port J: Port J is a 2-bit port that can be used for I/O or to generate interrupts. See Chapter 12 to learn more about interrupts.

11.4 Bidirectional I/O Ports

The parallel I/O ports in the port integration module have bidirectional capability. That is, they operate in either output or input mode.

Figure 11-3 shows a model of one bit of an 8-bit, bidirectional, parallel input/output port. It consists of an *output interface*, an *input interface*, *input/output direction control*, and an *address decoder/read and write control logic*.

The *output interface* consists of eight latches (only one of which is shown here [1]) and logic that provides the clock signal for the latch. During an output operation, say, by the instruction

```
staa      PTx
```

the address of PTx is placed on the internal address bus [2], the data are placed on the internal data bus [4], and then the R/W_L control signal [5] is asserted low. The *Bus Clock* signal [6] provides the correct timing for the latch clock. When the clock is asserted, the data bit is latched into the output port latch [1].

You must enable the output three-state gate [6] before data can appear at the output bit of the port. This is done by initializing the *Data Direction Register* [7] in the input/output direction control block. This register has a different address than the I/O data port and by writing a one into the DDR corresponding to the I/O bit, the output three-state gate [6] is enabled. You can control the direction of each bit individually in the I/O port in this way.

The *input interface* consists of eight three-state gates [8] (again, only one of the bits is shown) that connect the signal from an input device, say, a set of switches, to the internal data bus. The three-state gate's enable is asserted when the address decoder receives the correct address for this I/O port, the R/W_L control signal is high, and the Bus Clock transitions from low to high. The data that is input through the three-state gate [8] depends on the state of the data direction register bit. If DDRxn is zero, the port pin is read. If DDRxn is one (the bit is an output bit), an input from the port reads the state of the output latch.

Figure 11-3 shows a second input three-state gate [9]. This is the port input register, PTIx. Reading this port always returns the state of the port pin regardless of DDRxn.

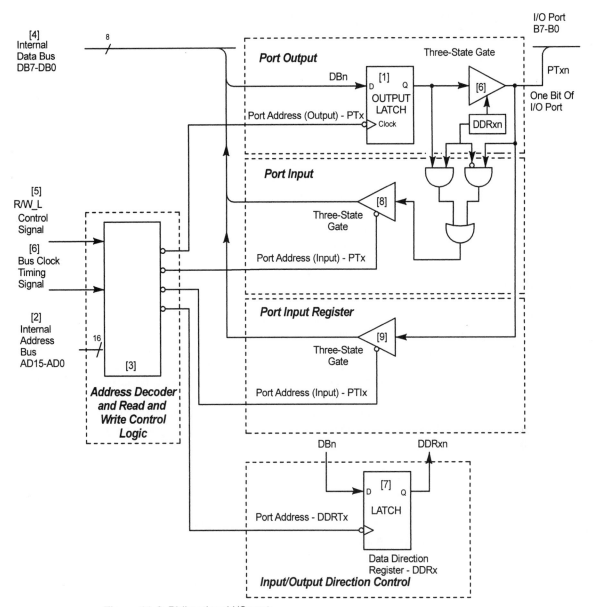

Figure 11-3 Bidirectional I/O port.

There are two features of this design. The first is that, even though the port may be configured as an output, you may read it. This operation reads what is on the output pins (by reading PTIx) or the output latch (by reading PTx). Another feature is that you can load the output latch before changing the direction of the port by writing to the data direction register.

One final note is that all I/O ports are active as inputs when the microcontroller is reset. This is a "safe" mode for the default state of the I/O line. A design consideration that we will consider in Chapter 18 is that any external device that expects the I/O line to be an output must

TABLE 11-1 MC9S12C Family Package Options

Port	48-LQFP	52-LQFP	80-QFP
		I/O Port Bits Available	
Port A	0	0, 1, 2	0–7
Port B	4	4	0–7
Port AD	0–7	0–7	0–7
Port E	0, 1, 4, 7	0, 1, 4, 7	0–7
Port J	None	None	6, 7
Port M	0–5	0–5	0–5
Port P	5	3, 4, 5	0–7
Port S	0, 1	0, 1	0–3
Port T	0–7	0–7	0–7

provide safe operation during the time the line is a high-impedance input before the Data Direction Register is programmed.

11.5 HCS12 Parallel I/O Ports

Figure 11-1 shows multiple I/O ports that can be used for either general purpose I/O or the specialized features associated with each module. However, we should note that not all of these I/O pins might be available in a particular version of the MC9S12C family of microcontrollers. At present, Freescale delivers the microcontroller in three different package versions. These are a 48-pin and 52-pin low profile quad flat pack (LQFP) and an 80-pin quad flat pack (QFP). These different packages allow the designer to minimize board space and pin connections by choosing one of the smaller chips, providing the chosen device has sufficient I/O. Table 11-1 shows the port I/O pins available for each of these packages. For devices assembled in 48- or 52-pin packages, all nonbonded out pins should be configured as outputs after reset to avoid current drain from floating inputs. Example 11-1 shows a short assembly language code segment to do this by setting the bits in the data direction registers high to force the pin into an output mode.

```
Example 11-1 Setting Nonbonded Pins to Output

Metrowerks HC12-Assembler
(c) COPYRIGHT METROWERKS 1987-2003

Rel. Loc   Obj. code   Source line

---- ----   ----------- ----------

     1                   ; ******************************
     2                   ; Assembly code segment to initialize
                           nonbonded out
     3                   ; pins in the 48-pin and 52-pin low profile
                           quad
     4                   ; flat packages
     5                   ; Illustrates conditional assembly
```

```
 6                          ;*******************************
 7                          ;*******************************
 8                          ; Register definitions
 9      0000 0002 DDRA:    EQU  $0002 ; Data Direction Registers
10      0000 0003 DDRB:    EQU  $0003
11      0000 0009 DDRE:    EQU  $0009
12      0000 026A DDRJ:    EQU  $026A
13      0000 025A DDRP:    EQU  $025A
14      0000 024A DDRS:    EQU  $024A
15                          ; *******************************
16                          ; Constants definitions
17      0000 0001 TRUE:    EQU  1
18      0000 0000 FALSE:   EQU  0
19                          ; Define the package version
20                          ; One of these must be true
21      0000 0001 PIN_48: EQU  TRUE
22      0000 0000 PIN_52: EQU  FALSE
23      0000 0000 PIN_80: EQU  FALSE
24                          ; *******************************
25                          ; Code Section
26                          MyCode: SECTION
27                          ; Place this code in the initialization
28                          ; section
29      0000 0001     IF PIN_48 = TRUE
30                          ; Set these DDR bits for 48-pin package
31 000000 4C02 FE              bset   DDRA,%11111110
32 000003 4C03 EF              bset   DDRB,%11101111
33 000006 4C09 6C              bset   DDRE,%01101100
34 000009 1C02 5ADF            bset   DDRP,%11011111
35 00000D 1C02 4AFC            bset   DDRS,%11111100
36                              ENDIF
37      0000 0000     IF PIN_52 = TRUE
44                              ENDIF
45                          ;
```

Ports A and B

Ports A and *B* may be used for I/O in single-chip mode but are used for the external address and data buses in expanded modes.

These ports are available as general purpose I/O ports in single-chip mode only. When in this mode, a data direction register for each, *DDRA* and *DDRB*, controls the direction, input or output, of each bit in each register. Any of the HCS12 memory addressing modes may be used when reading from or writing to these registers (see Figure 11-4). Example 11-2 illustrates the various addressing modes. See Section 11.6 for information

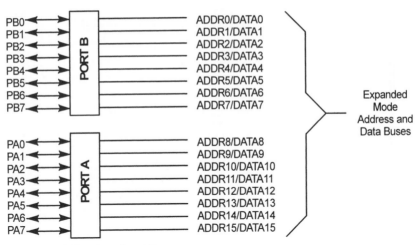

Figure 11-4 Port A and port B.

about the data direction registers for these ports and Section 11.7 on how to change the electronic properties for these bits.

When the HCS12 is in expanded mode, ports A and B are used for the external address and data buses and are not in the accessible register set.

PORTA—Base + $0000—Port A I/O Register[1]

	Bit 7	6	5	4	3	2	1	0
Read: Write:	PA7	PA6	PA5	PA4	PA3	PA2	PA1	PA0
Single- chip:	PA7	PA6	PA5	PA4	PA3	PA2	PA1	PA0
Expanded Wide[2]:	ADDR15/ DB15	ADDR14/ DB14	ADDR13/ DB13	ADDR12/ DB12	ADDR11/ DB11	ADDR10/ DB10	ADDR9/ DB9	ADDR8/ DB8
Expanded Narrow:	ADDR15/ DB15/ DB7	ADDR14/ DB14/ DB6	ADDR13/ DB13/ DB5	ADDR12/ DB12/ DB4	ADDR11/ DB11/ DB3	ADDR10/ DB10/ DB2	ADDR9/ DB9/ DB1	ADDR/8 DB8/ DB0

[1] The addresses of all I/O registers are shown in this way—Base + $XXXX—to denote that the actual register address is determined by offset ($XXXX) plus the register *base* address that may be programmed by the user after startup. The default base is $0000.

[2] The notation ADDRn/DBn and ADDRn/DBm/DBn means that these bits are multiplexed onto that register bit in expanded mode.

PORTB—Base + $0001—Port B I/O Register

	Bit 7	6	5	4	3	2	1	0
Read: Write:	PB7	PB6	PB5	PB4	PB3	PB2	PB1	PB0
Single-chip:	PB7	PB6	PB5	PB4	PB3	PB2	PB1	PB0
Expanded Wide:	ADDR7/ DB7	ADDR6/ DB6	ADDR5/ DB5	ADDR4/ DB4	ADDR3/ DB3	ADDR2/ DB2	ADDR1/ DB1	ADDR0/ DB0
Expanded Narrow:	ADDR7	ADDR6	ADDR5	ADDR4	ADDR3	ADDR2	ADDR1	ADDR0

See Also:

Topic	Register	Chapter/Section
Data Direction Registers	DDRA (Base + $0002), DDRB (Base + $0003)	11.6
Pull-up or Pull-down Enable	PUCR (Base + $000C)	11.7
Reduced Drive Control	RDRIV (Base + $000D)	11.7
Expanded Mode Operation		13

Example 11-2 Addressing Modes to Access I/O Ports

```
Metrowerks HC12-Assembler
(c) COPYRIGHT METROWERKS 1987-2003


Rel. Loc  Obj. code Source line
---- ------ --------- -----------
   1      0000 0000 BASE:   EQU   $0000        ; Base address
   2      0000 0001 PTB:    EQU   BASE+$0001   ; Address of
                                                 Port B
   3                        ;     . . .
   4 000000 CE00 00         ldx   #BASE
   5                        ;     . . .
   6                    ; Read Port B
   7 000003 9601             ldaa  PTB          ; Direct
                                                  Addressing
```

```
 8 000005 A601              ldaa  (PTB-BASE),x  ; Indexed
 9 000007 CExx xx           ldx   #VECTOR
10 00000A A6E3 0000         ldaa  [0,x]          ; Indexed-
                                                   Indirect
11                     ;      . . .
12 00000E 0001       VECTOR: DC.W PTB            ;Address of
                                                   Port B
```

Port AD

Port AD I/O pins PAD0–PAD7 may be used for analog input or for general purpose digital inputs or outputs (see Figure 11-5). Port AD pins not being used for analog input may be used for digital inputs (although port AD digital reads are not recommended during the sample period). These digital inputs (but not outputs) are through the *PORTAD Data Input Register*. See Chapter 17 for complete information on the A/D converter. In addition to analog and digital input, the port integration module port PTAD allows digital input and output from and to the PAD0–PAD7 pins.

There are three digital I/O registers associated with the ATD port:

PORTAD (Base + $008F): ATD port allowing analog or digital input only. Bits in the ATDDIEN register must be set and in the DDRAD register must be reset for digital input.

PTAD (Base + $0270): Port integration module ATD connection port allowing digital input and output. ATDDIEN must be set and DDRAD must be reset.

PTIAD (Base + $0271): Port integration module ATD connection port for reading PAD input pins. ATDDIEN must be set and DDRAD must be reset.

Figure 11-6 shows the initialization needed to read digital data from a PADn pin in registers PORTAD, PTAD, or PTIAD. The *ATD Digital Input Enable Register* (*ATDDIEN*) must be set and the *Data Direction Register AD* (*DDRAD*) must be cleared for each bit that is to be used as an input. The input data can be read from any of the three registers.

Figure 11-6 shows that to output from PTAD to a PADn pin, the data direction register bit associated with the output bit must be set. Only PTAD may be used to output data. See Example 11-3 and Section 11.8 for examples showing how to input from this port. (See also Example 11-4.)

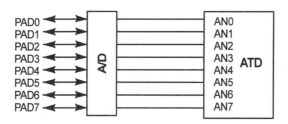

Figure 11-5 Port AD.

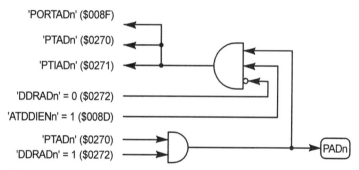

Figure 11-6 PORTAD and PTAD digital input and output.

PTAD—Base + $0270—Port AD I/O Register

	Bit 7	6	5	4	3	2	1	0
Read: Write:	PTAD7	PTAD6	PTAD5	PTAD4	PTAD3	PTAD2	PTAD1	PTAD0
Reset:	0	0	0	0	0	0	0	0

See Also:

Topic	Register	Chapter/Section
Port AD Input Register	PTIAD (Base + $0271)	11.7
Data Direction Register	DDRAD (Base + $0272)	11.6
ATD Digital Input Enable Register	ATDDIEN (Base + $008D)	17
Reduced Drive Control	RDRAD (Base + $0273)	11.7
Pull-up or Pull-down Enable	PERAD (Base + $0274)	11.7
Polarity Select	PPSAD (Base + $0275)	11.7
A/D Operation		17

ATDDIEN—Base + $008D—ATD Digital Input Enable Register

	Bit 7	6	5	4	3	2	1	0
Read: Write:	IEN7	IEN6	IEN5	IEN4	IEN3	IEN2	IEN1	IEN0
Reset:	0	0	0	0	0	0	0	0

Read: Anytime. Write: Anytime.

IEN7 – IEN0

ATD Digital Input Enable

0 = Disable digital input buffer (default).

1 = Enable digital input buffer.

These bits control the digital input buffer from the microcontroller's analog input pin to the PTADx data register. Setting this bit will enable the corresponding digital input buffer continuously. If this bit is set while simultaneously using it as an analog input port, there is a potential of increased power consumption because the input voltage may put the input electronics into a linear mode.

Example 11-3 Initializing PTAD I/O Direction

```
Metrowerks HC12-Assembler
(c) COPYRIGHT METROWERKS 1987-2003


Rel. Loc    Obj. code Source line
---- ------ --------- -----------
   1                  ; Set the register BASE address
   2    0000 0000 BASE:    EQU    $0000
   3    0000 0270 PTAD:    EQU    BASE+$0270   ; PTAD  I/O
   4    0000 0272 DDRAD:   EQU    BASE+$0272   ; DDR PTAD
   5    0000 008D ATDDIEN:EQU     BASE+$008D   ; ATD Digital
                                                 Input Enable
   6                  ; Define bits to be output and input
   7                  ;   1 = output, 0 = input
   8    0000 00F0 OBITS:   EQU    %11110000
   9    0000 000F INBITS:  EQU    %00001111
  10                  ;    . . .
  11                  ; Set direction register for PTAD for
                        outputs
  12 000000 1C02 72F0        bset   DDRAD,OBITS
  13                  ; Set digital input enable for Port AD for
                        the inputs
  14 000004 4C8D 0F          bset   ATDDIEN,INBITS
  15                  ; Reset DDRAD bits for the inputs
  16 000007 1D02 720F        bclr   DDRAD,INBITS
  17                  ;    . . .
  18                  ; Output data to bits 7 - 4
  19 00000B 86B0              ldaa   #%10110000
  20 00000D 7A02 70           staa   PTAD
  21                  ; Read data on bits 3 - 0
  22 000010 B602 70           ldaa   PTAD
```

Example 11-4 Initializing PTAD I/O Direction in C

```c
/*************************************************************
 * Initializing PTAD I/O Directions
 *************************************************************/
#include <mc9s12c32.h>              /* derivative information */
#define OBITS 0b11110000
#define INBITS 0b00001111
/*************************************************************/
void main(void) {
volatile in_data;
/*************************************************************/
  /* Set direction register for PTAD for outputs
   * and clear the bits for the inputs */
    DDRAD = OBITS;
  /* Set digital input enable for PTAD inputs */
    ATDDIEN = INBITS;
  /* Output data to bits 7 - 4 */
    PTAD = 0b10110000;
  /* Input data from bits 3 - 0 */
    in_data = PTAD;

  for(;;) {} /* wait forever */
}
```

EXPLANATION OF EXAMPLES 11-3 AND 11-4

As shown in Figure 11-6, both the DDRAD and the ATDDIEN registers must be initialized to be able to use PTAD bits as digital inputs or outputs or port AD as digital inputs (only). In these examples bits 7–4 are output and bits 3–0 are inputs. The outputs for PTAD are initialized by setting bits 7–4 in DDRAD (*line 12*). The inputs for PTAD are initialized by *setting* bits 3–0 in ATDDIEN (*line 14*) and *resetting* bits 3–0 in DDRAD (*line 16*).

Port E

Port E is a register in which the bits have a variety of functions in the expanded modes (see Figure 11-7). In single-chip mode, bits 2–7 may be input or output as controlled by the data direction register DDRE. Bits 0 and 1 are associated with the interrupt pins IRQ_L and XIRQ_L and may be used for general purpose input only if they are not being used for interrupt inputs. The *Port E Assignment Register* (*PEAR*) enables control signals when in expanded modes. See Chapter 21.

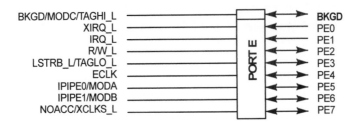

Figure 11-7 Port E.

PTE—Base + $0008—Port E I/O Register

	Bit 7	6	5	4	3	2	1	0
Read:	PE7	PE6	PE5	PE4	PE3	PE2	PE1	PE0
Write:								
Reset:	0	MODB	MODA	0	0	0	0	0
Single-chip Mode:	PE7	PE6	PE5	PE4	PE3	PE2	PE1/ XIRQ_L	PE0/ IRQ_L
Expanded Mode:	NOACC/ XCLKS	IPIPE1	IPIPE0	ECLK	LSTRB_L TAGLO_L	R/W_L	PE1/ XIRQ_L	PE0/ IRQ_L

▨ = reserved, unimplemented, or cannot be written to.

See Also:

Topic	Register	Chapter/Section
Data Direction Register	DDRE (Base + $0009)	11.6
Port E Assignment	PEAR (Base + $000A)	21
Pull-up or Pull-down Enable	PUCR (Base + $000C)	11.7
Reduced Drive Control	RDRIV (Base + $000D)	11.7
Mode Register	MODE (Base + $000B)	21
Expanded Mode Operation		21

Port J

Port J is a 2-bit, general purpose register with interrupt capabilities (see Figure 11-8). These pins are available only on the 80-pin QFP package.

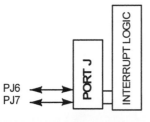

Figure 11-8 Port J.

PTJ—Base + $0268—Port J I/O Register

	Bit 7	6	5	4	3	2	1	0
Read:	PTJ7	PTJ6	0	0	0	0	0	0
Write:								
Reset:	0	0	—	—	—	—	—	—

▨ = reserved, unimplemented, or cannot be written to.

See Also:

Topic	Register	Chapter/Section
Port J Input Register	PTIJ (Base + $0269)	11.7
Data Direction Register	DDRJ (Base + $026A)	11.6
Reduced Drive Control	RDRJ (Base + $026B)	11.7
Pull-up or Pull-down Enable	PERJ (Base + $026C)	11.7
Port J Polarity Select	PPSJ (Base + $026D)	11.7
Port J Interrupt Enable	PIEJ (Base + $026E)	12
Port J Interrupt Flag	PIFJ (Base + $026F)	12

Port M

Port M is a 6-bit, general purpose I/O register when not being used for the synchronous peripheral interface (SPI) or the controller area network (CAN) interface (see Figure 11-9). When the SPI or the MSCAN interfaces are enabled, the port M bits are configured for input and output as needed by the device. The data direction register does not need to be configured.

Port P

Port P is an 8-bit, general purpose I/O port (see Figure 11-10). Alternatively, up to six pulse-width modulated waves can be output. In some versions of the microcontroller, such as the MC9S12C32 where only port P, bit 5 is available at the chip, five of the six PWM outputs can be routed to port T. The *Module Routing Register* (*MODRR*) is used to do this.

Port S

The serial communications interface (SCI) shares two of the four bidirectional port S pins (see Figure 11-11). If the SCI is enabled, bit 0 is assigned to received-data (RxD) and is configured as an input. Bit 1 is assigned to transmitted-data (TxD) and configured as an output. Bit 2 and bit 3 are not assigned and can be used to implement flow control in the serial interface.[3] See Chapter 15 for a complete discussion on the operation of the serial port.

---

[3] Port S, bit 2 and bit 3 are available only on the 80-pin QFP package.

PTM—Base + $0250—Port M I/O Register

	Bit 7	6	5	4	3	2	1	0
Read:	0	0	PTM5	PTM4	PTM3	PTM2	PTM1	PTM0
Write:								
Reset:	0	0	0	0	0	0	0	0
SPI and CAN:			SCK	MOSI	SS*	MISO	TXCAN	RXCAN

☐ = reserved, unimplemented, or cannot be written to.

See Also:

Topic	Register	Chapter/Section
Port M Input Register	PTIM (Base + $0251)	11.6
Data Direction Register	DDRM (Base + $0252)	11.7
Reduced Drive Control	RDRM (Base + $0253)	11.7
Pull-up or Pull-down Enable	PERM (Base + $00254)	11.7
Polarity Select Register	PPSM (Base + $0255)	11.7
Wired-OR Mode	WOMM (Base + $0256)	11.7
SPI Operation		15
CAN Operation		16

PTP—Base + $0258—Port P I/O Register

	Bit 7	6	5	4	3	2	1	0
Read:	PTP7	PTP6	PTP5	PTP4	PTP3	PTP2	PTP1	PTP0
Write:								
Reset:	0	0	0	0	0	0	0	0
PWM	—	—	PWM5	PWM4	PWM3	PWM2	PWM1	PWM0

See Also:

Topic	Register	Chapter/Section
Port P Input Register	PTIP (Base + $0259)	11.7
Data Direction Register	DDRP (Base + $025A)	11.6
Reduced Drive Control	RDRP (Base + $025B)	11.7
Pull-up or Pull-down Enable	PERP (Base + $025C)	11.7
Polarity Select Register	PPSP (Base + $025D)	11.7
Port P Interrupt Enable	PIEP (Base + $025E)	12
Port P Interrupt Flag	PIFP (Base + $025F)	12
Module Routing Register	MODRR (Base + $0247)	14
Pulse-Width Modulation		14

PTS—Base + $0248—Port S I/O Register

	Bit 7	6	5	4	3	2	1	0
Read:	0	0	0	0	PTS3	PTS2	PTS1	PTS0
Write:					PTS3	PTS2	PTS1	PTS0
Reset:	0	0	0	0	0	0	0	0
SCI	—	—	—	—	—	—	TXD	RXD

▓ = reserved, unimplemented, or cannot be written to.

See Also:

Topic	Register	Chapter/Section
Port S Input Register	PTIS (Base + $0249)	11.7
Data Direction Register	DDRS (Base + $024A)	11.6
Reduced Drive Control	RDRS (Base + $024B)	11.7
Pull-up or Pull-down Enable	PERS (Base + $024C)	11.7
Polarity Select Register	PPSS (Base + $024D)	11.7
Wired-OR Mode	WOMS (Base + $024E)	11.7
SCI Serial I/O		15

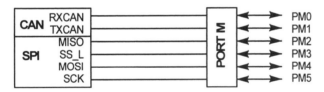

Figure 11-9 Port M.

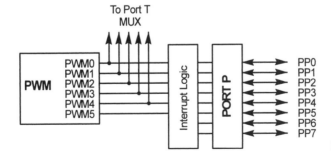

Figure 11-10 Port P.

Figure 11-11 Port S.

Port T

The timer port can provide a mixture of output compare and input capture functions (see Figure 11-12). It can also be used to output up to five channels of pulse-width modulated waveforms. We will cover each of these functions in Chapter 14.

PTT—Base + $0240—Port T I/O Register

	Bit 7	6	5	4	3	2	1	0
Read:	PTT7	PTT6	PTT5	PTT4	PTT3	PTT2	PTT1	PTT0
Write:								
Reset:	0	0	0	0	0	0	0	0
Timer	IOC7	IOC6	IOC5	IOC4	IOC3	IOC2	IOC1	IOC0
PWM				PWM4	PWM3	PWM2	PWM1	PWM0

MODRR—Base + $0247—Port T Module Routing Register

	Bit 7	6	5	4	3	2	1	0
Read:	0	0	0	MODRR4	MODRR3	MODRR2	MODRR1	MODRR0
Write:								
Reset:	0	0	0	0	0	0	0	0

▭ = reserved, unimplemented, or cannot be written to.

Read: Anytime. Write: Anytime.

MODRR4:MODRR0

Port T Module Selection Bits

0 = Associated pin is connected to the timer module (default).

1 = Associated pin is connected to the PWM module.

See Also:

Topic	Register	Chapter/Section
Port T Input Register	PTIT (Base + $0241)	11.7
Data Direction Register	DDRT (Base + $0242)	11.6
Reduced Drive Control	RDRT (Base + $0243)	11.7
Pull-up or Pull-down Enable	PERT (Base + $0244)	11.7
Polarity Select Register	PPST (Base + $0245)	11.7
Port T Module Routing (MUX Control)	MODRR (Base + $0247)	14
Timer Operation		14
Pulse-Width Modulation		14

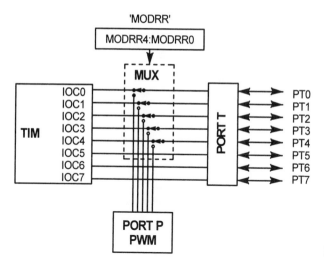

Figure 11-12 Port T.

11.6 Data Direction Registers

Each bit in the bidirectional data registers may be programmed to be either input or output. When the CPU is reset, all registers are placed in the input mode, and you must set bits in a *Data Direction Register* (*DDR*) to change input bits to be outputs. Each DDR has the following format, where x is A, B, AD, E, J, M, P, S, or T; the addresses for each are given in Table 11-2.

DDRx—Base + $(Table 11-2)—Port x Data Direction Register

	Bit 7	6	5	4	3	2	1	0
Read:	DDRx7	DDRx6	DDRx5	DDRx4	DDRx3	DDRx2	DDRx1	DDRx0
Write:								
Reset:	0	0	0	0	0	0	0	0
Default:	Input	Input	Input	Input	Input	Input	Input	Input

      ■ = reserved, unimplemented, or cannot be written to.

      Read: Anytime. Write: Anytime.

DDRx7:DDRx0

Data Direction Control Bits

0 = Associated pin is high-impedance input (default).
1 = Associated pin is an output.

Data direction register bits determine the direction of the corresponding data register. The direction can be set individually for each bit in the port.

TABLE 11-2 Data Direction Register Addresses

Data Direction Register	Address Base +
DDRA	$0002
DDRB	$0003
DDRAD	$0272
DDRE[a]	$0009
DDRJ	$026A
DDRM	$0252
DDRP	$025A
DDRS	$024A
DDRT	$0242

[a] You cannot configure bit 0 and bit 1 in port E to be output because they are associated with interrupt inputs.

Examples 11-5 and 11-6 show how to initialize the most significant nibble in port P for output. If a port has a mixture of input and output bits, writing to the port affects only those bits that are outputs. Reading the port returns the values on the input bits as well as the last values output to the output bits.

Example 11-5 Initializing the Data Direction Register

```
Metrowerks HC12-Assembler
(c) COPYRIGHT METROWERKS 1987-2003

Rel. Loc   Obj. code Source line

---- ----   --------- -----------

  1                   ; Set the register BASE address
  2         0000 0000 BASE:   EQU    $0000
  3         0000 0258 PTP:    EQU    BASE+$0258; Port P I/O
  4         0000 025A DDRP:   EQU    BASE+$025A; DDR Port P
  5                   ; Define bits to be output and input
  6                   ;   1 = output, 0 = input
  7         0000 00F0 OBITS:  EQU    %11110000
  8                   ;      . . .
  9                   ; Set direction register for Port P
 10 000000 1C02 5AF0          bset   DDRP,OBITS
 11                   ;      . . .
 12                   ; Output data to bits 7 - 4
 13 000004 86B0               ldaa   #%10110000
 14 000006 7A02 58            staa   PTP
 15                   ; Read data on bits 3 - 0
 16 000009 B602 58            ldaa   PTP
```

Example 11-6 Initializing the Data Direction Register with C

```
/****************************************************
* Define bits to be output and input
*     1 = output, 0 = input
****************************************************/
#define OBITS 0xf0 /* 240 */
void main(void) {
/****************************************************/
  /* Set direction register for Port P */
   DDRP = OBITS;
  /* . . . */
  /* Output data to bits 7 - 4 */
   PTP = 0xB0;
  /* . . . */
  /* Or here is another way */
  /* These compile to bit set and bit clear instructions */
  /* Set direction register for Port P */
    DDRP_DDRP7 = 1;
    DDRP_DDRP6 = 1;
    DDRP_DDRP5 = 1;
    DDRP_DDRP4 = 1;
  /* . . . */
  /* Output data to bits 7 - 4 */
    PTP_PTP7 = 1;
    PTP_PTP6 = 0;
    PTP_PTP5 = 1;
    PTP_PTP4 = 1;
}
```

11.7 I/O Port Bit Electronics

Reduced Drive Control

Drive refers to the capability of the output circuitry to source current to whatever is connected to the pin. High drive current is an advantage when the output must drive a capacitive load. High drive current results in higher speed switching between logic levels. Unfortunately, high drive current means higher power consumption and the increased likelihood of radio frequency interference (RFI). The *RDRIV, Reduced Drive of I/O Lines* register, allows you to reduce the drive level to reduce power consumption and RFI emissions for ports K[4], E, B, and A. The other

---

[4] Port K is not an available register in the MC9S12C32 version of the microcontroller.

general purpose I/O registers in the port integration module have individual registers as shown in Table 11-3. If any of the port bits are used for an input, the reduced drive control bit is ignored.

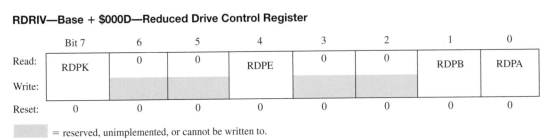

RDRIV—Base + $000D—Reduced Drive Control Register

	Bit 7	6	5	4	3	2	1	0
Read:	RDPK	0	0	RDPE	0	0	RDPB	RDPA
Write:								
Reset:	0	0	0	0	0	0	0	0

     = reserved, unimplemented, or cannot be written to.

Read: Anytime. Write: Anytime.

RDPK, RDPE, RDPB, RDPA

Reduced Drive for Port

0 = All output pins have full drive enabled (default).
1 = All output pins have reduced drive enabled.

RDRx—Base + $(Table 11-3)—Port x Reduced Drive Register

	Bit 7	6	5	4	3	2	1	0
Read:	RDRx7	RDRx6	RDRx5	RDRx4	RDRx3	RDRx2	RDRx1	RDRx0
Write:								
Reset:	0	0	0	0	0	0	0	0
RDRAD:	Full Drive	Full Drive	Full Drive	Full Drive	Full Drive	Full Drive	Full Drive	Full Drive
RDRJ:	Full Drive	Full Drive						
RDRM:			Full Drive	Full Drive	Full Drive	Full Drive	Full Drive	Full Drive
RDRP:	Full Drive	Full Drive	Full Drive	Full Drive	Full Drive	Full Drive	Full Drive	Full Drive
RDRS:					Full Drive	Full Drive	Full Drive	Full Drive
RDRT:	Full Drive	Full Drive	Full Drive	Full Drive	Full Drive	Full Drive	Full Drive	Full Drive

     = reserved, unimplemented, or cannot be written to.

Read: Anytime.   Write: Anytime.

RDRx7:RDRx0

Reduced Drive for Ports

0 = Full drive strength at output (default).
1 = Associated pin drives at about one-third the full drive output.

TABLE 11-3 Reduced Drive Enable Registers

Pull Device Enable Register	Address Base +
RDRAD	$0273
RDRJ	$026B
RDRM	$0253
RDRP	$025B
RDRS	$024B
RDRT	$0243

Pull-up or Pull-down Control

It is a good electronic design practice to tie unused input pins to either a high or a low logic level. In CMOS devices this reduces the chance for a potentially destructive condition called *latch-up* to occur. The HCS12 provides a variety of registers to enable pull-up or pull-down resistors on ports that are configured as inputs. The choice of implementing a pull-up or

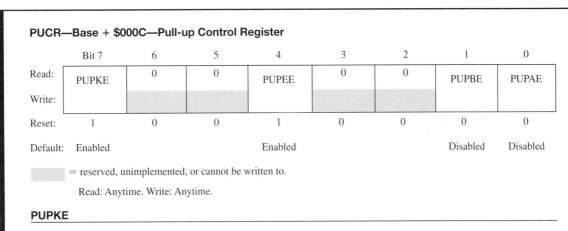

PUCR—Base + $000C—Pull-up Control Register

	Bit 7	6	5	4	3	2	1	0
Read:	PUPKE	0	0	PUPEE	0	0	PUPBE	PUPAE
Write:								
Reset:	1	0	0	1	0	0	0	0
Default:	Enabled			Enabled			Disabled	Disabled

= reserved, unimplemented, or cannot be written to.

Read: Anytime. Write: Anytime.

PUPKE

Pull Resistor Port K Enable

0 = Disable pull resistors.
1 = Enable pull resistors for port K input pins (default).

PUPEE

Pull Resistor Port E Enable

0 = Disable pull resistors.
1 = Enable pull resistors for port E input pins 7 and 4–0 (default).
Bits 5 and 6 of port E have pull resistors that are enabled only during reset. PUPEE has no effect on those pins.

PUPBE, PUPAE

Pull Resistor Ports B and A Enable

0 = Disable pull resistors (default).
1 = Enable pull resistors for port B and port A input pins.

PERx—Base + $(Table 11-4)—Port x Pull Device Enable Register

	Bit 7	6	5	4	3	2	1	0
Read:	PERx7	PERx6	PERx5	PERx4	PERx3	PERx2	PERx1	PERx0
Write:								

Reset:

PERAD:	Disabled	Disabled	Disabled	Disabled	Disabled	Disabled	Disabled	Disabled
PERJ:	Disabled	Disabled						
PERM:			Enabled	Enabled	Enabled	Enabled	Enabled	Enabled
PERP:	Disabled	Disabled	Disabled	Disabled	Disabled	Disabled	Disabled	Disabled
PERS:					Enabled	Enabled	Enabled	Enabled
PERT:	Disabled	Disabled	Disabled	Disabled	Disabled	Disabled	Disabled	Disabled

  ░ = reserved, unimplemented, or cannot be written to.

  Read: Anytime. Write: Anytime.

PERx7:PERx0

Pull Device Enable Bits

0 = Pull-up or pull-down device is disabled.

1 = Either a pull-up or pull-down device is enabled.

Port Specific Notes:

  PERAD (ATD): It is not possible to enable pull devices when an associated A/D channel is enabled simultaneously.

  PERJ (Port J): Configures pull-up or pull-down for bits 6 and 7 used as input or wired-OR output. It has no effect if the port is used as a normal, active pull-up (push-pull) output.

  PERM (Port M): Pull-up devices enabled at reset.

  PERP (Port P): Pull devices disabled at reset.

  PERS (Port S): Configures pull device for input or wired-OR output. Pull-up devices enabled at reset.

  PERT (Port T): Pull devices disabled at reset.

pull-down is controlled by the polarity selection registers, and the pull-ups and pull-downs are enabled by the PUCR register for K, E, A, and B and the PERx register for all other ports.

The other registers in the port integration module—ports AD, J, M, P, S, and T—have separate pull device enable registers that control each bit in the port individually. See Table 11-4. It is not possible to enable a pull-up or pull-down resister if the port bit is used as an output.

Polarity Selection

The polarity (pull-up or pull-down) of the pull devices enabled by the PERx registers is controlled by the *Polarity Select Registers*.

PPSx—Base + $(Table 11-5)—Port Polarity Select Register

	Bit 7	6	5	4	3	2	1	0
Read: Write:	PPSx7	PPSx6	PPSx5	PPSx4	PPSx3	PPSx2	PPSx1	PPSx0
Reset:	0	0	0	0	0	0	0	0
PPSAD:	Pull-up	Pull-up	Pull-up	Pull-up	Pull-up	Pull-up	Pull-up	Pull-up
PPSJ:	Pull-up	Pull-up						
PPSM:			Pull-up	Pull-up	Pull-up	Pull-up	Pull-up	Pull-up
PPSP:	Pull-up	Pull-up	Pull-up	Pull-up	Pull-up	Pull-up	Pull-up	Pull-up
PPSS:					Pull-up	Pull-up	Pull-up	Pull-up
PPST:	Pull-up	Pull-up	Pull-up	Pull-up	Pull-up	Pull-up	Pull-up	Pull-up

▨ = reserved, unimplemented, or cannot be written to.

Read: Anytime. Write: Anytime.

PPSx7:PPSx0

Pull Device Polarity Select

0 = Pull-up device is connected to the associated port pin if enabled by the associated bit in the PERx register and the port pin is used as an input (default).

1 = Pull-down device is connected to the associated port pin.

Port Specific Notes:

PPSP (Port P): The register selects both the polarity of the pull device and the active edge of associated interrupt request. (See Chapter 12.)

0 = Pull-up selected and falling edge on the associated port P pin sets the associated flag bit in the PIFP register.

1 = Pull-down selected and rising edge on the associated port P pin sets the associated flag bit in the PIFP register. (See Chapter 14.)

PPSJ (Port J): The register selects the polarity of both the pull device and the active edge of associated interrupt request.

0 = Pull-up selected and falling edge on the associated port J pin sets the associated flag bit in the PIFJ register.

1 = Pull-down selected and rising edge on the associated port J pin sets the associated flag bit in the PIFJ register.

TABLE 11-4 Pull Device Enable Register Addresses

Pull Device Enable Register	Address Base +
PERAD	$0274
PERJ	$026C
PERM	$0254
PERP	$025C
PERS	$024C
PERT	$0244

TABLE 11-5 Pull Device Polarity Select Register Addresses

Pull Device Enable Register	Address Base +
PPSAD	$0275
PPSJ	$026D
PPSM	$0255
PPSP	$025D
PPSS	$024D
PPST	$0245

Wired-OR

There are times when you would like to tie two outputs together. As we know, this is not a good idea when the outputs have an active-high pull-up, such as in a normal CMOS or TTL gate. This can be done when the outputs of the gates are open-drain with no active pull-up as shown in

---

WOMM—Base + $0256—Port M Wired-OR Mode Register

	Bit 7	6	5	4	3	2	1	0
Read:	0	0	WOMM5	WOMM4	WOMM3	WOMM2	WOMM1	WOMM0
Write:								
Reset:	0	0	0	0	0	0	0	0

 = reserved, unimplemented, or cannot be written to.

WOMS—Base + $024E—Port S Wired-OR Mode Register

	Bit 7	6	5	4	3	2	1	0
Read:	0	0	0	0	WOMS3	WOMS2	WOMS1	WOMS0
Write:								
Reset:								

 = reserved, unimplemented, or cannot be written to.

Read: Anytime. Write: Anytime.

WOMM5:WOMM0 and WOMS3:WOMS0

Configure Wired-OR Open-Drain Outputs

These bits configure the associated port outputs as wired-OR open drain. If enabled, the output is driven active-low only. A logic level of "1" is not driven. This has no effect on input pins.

0 = Output buffers are normal active-high (push-pull outputs) (default).
1 = Output buffers act as open-drain outputs.

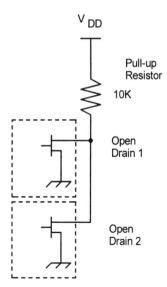

Figure 11-13 Open-drain wired-OR connection.

Figure 11-13. Each of the open-drain outputs can pull low but for a logic high to be realized, an external pull-up resistor is needed. This allows two sources to be tied together and if we guarantee that only one will be active-low at a time, each can drive the output line independently.

Port M and port S each have the capability to be configured as open-drain outputs to allow them to be connected in a wired-OR configuration. This is controlled by the *WOMM* and *WOMS* registers.

Input Registers

The ports AD, J, M, P, S, and T have separate input registers. The register always reads back the status of the associated pins and you cannot write to this register. This can be used to detect overload or short-circuit conditions on output pins. The addresses of these registers are shown in Table 11-6.

PTIx —Base + $(Table 11-6)—Port x Input Register

	Bit 7	6	5	4	3	2	1	0
Write:								
Read:								
PTIAD:	PTIAD7	PTIAD6	PTIAD5	PTIAD4	PTIAD3	PTIAD2	PTIAD1	PTIAD0
PTIJ:	PTIJ7	PTIJ6	0	0	0	0	0	0
PTIM:	0	0	PTIM5	PTIM4	PTIM3	PTIM2	PTIM1	PTIM0
PTIP:	PTIP7	PTIP6	PTIP5	PTIP4	PTIP3	PTIP2	PTIP1	PTIP0
PTIS:	0	0	0	0	PTIS3	PTIS2	PTIS1	PTIS0
PTIT:	PTIT7	PTIT6	PTIT5	PTIT4	PTIT3	PTIT2	PTIT1	PTIT0

▨ = reserved, unimplemented, or cannot be written to.

TABLE 11-6 Port Input Register Addresses

Port Input Register	Address Base +
PTIAD	$0271
PTIJ	$0269
PTIM	$0251
PTIP	$0259
PTIS	$0249
PTIT	$0241

11.8 Parallel I/O Programming

Parallel I/O Software Checklist

1. Initialize the data direction register (DDRx) for the port. The default direction is input signified by the data direction register bit equal to zero. Set the data direction register bits to one for any bits that are to be used for output.
2. If PORTAD is used for digital input, bits in the ATDDIEN register must be set and bits in the DDRAD register must be reset for the input bits.
3. If PTAD is used for digital input, ATDDIEN must be set and DDRAD must be reset.
4. If PTAD is used for digital output, DDRAD bits must be set.
5. If PTIAD is used for digital input, ATDDIEN must be set and DDRAD must be reset.
6. Initialize reduced drive control, pull-up or pull-down control, polarity selection, and wired-OR mode control registers as needed.

Parallel I/O Software Examples

See Example 11-7.

Example 11-7 Parallel I/O to Toggle LEDs

```
/****************************************************************
 * The program will alternately blink LED1 and LED2 on
 * the Student Learning Kit with a CSM-12C32 board
 * LED1 is connected to Port A, bit 0 and LED2 connected
```

```
     * to Port B, bit 4
     **********************************************************/
     #include <mc9s12c32.h>                    /* derivative
                                                   information */
     /*********************************************************/
     void main(void) {
     volatile int i;
     /*********************************************************/
       /* Initialize the I/O */
         DDRA_BIT0 = 1;  /* Make Port A bit-0 output */
         DDRB_BIT4 = 1;  /* Port B bit-4 output */
       /* Set LED1 on and LED2 off */
         PORTA_BIT0 = 0; /* Active low LEDs */
         PORTB_BIT4 = 1;
       /* DO */
         for(;;) {
           /* Delay for a while */
           for (i = 1; i < 30000; ++i ) {
           };
         /* Toggle the display bits */
           if (PORTA_BIT0 == 1) PORTA_BIT0 = 0;
           else PORTA_BIT0 = 1;
           if (PORTB_BIT4 == 1) PORTB_BIT4 = 0;
           else PORTB_BIT4 = 1;
         }
       /* FOREVER */
     }
```

EXPLANATION OF EXAMPLE 11-7

There are two LEDs connected to port A, bit 0 and port B, bit 4. The port A and port B data direction registers are set for output and then the active-low LEDs are turned on and off by outputting a zero and a one on the ports. A software delay loop generates a delay long enough so that we can see the LEDs flash and they are toggled on and off. We will see in Chapter 14 how to do accurate timing using the MC9S12C32's timer module.

Examples 11-8 and 11-9 show how, in C and assembly, to input 8-bit data from PORTAD0, swap the bits, and output to PORTT.

Example 11-8 C Parallel I/O Program

```
/*****************************************************************
 * C program to input 8-bit data from PORTAD0, reverse the
 * order of the bits, and output them to PORTT
 *****************************************************************/
#include <mc9s12c32.h>                /* derivative information */
/*****************************************************************/
/* Define a bitfield type as unsigned int */
typedef unsigned int BITFIELD;
/* Define a union of an unsigned char and 8 bit fields */
typedef union {
  unsigned char DataByte;
  struct {
    BITFIELD BIT0 : 1;
    BITFIELD BIT1 : 1;
    BITFIELD BIT2 : 1;
    BITFIELD BIT3 : 1;
    BITFIELD BIT4 : 1;
    BITFIELD BIT5 : 1;
    BITFIELD BIT6 : 1;
    BITFIELD BIT7 : 1;
  } ByteBits;
} TEST_DATA;
/*****************************************************************/
void main(void) {
  volatile TEST_DATA in_data, out_data;
/*****************************************************************/
  /* Set data direction register to make Port T output */
  DDRT = 0xFF;
  /* Initialize PORTAD to be used as digital input */
  ATDDIEN = 0xFF;        /* ATDDIEN bits are set */
  DDRAD = 0;             /* DDRAD bits are reset */
  /* DO */
  for (; ; ){
    /* Get a value from PORTAD0 */
      in_data.DataByte = PORTAD0;
    /* Swap the bits: out_data[7] = in_data[0] etc */
      out_data.ByteBits.BIT0 = in_data.ByteBits.BIT7;
      out_data.ByteBits.BIT1 = in_data.ByteBits.BIT6;
```

```
        out_data.ByteBits.BIT2 = in_data.ByteBits.BIT5;
        out_data.ByteBits.BIT3 = in_data.ByteBits.BIT4;
        out_data.ByteBits.BIT4 = in_data.ByteBits.BIT3;
        out_data.ByteBits.BIT5 = in_data.ByteBits.BIT2;
        out_data.ByteBits.BIT6 = in_data.ByteBits.BIT1;
        out_data.ByteBits.BIT7 = in_data.ByteBits.BIT0;
    /* Output the swapped data */
      PTT = out_data.DataByte;
  }
  /* FOREVER */
}
```

EXPLANATION OF EXAMPLE 11-8

This example defines a union between an unsigned char `DataByte` and an 8-bit structure `ByteBits`. Doing this allows both byte and bit addressing in the same data value. The data direction register for port T is set for output and the port AD ATDDIEN and DDRAD are set and reset, respectively. PORTAD0 is read as a byte and then each bit is individually transferred to the output `out_data` variable. Note that all addresses and definitions for the ports in this example are contained in *mc9s12c32.h*.

Example 11-9 Parallel I/O in Assembly

```
Metrowerks HC12-Assembler
(c) COPYRIGHT METROWERKS 1987-2003

 Rel. Loc    Obj. code Source line
 ----  ------  ---------  -----------
    1                     ;*******************************
    2                     ; Input 8-bit data from Port AD, swap the
                          bits
    3                     ; and output to Port T
    4                     ;*******************************
    5                     ; Define the entry point for the main program
    6                             XDEF   Entry, main
    7                             XREF   __SEG_END_SSTACK ; Note double
                                  underbar
    8                     ;*******************************
```

```
 9                  ; Include files
10                          include portt.inc
11                          include portad.inc
 5i                             INCLUDE "bits.inc"
12                  ;*******************************
13                  ; Code Section
14                  MyCode: SECTION
15                  Entry:
16                  main:
17                  ;*******************************
18                  ; Initialize stack pointer register
19 000000 CFxx xx           lds   #__SEG_END_SSTACK
20                  ;*******************************
21                  ; Initialize I/O
22                  ; Set data direction register to make Port
                    T output
23 000003 180B FF02         movb  #$FF,DDRT
   000007 42
24                  ; Initialize PORTAD to be used as digital
                    input
25 000008 180B FF00         movb  #$FF,ATDDIEN   ; Set ATDDIEN
                                                    bits
   00000C 8D
26 00000D 7902 72           clr   DDRAD           ; Clear DDRAD
                                                    bits
27                  main_loop:
28                  ; DO
29                  ;   Initialize the output value and set
                    only bits
30                  ;   that must be set
31 000010 C7               clrb
32                  ;   Check each bit in PORTAD0
33                  ;     IF PORTAD0[0] = 1
34 000011 4F8F 0102        brclr PORTAD0,%00000001,check_1
35                  ;     THEN Set bit 7
36 000015 CA80             orab  #%10000000
37                  check_1:
38                  ;     IF PORTAD0[1] = 1
39 000017 4F8F 0202        brclr PORTAD0,%00000010,check_2
40                  ;     THEN Set bit 6
41 00001B CA40             orab  #%01000000
```

```
42                      check_2:
43                      ;      IF PORTAD0[2] = 1
44  00001D 4F8F 0402           brclr PORTAD0,%00000100,check_3
45                      ;      THEN Set bit 5
46  000021 CA20                orab  #%00100000
47                      check_3:
48                      ;      IF PORTAD0[3] = 1
49  000023 4F8F 0802           brclr PORTAD0,%00001000,check_4
50                      ;      THEN Set bit 4
51  000027 CA10                orab  #%00010000
52                      check_4:
53                      ;      IF PORTAD0[4] = 1
54  000029 4F8F 1002           brclr PORTAD0,%00010000,check_5
55                      ;      THEN Set bit 3
56  00002D CA08                orab  #%00001000
57                      check_5:
58                      ;      IF PORTAD0[5] = 1
59  00002F 4F8F 2002           brclr PORTAD0,%00100000,check_6
60                      ;      THEN Set bit 2
61  000033 CA04                orab  #%00000100
62                      check_6:
63                      ;      IF PORTAD0[6] = 1
64  000035 4F8F 4002           brclr PORTAD0,%01000000,check_7
65                      ;      THEN Set bit 1
66  000039 CA02                orab  #%00000010
67                      check_7:
68                      ;      IF PORTAD0[7] = 1
69  00003B 4F8F 8002           brclr PORTAD0,%10000000,done
70                      ;      THEN Set bit 0
71  00003F CA01                orab  #%00000001
72                      done:
73  000041 7B02 40             stab  PTT
74                      ; FOREVER
75  000044 20CA               bra   main_loop
```

EXPLANATION OF EXAMPLE 11-9

Example 11-9 is an assembly language version of Example 11-8. The data direction registers are initialized in *lines 22–26*. Register B is to be used for the output value so it is cleared in *line 31*. Then each bit in checked in PORTAD0 using a branch-if-clear instruction. If the bit in PORTAD0 is set, then the corresponding swapped bit is set in the B register. The final value is output in *line 73*.

More Parallel I/O Programming and Interfacing Examples

See Chapter 18 for more examples showing I/O interfacing and programming.

11.9 Remaining Questions

- How do we interface real-world devices to the microcontroller? *We will be discussing that for a variety of applications in Chapter 18.*
- How are the interrupting capabilities of port J and port P used? *We will cover those interrupts as well as other capabilities in Chapter 12.*
- Where do I find out about expanded mode operation? *We will discuss expanded mode operation in Chapter 21.*
- How do I synchronize my I/O software with different I/O devices? *There are three ways to do this—polling, handshaking I/O, and interrupts. We cover interrupts in Chapter 12 and other I/O synchronization techniques in Chapter 18.*

11.10 Parallel I/O Register Address Summary

We present this summary in Table 11-7.

TABLE 11-7 Parallel I/O Registers

Name	Register	Address Base +
PORTA	Port A I/O Register	$0000
PORTB	Port B I/I Register	$0001
DDRA	Data Direction Register Port A	$0002
DDRB	Data Direction Register Port B	$0003
PTE	Port E I/O Register	$0008
DDRE	Data Direction Register Port E	$0009
PEAR	Port E Assignment Register	$000A
PUCR	Pull Device Enable Register	$000C
RDRIV	Reduced Drive Control	$000D
INITRM	Initialize RAM Location	$0010
INITRG	Initialize Register Location	$0011
INITEE	Initialize EEPROM Location	$0012
ATDDIEN	ATD Digital Input Enable Register	$008D
PORTAD0	ATD Digital Input Port	$008F
PTT	Port T I/O Register	$0240
PTIT	Port T Input Register	$0241
DDRT	Data Direction Register Port T	$0242
RDRT	Reduced Drive Register T	$0243
PERT	Pull-up or Pull-down Enable Port T	$0244
PPST	Port T Polarity Select	$0245
MODRR	Port T Module Routine Register	$0247
PTS	Port S I/O Register	$0248
PTIS	Port S Input Register	$0249
DDRS	Data Direction Register Port S	$024A

TABLE 11-7 Continued

Name	Register	Address Base +
RDRS	Reduced Drive Register S	$024B
PERS	Pull-up or Pull-down Enable Port S	$024C
PPSS	Port S Polarity Select	$024D
WOMS	Wired-OR Mode	$024E
PTM	Port M I/O Register	$0250
PTIM	Port M Input Register	$0251
DDRM	Data Direction Register Port M	$0252
RDRM	Reduced Drive Register M	$0253
PERM	Pull-up or Pull-down Enable Port M	$0254
PPSM	Port M Polarity Select	$0255
WOMM	Wired-OR Mode	$0256
PTP	Port P I/O Register	$0258
PTIP	Port P Input Register	$0259
DDRP	Data Direction Register Port P	$025A
RDRP	Reduced Drive Register P	$025B
PERP	Pull-up or Pull-down Enable Port P	$025C
PPSP	Port P Polarity Select	$025D
PIEP	Port P Interrupt Enable Register	$025E
PIFP	Port P Interrupt Flag Register	$025F
PTJ	Port J I/O Register	$0268
PTIJ	Port J Input Register	$0269
DDRJ	Data Direction Register Port J	$026A
RDRJ	Reduced Drive Register J	$026B
PERJ	Pull-up or Pull-down Enable Port J	$026C
PPSJ	Port J Polarity Select	$026D
PIEJ	Port J Interrupt Enable Register	$026E
PIFJ	Port J Interrupt Flag Register	$026F
PTAD	Port AD I/O Register	$0270
PTIAD	Port AD Input Register	$0271
DDRAD	Data Direction Register Port AD	$0272
RDRAD	Reduced Drive Register Port AD	$0273
PERAD	Pull-up or Pull-down Enable Port AD	$0274
PPSAD	Port AD Polarity Select	$0275

11.11 Conclusion and Chapter Summary Points

In this chapter we discussed the operation of the general purpose I/O ports. The following points summarize what we have covered in this chapter:

- Most of the ports have some secondary function that overrides their use as general purpose I/O. These are:
  - Port AD—A/D converter.
  - Port J—Interrupt inputs.
  - Port M—CAN and SPI serial I/O.
  - Port P—Pulse-width modulator.
  - Port S—SCI serial I/O.
  - Port T–Timer.
  - Port A and Port B—Expanded mode address and data buses.
  - Port E—Expanded mode control signals.

- When ports are not being used for expanded mode or other I/O, they may be used for general purpose parallel I/O.
- Ports A and B may not be used for I/O when in expanded mode.
- All ports have programmable functions.
- Bidirectional ports have data direction registers to specify the data flow direction.

11.12 Bibliography and Further Reading

PIM_9C32 (Port Integration Module) Block Guide, Document Number S12C32PIMV1/D, Motorola, 2001.

MC9S12C Family Device User Guide, Document Number 9S12C128DGV1/D, Motorola, 2002.

Multiplexed External Bus Interface (MEBI) Module V3 Block User Guide, Document Number S12MEBIV3, Motorola, 2003.

11.13 Problems

Basic

11.1 Give the data direction register addresses for ports A, B, AD, T, P, S, and M. **[a]**

11.2 Give the data register addresses for ports A, B, AD, T, P, S, M, and E. **[a]**

11.3 How do you control the direction of the bidirectional bits in the HCS12 I/O ports? **[a]**

11.4 What is the purpose of the MODDR register in the HCS12 I/O ports? **[a]**

Intermediate

11.5 In the HCS12, port T is a bidirectional port. Write a short segment of code that illustrates how to initialize port T so that bits 7, 4, and 3 may be used as outputs and bits 6, 5, 2, 1, and 0 may be used as inputs. **[c, k]**

11.6 The MC9S12C32 microcontroller has a variety of package options. Discuss why they have done this and what advantages and disadvantages each package has. **[g]**

11.7 In addition to analog input signals, port AD has digital input and output capability. How must bits in the DDRAD and ATDDIEN registers be initialized for the following? **[a, c]**

  a. PAD0–PAD3 input and PAD4–PAD7 output.

  b. PAD7 digital input and PAD0 analog input.

11.8 Write a small assembly program snippet to move the internal registers from their location at reset to start at $4000. **[c, k]**

12 HCS12 Interrupts

OBJECTIVES

This chapter shows how an important external or internal event can interrupt the normal flow of a program. We cover the interrupt system of the HCS12. You will learn about the vectors and the hardware prioritization that the program can modify dynamically. Assembly and C programming examples are given for interrupt service routines.

12.1 General Introduction

An *interrupt* is an important asynchronous event that requires immediate attention.

An *exception* is an event even more important than an interrupt.

An interrupt is a way for an *important, asynchronous event* to be recognized and taken care of (*serviced*) by the CPU executing instructions in a normal program. Consider, for example, a computer system controlling an oil refinery. It would have sensors measuring the chemical composition of the product being refined and outputs controlling the process. A typical process control software loop to do this is shown in Figure 12-1(a). The time taken to go around the loop depends on the complexity of the control algorithms and the speed of the processor. Now consider an important, external, asynchronous event. A fire breaks out in the oil refinery! If the control computer is responsible for activating fire suppression measures, the program should respond immediately and not wait for the software to come around the loop to check on the fire detection sensors. On the other hand, we do not want to write a program that is checking the fire sensors all the time, or even frequently, because it would take time away from the control calculations. This is an ideal application for an interrupt. The interrupt is caused by an external device, the fire sensor, generating a signal called *interrupt request*, or IRQ. The interrupt request is asynchronous. That is, it can happen any time, not necessarily corresponding to any particular time in the instruction execution sequence of the CPU. The IRQ requests the program to take immediate action, called an *interrupt service routine*, or *ISR*. Figure 12-1(b) shows an interrupt service routine added to the process control software of Figure 12-1(a). The interrupt service routine is executed whenever the interrupt occurs.

Some events that interrupt a processor's normal program flow are called *exceptions*. This terminology indicates that a higher priority is assigned to these events. An example of an exception is the system reset.

287

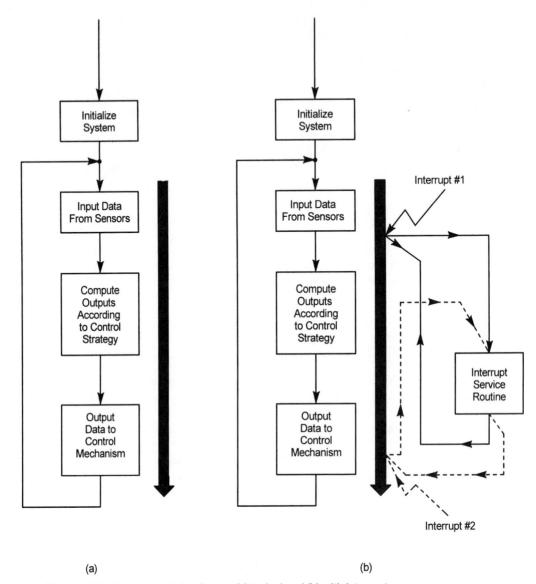

Figure 12-1 Process control software: (a) typical and (b) with interrupts.

A *real-time system* uses interrupts to control *when* things are done in a program.

Interrupts also can *synchronize* the operation of the computer with an external process. Consider sending data to a printer. Typically, computers are much faster than printers, so the data output must be synchronized to the speed of the printer. In addition to the I/O synchronization techniques such as polling or delay loops discussed in Chapter 18, interrupts may be used. In this case, the printer generates an interrupt to signify that it is ready for the next character or, perhaps more likely, the next block of characters. The input of data from a device such as an analog-to-digital (A/D) converter can be synchronized by the A/D converter generating an interrupt when the conversion is complete.

A term used to describe these systems is *real-time*. A real-time system is one that does some process, either at a specific time, say, midnight, or at specific intervals, say, every 10 milliseconds, or at a time required by some external device or event.

Interrupt System Specifications

Let us list some of the general specifications for an interrupt system. The system is to do the following:

- Allow asynchronous events to occur and be recognized.
- Wait for the current instruction to finish before taking care of any interrupt.
- Branch to the correct interrupt service routine to service the interrupting device.
- Return to the interrupted program at the point it was interrupted.
- Allow for a variety of interrupting signals, including levels and edges.
- Allow the programmer to selectively enable and disable all interrupts.
- Allow the programmer to selectively enable and disable selected interrupts.
- Disable further interrupts while the first is being serviced.
- Deal with multiple sources of interrupts.
- Deal with multiple, simultaneous interrupts.

Asynchronous Events and Internal Processor Timing

The current instruction must be finished before an interrupt request is acted upon.

Figure 12-2(a) shows a timeline of program execution. The ticks along the line represent the start of each instruction that a normal program executes in sequence. The normal program does not specify *when*, in a real-time sense, an instruction is to be executed, just the *sequence* of instructions. Asynchronous events, the IRQs, can occur any time.

Figure 12-2(b) shows an expanded timeline. We showed in Chapter 2 that an instruction execution cycle consists of the instruction fetch and instruction execution parts. The sequence controller can be modified to check for an interrupt request before it fetches the next instruction. More states are added to sample the interrupt request and generate more control signals, including one to acknowledge the interrupt. This change allows the CPU to finish the current instruction and then to service the interrupt by entering a special interrupt processing sequence; otherwise, it fetches the next instruction.

Enabling the Interrupts

The programmer of the HCS12 must have total control over the operation of the interrupt system. This is done in two ways.

A *global mask bit*, called the I bit in the condition code register, is used to mask (disallow) or *unmask* (allow) all interrupts. When the I bit is set, interrupts are masked and are not acted upon until the I bit is reset.

Each interrupting subsystem, such as each timer channel, also has an *enable bit* used to *enable* (allow) or *disable* (disallow) *that device* from interrupting. When the enable bit is set,

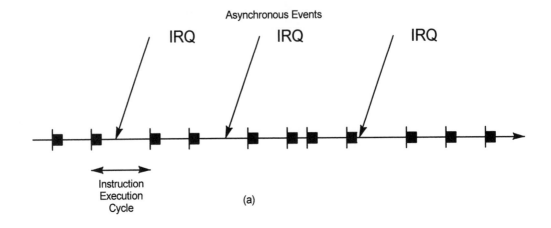

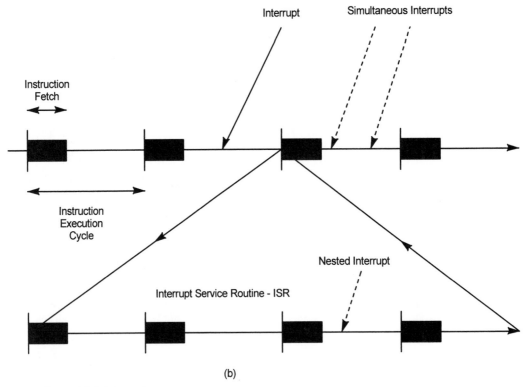

Figure 12-2 Interrupts.

the interrupt is enabled, just the reverse logic from the I bit. Figure 12-3 shows the hardware for this scheme.

Pending Interrupts

For a CPU to act upon a device interrupt, we see from Figure 12-3 that the I bit must be reset (low) to unmask interrupts and the device interrupt enable must be set (high). If these conditions are not

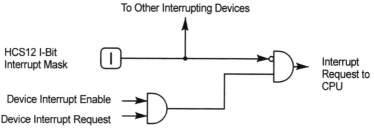

Figure 12-3 Interrupt enable mask and hardware.

A *pending* interrupt is one that will be taken care of after the current interrupt is done or after interrupts are unmasked and enabled.

met and the device asserts the interrupt request signal (high) (and keeps it asserted), the interrupt request is said to be *pending*. Unless the interrupting signal is deasserted, as soon as the I bit is reset and the device interrupt enable is set, the pending interrupt will be transferred through the logic to the CPU to be acted upon.

Multiple Sources of Interrupts

Our processor must deal with multiple devices generating interrupts. Allowing for these requires the system to do the following:

- Determine which of the multiple devices has generated the IRQ to be able to execute the correct interrupt service routine.
- Resolve simultaneous requests for interrupts with a prioritization scheme.

One of the most common ways to resolve which of several interrupting devices has generated the interrupt request is called *vectored interrupts*. A vector is simply an address, and in this case, it is the starting address of the interrupt service routine. In the case of modern microcontrollers such as the HCS12, a specific area of nonvolatile memory is reserved for and dedicated to the vectors (addresses) for all of the possible interrupting devices such as the timer, analog-to-digital converter, and external interrupting devices. When the interrupt request occurs, providing everything is enabled properly, the CPU fetches the address of the interrupt service routine from the vector location and branches to that address to start executing the interrupt service routine.

SEQUENTIAL AND NESTED INTERRUPTS

Nested interrupts are interrupts interrupting interrupts.

An interrupting system can resolve many interrupting sources by using multiple interrupting signals and vectors for determining where the correct interrupt service routine is located. If a subsequent interrupt occurs while another is being serviced—that is, when the interrupt service routine code shown in Figure 12-2(b) is being executed—the programmer may control whether or not the first interrupt service routine is interrupted by the second request. The interrupting system automatically sets the I bit to mask further interrupts just before entering the interrupt service routine. As we can see in Figure 12-3, this stops the second interrupt request from being passed to the CPU for service. The programmer may reset the I bit in the interrupt service routine if there are interrupts of higher importance than the first. This is optional and the programmer does not have to do this unless there are more important interrupts that may occur. Before clearing the I bit in an interrupt service routine, be sure to clear the flag associated with that interrupt. Failure to do so will cause an interrupt to interrupt itself over and over again until the dedicated stack space is overrun. If the mask bit is not reset, the second interrupt remains pending until the interrupt service routine completes and returns to the interrupted program. As part of the return-from-interrupt (RTI) instruction, the I bit is reset automatically and further interrupts are unmasked at that time, allowing the pending interrupt to be serviced.

SIMULTANEOUS INTERRUPTS

Simultaneous interrupts are interrupts that occur within the same instruction execution cycle. See Figure 12-2(b). In this case, both need to be serviced and there must be a prioritization mechanism to resolve the conflict. We will discuss the prioritization scheme used in the HCS12 in Section 12.2.

INTERRUPT TERMINOLOGY

> Foreground jobs may be interrupted by background jobs or vice versa!

There are a variety of terms used to describe these interrupt processes. In some systems, the terms *foreground* and *background* jobs are used. The foreground job is usually the "main" program that is interrupted by the background job. In some real-time systems, these definitions are reversed.

Interrupt latency is the time delay from the initiation of the interrupt request by the hardware to the start of the interrupt service routine. Elements contributing to interrupt latency are the time to complete the current instruction, time to save the machine context and return address on the stack, and time to find the correct interrupt service routine.

12.2 HCS12 Interrupts

The HCS12 microcontroller contains vectored interrupts with hardware priority resolution that we can customize with software. It has two, dedicated, external interrupt inputs. These are IRQ_L, a maskable, general purpose, external interrupt request, and XIRQ_L, a nonmaskable (after the programmer enables it) interrupt. Up to 12 other signals associated with the timer subsystem (a Real-Time Interrupt, eight Timer Channels, a Timer Overflow, and two Pulse Accumulator signals) generate interrupts, and we will discuss these in Chapter 14. The serial interface has two interrupts and the analog-to-digital converter one. The serial interface and its interrupts are covered in Chapter 15 and the A/D in Chapter 17. In addition to external interrupts IRQ_L and XIRQ_L, there are I/O ports with interrupt capability, and we discuss these later in this chapter. There are four other special interrupts, or exceptions, including a software interrupt, unimplemented opcode trap, a watchdog timer, and a clock failure interrupt.

12.3 The Interrupt Process

The Interrupt Enable

> The condition code register contains bits to globally *mask* and *unmask* interrupts.

Two bits in the condition code register give overall control of the interrupt system. The I and X bits are *mask* bits that, when set, disable the interrupt system. The I bit is controlled by the instructions ORCC #%00010000, or preferably SEI, to set the interrupt mask and ANDCC #%11101111 (CLI) to clear it. The I bit can be thought of as a shade in a window that looks out on the interrupting

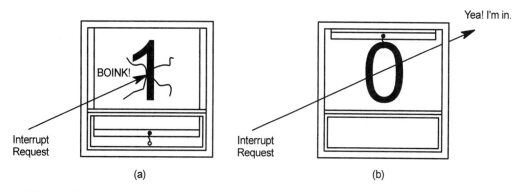

Figure 12-4 (a) I bit = 1 to mask; (b) I bit = 0 to unmask.

world. If the blind is pulled, I = 1, and interrupt requests cannot get through. An open window, I = 0, lets the CPU "see" the interrupts. See Figure 12-4. As we discovered in Chapter 4, when the CPU is reset the I and X bits are set so that interrupts will not be acted on until the program is ready for them.

The I bit acts *globally* and allows or disallows *all* interrupts except for a few that are *unmaskable*. We will discuss these exceptions in Section 12.6. A second level of interrupt control is available. Each interrupt source may be enabled or disabled individually by setting a bit in a control register. As we will see in more detail later, when the I/O device's enable bit is set and an interrupt is to be generated, a flag is set in the I/O device triggering the interrupt request. In most of the I/O devices in the HCS12, you must reset this flag bit in the interrupt service routine to avoid another interrupt request being generated immediately.

The Interrupt Disable

When an interrupt occurs the I bit is automatically set, masking further interrupts. Nested interrupts, which should be avoided, are allowed if the I bit is cleared in the interrupt service routine. Before doing this, you must disable the interrupting source or clear its interrupting flag so that it does not immediately generate another interrupt, resulting in an infinite loop and a locked-up program.

All interrupts are disabled when the HCS12 is reset. Interrupts may be globally masked (disabled) in your program at any time by setting the interrupt mask with the SEI instruction. Individual interrupts are disabled by clearing the enable bit associated with the device.

The Interrupt Request

The HCS12 has both internal and external sources of interrupts. The internal requests come from the internal systems and from exceptions or error conditions. Each interrupt is serviced through its own vector as described in Section 12.4. The two external interrupt request signals, IRQ_L and XIRQ_L, are active-low, edge- or level-sensing. Figure 12-5 shows how to interface these signals to the HCS12. Multiple interrupting devices may pull the request line low using wired-OR (wired-AND), open-drain, or open-collector gates. The HCS12 must poll the devices to determine which generated the interrupt because there is only one vector associated with the IRQ_L signal.

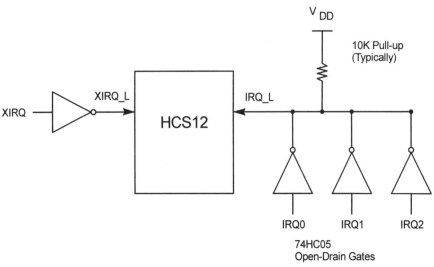

Figure 12-5 Hardware interface for IRQ_L and XIRQ_L signals.

The Interrupt Sequence

All registers, including the *condition code register*, are pushed onto the stack at the start of the interrupt service.

The following events take place when interrupts have been unmasked and enabled, and a request has been generated:

1. Generally, the CPU waits until the currently executing instruction finishes before servicing the interrupt. This component of interrupt latency will depend on the instruction being executed. Most instructions are two to four cycles, but the EDIVS instruction takes 12 clock cycles. Some of the longer HCS12 instructions can be interrupted before they complete.

Interrupt latency is the time between the assertion of the interrupt request and when the interrupt service routine starts to execute.

2. The CPU then determines the address of the interrupt service routine to be executed through a vectoring system.[1]

3. The return address is pushed onto the stack.

4. All CPU registers, including X, Y, A, B, and CCR, are pushed onto the stack.

5. After the CCR is pushed onto the stack, the I bit is set, masking further interrupts.

6. The CPU branches to the interrupt service routine.

The Interrupt Return

If an interrupt is generated by an internal source, the *flag* causing the interrupt must be *reset* in the interrupt service.

Before returning to the interrupted program, you must *reenable the interrupting device's interrupt capability*. This is usually done by *resetting the flag* that caused the interrupt. If this is not done correctly, another interrupt will immediately occur. The RTI—return from interrupt instruction—is used to return to the interrupted program.

You do not have to, and in general you should not, unmask global interrupts in the interrupt service routine.[2] The RTI pulls all registers that were pushed onto the

---

[1] The HCS12 chooses the interrupt source and fetches the vector at the start of the interrupt sequence. This allows it to interleave three program word fetches in the interrupt sequence to refill the instruction queue.

[2] Exception: You will have to unmask interrupts if you need to allow nested interrupts in case there is a higher priority device that may need service.

The *RTI* instruction is used at the end of the interrupt service.

stack at the start of the interrupt service routine, including the CCR, which had the mask bit cleared. Beginning programmers often put a redundant CLI instruction in their interrupt service routines.

12.4 Interrupt Vectors

In any interrupt driven system, the CPU must somehow branch to and start executing the correct interrupt service routine for that request. The HCS12 uses *vectored* interrupts. Each interrupt source has its own vector in memory. This vector is simply the *address* of the interrupt service routine to be executed.

HCS12 System Vectors

An *interrupt vector* is the address of the start of the interrupt service routine.

The HCS12 CPU can handle up to 128 interrupt vectors, but the number implemented in each version of the CPU varies from device to device. There are six resets and nonmaskable interrupts. These are used to reset the processor and for other critical exceptional conditions. The rest of the hardware interrupts are masked by the I bit in the condition code register and are further enabled by bits locally within the particular I/O system generating the interrupt. Each of these interrupting sources has dedicated vector locations as shown in Table 12-1 for the MC9S12C32. The CPU fetches the address of the interrupt service routine from the vector location when the hardware generates an interrupt request. The table also shows the vector locations for the nonmaskable interrupt XIRQ_L, a software interrupt, SWI, an unimplemented opcode trap interrupt, interrupts associated with the computer operating properly (COP) timer and the clock monitor circuit, and the hardware RESET_L.

TABLE 12-1 Interrupt Vector Assignments

Priority[a]	Vector Address	Interrupt Source	Local Enable Bit	See Register	See Chapter	HPRIO Value to Promote
—	$FF80:FF89	Reserved	—	—	—	—
58	$FF8A:FF8B	VREG LVI	LVIE	CTRL0	—	$8A
57	$FF8C:FF8D	PWM Emergency Shutdown	PWMIE	PWMSDN	14	$8C
56	$FF8E:FF8F	Port P	PIEP[7:0]	PIEP	—	$8E
33	$FF90:FFAF	Reserved	—	—	—	—
39	$FFB0:FFB1	CAN Transmit	TXEIE[2:0]	CANTIER	16	$B0
38	$FFB2:FFB3	CAN Receive	RXFIE	CANRIER	16	$B2
37	$FFB4:FFB5	CAN Errors	CSCIE, OVRIE	CANRIER	16	$B4
36	$FFB6:FFB7	CAN Wake-up	WUPIE	CANRIER	16	$B6
35	$FFB8:FFB9	Flash	CCIE, CBEIE	FCNFG	—	$B8
—	$FFBA:FFC3	Reserved	—	—	—	—
29	$FFC4:FFC5	CRG Self Clock Mode	SCMIE	CRGINT	20	$C4
28	$FFC6:FFC7	CRG PLL Lock	LOCKIE	CRGINT	20	$C6
—	$FFC8:FFCD	Reserved	—	—	—	—
24	$FFCE:FFCF	Port J	PIEJ[7:6]	PIEJ	11	$CE
—	$FFD0:FFD1	Reserved	—	—	—	—
22	$FFD2:FFD3	A/D Converter	ASCIE	ATDCTL2	17	$D2
—	$FFD4:FFD5	Reserved	—	—	—	—

TABLE 12-1 Continued

Priority[a]	Vector Address	Interrupt Source	Local Enable Bit	See Register	See Chapter	HPRIO Value to Promote
20	$FFD6:FFD7	SCI Serial System	TIE, TCIE, RIE, ILIE	SCICR2	15	$D6
19	$FFD8:FFD9	SPI Serial Peripheral System	SPIE, SPTIE	SPICR1	15	$D8
18	$FFDA:FFDB	Pulse Accumulator Input Edge	PAI	PACTL	14	$DA
17	$FFDC:FFDD	Pulse Accumulator Overflow	PAOVI	PACTL	14	$DC
16	$FFDE:FFDF	Timer Overflow	TOI	TSCR2	14	$DE
15	$FFE0:FFE1	Timer Channel 7	C7I	TIE	14	$E0
14	$FFE2:FFE3	Timer Channel 6	C6I	TIE	14	$E2
13	$FFE4:FFE5	Timer Channel 5	C5I	TIE	14	$E4
12	$FFE6:FFE7	Timer Channel 4	C4I	TIE	14	$E6
11	$FFE8:FFE9	Timer Channel 3	C3I	TIE	14	$E8
10	$FFEA:FFEB	Timer Channel 2	C2I	TIE	14	$EA
9	$FFEC:FFED	Timer Channel 1	C1I	TIE	14	$EC
8	$FFEE:FFEF	Timer Channel 0	C0I	TIE	14	$EE
7	$FFF0:FFF1	Real-Time Interrupt	RTIE	CRGINT	14	$F0
6	$FFF2:FFF3	IRQ_L Pin	IRQEN	INTCR	12	$F2
5	$FFF4:FFF5	XIRQ_L Pin	X bit	CCR	12	—
4	$FFF6:FFF7	SWI	None	—	—	—
3	$FFF8:FFF9	Unimplemented Instruction Trap	None	—	—	—
2	$FFFA:FFFB	COP Failure Reset	CR2:CR1:CR0	COPCTL	—	—
1	$FFFC:FFFD	Clock Monitor Fail Reset	CME, SCME	PLLCTL	—	—
0	$FFFE:FFFF	External Reset	None	—	—	—

[a] The numbers given for the priority show zero as the highest priority. This numbering scheme is also used by the CodeWarrior linker to locate the vector in the proper place in memory. See Example 12-3.

Initializing the Interrupt Vectors

In an embedded system the interrupt vector locations shown in Table 12-1 must be initialized to the start of the interrupt service routine. This is easy to do in assembly language, and Example 12-1 shows how to do this by using a relocatable assembler such as CodeWarrior. Example 12-2 shows an equivalent C routine.

Example 12-1 An Interrupt Program with an Interrupt Service Routine

```
Metrowerks HC12-Assembler
(c) COPYRIGHT METROWERKS 1987-2003

Rel. Loc Obj. code Source line
---- --- --------- -----------
   1            ;**********************************************
   2            ; Sample program to initialize interrupt vectors.
   3            ; Define entry point etc.
   4                    XDEF   Entry,main
```

```
  5                          XREF    __SEG_END_SSTACK
  6                          XDEF    IRQISR    ; ISR Address
  7                ; IRQ_L system equates
  8     0000 001E INTCR: EQU   $001e      ; IRQ_L control reg
  9     0000 0080 IRQE:  EQU   %10000000 ; IRQ_L edge bit
 10     0000 0040 IRQEN: EQU   %01000000 ; IRQ_L enable
 11                MyCode: SECTION
 12                main:
 13                Entry:
 14 000000 CFxx xx       lds   #__SEG_END_SSTACK
 15                ; Initialize interrupt system
 16                ; Set low level interrupt on IRQ
 17 000003 4D1E 80       bclr  INTCR,IRQE
 18                ; Enable the IRQ_L interrupt
 19 000006 4C1E 40       bset  INTCR,IRQEN
 20                ; Unmask interrupts
 21 000009 10EF          cli
 22                ; The main process is a loop that runs forever.
 23                ; DO
 24                loop:
 25                ; blah, blah, blah whatever
 26                ; WHILE (forever)
 27 00000B 20FE          bra   loop;
 28                ;*********************************************
 29                ; Here is the interrupt service routine
 30                ;*********************************************
 31                IRQISR:
 32                ; blah, blah, blah whatever
 33 00000D 0B            rti   ; Special return (not rts)
 34                ;*********************************************
```

EXPLANATION OF EXAMPLE 12-1

In *line 6*, the address of the IRQ_L interrupt service routine is made visible to the linker by the XDEF directive. The program initializes the system, enables the interrupts, and unmasks the I bit in *lines 15–21*. The main, or foreground job, is simply a loop, hopefully doing more than this simple example. The interrupt service routine starts with the label IRQISR at *line 31* and ends with the RTI—Return from Interrupt—instruction at *line 33*. To place the interrupt service routines address in the correct vector location, the linker parameter file shown in Example 12-3 is used.

Example 12-2 Initializing Interrupts with C

```c
/******************************************************************
 * IRQ_L Interrupt Example
 * Turns on the LED1 on the CSM-12C32 CPU module
 * and enables IRQ_L for falling edge interrupts.
 * Each time the interrupt occurs, the display LED
 * is toggled.
 ******************************************************************/
#include <mc9s12c32.h>                    /* derivative information
*/
/*****************************************************************/
void main(void) {
/*****************************************************************/
  /* Initialize I/O */
  /* Make Port A bit-0 output */
    DDRA_BIT0 = 1;
  /* Turn LED1 on (active low) */
    PORTA_BIT0 = 0;
  /* Initialize INTCR:IRQE for falling edge interrupts */
    INTCR_IRQE = 1;
  /* Enable the IRQ_L Interrupt */
    INTCR_IRQEN = 1;
  /* Enable Interrupts */
    EnableInterrupts;
  /* DO Nothing */
    for(;;) {
    }
  /* FOREVER */
}
/******************************************************************
 * Interrupt handler for the IRQ_L interrupt
 ******************************************************************/
void interrupt IRQISR( void ) {
/*****************************************************************/
  /* Complement the display bits */
    if (PORTA_BIT0 == 1)
      PORTA_BIT0 = 0;
    else
      PORTA_BIT0 = 1;
  /* There may need to be some debouncing done here! */
} /* End of IRQ_L interrupt handler */
```

Example 12-3 Linker Parameter File for Linking the Vectors

```
1.  // The linker parameter file will locate the
2.  // code in ROM and the interrupt service routine vector
3.  // at $FFF2
4.  SEGMENTS
5.      RAM = READ_WRITE    0x0800 TO 0x0BFF;
6.      FLASH = READ_ONLY   0xC000 TO 0xFEFF;
7.  //  EEPROM = READ_WRITE 0x0D00 TO 0x0FFF;
8.  END
9.
10. PLACEMENT
11.     DEFAULT_ROM, NON_BANKED, ROM_VAR, STRINGS INTO  FLASH;
12.     DEFAULT_RAM                                INTO  RAM;
13. END
14.
15. STACKSIZE 0x100
16.
17. VECTOR 0 Entry  /* Reset vector. */
18. INIT  Entry     /* Include for assembly applications*/
19. VECTOR ADDRESS 0xFFF2 IRQISR  /* IRQISR Vector location */
```

EXPLANATION OF EXAMPLE 12-2

The *mc9s12c32.h* header file defines all registers and bits in registers. The header defines memory addresses for the control registers, such as INTCR, and assignment to these are unsigned char types. The header file defines bits within registers as, for example, INTCR_IRQE. Assignments to these are bit assignments use BSET or BCLR instructions. Your C program can access registers as bytes or individual bits.

The declaration of the interrupt handler *void interrupt IRQISR (void)* allows the compiler to use a return from interrupt—RTI—instruction instead of the return from subroutine—RTS—instruction. It also passes the label IRQISR to the linker to allow the interrupt vector to be set. See Example 12-3.

EXPLANATION OF EXAMPLE 12-3

The linker parameter file controls where all parts of the program are located in the microcontroller's memory. We discussed it fully in Chapter 6. The important part here are the *lines 17–19*. In *line 17*, the reset vector for the power-on reset is initialized. The linker locates the address given by the label *Entry*. The syntax

```
VECTOR 0 Entry
```

uses a numbering scheme for the vectors starting with zero for the highest priority interrupt vector as shown in Table 12-1. *Line 18* is included to be able to run assembly applications in the CodeWarrior system. *Line 19*,

```
VECTOR ADDRESS 0xFFF2 IRQISR
```

illustrates another way to specify the vector location specifically. An alternative to *line 19* would be

```
VECTOR 6 IRQISR
```

INITIALIZING THE INTERRUPT VECTOR IN C

A C interrupt program is shown in Example 12-2 with the linker parameter file shown in Example 12-3 initializing the interrupt vector locations. You may also include interrupt vector number in the function definition as follows:

```
/********************************************************/
void interrupt 6 IRQISR( void ) {
/********************************************************/
```

12.5 Interrupt Priorities

Hardware Prioritization

Hardware must be used to resolve simultaneous interrupts in a vectored system. The priorities in the HCS12 are fixed in hardware as shown in Table 12-1, where the higher vector addresses have higher priority. The programmer can dynamically change these fixed priorities. Any single interrupting source can be elevated to the highest priority position. The rest of the order remains fixed as given in Table 12-1. The *HPRIO* (*Highest Priority Interrupt Register*) contains bits *PSEL7–PSEL1* to select which device has the highest priority as shown in Table 12-1. To promote any of the interrupts to the highest priority, write the low-byte of its vector to HPRIO. Do this only when interrupts are masked with the I bit set. See Example 12-4. When the CPU is reset, HPRIO = $F2 giving IRQ_L the highest priority. Note that this prioritization is working at the interrupt request level and does not imply that if you are in an interrupt service routine a higher priority interrupt can take control. This must be done with software prioritization as discussed in the next section.

HPRIO—Base³ + $001F—Highest Priority Interrupt Register

	Bit 7	6	5	4	3	2	1	0
Read:	PSEL7	PSEL6	PSEL5	PSEL4	PSEL3	PSEL2	PSEL1	0
Write:								
Reset:	1	1	1	1	0	0	1	0

    = reserved, unimplemented, or cannot be written to.

    Read: Anytime. Write: Only if I bit in CCR is set.

PSEL[7:1]

Write the value shown in Table 12-1 to promote a particular interrupt to the highest priority.

---

³ The addresses of all I/O registers are shown in this way—Base + $XXXX—to denote that the actual register address is determined by offset ($XXXX) plus the register *base* address that may be programmed by the user after startup. The default base is $0000 and may be changed by writing to the INITRG register.

Example 12-4 Raising an Interrupt to the Highest Priority
Write a small segment of code to raise Timer Channel 2 to the highest priority position.

Solution:

```
Metrowerks HC12-Assembler
(c) COPYRIGHT METROWERKS 1987-2003

Rel. Loc  Obj. code Source line
---- ---  --------- -----------
 1         0000 001F HPRIO:  EQU    $1F    ; Address of HPRIO reg
 2         0000 FFEA TC2VECT: EQU   $FFEA  ; Channel 2 vector
 3                        ;
 4                        ; Masks interrupts while setting HPRIO
 5 000000 1410                sei         ; Set I bit
 6                        ; Raise timer channel 2 to the
 7                        ; highest priority
 8 000002 CCFF EA            ldd    #TC2VECT
 9 000005 5B1F              stab   HPRIO
10 000007 10EF              cli         ; Clear interrupt mask
```

Solution in C:
```c
void main(void) {
  /* Mask interrupt while setting HPRIO */
  DisableInterrupts;
  /* Raise timer channel 2 to the highest priority */
  HPRIO = Vtimch2;
  /* Reenable interrupts */
  EnableInterrupts;
}
```

Software Prioritization

Software prioritization can allow higher priority interrupts to interrupt a lower priority one.

Although the resolution of the *simultaneous* interrupts shown in Figure 12-2(b) requires hardware, and although the HPRIO register allows us some control over what is currently the most important or highest priority interrupt source, the system allows us total control over the *sequential* or *nested* interrupts (interrupts interrupting interrupts) shown in Figure 12-2(b). This is done in the following way:

- When the first interrupt service routine is entered, the I bit is set, masking any further interrupts.
- If an interrupt does occur while executing the interrupt service routine, it will remain pending until the current ISR is finished and control returns to the interrupted program.

- If higher priority interrupts must be allowed, the programmer must do the following:

  1. Disable the interrupt enable bits in all lower priority interrupting devices leaving higher priority interrupts enabled. (Note that you may or may not leave the current interrupt enabled. If you do, you must allow for it to interrupt itself.)
  2. Clear any interrupt flag associated with the current interrupt.
  3. Unmask interrupts by clearing the I bit.
  4. Proceed with the interrupt service routine for the current interrupt.
  5. When the current interrupt service routine is completed, mask interrupts by setting the I bit and reenable those interrupts that were disabled.
  6. Execute the return from interrupt instruction that unmasks interrupts again and allows any pending interrupts to be serviced.

12.6 Nonmaskable Interrupts

In any system there are events that are so important that they should never be masked. These are sometimes called *exceptions* and a good example is the RESET_L signal. When this is asserted, everything stops and the processor is reset. These very important events are called *nonmaskable interrupts*.

Table 12-2 shows six nonmaskable interrupt sources. These can always interrupt the CPU and thus have higher priority than any of the maskable interrupts.

System Reset

An *internally generated CPU reset* asserts the external RESET_L signal, which can be used to reset external hardware.

The system reset vector is at $FFFE:FFFF.

This is the hardware *power-on reset* (POR) normally done when powering up the microcontroller. It can be accomplished also by asserting the RESET_L signal and has the highest priority of all. The HCS12 has internal hardware that detects a positive transition in the V_{DD} supply and initializes the device during cold starts by asserting the reset signal internally. An external RESET_L signal may be applied also. The microcontroller distinguishes between internal and external resets by sensing how quickly the signal on RESET_L rises to a high logic level after it has been asserted. When any of the internally generated resets are triggered (power-on, clock monitor, or COP), the microcontroller drives the RESET_L low for a few clock cycles. After a few more clock cycles it checks the state of the RESET_L pin.

TABLE 12-2 Nonmaskable Interrupt Priorities

Priority	Nonmaskable Interrupt Source	Vector Address	Enable Bit	See Register
5	XIRQ_L	$FFF4:FFF5	X	CCR
4	Software Interrupt Instruction (SWI)	$FFF6:FFF7	None	None
3	Unimplemented Opcode Trap	$FFF8:FFF9	None	None
2	COP Reset	$FFFA:FFFB	CR2, CR1, CR0	COPCTL
1	Clock Monitor Reset	$FFFC:FFFD	CME, SCME	PLLCTL
0	System Reset (RESET_L)	$FFFE:FFFF	None	None

If it is still low, it assumes that an external signal has supplied the reset signal. To properly apply an external reset signal, ensure your circuitry asserts RESET_L for more than 32 E-clock cycles.

When an internally generated reset is asserted, other external hardware can be reset when the microcontroller drives the external RESET_L low.

Clock Monitor Reset

If the CPU's clock signals slow down or fail, and the *CME* (*Clock Monitor Enable*) bit in the *PLLCTL* (*CRG PLL Control*) *Register* is set and the *SCME* (*Self Clock Mode Enable*) bit is reset, the clock monitor will detect the problem and issue a CME RESET signal. A vector at $FFFC:FFFD is available in case something special should be done if this occurs. Note that the HCS12 cannot complete the reset sequence, including the low-drive on RESET_L and fetching the vector until clocks resume. See Chapter 21 for more information on the clock generator circuitry and features.

PLLCTL—Base + $003A—CRG PLL Control Register

	Bit 7	6	5	4	3	2	1	0
Read:	CME	PLLON	AUTO	ACQ	0	PRE	PCE	SCME
Write:								
Reset:	1	1	1	1	0	0	0	1

☐ = reserved, unimplemented, or cannot be written to.

CME

Clock Monitor Enable Bit

Read: Anytime. Write: Anytime except when SCME = 1.

0 = Clock monitor is disabled.

1 = Clock monitor is enabled. Slow or stopped clocks will cause a clock monitor reset sequence of Self Clock Mode (default).

PLLON, AUTO, ACQ, PRE, PCE

PLLON—Phase Lock Loop On Bit. See Chapter 21.

AUTO—Automatic Bandwidth Control Bit. See Chapter 21.

ACQ—Acquisition Bit. See Chapter 21.

PRE—RTI Enable during Pseudo-Stop Bit. See Chapters 14 and 21.

PCE—COP Enable during Pseudo-Stop Bit. See Chapter 21.

SCME

Self Clock Mode Enable Bit

Read: Anytime. Write: Once.

0 = Detection of crystal clock failure causes clock monitor reset.

1 = Detection of crystal clock failure forces the MCU into Self Clock Mode (default). See Chapter 21.

Computer Operating Properly (COP) Reset

The *Computer Operating Properly* function is a *watchdog timer*. It can reset the microcontroller if the program gets lost.

A COP, or watchdog, system is a vital part of computers used in embedded applications. The system must have some way to recover from unexpected errors that may occur. Power surges or programming errors may cause the program "to get lost" and to lose control of the system. This could be disastrous and so the watchdog timer is included to help the program recover. When in operation, the program is responsible for pulsing the COP at specific intervals. This is accomplished by choosing a place in the program to pulse the watchdog timer regularly. Then, if the program fails to do this, the COP automatically provides a hardware RESET_L to begin the processing again.[4]

The COP failure interrupt vector is at $FFFA:FFFB.

For the HCS12 operating in normal modes, the COP is disabled when the CPU is reset.[5] The COP time-out period is controlled by the *CR2:CR1:CR0* bits in the *COPCTL* register, and the COP is enabled by writing a nonzero value to these bits. CR2:CR1:CR0 may be programmed to give a COP time-out period ranging from 1.024 ms to 1.049 seconds when the OSCCLK is 16 MHz. After the COP timer has started, the program must write first $55 followed by $AA to the *ARMCOP (Arm COP) Register* before the COP times out. During each COP time-out period, this sequence ($55 followed by $AA) must be written. Other instructions can be executed between the $55 and the $AA but both must be completed in the time-out period to avoid a COP reset. When a COP time-out occurs, or if the program writes anything other than $55 or $AA to the ARMCOP register, a COP RESET is generated and the program restarts at the program location given by the COP failure vector. See Examples 12-5 to 12-7.

A *windowed* COP operation is available. When the *WCOP* bit in the COPCTL register is set, writes to the ARMCOP register to clear the COP timer must occur in the last 25% of the selected time-out period. A premature write will immediately reset the microcontroller.

COPCTL—Base + $003C—COP Control Register

	Bit 7	6	5	4	3	2	1	0
Read:	WCOP	RSBCK	0	0	0	CR2	CR1	CR0
Write:								
Reset:	0	0	0	0	0	0	0	0

       = reserved, unimplemented, or cannot be written to.

---

[4] A particularly good article on watchdog timers can be found at http://www.ganssle.com/watchdogs.htm.

[5] The older HC12 versions enabled the COP on reset and the programmer could turn it off by writing CR2:CR1:CR0 = %000.

WCOP

Windowed COP Mode Bit

Read: Anytime. Write: Once in user mode, anytime in special mode.

0 = Normal COP operation (default).
1 = Windowed COP operation.

When set, a write to the ARMCOP register must occur in the last 25% of the selected period. A write during the first 75% will reset the microcontroller. As long as all writes occur during this window, $55 can be written as often as desired. Once $AA is written after the $55, the time-out logic restarts and the user must wait until the next window before writing to ARMCOP again.

RSBCK

COP and RTI Stop in Active BDM Mode Bit

Read: Anytime. Write: Once.

0 = Allows the COP and real-time interrupt to keep running in active background debugger mode (default).
1 = Stops the COP and RTI counter whenever the microcontroller is in active BDM mode.

See Chapters 14 and 20.

CR2, CR1, CR0

COP Watchdog Timer Rate Select

Read: Anytime. Write: Once in user mode, anytime in special mode.
See Table 12-3.

TABLE 12-3 COP Watchdog Rates

CR2	CR1	CR0	OSSCLK Cycles to Time-out	Time to Time-out with 16-MHz Clock
0	0	0	COP disabled	
0	0	1	2^{14}	1.024 ms
0	1	0	2^{16}	4.096 ms
0	1	1	2^{18}	16.384 ms
1	0	0	2^{20}	65.536 ms
1	0	1	2^{22}	262.144 ms
1	1	0	2^{23}	524.288 ms
1	1	1	2^{24}	1.049 s

Note: The very first COP time-out is only half of the nominal COP time-out period. This is because the counter starts at 0000 and the time-out becomes active when the MSB goes high. The period is from a rising edge to the next rising edge.

ARMCOP—Base + $003F—Arm COP Register

	Bit 7	6	5	4	3	2	1	0
Read:	0	0	0	0	0	0	0	0
Write:	Bit 7	Bit 6	Bit 5	Bit 4	Bit 3	Bit 2	Bit 1	Bit 0
Reset:	0	0	0	0	0	0	0	0

Read: Always reads $00.

Write: Anytime.

When the COP is disabled (CR[2:0] = "000") writing to this register has no effect.

When the COP is enabled by setting CR[2:0] to nonzero, the following applies:

Writing any value other than $55 or $AA causes a COP reset. To restart the COP time-out period you must write $55 followed by a write of $AA. Other instructions may be executed between these writes but the sequence $55 $AA must be completed prior to the COP end of time-out period to avoid a COP reset. Sequences of $55 or $AA writes are allowed. When the WCOP bit is set, $55 and $AA must be done in the last 25% of the selected time-out period; writing any value in the first 75% of the selected period will cause a COP reset.

Example 12-5 COP Initialization

Show a segment of code that initializes the COP time-out period to 1.049 seconds and then, in the main processing loop, resets the COP timer.

Solution:

```
Metrowerks HC12-Assembler
(c) COPYRIGHT METROWERKS 1987-2003
Rel. Loc  Obj. code Source line
----  ---  --------- -----------
    1                 ; Example program to initialize and start
    2                 ; the COP and then to reset it in the main
    3                 ; control loop.
    4                       XDEF COP_Restart
    5      0000 0000 BASE:   EQU   0      ; Registers base address
    6      0000 003C COPCTL: EQU   BASE+$003C  ; COP Control reg
    7      0000 003F ARMCOP: EQU   BASE+$003F  ; COP Arm register
    8      0000 0007 SEC_1   EQU   %00000111   ; 1.049 s time-out
    9
   10                 Entry:
   11                 MyCode: SECTION
   12                 ; All I/O initialization
   13                 ;
```

```
14                      ; Set up the COP
15 000000 8607                    ldaa    #SEC_1
16 000002 5A3C                    staa    COPCTL
17                      ;
18
19                      loop:
20                      ; Main processing loop
21                      ; Blah, blah, blah
22                      ; Now reset the COP before it times out
23 000004 180B 5500             movb    #$55,ARMCOP
   000008 3F
24 000009 180B AA00             movb    #$AA,ARMCOP
   00000D 3F
25                      ; Continue with the main processing loop
26 00000E 20F4                    bra    loop
27
28                      ; COP restart code
29                      ; This code is executed when the COP reset
30                      ; occurs.
31                      COP_Restart:
32 000010 06xx xx             jmp    Entry
```

Example 12-6 COP Initialization in C

```
/****************************************************************
 * COP timer initialization and reset
 ****************************************************************/
#include <mc9s12c32.h>     /* derivative information */
/****************************************************************/
void main(void) {
/****************************************************************/
  /* Set up the COP for 1.049 s time-out */
    COPCTL = 0x07;
  /* DO */
  for ( ; ; ) {
```

```
        /* Do whatever main processing is to be done */

      /* Before the COP times out, reset it */
        ARMCOP = 0x55;
        ARMCOP = 0xAA;
    }
  }
```

Example 12-7 Linker Parameter Code to Set Up COP Vector

Show the addition needed in the linker parameter file to set the COP time-out vector to restart the C program.

Solution:

A complete restart with the C startup code is necessary. The vector is located at 0xFFFA and is priority 2. The code to be added to the linker parameter file is

```
        VECTOR 2 _Startup   /* COP time-out reset vector */
```

Unimplemented Instruction Opcode Trap

If the program somehow gets lost and starts executing data, it is likely to encounter an unimplemented opcode. Executing data is a disaster and executing an illegal opcode even more of a disaster. The CPU can detect an unimplemented opcode and will vector itself to the address in $FFF8:FFF9.

Software Interrupt (SWI)

The SWI is, in effect, a 1-byte, indirect branch to a subroutine whose address is at the vector location $FFF6:FFF7. It operates like the rest of the interrupt system in that all registers are pushed onto the stack, making it ideal for debugging.

Nonmaskable Interrupt Request XIRQ_L

XIRQ_L is an external, nonmaskable interrupt input. The system designers have included the X bit in the condition code register to mask interrupts on this pin until the program has initialized the stack pointer and other critical program elements (and the vector jump table, if required). The X bit is similar to the I bit in that it masks when set and unmasks when cleared. The ANDCC #%10111111 instruction is used to reset the X bit and to unmask the interrupt.

Once the X bit is reset, the program *absolutely cannot* mask it again. However, when an XIRQ_L occurs, the X bit is set, just like the I bit, so that nested XIRQ_Ls cannot occur. On leaving the interrupt service routine, the bit is reset and further XIRQ_L interrupts can then occur. Figure 12-5 shows how XIRQ_L is to be interfaced to the HCS12. The XIRQ_L vector is at $FFF4:FFF5.

12.7 External Interrupt Sources

Chapter 11 did not cover the interrupts of the parallel I/O system. Now that we know more about the interrupt system, let us look at the details we postponed.

IRQ_L

The IRQ_L vector is at $FFF2:FFF3.

IRQ_L is an interrupt request that is generated by some external device. For example, you may have a sensor that generates an interrupt every time a vehicle passes a traffic counter. IRQ_L is an *active-low* signal and you may choose a level-activate (the default) or a negative-edge-sensitive response. The *INT Control Register* (*INTCR*) contains a bit to select either of these responses and to enable/disable the IRQ_L signal.

INTCR—Base + $001E—Interrupt Control Register

	Bit 7	6	5	4	3	2	1	0
Read:	IRQE	IRQEN	0	0	0	0	0	0
Write:								
Reset:	0	1	0	0	0	0	0	0

▨ = reserved, unimplemented, or cannot be written to.

IRQE

IRQ_L Select Edge Sensitivity

Special modes: Read or write anytime.

Normal and Emulation modes: Read anytime, write once.

0 = IRQ_L configured for low level recognition (default).

1 = IRQ_L configured for falling edges. Falling edges will be detected anytime IRQE = 1 and will be cleared only upon a reset or the servicing of the IRQ_L interrupt.

IRQEN

External IRQ_L Enable

Read or write anytime.

0 = External IRQ_L is disabled (default).

1 = External IRQ_L is enabled.

Port P

Port P, whose I/O capabilities were covered in Chapter 11, can also generate interrupts. Each of the 8 bits can be an interrupt source and there are four registers that allow programmer control over the interrupts. See Example 12-8.

PERP—Base + $025C—Port P Pull Device Enable Register

	Bit 7	6	5	4	3	2	1	0
Read:	PERP7	PERP6	PERP5	PERP4	PERP3	PERP2	PERP1	PERP0
Write:								
Reset:	0	0	0	0	0	0	0	0

Read: Anytime. Write: Anytime.

PERP[7:0]

Pull Device Enable Bits

0 = Pull-up or pull-down device is disabled (default).
1 = Either a pull-up or pull-down device is enabled depending on the associated bit in PPSP.

PPSP—Base + $025D—Port P Polarity Select Register

	Bit 7	6	5	4	3	2	1	0
Read:	PPSP7	PPSP6	PPSP5	PPSP4	PPSP3	PPSP2	PPSP1	PPSP0
Write:								
Reset:	0	0	0	0	0	0	0	0

Read: Anytime. Write: Anytime.
This register provides a dual purpose by selecting the polarity of the active interrupt edge as well as selecting a pull-up or pull-down device if enabled.

PPSP[7:0]

Port P Polarity Select Bits

0 = Falling edge on the associated port P pin sets the associated flag bit in the PIFP register. A pull-up device is connected to the associated pin if enabled by the associated bit in register PERP and if the pin is used as input (default).
1 = Rising edge on the associated port P pin sets the associated flag bit in the PIFP register. A pull-down device is connected to the associated pin if enabled by the associated bit in register PERP and if the pin is used as input.

PIEP—Base + $025E—Port P Interrupt Enable Register

	Bit 7	6	5	4	3	2	1	0
Read:	PIEP7	PIEP6	PIEP5	PIEP4	PIEP3	PIEP2	PIEP1	PIEP0
Write:								
Reset:	0	0	0	0	0	0	0	0

Read: Anytime. Write: Anytime.

PIEP[7:0]

Port P Interrupt Enable Bits

0 = Interrupt on the associated bit in port P is disabled (default).

1 = Interrupt on the associated bit in port P is enabled. When an edge as defined by the associated bit in PPSP occurs, the associated interrupt flag in PIFP is unmasked.

PIFP—Base + $025F—Port P Interrupt Flag Register

	Bit 7	6	5	4	3	2	1	0
Read:	PIFP7	PIFP6	PIFP5	PIFP4	PIFP3	PIFP2	PIFP1	PIFP0
Write:								
Reset:	0	0	0	0	0	0	0	0

Read: Anytime. Write: Anytime. Writing a 1 clears the flag; writing a 0 has no effect.

PIFP[7:0]

Port P Interrupt Flags

0 = No pending interrupt (default).

1 = Active edge on the associated bit has occurred and an interrupt will occur if the associated enable bit is set in the PIEP register and the I bit in the CCR is cleared.

See Also:

Topic	Register	Chapter/Section
Port P Input Register	PTIP (Base + $0259)	11
Data Direction Register	DDRP (Base + $025A)	11
Reduced Drive Control	RDRP (Base + $025B)	11
Pull-up or Pull-down Enable	PERP (Base + $025C)	11
Polarity Select Register	PPSP (Base + $025D)	11
Port P Interrupt Vector	$FF8E:FF8F	12.4
Pulse-Width Modulation	—	14

Example 12-8 Port P I/O and Interrupts

```
Metrowerks HC12-Assembler
(c) COPYRIGHT METROWERKS 1987-2003


Rel. Loc    Obj. code Source line
---- ---     --------- -----------
   1                     ; Example showing how to initialize
   2                     ; Port P for I/O and interrupts.
   3                     ; Use bits 7-4 for interrupt inputs and
   4                     ; bits 3-0 for output.
   5                             INCLUDE base.inc
   6                             INCLUDE bits.inc
   7                             INCLUDE portp.inc ; Port P defn's
   8         0000 000F OUT_BITS:  EQU  %00001111  ; Bits 3-0
   9         0000 00F0 IN_BITS:   EQU  %11110000  ; Bits 7-4
  10                     ; Bits 5 and 4 used for falling edge
  11                     ;    interrupts
  12         0000 0030 FALLING:   EQU  BIT5|BIT4
  13                     ; Bits 7 and 6 used for rising edge
  14                     ;    interrupts
  15         0000 00C0 RISING:    EQU  BIT7|BIT6
  16
  17                     ; export symbols
  18                             XDEF   Entry, main
  19                             XREF   __SEG_END_SSTACK
  20                             XDEF   Port_P_ISR
  21                     ; code section
  22                     MyCode:   SECTION
  23                     main:
  24                     Entry:
  25 000000 CFxx xx            LDS   #__SEG_END_SSTACK
  26                     ; Initialize data direction register
  27 000003 1C02 5A0F          bset  DDRP,OUT_BITS ; Set outputs
  28 000007 1D02 5AF0          bclr DDRP,IN_BITS    ; Set inputs
  29                     ; Enable pull devices on inputs
  30 00000B 1C02 5CF0          bset  PERP,IN_BITS
  31                     ; Select the falling edge bits
  32 00000F 1D02 5D30          bclr  PPSP,FALLING
  33                     ; Select the rising edge bits
```

```
34 000013 1C02 5DC0            bset   PPSP,RISING
35                        ; Clear any pending flags
36 000017 86F0                 ldaa   #IN_BITS
37 000019 7A02 5F              staa   PIFP
38                        ; Now it is safe to enable the interrupts
39 00001C 1C02 5EF0            bset   PIEP,IN_BITS
40                        ; Unmask I bit
41 000020 10EF                 cli
42                        ; Do the foreground job
43                        loop:
44 000022 3E                   wai            ; Wait for interrupt
45 000023 20FD                 bra    loop
46                        ;*****************************************
47                        ; Interrupt Service Routine
48                        Port_P_ISR:
49                        ; Find out which bit set the flag
50 000025 B602 5F              ldaa   PIFP
51                        ; IF bit 4
52 000028 8510                 bita   #BIT4
53 00002A 2707                 beq    check_bit_5
54                        ; THEN DO the bit 4 ISR
55                        ;   . . .
56                        ; Reset the bit 4 interrupt flag
57 00002C 8610                 ldaa   #BIT4
58 00002E 7A02 5F              staa   PIFP
59 000031 2021                 bra    done
60                        ; ELSE IF bit 4
61                        check_bit_5:
62 000033 8520                 bita   #BIT5
63 000035 2707                 beq    check_bit_6
64                        ; THEN DO the bit 5 ISR
65                        ;   . . .
66                        ; Reset the bit 5 interrupt flag
67 000037 8620                 ldaa   #BIT5
68 000039 7A02 5F              staa   PIFP
69 00003C 2016                 bra    done
70                        ; ELSE IF bit 5
71                        check_bit_6:
72 00003E 8540                 bita   #BIT6
73 000040 2707                 beq    check_bit_7
74                        ; THEN DO the bit 6 ISR
```

```
75                          ; . . .
76                          ; Reset the bit 6 interrupt flag
77 000042 8640                     ldaa    #BIT6
78 000044 7A02 5F                  staa    PIFP
79 000047 200B                     bra     done
80                          ; ELSE IF bit 6
81                          check_bit_7:
82 000049 8580                     bita    #BIT7
83 00004B 2707                     beq     done
84                          ; THEN DO the bit 7 ISR
85                          ; . . .
86                          ; Reset the bit 7 interrupt flag
87 00004D 8680                     ldaa    #BIT7
88 00004F 7A02 5F                  staa    PIFP
89 000052 2000                     bra     done
90                          ;
91                          done:
92                          ; You may need some debouncing here
93                          ; Return to interrupted program
94 000054 0B                       rti
```

EXPLANATION OF EXAMPLE 12-8

This example illustrates how to choose port P bits for input and output operation (*lines 27* and *28*) and to enable pull-up or pull-down devices (*line 30*). Falling edge and rising edge interrupts are chosen in *lines 32* and *34*. The interrupts on port P are enabled in *line 39*, interrupts are unmasked in *line 41*, and a wait for interrupt instruction is entered in line 44. The interrupt service routine starts on *line 50*. Because in this example we can have up to four different interrupting sources but with a single vector for all four, the interrupt service routine must poll the flags in the PIFP register to determine which has generated the interrupt. The code in *lines 50–83* does this and this code checks bit 4 through bit 7 in order. It services the first interrupt whose flag is set and then returns. If more than one flag has been set, it will remain as a pending interrupt and be serviced next. Example 12-9 shows a C version of this program.

Example 12-9 Port P Interrupts with C

```
/******************************************************************
 * Initialize Port P for I/O and interrupts.
 * Use bits 7-4 for input with interrupts and
```

```c
 * bits 3-0 for general purpose output.
 * Bits 5 and 4 are falling edge interrupts and
 * bits 7 and 6 are rising edge interrupts.
 ****************************************************************/
#include <mc9s12c32.h>              /* derivative information */
#define IN_BITS   0xF0  /* Bits 7 - 4 */
#define OUT_BITS  0x0F  /* Bits 3 - 0 */
/****************************************************************/
void main(void) {
/****************************************************************/
  /* Initialize data direction register */
    DDRP = OUT_BITS;
  /* Enable pull devices on inputs */
    PERP = IN_BITS;
  /* Select the falling edge bits */
    PPSP_PPSP5 = 0;
    PPSP_PPSP4 = 0;
  /* Select the rising edge bits */
    PPSP_PPSP7 = 1;
    PPSP_PPSP6 = 1;
  /* Clear any pending flags */
    PIFP = IN_BITS;
  /* Now it is safe to enable the Port P interrupts */
    PIEP = IN_BITS;
  /* Now can unmask interrupts */
    EnableInterrupts;
  /* Do the foreground job */
    for(;;) {
      asm (wai);   /* Wait for interrupt */
      } /* wait forever */
}
/****************************************************************
 * Interrupt handler.
 * Polls the PIFP register to find the interrupt source.
 * Software prioritization is used by the test order
 * 4, 5, 6, and 7. The first one found is serviced and
```

```
    * any others remain as pending interrupts.
    ****************************************************/
void interrupt 56 Port_P_ISR( void ){
/******************************************************/
  /* Find out which bit set the flag */
    if (PIFP_PIFP4 == 1) {
        /* THEN DO the bit 4 ISR */
        /* Reset the bit 4 flag */
        PIFP = PIFP_PIFP4_MASK;
    }
     else {
      if (PIFP_PIFP5 == 1) {
        /* THEN DO the bit 5 ISR */
        /* Reset the bit 5 flag */
        PIFP = PIFP_PIFP5_MASK;
      } else {
        if (PIFP_PIFP6 == 1) {
          /* Then DO the bit 6 ISR */
          /* Reset the bit 6 flag */
          PIFP = PIFP_PIFP6_MASK;
        } else {
          if (PIFP_PIFP7 == 1) {
            /* Then DO the bit 7 ISR */
            /* Reset the bit 7 flag */
            PIFP = PIFP_PIFP5_MASK;
          }
        }
      }
    }
}
```

Port J

Port J's interrupting capabilities are similar to those found in port P, although only two bits are available.

PERJ—Base + $026C—Port J Pull Device Enable Register

	Bit 7	6	5	4	3	2	1	0
Read:	PERJ7	PERJ6	0	0	0	0	0	0
Write:								
Reset:	0	0	—	—	—	—	—	—

Read: Anytime. Write: Anytime.

PERJ[7:6]

Port J Pull Device Enable Bits

0 = Pull-up or pull-down device is disabled (default).
1 = Either a pull-up or pull-down device is enabled depending on the associated bit in PPSJ.

PPSJ—Base + $026D—Port J Polarity Select Register

	Bit 7	6	5	4	3	2	1	0
Read:	PPSJ7	PPSJ6	0	0	0	0	0	0
Write:								
Reset:	0	0	—	—	—	—	—	—

Read: Anytime. Write: Anytime.
This register provides a dual purpose by selecting the polarity of the active interrupt edge as well as selecting a pull-up or pull-down device if enabled.

PPSJ[7:6]

Port J Polarity Select Bits

0 = Falling edge on the associated port J pin sets the associated flag bit in the PIFJ register. A pull-up device is connected to the associated pin if enabled by the associated bit in register PERJ and if the pin is used as input (default).
1 = Rising edge on the associated port P pin sets the associated flag bit in the PIFJ register. A pull-down device is connected to the associated pin if enabled by the associated bit in register PERJ and if the pin is used as an input.

PIEJ—Base + $026E—Port J Interrupt Enable Register

	Bit 7	6	5	4	3	2	1	0
Read:	PIEJ7	PIEJ6	0	0	0	0	0	0
Write:								
Reset:	0	0	—	—	—	—	—	—

Read: Anytime. Write: Anytime.

PIEJ[7:6]

Port J Interrupt Enable Bits

0 = Interrupt on the associated bit in port J is disabled (default).

1 = Interrupt on the associated bit in port J is enabled. When an edge as defined by the associated bit in PPSJ occurs, the associated interrupt flag in PIFJ is unmasked.

PIFJ—Base + $026F—Port J Interrupt Flag Register

	Bit 7	6	5	4	3	2	1	0
Read:	PIFJ7	PIFJ6	0	0	0	0	0	0
Write:								
Reset:	0	0	—	—	—	—	—	—

Read: Anytime. Write: Anytime. Writing a 1 clears the flag; writing a 0 has no effect.

PIFJ[7:6]

Port J Interrupt Flags

0 = No pending interrupt (default).

1 = Active edge on the associated bit has occurred and an interrupt will occur if the associated enable bit is set in the PIEJ register and the I bit in the CCR is cleared.

See Also:

Topic	Register	Chapter/Section
Port J Input Register	PTIJ (Base + $0269)	11
Data Direction Register	DDRJ (Base + $026A)	11
Reduced Drive Control	RDRJ (Base + $026B)	11
Pull-up or Pull-down Enable	PERJ (Base + $026C)	11
Port J Polarity Select	PPSJ (Base + $026D)	11
Port J Interrupt Vector	$FFCE:FFCF	12.4

12.8 Interrupt Flags

Resetting Interrupt Flags

> Most interrupt flags are reset by writing a *one* to the bit. Writing a zero *has no effect.*

Up to eight interrupting sources can be connected to port P and two to port J. When an interrupt occurs on any of these bits, the associated bit in the flags register is set. This generates an interrupt if that bit is enabled (in the interrupt enable register) and the I bit is cleared to unmask interrupts. This is also true for all other interrupting devices such as the various timer interrupts. The flags must be reset, or cleared, in the interrupt service routine before interrupts are unmasked by the RTI instruction. Flags are cleared *by writing a one* to the flag that is set. *Writing a zero to a flag bit has no effect.* For example, resetting the bit-0 flag in port P can be done with the following code sequences:

```
ldaa  #%00000001
staa  PIFP
```

An alternative is

```
bclr  PIFP,#%11111110
```

The bit clear (BCLR) instruction has a mask byte with ones in the bit positions where zeros are to be written (bits cleared). Here is how this instruction works:

The data byte is read from PIFP, say, %00000011 (bit-1 and bit-0 flags both set).

The mask byte is complemented %11111110 → %0000001.

The complemented mask byte is ANDed with the data and written back to PIFP; that is, PIFP is written with %000000001, resetting bit-0 *and leaving the bit-1 flag set.*

The bit set instruction (BSET) *will not work.* The instruction

```
bset  PIFP,#%000000001
```

operates this way:

The data byte, say, %00000011 (bit-1 and bit-0 flags both set), is read from PIFP.

The mask byte is ORed with the data byte and written back to PIFP; that is, PIFP is written with %00000011! This resets *both* bit 1 and bit 0.

Resetting Interrupt Flags in C

As described previously, we must avoid the bset and bclr instructions when resetting the interrupt flags. Example 12-10 shows a code segment to use the store instruction.

The CodeWarrior system gives us a header file that defines all registers and bits within registers. We should use unsigned char assignments to the registers to reset the flags as shown in Example 12-11.

Example 12-10 Resetting Flags in C

```
1:  /****************************************************
2:   * Resetting flags in C
3:   ****************************************************/
4:  volatile unsigned char TFLG1 @ (0x0043); /* TFLG1 location */
5:  #define C0F 1   /* Timer flag channel 0 */
```

```
 6:   #define C1F 2   /* Timer flag channel 1 */
 7:   #define C2F 4   /* Timer flag channel 2 */
 8:                   /* etc */
 9:   /***********************************************/
10:   void main(void) {
11:   /***********************************************/
12:     /* . . . */
13:     /* Reset bit-0 in TFLG1 */
14:       TFLG1 = C0F;
0000 c601          LDAB   #1
0002 5b00          STAB   TFLG1
15:     /* Reset bit-0 and bit-2 in TFLG1 */
16:       TFLG1 = C0F|C2F;
0004 8605          LDAA   #5
0006 5a00          STAA   TFLG1
17:     /* . . . */
18:   }
0008 3d            RTS
```

Example 12-11 Resetting Flags with CodeWarrior C

```
 4:   /************************************************
 5:   * Resetting flags in C
 6:   ************************************************/
 7:   #include <mc9s12c32.h>        /* derivative information */
 8:   /***********************************************/
 9:   void main(void) {
10:   /***********************************************/
11:     /* . . . */
12:     /* Reset bit-0 in PIFP */
13:       PIFP = PIFP_PIFP0_MASK;
0000 c601          LDAB   #1
0002 7b0000        STAB   _PIFP
14:     /* Reset bit-7 and bit-6 in PIFP */
15:       PIFP = PIFP_PIFP7_MASK | PIFP_PIFP6_MASK;
0005 86c0          LDAA   #192
0007 7a0000        STAA   _PIFP
16:     /* . . . */
17:   }
000a 3d            RTS
```

Multiple Flags with a Single Interrupt Vector

There is a single interrupt vector each for port P and port J even though port P can provide up to eight separate interrupt sources and port J two. If you have interrupts from either of these two ports, you must determine which of the bits has caused the interrupt in the interrupt service routine. Examples 12-8 and 12-9 show how to do this.

Internal Interrupt Sources

Most internal interrupts are generated by a flag that must be reset in the ISR.

Table 12-1 shows several other internal interrupt sources, such as timer channels, that are generated within the HCS12. These all operate like external I/O interrupts. An enable bit must be set, and when an interrupt occurs, a flag is set in a flags register. The flag must be reset in the interrupt service routine before the return from interrupt instruction is executed. Flags that are cleared by writing a one to them are called direct clearing flags. Some of the other internal systems, for example, the serial communications interface (SCI) has flags that are cleared when reading the data that generated the interrupt. We will see examples of these other interrupts when we study the devices generating them.

12.9 Advanced Interrupts

Selecting Edge or Level Triggering

The external IRQ_L interrupt is normally a low-level sensitive input. This is suitable for use in a system with several devices whose interrupt request lines may be tied in a wired-OR configuration as shown in Figure 12-5. The reason for this is that if more than one device interrupts, it can keep its interrupt request asserted so that it will be a pending interrupt when another is finished. You may choose to have a negative-edge-sensitive interrupt request by programming the IRQE bit in the INTCR register. An edge-sensitive interrupt is only appropriate if there is only one interrupt source connected to IRQ_L. Another reason to use the edge triggering for the IRQ_L interrupt is that in this mode the interrupt request is latched. If the interrupt signal does not stay low for a long enough time, this mode should be used.

What to Do While Waiting for an Interrupt

There are three ways to make the HCS12 spin its wheels while waiting for an interrupt to occur. These are *spin loops*, and the WAI, *wait for interrupt*, and STOP, *stop clocks*, instructions.

Spin Loop: The simplest way to make the CPU wait is the spin loop. You make the processor branch to itself with the code shown in Example 12-12.

When an interrupt occurs, the CPU will finish executing the instruction, which is, of course, a branch to the same instruction. Before executing it again, the interrupt will be acknowledged and the interrupt service routine executed. The program will fall back into the spin loop when it returns. When using the spin loop, all processing is done in the interrupt service routine.

Spinning your wheels is not normally done except, perhaps, during the debugging phase of your software development.

Example 12-12 Using a Spin Loop to Wait for an Interrupt

```
Metrowerks HC12-Assembler
(c) COPYRIGHT METROWERKS 1987-2003


Rel. Loc     Obj. code Source line
---- ---     --------- -----------
  1  000000 20FE        spin:   bra     spin    ; Wait for interrupt
  2                     ; An equivalent instruction is
  3  000002 20FE                bra     *
  4                     ; The * stands for the current PC location
```

The equivalent code in C is

```
for(;;) {} /* wait forever */
```

WAI—Wait for Interrupt: The WAI instruction performs two functions. First, it pushes all the registers onto the stack in preparation for a subsequent interrupt. This reduces the delay (the *latency*) in executing the interrupt service routine. This could be important in time-critical applications. Second, the WAI places the CPU into the WAIT mode. This is a reduced power consumption, standby state that will be discussed further in Chapter 21. See Example 12-13.

Example 12-13 The WAI Instruction

Show a short code example of how to use the WAI instruction to wait for an interrupt after doing what needs to be done in a foreground job.

Solution:

```
Metrowerks HC12-Assembler
(c) COPYRIGHT METROWERKS 1987-2003


Rel. Loc     Obj. code Source line
---- ---     --------- -----------
  1                     ; Example of the WAI instruction in
  2                     ; a foreground job.
  3                     ;
  4                     foreground:
  5                     ; Here is the code to be done in the
  6                     ; foreground. When all is complete,
```

```
   7                              ; wait for the next interrupt.
   8                              ; . . .
   9 000000 3E                        wai
  10                              ; When you come out of the interrupt,
  11                              ; branch back to the foreground job.
  12 000001 20FD                      bra    foreground
```

The equivalent code in C is

```
    for ( ;  ;  ) {
      /* Here is the foreground code. When complete,
       * wait for the next interrupt
       */
      asm( wai );  /* In-line assembly */
    }
```

STOP—Stop Clocks: The STOP instruction stops all HCS12 clocks, thus dramatically reducing the power consumption. The S bit in the condition code register must be zero for the instruction to operate. See Chapter 21.

Initializing Unused Interrupt Vectors

Whenever you unmask interrupts, you are vulnerable to getting interrupts from any interrupting source. To prevent unfortunate things from happening should an unexpected interrupt occur, you should always initialize all interrupt vectors to point to a dummy interrupt service routine, which is just a return from interrupt instruction. See Example 12-14.

Example 12-14 Initializing Unused Interrupt Vectors

```
Metrowerks HC12-Assembler
(c) COPYRIGHT METROWERKS 1987-2003


Rel. Loc    Obj. code Source line
---- ---    --------- -----------
   1                              ; Initialization code for unused interrupt
   2                              ; vectors.
   3
   4                                  XDEF  Port_P_ISR, dummy_isr
   5                                  XDEF  Entry, main
```

```
 6                        XREF    __SEG_END_SSTACK
 7      0000 0000  BASE:   EQU    0
 8      0000 025F  PIFP:   EQU    BASE+$025F ; Port P Flags
 9                 main:
10                 Entry:
11                 MyCode: SECTION
12 000000 CFxx xx         lds    #__SEG_END_SSTACK
13                 ; This is the main process
14                 ; . . .
15                 ; This is the ISR for Port P, bit 0
16                 Port_P_ISR:
17                 ; . . .
18                 ; Reset the interrupt flag
19 000003 8601            ldaa   #%00000001
20 000005 7A02 5F         staa   PIFP
21                 ; Return to interrupted program
22 000008 0B              rti
23                 ;******************************
24                 ; Dummy interrupt service routine
25                 dummy_isr:
26 000009 0B              rti
```

In C, the dummy interrupt handler is simply

```
void interrupt dummy_isr( void ){
}
```

The linker parameter file must initialize the vectors by including the following commands:

```
// The interrupt vector addresses are set by the following lines:
VECTOR 0 Entry       /* Reset vector */
VECTOR 1 dummy_isr   /* Clock monitor fail reset */
VECTOR 2 dummy_isr   /* COP failure reset */
VECTOR 3 dummy_isr   /* Unknown opcode trap */
. . .
VECTOR 56 Port_P_ISR /* Port P interrupt */
VECTOR 57 dummy_isr  /* PWM emergency shutdown */
VECTOR 58 dummy_isr  /* VREG LVI interrupt */
```

Other Interrupt Registers

There are some other registers associated with the interrupt module. These are the *Interrupt Test Control Register* (*ITCR*) and *Interrupt Test Registers* (*ITEST*).

ITCR—Base + $0015—Interrupt Test Control Register

	Bit 7	6	5	4	3	2	1	0
Read:	0	0	0	WRINT	ADR3	ADR2	ADR1	ADR0
Write:								
Reset:	0	0	0	0	1	1	1	1

▨ = reserved, unimplemented, or cannot be written to.

Read: Anytime. Write: Only in special modes and with the I-bit mask and X-bit mask set.

WRINT

Write to the Interrupt Test Registers

0 = Disables writes to the test registers; reads of the test registers will return the state of the interrupt inputs (default).

1 = Disconnect the interrupt inputs from the priority decoder and use the values written into the ITEST registers instead.

ADR3–ADR0

Test Register Select Bits

These bits determine which test register is selected on a read or write. The hexadecimal value written here will be the same as the upper nibble of the lower byte of the vector selects. That is, an $F written into ADR3:ADR0 will select vectors $FFF0–$FFFE while a $7 will select vectors $FF70–$FF7E.

The sixteen ITEST registers are used in special modes for testing the interrupt logic and priority independent of the system configuration. Each bit is used to force a specific interrupt vector by writing to it a logic one. Bits are named INT0 through INTE to indicate vectors $FFx0 through $FFxE. These bits can be written only in special modes and only with the WRTINT bit in the ITCR register set to one. The interrupts must be masked by setting the I bit in the CCR. In this state the interrupt lines to the interrupt subblock will be disconnected and interrupt requests will be generated only by these registers. These bits can be read in special modes to verify that an interrupt request has reached the interrupt module.

ITEST—Base + $0016—Interrupt Test Registers

	Bit 7	6	5	4	3	2	1	0
Read:	INTE	INTC	INTA	INT8	INT6	INT4	INT2	INT0
Write:								
Reset:	0	0	0	0	0	0	0	0

INTE–INT0

Read: Only in special modes. Reads will return either the state of the interrupt inputs of the Interrupt subblock (WRTINT = 0) or the values written into the TEST registers (WRINT = 1). Reads will always return zeros in normal modes.

Write: Only in special modes and with WRINT = 1 and CCR I mask = 1.

There is an ITEST register for every eight interrupts in the overall system. All ITEST registers share the same address and are individually selected using the value stored in the ADR3:ADR0 bits of the interrupt control register.

When ADR3:ADR0 have the value of $F, only bits 2–0 in the ITEST register will be accessible because vectors higher than $FFF4 cannot be tested using the test registers. If ADR3:ADR0 point to an unimplemented test register, writes will have no effect and reads will always return a logic zero.

12.10 The Interrupt Service Routine or Interrupt Handler

The interrupt service routine is called an *ISR*, in assembly, and interrupt handler in C.

The interrupt service routine, *ISR*, or interrupt handler is executed when the vector has been initialized properly, interrupts have been unmasked and enabled, an interrupt has occurred, the CPU registers have been pushed onto the stack, and the vector has been fetched. Here are some hints for HCS12 interrupt service routines.

Interrupt Service Routine Hints

Reenable interrupts in the ISR only if you need to: If there are higher priority interrupts that must be serviced, you must unmask interrupts by clearing the I bit.

Do not use nested interrupts: Unless you have to.

Reset any interrupt generating flags in I/O devices: Each device is different and requires somewhat different procedures. If you do not reset the flag, interrupts will be generated continuously.

Do not worry about using registers in the ISR: The HCS12 automatically pushes all registers and the CCR onto the stack before entering the ISR. Remember to use the RTI instruction to restore them automatically at the end of the ISR.

Do not assume any register contents: Never assume the registers contain a value needed in the interrupt service routine unless you have full control over the whole program and can guarantee the contents of a register never changes in the program that is interrupted.

Keep it simple to start: Learning how to use an interrupt can be frustrating if you try to do too much in the ISR. The first step should be to see if the interrupts are occurring and if the interrupt service routine is being entered properly. After that is working, you can make the ISR do what it is supposed to do.

Keep it short: Do as little as possible in the ISR. This reduces the latency in servicing other interrupts should they occur while in the current ISR.

Interprocess Communication

Frequently, an interrupt service routine and another part of the program must exchange information. For example, an ISR may be incrementing a counter each time a product goes by on an assembly line. Another part of the program may be monitoring this counter to package the product when the counter reaches a certain value. The only interprocess data exchange technique appropriate for interrupt service routines uses a global or local data element as shown in Example 12-15. Clearly, except for only the simplest programs, registers cannot pass information back and forth.

In addition to the method of data transfer, we must be concerned about the timing of the data exchange. In normal program flow, we have some control over when data are written to data elements. With interrupts, however, the interrupt can occur at any time, and we must ensure that data is not changed while it is being used. Consider the situation in Example 12-15 where both the main program and the interrupt service routine must read, modify, and write the data. There is no problem if the interrupt does not occur while the main program is reading, modifying, and writing the data. If it does, the data modification that the ISR produced may be lost. *A critical region* in a program is one in which the interrupted program takes more than one instruction to read, modify, and write data. A solution to this problem is to disable the interrupt just before the critical region and reenable it just after.

Example 12-15 Interprocess Communication and Critical Code

```
Metrowerks HC12-Assembler
(c) COPYRIGHT METROWERKS 1987-2003

Rel. Loc    Obj. code Source line
---- ---     --------- -----------
  1                     ; Sampling listing showing a section of
  2                     ; critical code.
  3                     ; Define entry point etc.
  4                            XDEF   Entry,main
  5                            XREF   __SEG_END_SSTACK
  6                            XREF   do_something
  7                            XDEF   IRQISR
  8
  9                     ; IRQ_L system equates
 10          0000 001E INTCR:  EQU    $1E        ; IRQ_L control
                                                   reg
 11          0000 0080 IRQE:   EQU    %10000000 ; IRQ_L edge bit
 12          0000 0040 IRQEN:  EQU    %01000000 ; IRQ_L enable
 13                     ;
 14                            Entry:
 15                            main:
 16                            MyCode: SECTION
 17 000000 CFxx xx             lds    #__SEG_END_SSTACK
 18                     ; Initialize the counter
 19 000003 1803 0000           movw   #0,count
    000007 xxxx
 20                     ; Initialize interrupt system details
 21                     ;   Set low level interrupt on IRQ
```

```
22 000009 4D1E 80             bclr  INTCR,IRQE
23                       ;   Enable the IRQ_L interrupt
24 00000C 4C1E 40             bset  INTCR,IRQEN
25                       ;   Unmask interrupts
26 00000F 10EF                cli
27                       ; The main process is a loop that runs
                         forever
28                       ; DO
29                       loop:
30                       ; . . .
31                       ; Get the current counter value
32 000011 FExx xx             ldx   count        ;-- Critical
                                                  code
33                       ; Do whatever you need to do with the
                         count
34 000014 16xx xx             jsr   do_something ;-- Critical
                                                  code
35                       ; Increment the counter
36 000017 08                  inx                ;-- Critical
                                                  code
37 000018 7Exx xx             stx   count        ;-- Critical
                                                  code
38                       ; . . .
39                        ; WHILE (forever)
40 00001B 20F4                bra   loop
41                       ;****************************************
42                       ; Here is the interrupt service routine
43                       IRQISR:
44                       ; Each time the interrupt occurs, clear
45                       ; the counter
46                       ;****************************************
47 00001D 1803 0000          movw  #0,count
   000021 xxxx
48 000023 0B                  rti
49                       ;****************************************
50                       MyData: SECTION
51                       ; Data shared between the main process and
52                       ; the interrupt service routine
53 000000                count: DS.W  1   ; Shared data
54                       ;****************************************
```

EXPLANATION OF EXAMPLE 12-15

This program uses a shared data word, a 16-bit counter, that is incremented by the main program each time it is used, *lines 32–37*, and reset by the interrupt service routine in *line 47*.

This program also illustrates a critical code region. If the program is executing the code in *lines 32–37* and the interrupt occurs, the resetting of the counter by the interrupt service routine will not take effect. The solution to this problem is to precede the critical code with an interrupt disable and to follow it with an interrupt enable. See Examples 12-16 and 12-17.

Example 12-16 Critical Code

Show how to modify the code in Example 12-15 to avoid the problems associated with critical code.

Solution:

```
; Get the current counter value
; First, disable the IRQ_L interrupt
        bclr    INTCR,IRQE
; Get the current counter value
        ldx     count           ;-- Critical code
; Do whatever you need to do with the count
        jsr     do_something ;-- Critical code
; Increment the counter
        inx                     ;-- Critical code
        stx     count           ;-- Critical code
; Now you can reenable the IRQ_L interrupt
        bset    INTCR,IRQE
```

In C, critical code can be protected in the following way:

```
INTCR_IRQE = 0;    /* Disable the IRQ_L interrupt */
do_something( count );
count++;   /* Increment the counter*/
INTCR_IRQE = 1;    /* Reenable the IRQ_L interrupt */
```

Example 12-17 Critical Code

Example 12-16 shows how to disable the IRQ_L interrupt before entering a critical code section. How else might you ensure interrupts do not occur during critical code?

Solution: All interrupts can be masked by preceding the critical code with a SEI instruction and following it with a CLI. In C, in-line assembly code must be used:

```
asm( sei );
do_something( count );
count++;   /* Increment the counter*/
asm( cli );
```

12.11 An Interrupt Program Template

All interrupt programs can start with the same basic format. A template you can use for your interrupt programs is shown in Example 12-18.

Example 12-18 Interrupt Program Template

```
; Interrupt Template
;*******************************
; Define the entry point for the main program
        XDEF   Entry, main, YOUR_ISR
        XREF   __SEG_END_SSTACK ; Note double underbar
;*******************************
; Include files
;*******************************
; Register definitions
;*******************************
; Constants definitions
;*******************************
; Code Section
MyCode: SECTION
Entry:
main:
;*******************************
; Initialize stack pointer register
        lds    #__SEG_END_SSTACK
;*******************************
; Initialize I/O
; 1. Initialize all necessary hardware.
;    I.e. the timer, A/D etc.
;*******************************
; 2. Clear any flags that could cause interrupts

;*******************************
; 3. Enable the specific interrupt source
;*******************************
; 4. Unmask HCS12 interrupts
        cli    ; Unmask I-bit
;*******************************
; 5. Go into the foreground job
foreground:
```

```
; DO

; FOREVER
        bra    foreground
;*******************************
; 6. Here is the background job or
;    interrupt service routine (ISR)
YOUR_ISR:
;*******************************
; 6A. Reset the interrupt flag

;*******************************
; 6B. Disable lower priority interrupts
;     and unmask (CLI) higher priority as
;     needed.
;*******************************
; 6C. Do the background specific task

;*******************************
; 6D. Reenable lower priority interrupts
;     disabled in 6A.

;*******************************
; 6E. Return from the ISR
        rti
;*******************************
MyConst:SECTION
; Place constant data here
;*******************************
MyData: SECTION
; Place variable data here
```

12.12 Remaining Question

- How do the interrupts work in the other subsystems, such as the timer? *Each of the chapters that cover these subsystems will show how to use the interrupts.*

12.13 Interrupt System Register Address Summary

We present this summary in Table 12-4.

TABLE 12-4 Interrupt System Registers

Name	Register	Address Base+
ARMCOP	Arm COP Register	$003F
COPCTL	COP Control Register	$003C
HPRIO	Highest Priority Interrupt Register	$001F
INTCR	Interrupt Control Register	$001E
ITCR	Interrupt Test Control Register	$0015
ITEST	Interrupt Test Registers	$0016
PERJ	Pull-up or Pull-down Enable Port J	$026C
PERP	Pull-up or Pull-down Enable Port P	$025C
PIEJ	Port J Interrupt Enable Register	$026E
PIEP	Port P Interrupt Enable Register	$025E
PIFJ	Port J Interrupt Flag Register	$026F
PIFP	Port P Interrupt Flag Register	$025F
PLLCTL	CRG PLL Control Register	$003A
PPSJ	Port J Polarity Select	$026D
PPSP	Port P Polarity Select	$025D

12.14 Conclusion and Chapter Summary Points

In this chapter we discussed interrupts in general and showed how the HCS12 processes interrupt requests. The following points summarize what we have covered in this chapter.

- Interrupts are important, asynchronous events that require immediate attention.
- Interrupts are globally masked and unmasked by the I bit in the condition code register.
- Masking means to disallow interrupts and unmasking means allowing them.
- Interrupts may be selectively enabled and disabled by individual bits in control registers.
- The routine that is executed when an interrupt occurs is called an interrupt service routine or interrupt handler.
- The HCS12 determines the correct interrupt service routine using a vector stored in memory.
- A COP, or Computer Operating Properly, reset may be used to reset the CPU if the program misbehaves and runs away.
- There are a variety of interrupting sources including externally generated ones on I/O ports and internally generated ones such as from the timer subsystem.
- A wait for interrupt instruction, WAI, and a stop instruction, STOP, can put the HCS12 into a power-saving mode until an interrupt occurs to wake-up the processor.

12.15 Bibliography and Further Reading

CPU12 Reference Manual, S12CPUV2.PDF, Freescale Semiconductor, Inc.

CPU12 Reference Manual, *M68HC12 & HCS12 Microcontrollers*, CPU12RM/AD, Rev. 3, Freescale Semiconductor, Inc., April 2002.

CRG Block User Guide V04.05, S12CRGV4.PDF, Freescale Semiconductor, Inc., August 2002.

Ganssle, J. G., *Great Watchdogs*, Version 1.2, January 2004. http://www.ganssle.com/watchdogs.htm.

INT Module Block Guide, S12INTV1.PDF, Freescale Semiconductor, Inc., May 2003.

More, G. M. and D. Malik, *A Software Interrupt Priority Scheme for HCS12 Microcontrollers*, AN2617, Freescale Semiconductor, Inc., February 2004.

12.16 Problems

Basic

12.1 What are polled interrupts? **[a, g]**

12.2 "An interrupt system must allow asynchronous events to interrupt an ongoing process." Give at least five more attributes of an interrupt system. **[a]**

12.3 The HCS12 is a vectored interrupt processor. Describe how the HCS12 uses a vector to find the correct interrupt service routine after an interrupt occurs. **[a, g, k]**

12.4 What advantage does a vectored interrupt system have over a polled interrupt system? **[a]**

12.5 Why, in most processors with interrupts, are further interrupts disabled when the processor reaches the interrupt service routine? **[a, k]**

12.6 What features does the HCS12 have to allow you to manage multiple sources of interrupts? **[a, g]**

12.7 Describe how the HCS12 manages to transfer control to the correct interrupt service routine for any particular interrupt in an embedded system. **[a, g]**

12.8 In the HCS12, why does the CPU automatically push all registers on the stack before transferring control to the user's interrupt service routine? **[a]**

12.9 When the I bit in the condition code register is set to 1: **[a]**

  a. Interrupts are masked.

  b. Interrupts are unmasked.

  c. The I bit doesn't affect interrupts.

12.10 Which instruction is used to globally mask interrupts? **[k]**

12.11 Which instruction is used to globally unmask interrupts? **[k]**

12.12 Interrupts are masked when you get to the interrupt service routine—true or false? **[a]**

12.13 In the HCS12 interrupt service routine, you *must* unmask interrupts with the CLI instruction before returning—true or false? **[a]**

12.14 How are interrupts unmasked if the CLI instruction is not executed in the interrupt service routine? **[g]**

12.15 In the HCS12, what is the difference between an interrupt mask bit and an interrupt enable bit? **[a, g]**

12.16 What address does the HCS12 use to find the address of an interrupt service routine for a timer overflow? **[k]**

12.17 How can the priority order of interrupts be changed? **[g]**

12.18 The Timer Channel 3 interrupt and the Real-Time Interrupt happen to occur simultaneously. Which is serviced first? **[k]**

12.19 What instructions can be used to reduce power consumption when waiting for an interrupt to occur? **[k]**

Intermediate

12.20 How many bytes are pushed onto the stack when the HCS12 processes an interrupt request? **[a]**

12.21 For the interrupt service routine in Example 12-8, where would you put a breakpoint to find out if you are getting to the interrupt service routine? **[k]**

12.22 Define interrupt latency. **[g]**

12.23 Give at least two components of interrupt latency in the HCS12. **[a]**

12.24 Describe how the microcontroller finds its way to your program when RESET_L is asserted. **[a]**

12.25 Describe the operation of the last instruction executed in an interrupt service routine. **[a]**

Advanced

12.26 Assume an embedded application system with ROM at $C000–$F000 and RAM at $0800–$0FFF. Show how to initialize the interrupt vectors for the IRQ_L and Timer Channel 1 interrupts. Assume IRQISR and TC1ISR are labels on the respective interrupt service routines. **[k]**

12.27 Write a complete HCS12 program in assembly language for an interrupt occurring on the external IRQ_L source. The interrupt vector is to be at $FFF2:FFF3. When the interrupt occurs, the ISR is to increment an 8-bit memory location "COUNT" starting from $00. The foreground job is to be a spin loop "SPIN BRA SPIN." Assume the following. [c, k]

    a. Code is to be located in ROM at $C000.
    b. RAM is available between $0800 and $0BFF.

12.28 Describe how you could measure interrupt latency in the lab using a lab processor board and other lab instrumentation. [c]

12.29 Describe how to modify the code in Example 12-9 to have the ISR service all interrupts generated by any of the four port P bits. [c, g]

12.30 Write a short section of code demonstrating how to reset the COP timer. [c, k]

13 | HCS12 Memories

OBJECTIVES

This chapter describes the types of memories in the MC9S12C32 microcontroller. Although there are several members of the HCS12 family with differing amounts of memory, we restrict our discussion in this chapter mostly to the MC9S12C32 version.

13.1 Introduction

The HCS12 family includes on-chip static *RAM* for program variables and the stack, electrically erasable *EEPROM* memory for nonvolatile variable storage, and *Flash EEPROM* for program code. The amount of each depends on the family member and can range from 2K to 12K bytes of RAM, 0 to 4K bytes of EEPROM, and 32K to 512K bytes of Flash EEPROM. In addition to the RAM and program memory, there are 1K to 2K bytes of the memory map allocated to I/O and control registers. The MC9S12C32 has 1K bytes for I/O, 2K bytes of RAM, and 32K bytes of Flash but does not have any EEPROM.

Flash and EEPROM

The two kinds of nonvolatile memory in the HCS12 are *Flash EEPROM* and *EEPROM*. The Flash memory is electrically erasable but it is faster to program (hence its name) and must be erased and programmed in sectors of 512 to 1024 bytes. Flash memory holds our program code because it is programmed and erased in large sectors. EEPROM can be erased and programmed in single bytes and is thus useful for program storing variables that you wish to retain if the power is removed from the microcontroller. We will use the term Flash to refer to the MC9S12C32 Flash EEPROM program memory.

13.2 Remapping Memory Resources

The memory locations of the internal control registers, the internal RAM, and the EEPROM (in some versions of the HCS12 family) can be remapped, or changed, from the default positions established at reset. There are a couple of reasons why we might want to do this.

- The direct addressing mode, which accesses the first 256 bytes of memory, allows a program to be written with fewer bytes of code and to run faster. In an application that frequently uses RAM, it is sensible to remap the RAM to these memory locations and to locate the registers somewhere else.
- Some members of the HCS12 family have overlapping memory locations because they have more memory than can be addressed with 16 bits. In this case, the user can select alternative locations for the registers, RAM, and EEPROM to avoid or minimize the overlap.

Each member of the HCS12 family has its own memory resources. You should refer to the device guide for that family for complete details.

13.3 MC9S12C32 Memory Map

The MC9S12C32 remappable resources are the 1K bytes used for I/O and control registers and 2K bytes of RAM. The 32K bytes of Flash are fixed at addresses $8000–$FFFF. These HCS12C family members do not have EEPROM.

The MC9S12C32 memory includes on-chip static *RAM* for program variables and the stack, and *Flash* memory for program code. In the "C" version of the MC9S12 family there is no standard EEPROM that some other family members have. Figure 13-1 shows the memory map for an MC9S12C32. We can see that the microcontroller has I/O registers to control hardware features of the microcontroller, RAM for variable data storage and the stack, and Flash for program and constant data storage. Figure 13-1 shows the location of the memory components at the time the microcontroller is reset. As we will show next, the locations of the registers and the RAM can be changed to be more convenient for any particular programming application.

Figure 13-1 shows areas of the memory with "Nothing"—no memory or resources are allocated to these addresses. We might ask: "What happens if we read from or write to these addresses?" The answer is: "Nothing." Reading from this address space will return an unspecified result. Writing does not affect anything.

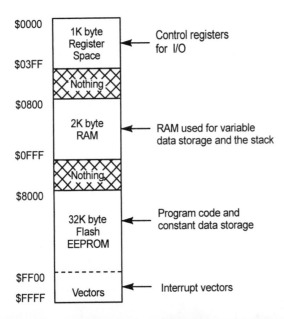

Figure 13-1 MC9S12C32 memory map at reset time.

Example 13-1 Remapping Internal RAM

What value must be written into the INITRM register to move the internal 2K-byte RAM from its reset location to a location so that the highest byte's address is $3FFF?

Solution: For the high byte of the address to be $3F, RAMHAL = 1 and thus INITRM should be %0011 1001 = $39.

13.4 Remapping Registers and Memory Locations

The memory addresses of the I/O registers and the RAM in the MC9S12C32 are initially at the locations shown in Figure 13-1. The members of the HCS12 family who have EEPROM will show its location in a reset memory map similar to Figure 13-1. The programmer can alter these locations by programming the *INITRM*, *INITRG*, and *INTEE* registers.

Remapping RAM

The 2K-byte block of RAM in the MC9S12C32 processor is initially in memory addresses $0800–$0FFF. The *INITRM* register initializes the position of the RAM in the internal address memory map. (See Examples 13-1 and 13-2.)

INITRM—Base[1] + $0010—Initialize Internal RAM Position Register

	Bit 7	6	5	4	3	2	1	0
Read:	RAM15	RAM14	RAM13	RAM12	RAM11	0	0	RAMHAL
Write:								
Reset:	0	0	0	0	1	0	0	1

⬜ = reserved, unimplemented, or cannot be written to.

Read: Anytime. Write: Once in normal and emulation modes, anytime in special modes.

RAM15:RAM11

Base Address of the Internal RAM

These bits set the upper five bits of the internal RAM memory.

RAMHAL

RAM High-Align

0 = Aligns RAM to the lowest address of the mappable space.

1 = Aligns RAM to the highest address of the mappable space (default).

See Examples 13-1 and 13-4.

---

[1] The addresses of all I/O registers are shown in this way—Base + $XXXX—to denote that the actual register address is determined by offset ($XXXX) plus the register *base* address that may be programmed by the user after startup. The default base is $0000 and may be changed by writing to the INITRG register.

Example 13-2 Remapping Internal RAM

What value must be written into the INITRM register to move the internal 2K-byte RAM from its reset location to a location so that the lowest byte's address is $1000?

Solution: For the low byte of the address to be $1000, RAMHAL = 0, and thus INITRM should be %0001 000 = $10.

Remapping I/O Registers

The *INITRG* register remaps the internal I/O registers. The registers can be mapped to any 2K-byte space in the first 32K bytes of the system memory map. (See Example 13-3.)

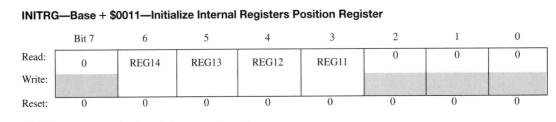

INITRG—Base + $0011—Initialize Internal Registers Position Register

	Bit 7	6	5	4	3	2	1	0
Read:	0	REG14	REG13	REG12	REG11	0	0	0
Write:								
Reset:	0	0	0	0	0	0	0	0

= reserved, unimplemented, or cannot be written to.

Read: Anytime. Write: Once in normal and emulation modes, anytime in special modes.

REG14:REG11

Base Address of I/O Registers

These bits establish the base address of the microcontroller's internal registers. See Example 13-4.

Example 13-3 Remapping Internal Registers

What value must be written into the INITRG register to remap the starting address of the registers to $1000?

Solution: The registers may be mapped into any 2K-byte space in the first 32K bytes of the system memory map. $1000 is in the allowable address space and bits 15–11 of this address are %00010. Thus %0001 0000 = $10 is the correct 8-bit answer.

Remapping EEPROM

The *INITEE* register remaps the internal EEPROM memory that is available in some versions of the HCS12 microcontroller. Unfortunately, the MC9S12C32 microcontroller does not have internal, remappable EEPROM. If you are using other versions of the processor, check with the device user guide for that chip.

INITEE—Base + $0012—Initialize Internal RAM Position Register

	Bit 7	6	5	4	3	2	1	0
Read:	EE15	EE14	EE13	EE12	EE11	0	0	EEON
Write:								

Reset: Reset state of these bits is controlled at chip integration and is different for various family members.

 = reserved, unimplemented, or cannot be written to.

Read: Anytime. Write: Once in normal and emulation modes, anytime in special modes.

EE15:EE11

Base Address of the Internal EEPROM

These bits set the upper 5 bits of the internal EEPROM available in some devices. In some devices, these can be written at any time.

EEON

Enable EEPROM

0 = Disables the EEPROM from the memory map.
1 = Enables the EEPROM at the base address given by EE15:EE11.

Example 13-4 Remapping RAM and Register Locations
Show example code to remap the I/O registers from $0000–$03FF to $0800–$0BFF and the RAM from $0800–$0FFF to $0000–$07FF.

Solution:

```
Metrowerks HC12-Assembler
(c) COPYRIGHT METROWERKS 1987-2003


Rel. Loc    Obj. code Source line
---- ------ --------- -----------
1                     ; Code example showing remapping
2                     ; RAM to $0000 and registers to $0800
3                             INCLUDE bits.inc
4
```

```
 5          0000 0000 RAMLOC: EQU   $0000  ; Locate 2K RAM
 6          0000 0800 REGLOC: EQU   $0800  ; Locate 1K registers
 7          0000 0010 INITRM: EQU   $0010  ; Init RAM register
 8          0000 0011 INITRG: EQU   $0011  ; Init registers reg
 9          0000 0001 RAMHAL: EQU   BIT0   ; RAM high-align
10                    ; Use expressions to calculate the values for
11                    ; INITRM and INITRG registers
12                    ; Make the INITRM register = %00000000
13          0000 0000 RAM_LOC:EQU   RAMLOC >> 8 & ~RAMHAL
14                    ; Make the INITRG register = %00001000
15          0000 0008 REG_LOC:EQU   REGLOC >> 8
16
17                    ; . . .
18                    ; Remap the RAM and registers.
19                    ; This can be done only once in the program
20 000000 180B 0000       movb  #RAM_LOC,INITRM ; Remap the RAM
   000004 10
21 000005 180B 0800       movb  #REG_LOC,INITRG ; Remap reg
   000009 11
```

Show how to relocate RAM and the I/O registers in C.

Solution:

```
/*****************************************************************/
/* Remap the RAM to 0x0000 - 0x0FFF and the registers
 * to 0x0800 - 0x07FF
/*****************************************************************/
#define INITRM  (*(volatile unsigned char *) 0x0010)
#define INITRG  (*(volatile unsigned char *) 0x0011)

void main(void) {
  /* . . . */
  INITRM = 0x00;  /* Remap the RAM start to 0x0000 */
  INITRG = 0x08;  /* Remap the registers to start at 0x0800 */
}
```

EXPLANATION OF EXAMPLE 13-4

In addition to showing how to remap the registers and RAM by using the MOVB instruction, this example shows a useful assembler expression evaluation. Each position register, INITRG and INITRM, is 8 bits. *Lines 5* and *6* define the locations of the RAM and registers as 16-bit addresses, which is a good documentation technique. In *lines 13* and *15* the assembler evaluates the immediate operand for the MOVB instructions at *lines 20* and *21* by *logical right shifting* the 16-bit RAMLOC and REGLOC values 8-bit positions. The RAMHAL bit is reset

TABLE 13-1 Memory Addressing Priorities

Priority	Address Space
Highest	Background Debugger Module (BDM); internal to Core firmware or register space
—	Internal registers
—	RAM
—	EEPROM
—	On-chip Flash or ROM
Lowest	Remaining external space

(by ANDing the shifted RAMLOC value with $FE) to indicate that the address in INITRM specifies the lowest address of the new position of the RAM. To review assembler expressions, see Chapter 8.

Memory Priorities

We notice in Example 13-4 that the RAM is first mapped into memory locations $0000–$07FFF before the registers are mapped to their final addresses $0800–$0BFF. If this is the case, how can the instruction at *line 21* write the correct data into the INITRG register at $0011? Table 13-1 shows how overlapping memory space addressing is resolved. Even though the RAM and register addresses overlap, at least for a short time, the instruction at *line 21* writes the value into the correct address.

13.5 Memory Expansion

The HCS12's 16-bit address limits the physical address space to 64K bytes. To allow for larger programs using more memory, some members of the HCS12 family allow up to 64, 16K-byte blocks (1M bytes) of Flash memory or ROM to be accessed by a 16K-byte memory "page." Figure 13-2 shows the memory map of the MC9S12DP256 microcontroller. At reset, the 1K-byte register block is at $0000–$03FF and is overlapped by a 4K-byte EEPROM block from $0000 to $0FFF. As shown in Table 13-1, any register access will take priority over an overlapped EEPROM address. A 16K-byte page window is located at $8000–$BFFF. Access to this expanded memory space is through the *Program Page Index Register* (*PPAGE*).

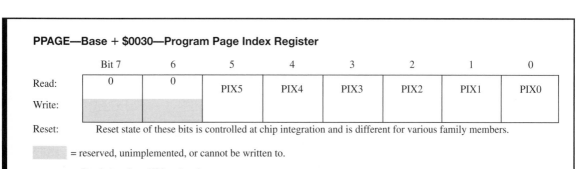

PPAGE—Base + $0030—Program Page Index Register

	Bit 7	6	5	4	3	2	1	0
Read:	0	0	PIX5	PIX4	PIX3	PIX2	PIX1	PIX0
Write:								

Reset: Reset state of these bits is controlled at chip integration and is different for various family members.

▨ = reserved, unimplemented, or cannot be written to.

Read: Anytime. Write: Anytime.

PIX5:PIX0

Program Page Index Bits

The page index register is used to select which of the 64 Flash pages is to be accessed by an address in the range $8000–$BFFF. (See Table 13-2.)

TABLE 13-2 Program Page Index Register Bits

PIX5	PIX4	PIX3	PIX2	PIX1	PIX0	Program Space Selected
0	0	0	0	0	0	16K page 0
0	0	0	0	0	1	16K page 1
0	0	0	0	1	0	16K page 2
—	—	—	—	—	—	—
1	1	1	1	0	1	16K page 61
1	1	1	1	1	0	16K page 62
1	1	1	1	1	1	16K page 63

Programming and the Program Page Window

The actual program memory page accessed when a memory address in the range of $8000–$BFFF is generated is controlled by the PPAGE register as shown in Table 13-2. The

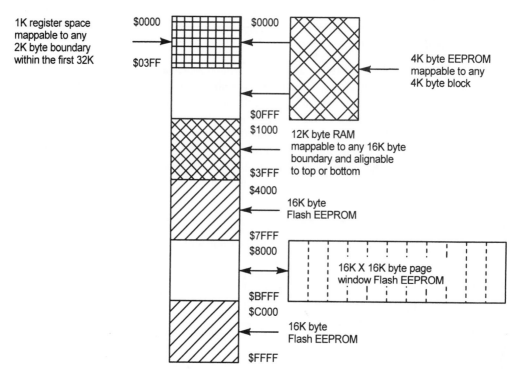

Figure 13-2 MC9S12DP256 memory map.

value in PPAGE may be read or written by normal instructions, although some devices allow writes only in special modes. Control registers, interrupt vectors, and I/O registers should not be located in the page window so that they can be accessed from any page in the window. The starting address of an interrupt service routine must be in unpaged memory because all interrupt vectors are 16 bits only.

The HCS12 instruction set includes the CALL and RTC (return from call) instructions to transfer to subroutines located in one of the expansion memory pages. CALL is similar to a JSR instruction except that the subroutine can be located anywhere in the 64K memory space or on any page of the expansion memory. The return from call (RTC) instruction must be used when CALL is used. It returns correctly taking into account page changes.

13.6 Other Memory Mapping Registers

Two registers are visible to the program but are fixed at the time the microcontroller is manufactured. Reading these registers allows you to determine the memory configuration of the processor.

MEMSIZ0—Base + $001C—Memory Size Register 0

	Bit 7	6	5	4	3	2	1	0
Read:	REG_SW	0	EEP_SW1	EEP_SW0	0	RAM_SW2	RAM_SW1	RAM_SW0
Write:								

Reset: Defined at chip integration; see chip level documentation.

= reserved, unimplemented, or cannot be written to.

Read: Anytime. Write: Has no effect.

REG_SW

Allocated System Register Space

0 = Allocated system register space is 1K bytes.
1 = Allocated space is 2K bytes.

EEP_SW1:EEP_SW0

Allocated System EEPROM Memory Space in Family Members with EEPROM

(See Table 13-3.)

TABLE 13-3 Allocated EEPROM Memory Space

EEP_SW1	EEP_SW0	EEPROM Space
0	0	0 byte
0	1	2K bytes
1	0	4K bytes
1	1	8K bytes

RAM_SW2:RAM_SW0

Allocated RAM Memory Space

(See Table 13-4.)

TABLE 13-4 RAM Memory Allocated Space

RAM_SW 2	1	0	Allocated RAM Space	RAM Mappable Region	INITRM Bits Used	RAM Reset Base Address
0	0	0	2K bytes	2K bytes	RAM15:11	$0800
0	0	1	4K bytes	4K bytes	RAM15:12	$0000
0	1	0	6K bytes	8K bytes	RAM15:13	$0800
0	1	1	8K bytes	8K bytes	RAM15:13	$0000
1	0	0	10K bytes	16K bytes	RAM15:14	$1800
1	0	1	12K bytes	16K bytes	RAM15:14	$1000
1	1	0	14K bytes	16K bytes	RAM15:14	$0800
1	1	1	16K bytes	16K bytes	RAM15:14	$0000

MEMSIZ1—Base + $001D—Memory Size Register 1

	Bit 7	6	5	4	3	2	1	0
Read:	ROM_SW1	ROM_SW0	0	0	0	0	PAG_SW1	PAG_SW0
Write:								

Reset: Defined at chip integration; see chip level documentation.

     = reserved, unimplemented, or cannot be written to.

Read: Anytime. Write: Has no effect.

ROM_SW1:ROM_SW0

Allocated System Flash or ROM Physical Memory Space

(See Table 13-5.)

TABLE 13-5 Allocated Flash/ROM Physical Memory Space

ROM_SW1	ROM_SW0	Flash or ROM Space
0	0	0K bytes
0	1	16K bytes
1	0	48K bytes
1	1	64K bytes

PAG_SW1:PAG_SW0

Allocated Off-Chip Flash Memory or ROM Space

(See Table 13-6.)

TABLE 13-6 Allocated Off-Chip Memory Options

PAG_SW1	PAG_SW 0	Off-Chip Space	On-Chip Space
0	0	876K bytes	128K bytes
0	1	768K bytes	256K bytes
1	0	512K bytes	512K bytes
1	1	0 byte	1M bytes

MISC—Base + $0013—Miscellaneous System Control Register

	Bit 7	6	5	4	3	2	1	0
Read:	0	0	0	0	EXSTR1	EXSTR0	ROMHM	ROMON
Write:								
Reset:	0	0	0	0	1	1	0	1

▨ = reserved, unimplemented, or cannot be written to.

Read: Anytime. Write: Once in normal and emulation modes and anytime in special modes.

EXSTR1:EXSTR0

External Access Stretch Bits 1 and 0

These bits control the amount of clock stretch on accesses to the external address space in expanded modes. They have no meaning in single-chip and peripheral modes. (See Table 13-7.)

TABLE 13-7 External Stretch Bits

EXSTR1	EXSTR0	Number of Bus Clocks Stretched
0	0	0
0	1	1
1	0	2
1	1	3

ROMHM

Flash EEPROM or ROM Only in Second Half of Memory

0 = The fixed pages of Flash EEPROM or ROM in the lower half of the memory map can be accessed (default).

1 = Disables direct access to the Flash EEPROM or ROM in the lower half of the memory map. These physical locations remain accessible through the program page window.

ROMON

Enable Flash EEPROM or ROM

0 = Disables the Flash EEPROM or ROM from the memory map.

1 = Enables the Flash EEPROM or ROM in the memory map (default).

TABLE 13-8 Memory System Registers

Name	Register	Address Base +
INITRM	Initialize RAM Location	$0010
INITRG	Initialize Register Location	$0011
INITEE	Initialize EEPROM Location	$0012
MISC	Miscellaneous System Control	$0013
MEMSIZ0	Memory Size Register 0	$001C
MEMSIZ1	Memory Size Register 1	$001D
PPAGE	Program Page Index Register	$0030

13.7 Remaining Question

- Are there HCS12 versions to which external memory and I/O can be added? *Yes there are. The HCS12 microcontroller can be reset into an expanded mode operation where port A and port B are used as external address and data buses. That is outside the scope of what we want to cover in this text, so to find out more details, please refer to* HCS12 External Bus Design, *AN2287, Freescale Semiconductor, Inc., 2004.*

13.8 Memory System Register Address Summary

We present this summary in Table 13-8.

13.9 Conclusion and Chapter Summary Points

In this chapter we discussed the memory capabilities of the MC9S12C32 microcontroller and other members of the HCS12 family.

- There are two kinds of memory in a microcontroller system—RAM and ROM.
- ROM usually denotes program memory and in the 9S12C family is implemented as Flash EEPROM.
- EEPROM is available in some other members of the HCS12 family.
- Flash is faster to erase and program than EEPROM.
- Flash EEPROM is erased and programmed in 512- or 1024-byte sectors.
- The RAM, registers, and EEPROM (in some versions) can be moved around in the memory map.
- Some versions of the HCS12 can have up to 1M bytes of program memory that is accessed in 16K-byte pages.

13.10 Bibliography and Further Reading

HCS12 External Bus Design, AN2287, Freescale Semiconductor, Inc., 2004.

Mapping Memory Resources on HCS12 Microcontrollers, AN2881, Freescale Semiconductor, Inc., 2005.

MC9S12C Family Device User Guide V01.05, 9S12C128DGV1/D, Freescale Semiconductor, Inc., 2004.

Module Mapping Control Manual, S12MMCV4/D, Rev. 4.00, Freescale Semiconductor, Inc., 2003.

Multiplexed External Bus Interface (MEBI) Module V3 Block User Guide, S12MEBIV3/D, Rev. 3.00, Freescale Semiconductor, Inc., 2003.

13.11 Problems

Basic

13.1 Give the value to which the INITRM register must be written to remap the internal 2K bytes of RAM to $1800. **[k]**

13.2 Give the value to which the INITRG register must be written to remap the internal registers to $1800. **[k]**

13.3 What registers are used to remap the data RAM, EEPROM, and control registers? **[k]**

13.4 The internal registers and the RAM have both been mapped to start at $0800. When a byte is written to $0800, which resource receives it—the RAM or the register? **[k]**

13.5 What register is used to control which expanded memory page is accessed in the MC9S12DP256? **[k]**

13.6 When using a microcontroller with expansion memory such as the MC9S12C256, what instructions must be used to access and return from subroutines located in an expansion memory page? **[k]**

Intermediate

13.7 On reset, the RAM in the MC9S12C32 is mapped to $0800 and the 1K-byte register block to $0000. The locations of these can be changed. Describe how this is done. **[g, k]**

13.8 Give the value to which the INITRG register must be written to remap the internal registers to $8000. **[k]**

13.9 When internal registers and RAM are mapped to the same address space, which has access priority? **[k]**

13.10 When RAM and on-chip Flash memory are mapped to the same address space, which has access priority? **[k]**

13.11 The internal registers and the RAM have both been mapped to start at $0800. When a byte is written to $0C00, which resource receives it—the RAM or the register? **[k]**

14 HCS12 Timer

OBJECTIVES

This chapter describes the HCS12 timer. The timer system includes a *free-running counter* and eight timer channels. These channels may be configured in any combination of timer comparison channels, called *output compares*, or *input capture* channels, which capture the time when an external event occurs. There is a *real-time periodic interrupt* and a counter for external events called the *pulse accumulator*. The interrupting capabilities of the timer are covered and programming examples are given. We also describe the *real-time periodic interrupt* and the *pulse-width modulated* waveform generator although they are separate hardware modules.

14.1 Introduction

The timer section in the HCS12 is based on a 16-bit counter operating from a system clock called the *bus clock*. We will discuss how this clock is generated in Chapter 21 and concentrate now on learning about the timer capabilities. The 16-bit counter provides basic, real-time functions with the following features:

- A *timer overflow* to extend the 16-bit capability of the timer section counter.
- Up to eight *output compare functions* that can generate a variety of output waveforms by comparing the 16-bit timer counter with the contents of a programmable register.
- Up to eight *input capture functions* that can latch the value of the 16-bit counter on selected edges of eight timer input pins.
- A 16-bit *pulse accumulator* to count external events or act as a gated timer counting internal clock pulses.

The timer system is by far the most complex subsystem in the HCS12. It uses many I/O control registers and many control bits. All timer functions have interrupt controls and separate interrupt vectors. Figures in each of the following sections show block diagrams of each timer subsystem.

All timer functions have similar programming and operational characteristics. They all have flags in a control register that are set when some programmable condition is satisfied and that must be reset by the program. They all have interrupts that are enabled or disabled by a bit in a control register. Thus, when the operation of one timer function has been learned, the procedures are easily transferred to the others.

14.2 Basic Timer

A 16-bit *TCNT* register forms the basis of all timer functions.

The key to the operation of the HCS12 timer is the 16-bit, free-running counter called *TCNT* shown in Figure 14-1. Its input clock is from the system bus clock, which may be prescaled by dividing it by 1, 2, 4, 8, 16, 32, 64, or 128. A typical bus clock frequency is 8 MHz. The counter starts at $0000 when the microcontroller is reset and runs continuously after that unless you turn it off or cause it to stop in WAIT or background debug mode. The counter cannot be set to a particular value by the program when the CPU is operating in a normal mode, but it can be written in special modes. It can also be reset when a successful output compare occurs on timer channel 7. The TCNT register's current value can be read anytime. Every 65,536 pulses the counter reaches a maximum and overflows. When this occurs, the counter is reset to $0000 and a *Timer Overflow Flag* is set. This flag can extend the counter's range.

As shown in Figure 14-1, the clock for the TCNT counter may be selected from a prescaled bus clock or from other sources generated in the pulse accumulator. Normally the prescaled bus clock is used but we will discuss the other clock sources in Section 14.7.

Prescaler

The clock source for the TCNT counter is the system bus clock, which can be prescaled by a programmable divider, or three other sources from the pulse accumulator subsystem. The prescaler shown in Figure 14-1 is controlled by the three PR2:PR1:PR0 bits in the *Timer System Control Register 2 (TSCR2)*.

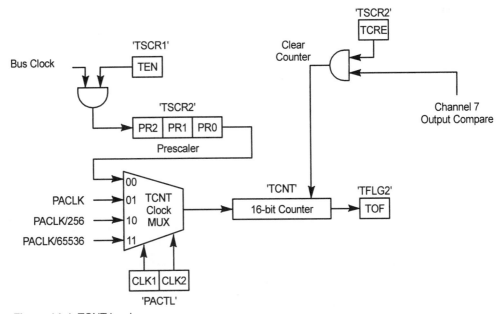

Figure 14-1 TCNT hardware.

TSCR2—Base[1] + $004D—Timer System Control Register 2

	Bit 7	6	5	4	3	2	1	0
Read:	TOI	0	0	0	TCRE	PR2	PR1	PR0
Write:								
Reset:	0	0	0	0	0	0	0	0

        = reserved, unimplemented, or cannot be written to.

        Read: Anytime. Write: Anytime.

TOI

Timer Overflow Interrupt Enable

0 = Interrupt inhibited (default).

1 = Hardware interrupt requested when the timer overflows setting the TOF.

For more information, see Section 14.3.

TCRE

Timer Counter Reset Enable

0 = Counter reset inhibited and TCNT free runs (default).

1 = TCNT is reset to $0000 when a successful output compare occurs on timer channel 7.

PR2:PR1:PR0

Timer Prescaler Select

(See Table 14.1.)

TABLE 14-1 Timer Clock Selection

PR2	PR1	PR0	Bus Clock Divisor
0	0	0	1
0	0	1	2
0	1	0	4
0	1	1	8
1	0	0	16
1	0	1	32
1	1	0	64
1	1	1	128

Sixteen-Bit Free-Running TCNT Register

The TCNT free-running counter is the heart of the timer system.

The timer must be enabled after reset to start the TCNT register counting. This is done by setting the *Timer Enable (TEN)* bit in the *Timer System Control Register 1* register.

---

[1] The addresses of all I/O registers are shown in this way—Base + $XXXX—to denote that the actual register address is determined by offset ($XXXX) plus the register *base* address that may be programmed by the user after startup. The default base is $0000 and may be changed by writing to the INITRG register.

TSCR1—Base + $0046—Timer System Control Register 1

	Bit 7	6	5	4	3	2	1	0
Read:	TEN	TSWAI	TSFRZ	TFFCA	0	0	0	0
Write:								
Reset:								

= reserved, unimplemented, or cannot be written to.

Read: Anytime. Write: Anytime.

TEN

Timer Enable

0 = Disables the main timer, including the counter. This can be used to reduce power consumption (default).

1 = Enables the timer to run normally.

TSWAI

Timer Module Stops While in Wait Mode

0 = Allows the timer to continue running during wait mode (default).

1 = Disables the timer when the microcontroller is in wait mode. Timer interrupts cannot be used to exit wait mode.

TSFRZ

Timer Stops While in Freeze Mode

0 = Allows the timer to continue running during freeze mode (default).

1 = Disables the timer whenever the microcontroller is in freeze mode. This is useful for emulation.

TFFCA

Timer Fast Flag Clear All

0 = Allows the timer flag clearing to function normally (writing a one to the flag bit) (default).

1 = When this bit is set, flags are cleared by reading from the register that set the flag.

For more information, see Section 14.8.

TCNT REGISTER OPERATION

The *TCNT* register starts at $0000 when the processor is reset and counts continuously until it reaches the maximum count of $FFFF. On the next pulse, the counter rolls over to $0000, sets the *Timer Overflow Flag* (*TOF*), which is in the TFLG2 register, and continues to count.

The timer counter is designed to be read with a 16-bit read instruction such as `ldd $44` or `ldx $44`. If you were to do two, 8-bit read instructions, for example,

```
ldaa    $44     ; Get the high 8 bits
ldab    $45     ; Get the low 8 bits
```

the low 8 bits of the counter will be incremented and will be different by the time the second load instruction is executed.

TCNT—Base + $0044, $0045—Timer Counter Register

Bit 15	14	13	12	11	10	9	8
TCNT15	TCNT14	TCNT13	TCNT12	TCNT11	TCNT10	TCNT9	TCNT8

Reset: 0   0   0   0   0   0   0   0

Bit 7	6	5	4	3	2	1	0
TCNT7	TCNT6	TCNT5	TCNT4	TCNT3	TCNT2	TCNT1	TCNT0

Reset: 0   0   0   0   0   0   0   0

Read: Anytime. Write: Only in special modes.

Timer Overflow Flag

The *TOF* bit is set when the TCNT register overflows and is reset by *writing a one* to bit 7 in the *TFLG2* register.

The timer overflow flag, *TOF*, is set when the timer rolls over from $FFFF to $0000. The programmer can extend the range of the count by detecting the overflow and incrementing another counter in the program. You may also achieve longer time intervals between overflows by setting a different prescaler value but this, of course, alters all timing done by the timer. The timer overflow flag is in the *TFLG2* register. Figure 14-2 shows the timer overflow hardware in the HCS12.

TFLG2—Base + $004F—Timer Interrupt Flag Register 2

	Bit 7	6	5	4	3	2	1	0
Read:	TOF	0	0	0	0	0	0	0
Write:								
Reset:	0	0	0	0	0	0	0	0

    = reserved, unimplemented, or cannot be written to.

Read: Anytime. Write: Used in the clearing process (writing a one clears the bit).

TOF

Timer Overflow Flag

This flag is set when the TCNT register overflows from $FFFF to $0000. The bit is cleared by writing a one to bit 7. It may be cleared by any access to TCNT if the fast flag clear all bit (TFFCA) in TSCR1 is set.

For more information, see Section 14.8.

TABLE 14-2 Pseudocode Design for a Timer Overflow Program

Program Requirements: Alternately flash LED1 and LED2 at about 0.5 Hz using the timer overflow bit.

```
; Initialize the I/O data direction for the output bits.
; Initialize the initial LED display values.
; Enable the timer.
; Clear the timer overflow flag (TOF)
; DO
;   Display the current value on the LEDs
;   Delay approximately 1 second
;   Complement the LED display values
; FOREVER

; Subroutine tof_1_sec_delay
; Save the registers
; Initialize a counter for the timer overflows
; DO
;   Wait until the timer overflows
;   Reset the timer overflow flag
;   Decrement the timer overflow counter
; ENDO
; WHILE Timer overflow counter > 0
; Restore the registers
; Return
```

Basic Timer Overflow Programming

The TOF can be used in two ways—polling or interrupting. In polling, the program is responsible for watching the TOF (by reading bit 7 in the TFLG2 register). When the flag is set, the program can increment its local counter. If you are using the timer overflow flag, it *must* be reset by the program each time it is set by the counter. This is done by *writing* a *one* to bit 7 in the TFLG2 register. Example 14-1 shows how to use the timer overflow bit to generate a delay in increments of 8.192 ms when using an 8-MHz bus clock. The pseudocode design of Example 14-1 is given in Table 14-2.

Example 14-1 Polling the Timer Overflow Flag

```
Metrowerks HC12-Assembler
(c) COPYRIGHT METROWERKS 1987-2003
 Rel. Loc    Obj. code Source line
 ---- ------ --------- -------------
     1                  ; This is an example of using the timer
     2                  ; overflow flag to generate a delay of
     3                  ; approximately 1 second (The delay will
     4                  ; be 122 timer overflow flags which is
     5                  ; 0.999 second with an 8 MHz bus clock.
     6                  ; The program will alternately blink
     7                  ; LED1 and LED2 on the Student Learning
```

```
 8                       ; Kit with a CSM-12C32 board
 9                       ;*******************************
10                       ; Define the entry point for the main
                         program
11                           XDEF   Entry, main
12                           XREF   __SEG_END_SSTACK
13                           INCLUDE  timer.inc   ; Timer defns
14                           INCLUDE  base.inc    ; Reg base defn
15                           INCLUDE  bits.inc    ; Bit defns
16                           INCLUDE  porta.inc   ; Port adr defns
17                           INCLUDE  portb.inc
18                       ;*******************************
19                       ; Constants
20         0000 0001 LED1:    EQU    %00000001 ; LED1 on Port A-0
21         0000 0010 LED2:    EQU    %00010000 ; LED2 on Port B-4
22                       ; Code Section
23                       MyCode: SECTION
24                       Entry:
25                       main:
26 000000 CFxx xx            lds    #__SEG_END_SSTACK
27                       ; Initialize the I/O
28 000003 4C02 01            bset   DDRA,LED1 ; Make PA-0 output
29 000006 4C03 10            bset   DDRB,LED2 ; Make PB-4 output
30 000009 8601              ldaa   #LED1     ; Initial LED
                                                display
31 00000B C600              ldab   #!LED2
32                       ; Enable the timer
33 00000D 4C46 80            bset   TSCR1,TEN
34                       ; Reset the timer overflow flag
35 000010 8680              ldaa   #TOF
36 000012 5A4F              staa   TFLG2
37                       main_loop:
38                       ; DO
39                       ; Display current value on the LEDs
40 000014 5A00              staa   PTA
41 000016 5B01              stab   PTB
42                       ; Delay approximately one sec
43 000018 16xx xx            jsr   tof_1_sec_delay
44                       ; Complement the display values
45 00001B 41                coma          ; Next display value
46 00001C 51                comb
```

```
47                      ; FOREVER
48 00001D 20F5              bra    main_loop
49                      ;*******************************
50                      ; Timer Overflow 1 second delay
51                      ; Registers modified: CCR
52                      ;*******************************
53                      ;
54                      ; Constant Equates
55        0000 007A SEC_1:  EQU    122   ; Number of TOFs for 1
                                               second
56
57                      SubCode: SECTION
58                      ; One second delay using the TOF
59                      tof_1_sec_delay:
60                      ; Save the registers
61 000000 36                psha
62 000001 34                pshx
63                      ; Initialize the TOF counter
64 000002 CE00 7A            ldx    #SEC_1
65                      ; DO
66                      ;   Wait until the TOF occurs
67                      spin:
68 000005 F700 4F           tst    TFLG2
69 000008 2AFB              bpl    spin
70                      ; Reset the TOF
71 00000A 8680              ldaa   #TOF
72 00000C 5A4F              staa   TFLG2
73                      ; Decrement the TOF counter
74 00000E 09                dex
75                      ; ENDO
76                      ; WHILE (number overflows < SEC_1)
77 00000F 26F4              bne    spin
78                      ; Now done, return
79                      ; Restore the registers
80 000011 30                pulx
81 000012 32                pula
82 000013 3D                rts
83                      ;*******************************
84                      ; No constant or variable data section
                        needed
```

EXPLANATION OF EXAMPLE 14-1

Example 14-1 shows how to use the timer overflow occurring at intervals of 8.192 ms to delay for longer periods. The initialization section (*lines 28–33*) sets the data direction registers for the bits in port A and port B that are connected to the LEDs, the timer is enabled in *line 33*, and the timer overflow flag is reset in *lines 35* and *36*. An overflow counter using the X register is initialized in *line 64*. *Lines 68* and *69* are a spin loop waiting for the TOF bit to be set. When it is, the flag is reset in *lines 71* and *72* and the counter is decremented (*line 74*). If the counter is not zero, the spin loop is reentered; otherwise, the subroutine returns to the calling program. Example 14-2 shows a C program version of this example.

The delay in Example 14-1 has a resolution one-half of the period of the timer overflow, ±4.096 ms. If you would like to generate an exact delay, to the resolution of the bus clock (125 ns), you could count the extra clock cycles needed to make up the delay. A better and easier way is to use the *output compare* function discussed in Section 14.4.

Example 14-2 C Program Version of the Timer Overflow Flag Program

```
/*****************************************************************
 * This is an example of using the timer overflow flag
 * to generate a delay of approximately 1 sec. (The delay
 * will be 122 timer overflow flags which is 0.999 second
 * with an 8 MHz bus clock.
 * The program will alternately blink LED1 and LED2 on
 * the Student Learning Kit with a CSM-12C32 board
 *****************************************************************/
#include <mc9s12c32.h>2              /* derivative information */
void tof_1_sec_delay( void );
void main(void) {
  /* Initialize the I/O */
     DDRA_BIT0 = 1;   /* Make Port A bit-0 output */
     DDRB_BIT4 = 1;   /* Port B bit-4 output */
  /* Enable the timer */
     TSCR1_TEN = 1;
  /* Reset the timer overflow flag */
     TFLG2 = TFLG2_TOF_MASK;
  /* Set LED1 on and LED2 off */
```

---

[2] All registers and bits in registers are defined in the mc9s12c32.h header file supplied by the CodeWarrior® environment. The header defines the memory addresses for the control registers, such as DDRA, and assignment to these are unsigned char types. The header file defines bits within registers as, for example, DDRA_BIT0. Assignments to these are bit assignments using bset or bclr instructions. Your C program can access registers as bytes or individual bits.

```
            PORTA_BIT0 = 0; /* Active low LEDs */
            PORTB_BIT4 = 1;
    /* DO */
        for(;;) {
          /* Delay 1 second */
            tof_1_sec_delay();
          /* Toggle the display bits */
            if (PORTA_BIT0 == 1) PORTA_BIT0 = 0;
            else PORTA_BIT0 = 1;
            if (PORTB_BIT4 == 1) PORTB_BIT4 = 0;
            else PORTB_BIT4 = 1;
        } /* wait forever */
    /* FOREVER */
}
/*****************************************************************
 * Timer overflow 1 second delay
 ****************************************************************/
#define SEC_1  122  /* Number of TOFs per second */
void tof_1_sec_delay( void ) {
    int counter;
/*****************************************************************/
    /* Initialize the counter */
        counter = SEC_1;
        do {
          /* Wait until the TOF */
          while (TFLG2_TOF == 0){/* spin */}
          /* Reset the TOF */
          TFLG2 = TFLG2_TOF_MASK;
        }
        while (--counter > 0);
    }
```

14.3 Timer Overflow Interrupts

Timer interrupts allow your program to do other things while waiting for a timing event to occur.

A shortcoming of the program in Example 14-1 is that the TOF bit must be polled until 122 overflows have occurred. During this time the program could be doing other things but an overflow might be missed. An interrupt can allow the program to go about some other business while waiting for an event—the timer overflow, for example—to occur.

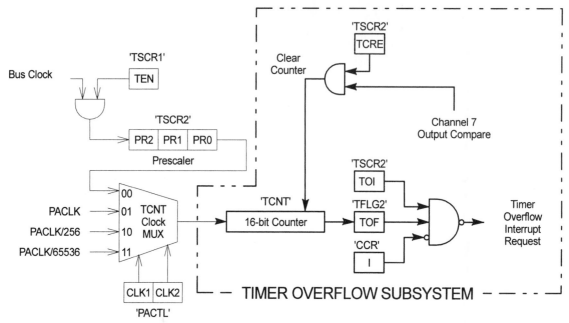

Figure 14-2 Timer overflow interrupt hardware.

To use the timer overflow interrupt, the *TOI* bit in *TSCR2* must be enabled, the interrupt vector must be initialized as described in Chapter 12, and the I bit in the condition code register must be unmasked.

The *timer overflow enable* bit must be set to allow the interrupt request to be generated.

The timer overflow flag (TOF) is ANDed with the timer overflow interrupt enable bit (TOI) to generate the interrupt request as shown in Figure 14-2. This request is further qualified by the interrupt mask bit (I bit) in the condition code register as we discussed in Chapter 12. The timer overflow interrupt vector must be initialized properly to be able to transfer to the interrupt service routine (see Table 14-3). An interrupt service routine using the timer overflow flag is given in pseudocode in Table 14-4, an assembly program in Example 14-3, and a C program in Example 14-6.

TABLE 14-3 Timer Overflow Interrupt Vector

Priority	Vector Address	Interrupt Source	Local Enable Bit	See Register	HPRIO Value to Promote
16	$FFDE:FFDF	Timer Overflow	TOI	TSCR2	$DE

TABLE 14-4 Pseudocode for Timer Overflow Interrupt Program

Program Requirements: Use the timer overflow interrupt to flash LED1 on the CSM-12C32 CPU board at 1 Hz.

```
; Initialize I/O to make PA-0 output
; Display the initial value on LED1
; Initialize a counter for 1/2 second
; Enable the timer
; Reset the timer overflow flag
; Enable timer overflow interrupts
; Unmask interrupts
; DO
;    Wait for an interrupt
;    IF it is time to toggle the display
;    THEN
;        Toggle the display bit
;        Clear the toggle flag
;    ENDIF
; FOREVER

; Timer Overflow Interrupt Service Routine
; Decrement the 1/2 second counter
; IF the counter is zero
; THEN
;    Set the display toggle flag
;    Reset the 1/2 second counter
; ENDIF
; Clear the Timer Overflow Flag
; Return from interrupt
```

Example 14-3 Timer Overflow Interrupts

```
Metrowerks HC12-Assembler
(c) COPYRIGHT METROWERKS 1987-2003

Rel. Loc    Obj. code Source line

---- ------ --------- -----------

    1                    ; This is a test program showing the use
    2                    ; of the Timer Overflow Interrupts.
    3                    ; The program uses the WAI instruction
    4                    ; while waiting for an interrupt.
    5                    ; After 1/2 second has elapsed, LED1 on
    6                    ; the Student Learning Kit CSM-12C32
    7                    ; board is toggled.
    8                    ;*******************************
    9                    ; Define the entry point for the main
                         program
   10                        XDEF    Entry, main
   11                        XDEF    tof_pa0_isr
   12                        XREF    __SEG_END_SSTACK
```

```
13                        INCLUDE   timer.inc  ; Timer defns
14                 .      INCLUDE   bits.inc   ; Bit defns
15                        INCLUDE   base.inc   ; Reg base defn
16                        INCLUDE   porta.inc  ; Port A defn
17                 ;*******************************
18                 ; Constants
19      0000 0001 LED1:    EQU    BIT0             ; LED1 on Port
                                                     A - 0
20      0000 003D HALF_SEC: EQU 61                 ; TOFs for 1/2
                                                     second
21                 ; Code Section
22                 MyCode: SECTION
23                 Entry:
24                 main:
25 000000 CFxx xx         lds    #__SEG_END_SSTACK
26                 ; DO Initialize the I/O
27 000003 4C02 01         bset   DDRA,LED1     ; Make PA-0
                                                  output
28 000006 8601            ldaa   #LED1         ; Initial LED
                                                  display
29                 ;    Display on the LEDs
30 000008 5A00            staa   PTA
31                 ;    Initialize the 1/2 second counter
32 00000A 180B 3Dxx       movb   #HALF_SEC,half_sec_count
   00000E xx
33                 ;    Enable the Timer
34 00000F 4C46 80         bset   TSCR1,TEN
35                 ;    Reset timer overflow flag
36 000012 8680            ldaa   #TOF
37 000014 5A4F            staa   TFLG2
38                 ;    Enable the TOF interrupts
39 000016 4C4D 80         bset   TSCR2,TOI
40                 ;    Unmask interrupts
41 000019 10EF            cli
42                 ; ENDO Initialize the I/O
43                 ; DO
44                 main_loop:
45 00001B 3E              wai                      ; Wait for
                                                     interrupt
46                 ;    IF the toggle flag is set
47 00001C 1Fxx xx80       brclr toggle_flag,BIT7,end_if
   000020 0A
```

```
48                          ;    THEN toggle the bit and clear the flag
49 000021 9600                          ldaa    PTA
50 000023 8801                          eora    #LED1              ; Toggle the
                                                                      bit
51 000025 5A00                          staa    PTA
52 000027 1Dxx xx80                     bclr    toggle_flag,BIT7 ; Clear flag
53                          end_if:
54 00002B 20EE                          bra     main_loop
55                          ; FOREVER
56                          ;*****************************
57                          ; Timer overflow ISR
58                          ;*****************************
59                          tof_pa0_isr:
60                          ; Blinks the LED1 on Port A, Bit 0 at 1 Hz
61                          ; Decrement the counter
62 00002D 73xx xx                       dec     half_sec_count
63                          ; IF the counter is zero
64 000030 2609                          bne     endif
65                          ; THEN set the flag to toggle the LED
66 000032 1Cxx xx80                     bset    toggle_flag,BIT7
67                          ;    Initialize the counter
68 000036 180B 3Dxx                     movb    #HALF_SEC,half_sec_count
00003A xx
69                          ; ENDIF counter is zero
70                          endif:
71                          ; Clear the TOF
72 00003B 8680                          ldaa    #TOF
73 00003D 5A4F                          staa    TFLG2
74 00003F 0B                            rti
75                          ;*****************************
76                          ;
77                          Data:  SECTION
78 000000                   half_sec_count: DS.B  1
79 000001                   toggle_flag:    DS.B  1
```

EXPLANATION OF EXAMPLE 14-3

Interrupts are used here as we described in Chapter 12. The main program consists of an initialization section, *lines 26–42*, where the port A direction is set and the LED turned on (*lines 27, 28, and 32*), a counter used in the interrupt service routine is initialized (*line 32*), and the timer is enabled (*line 34*). Then the timer overflow flag is reset (*line 37*), the TOF interrupts are enabled (*line 39*), and interrupts are unmasked (*line 41*). The main loop waits for an interrupt with the WAI instruction at *line 45* and then checks to see if 0.5 second has elapsed to toggle the LED (*lines 47–52*).

The interrupt service routine is entered each time the timer overflows, at intervals of 8.192 ms. At half-second intervals, or 61 timer overflows, a *toggle_flag* is set to be used in the main program to toggle LED. Before returning from the interrupt service routine, the timer overflow flag is reset in *lines 72* and *73*. See Examples 14-4 and 14-5.

Example 14-4 Clearing the TOF

Why is the TOF bit cleared in *line 37* in Example 14-3?

Solution: If the TOF bit is set when the timer interrupts are enabled and interrupts are unmasked, an interrupt will occur immediately. This may upset the required timing for the first iteration of the program.

Example 14-5 Clearing the TOF

Why is the TOF bit cleared in *line 73* in Example 14-3?

Solution: If the TOF bit is not cleared in the interrupt service routine, as soon as the RTI instruction is executed the TOF bit will immediately generate another interrupt request. The program will appear to never get out of the ISR.

Example 14-6 Timer Overflow Interrupts in C

```
/****************************************************************
 * This is a test program showing the use of the Timer
 * Overflow interrupt. The program uses the WAI instruction
 * while waiting for an interrupt. After 1/2 second has elapsed,
 * LED1 on the Student Learning Kit CSM-12C32 board
 * is toggled.
 ***************************************************************/
#include <mc9s12c32.h>      /* derivative information */
#define HALF_SEC  61        /* TOFs per 1/2 sec */
static int half_sec_counter;
void main(void) {
/***************************************************************/
  /* Initialize the I/O */
    DDRA_BIT0 = 1;  /* Make Port A bit-0 output */
```

```
        /* Set LED1 on */
           PORTA_BIT0 = 0; /* Active low LEDs */
        /* Initialize 1/2 second counter */
           half_sec_counter = HALF_SEC;
        /* Clear the timer overflow flag */
           TFLG2 = TFLG2_TOF_MASK;
        /* Enable the timer */
           TSCR1_TEN = 1;
        /* Enable the TOF interrupts */
           TSCR2_TOI = 1;
        /* Unmask interrupts */
           EnableInterrupts;

        /* DO */
          for(;;) {
            /* Wait for the interrupt */
               asm (wai );
            /* If the half_sec_counter is zero */
            if (half_sec_counter == 0) {
              /* Then toggle the bit and reset the counter */
                 if (PORTA_BIT0 == 1) PORTA_BIT0 = 0;
                 else PORTA_BIT0 = 1;
                 half_sec_counter = HALF_SEC;
            }
        /* FOREVER */
          }
}


/****************************************************************
 * Timer overflow interrupt service routine

 ****************************************************************/
#define SEC_1  122  /* Number of TOFs per second */
void interrupt 16 tof_pa0_isr( void ) {
  /* Decrement the counter */
     --half_sec_counter;
  /* Reset the TOF */
  TFLG2 = TFLG2_TOF_MASK;
}
```

14.4 Output Compare

The timer overflow flag and interrupt discussed in the last section is suitable for timing to a resolution of $\pm 2^{15}$ clock cycles (± 4.096 ms for an 8-MHz clock). This may be sufficiently accurate for many applications, but when timing that is more precise is needed, the *output compare* features of the HCS12 timer can be used.

The output compare hardware is shown in Figure 14-3. The 16-bit TCNT register is clocked by the bus clock and operates as described in the previous section. A 16-bit *Timer Input Capture/Output Compare* register, *TCn*,[3] may be loaded by the program with a 16-bit

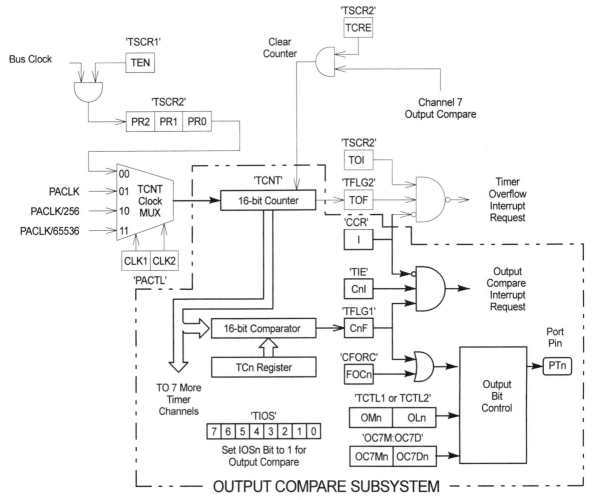

Figure 14-3 Timer output compare hardware.

---

[3] The n is 0–7 for each of the timer channels. Any of the channels can be input capture or output compare.

load instruction. This register is compared with the current value of the TCNT register at every bus clock cycle, and when the TCn is identical to TCNT, the *Input Capture/Output Compare Channel Flag*, *CnF*, is set. Tracing through Figure 14-3, you can see that CnF is ANDed with a *Timer Interrupt Enable* bit (*CnI*) to generate an interrupt when both the flag and the enable bit are set and interrupts are unmasked. CnF is also ORed with a bit called *FOCn* (*Timer Compare Force* bit-n) and routed to the output bit control logic. Two bits, *OMn* and *OLn*, control the action on the port T output bit when a successful output compare occurs.

Each of the timer channels may be configured to give an input capture or an output compare function. Port T bits can be controlled by the output compare functions or can serve as the input signals for the input captures. As we saw in Chapter 11, port T can be used for general purpose I/O if the timer functions are not being used. The choice is made by programming the *Timer Input Capture/Output Compare Select* (*TIOS*) register. When bits in TIOS are set, the output compare bit is controlled by the OMn:OLn bits in TCTL1 and TCTL2. When bits in TIOS are zero, the associated port T bit can be an input capture or an I/O bit whose direction is controlled by DDRT.

After you have initialized the TIOS register to select which channel is to be an Output Compare, and have initialized the corresponding Timer Input/Output Compare (TCn) register with a 16-bit value, and have reset the Input Capture/Output Compare Channel Flag (CnF), you may start to generate very accurate timing signals. When the TCNT and TCn registers are identical, the CnF is set. You can use this event in one of two ways. You can *poll* the flag and when it is set accomplish the timing that you want, or you can use an *interrupt* that can be generated. You can use the timing information in your program, say, for generating a time delay for some purpose, or you can allow the flag to generate an action on the associated port T output bit.

TIOS—Base + $0040—Timer Input Capture/Output Compare Select

	Bit 7	6	5	4	3	2	1	0
Read:	IOS7	IOS6	IOS5	IOS4	IOS3	IOS2	IOS1	IOS0
Write:								
Reset:								

Read: Anytime. Write: Anytime.

IOS7:IOS0

Input Capture or Output Compare Channel Configuration
0 = The corresponding channel acts as input capture or an I/O bit (default).
1 = The channel acts as an output compare.
When IOS[7:0] is zero and EDGnB:EDGnA = %00 (see Section 14.5), the port T bit can act as an I/O pin whose direction is controlled by the state of the data direction register DDRT.

TC0—Base + $0050, $0051—Timer Input Capture/Output Compare Register 0
TC1—Base + $0052, $0053—Timer Input Capture/Output Compare Register 1
TC2—Base + $0054, $0055—Timer Input Capture/Output Compare Register 2
TC3—Base + $0056, $0057—Timer Input Capture/Output Compare Register 3
TC4—Base + $0058, $0059—Timer Input Capture/Output Compare Register 4

TC5—Base + $005A, $005B—Timer Input Capture/Output Compare Register 5

TC6—Base + $005C, $005D—Timer Input Capture/Output Compare Register 6

TC7—Base + $005E, $005F—Timer Input Capture/Output Compare Register 7

Bit 15	14	13	12	11	10	9	8
TC15	TC14	TC13	TC12	TC11	TC10	TC9	TC8

Reset: 0 (Bit 15), 0 (14), 0 (13), 0 (12), 0 (11), 0 (10), 0 (9), 0 (8)

Bit 7	6	5	4	3	2	1	0
TC7	TC6	TC5	TC4	TC3	TC2	TC1	TC0

Reset: 0 (Bit 7), 0 (6), 0 (5), 0 (4), 0 (3), 0 (2), 0 (1), 0 (0)

Read: Anytime. Write: Anytime for output compare operations. Writes have no effect in pulse accumulator mode.

TFLG1—Base + $004E—Timer Interrupt Flag 1

	Bit 7	6	5	4	3	2	1	0
Read:	C7F	C6F	C5F	C4F	C3F	C2F	C1F	C0F
Write:								
Reset:	0	0	0	0	0	0	0	0

Read: Anytime. Write: Used in the clearing process (writing a one clears the bit).

CnF

Timer Interrupt Flags

When the corresponding timer channel is configured for output compare, the flag is set when the TCNT and TCn register contents are identical.

When the corresponding timer channel is configured for input capture, the flag is set when a selected edge is detected at the timer input pin. See Section 14.5.

The flag is reset by writing a one to the register. See Section 14.8.

Output Compare Time Delays

The output compare can generate timing delays with much higher accuracy than the timer overflow flag. Consider generating a delay that is less than 8.192 ms, for example, 1 ms. In a system with an 8-MHz bus clock, a 1-ms delay is 8000 bus clock cycles. In Example 14-7, a 1-ms delay is generated by reading the current contents of the TCNT register (*line 32*), adding 8000 cycles to it (*line 34*), and storing this value into the TC1 register (*line 35*). The C1F bit is reset in *lines 37* and *38* by writing a one to the register. The program then waits for C1F to be set again in a spin loop at *line 42*. One millisecond after the TCNT register was read in *line 32* the C1F bit is set and the program will drop out of the spin loop. Table 14-5 shows the pseudocode and Example 14-7 the assembly program. See Example 14-8 for the same program in C.

TABLE 14-5 Pseudocode Design for 500-Hz Square Wave

Program Requirements: Generate a 500-Hz square wave on port T, bit 0 using the output compare.

```
; Initialize Port T bit-0 for output
; Enable the timer
; Enable Output Compare Channel 1
; Set the output bit high
; Delay 1 ms
;     Get the current value of the TCNT register
;     Add the number of clocks for 1 ms delay
;     Store in TC1
;     Reset the Output Compare channel 1 flag
; DO
;     Wait until the flag is set again
;     Toggle the output bit
;     Add the number of clocks for 1 ms delay to TC1
;     Reset the Output Compare channel 1 flag
; FOREVER
```

Example 14-7 A 1-ms Delay Program Using the Output Compare

```
Metrowerks HC12-Assembler
(c) COPYRIGHT METROWERKS 1987-2003

Rel. Loc    Obj. code Source line
---- ------ --------- -----------

    1                     ; This is a test program showing how to
    2                     ; use the Output Compare to generate time
    3                     ; delays. It outputs a 500 Hz square
    4                     ; wave on Port T, Bit 0.
    5                     ;********************************
    6                     ; Define the entry point for the main
                            program
    7                             XDEF     Entry, main
    8                             XREF     __SEG_END_SSTACK
    9                             INCLUDE  timer.inc   ; Timer defns
   10                             INCLUDE  base.inc    ; Reg base defn
   11                             INCLUDE  bits.inc    ; Bit defns
   12                             INCLUDE  portt.inc   ; Port T defns
   13                     ;********************************
   14                     ; Constants
   15    0000 1F40 MS_1:   EQU     8000        ; # Clocks for a ms
   16    0000 0001 LED1:   EQU     BIT0        ; LED1 on Port A - 0
   17                     ; Code Section
   18                     MyCode: SECTION
```

```
19                     Entry:
20                     main:
21 000000 CFxx xx          lds    #__SEG_END_SSTACK
22                     ; Initialize the I/O
23 000003 1C02 4201        bset   DDRT,BIT0 ; Make PT-0 output
24                     ; Enable the timer
25 000007 4C46 80          bset   TSCR1,TEN
26                     ; Enable Output Compare Channel 1
27 00000A 4C40 02          bset   TIOS,IOS1
28                     ; Set the output bit high
29 00000D 1C02 4001        bset   PTT,BIT0
30                     ; Delay 1 ms
31                     ; Grab the value of the TCNT register
32 000011 DC44            ldd    TCNT
33                     ; Add the number of clocks for 1 ms delay
34 000013 C31F 40         addd   #MS_1
35 000016 5C52            std    TC1   ; Set up the next compare
36                     ; Now reset the flag
37 000018 8602            ldaa   #C1F
38 00001A 5A4E            staa   TFLG1
39                     ; and wait until it is set
40                     ; DO
41                     spin:
42 00001C 4F4E 02FC        brclr TFLG1,C1F,spin
43                     ;   Timed out, so toggle the bit
44 000020 B602 40         ldaa   PTT
45 000023 8801            eora   #BIT0
46 000025 7A02 40         staa   PTT
47                     ;   Get the current time
48 000028 DC52            ldd    TC1
49                     ;   Add the number of clocks for 1 ms delay
50 00002A C31F 40         addd   #MS_1
51 00002D 5C52            std    TC1
52                     ;   Now reset the output compare 1 flag
53 00002F 8602            ldaa   #C1F
54 000031 5A4E            staa   TFLG1
55                     ;   and repeat
56 000033 20E7             bra    spin
57                     ; FOREVER
58                     ;*******************************
```

Example 14-8 A 1-ms Delay Program in C

```
/****************************************************************
 * This program shows how to use the output compare
 * channel 1 to generate time delays. It outputs a
 * 500 Hz squarewave on Port T, bit-0
 ****************************************************************/
#include <mc9s12c32.h>            /* derivative information */
#define  MS_1  8000               /* Number of clocks per ms */
/****************************************************************/
void main(void) {
/****************************************************************/
    /* Initialize the I/O */
    DDRT_DDRT0 = 1;  /* Make PT-0 output */
    TSCR1_TEN = 1;  /* Enable the timer */
    TIOS_IOS1 = 1;  /* Enable Output Compare 1 */
  /* set the output bit high */
    PTT_PTT0 = 1;
  /* Grab the value of the TCNT register and
   * Add the clocks for 1 ms delay */
    TC1 = TCNT + MS_1;
  /* DO */
    for(;;) {
    /* Reset the OC flag */
       TFLG1 = TFLG1_C1F_MASK;
    /* Wait until it is set */
       while (TFLG1_C1F == 0) {}
    /* Timed out so toggle the bit */
       if (PTT_PTT0 == 1) PTT_PTT0 = 0;
       else PTT_PTT0 = 1;
    /* Set up the next output compare time */
       TC1 += MS_1;
    }
  /* FOREVER */
}
```

CHANGING THE TIMER PRESCALER

Example 14-7 shows an output compare time delay that can have at most an 8.192-ms delay when an 8-MHz bus clock is used. Figure 14-3 shows that the TSCR2 register has three prescaler bits PR2:PR1:PR0. These can be changed to select a prescaler of 1, 2, 4, 8, 16, 32, 64, or 128. Example 14-9 shows a program to generate a 10-ms delay. *Lines 24* and *25* set the prescaler bits to 010 to divide by four. Now, the clock period is 0.5 μs and a 10-ms delay is represented by 20,000 counts. Previously, each clock period was 0.125 μs and a 10-ms delay was impossible to achieve.

If you use this technique to generate longer time delays, remember that any changes you make will affect all timing being done by the timer system in other parts of the program.

Example 14-9 Changing the Timer Prescaler

```
Metrowerks HC12-Assembler
(c) COPYRIGHT METROWERKS 1987-2003
Rel. Loc    Obj. code Source line
---- ------ --------- -----------
   1                   ; This is a test program showing how to
   2                   ; use the Output Compare to generate time
   3                   ; delays longer than 8.192 ms by changing
   4                   ; the TCNT register prescaler.
   5                   ; It outputs a 10 ms wide pulse
   6                   ; Port T, Bit 0.
   7                   ;*******************************
   8                   ; Define the entry point for the main
                         program
   9                          XDEF     Entry, main
  10                          XREF     __SEG_END_SSTACK
  11                          INCLUDE  timer.inc   ; Timer defns
  12                          INCLUDE  base.inc    ; Reg base defn
  13                          INCLUDE  bits.inc    ; Bit defns
  14                          INCLUDE  portt.inc
  15                   ;*******************************
  16                   ; Constants
  17        0000 4E20 MS_10:  EQU    20000         ; # Clocks
                                                     for 10 ms
  18                   ; Code Section
  19                   MyCode: SECTION
  20                   Entry:
  21                   main:
  22 000000 CFxx xx           lds    #__SEG_END_SSTACK
  23                   ; Set the prescaler to divide by 4
  24 000003 4C4D 02           bset   TSCR2,PR1     ; PR1
  25 000006 4D4D 05           bclr   TSCR2,PR2|PR0 ; PR2, PR0
  26                   ; Initialize the I/O
  27 000009 1C02 4201         bset   DDRT,BIT0     ; Make PT-0
                                                     output
  28                   ; Enable the timer
```

```
29 00000D 4C46 80              bset   TSCR1,TEN
30                        ; Enable Output Compare Channel 1
31 000010 4C40 02              bset   TIOS,IOS1
32                  main_loop:
33                ; DO
34                ;   Set the output bit high
35 000013 1C02 4001     bset   PTT,BIT0
36                ;   Delay 10 ms
37                ;   Grab the value of the TCNT register
38 000017 DC44          ldd    TCNT
39                ;   Add the number of clocks for 10 ms delay
40 000019 C34E 20       addd   #MS_10
41 00001C 5C52          std    TC1   ; Set up the next compare
42                ;   Now reset the flag
43 00001E 8602          ldaa   #C1F
44 000020 5A4E          staa   TFLG1
45                ;   and wait until it is set
46                  spin:
47 000022 4F4E 02FC     brclr TFLG1,C1F,spin
48                ;   Timed out, so set the bit low
49 000026 1D02 4001     bclr   PTT,BIT0
50                ;   Now reset the flag
51 00002A 8602          ldaa   #C1F
52 00002C 5A4E          staa   TFLG1
53                ;   and repeat
54 00002E 20E3          bra    main_loop
55                ; FOREVER
56                ;******************************
```

C code to change the prescaler bits is

```
/* . . . */
/* PR2:PR1:PR0 = "010" = divide by 4 */
TSCR2_PR2 = 0;
TSCR2_PR1 = 1;
TSCR2_PR0 = 0;
/* . . . */
```

Output Compare Interrupts

An interrupt can be generated by the output compare flag if, like the timer overflow flag, the *Timer Interrupt* enable bit (*CnI*) is set. The enable bits for all timer functions are in the *Timer Interrupt Enable* (*TIE*) register.

TIE—Base + $004C—Timer Interrupt Enable Register

	Bit 7	6	5	4	3	2	1	0
Read:								
	C7I	C6I	C5I	C4I	C3I	C2I	C1I	C0I
Write:								
Reset:	0	0	0	0	0	0	0	0

Read: Anytime. Write: Anytime.

CnI

Timer Interrupt Flags

0 = The corresponding bit in TFLG1 is disabled from generating interrupts (default).

1 = The corresponding bit in TFLG1 may generate an interrupt.

Delays longer than 8.192 ms can be generated by changing the prescaler as shown in Example 14-9 or by waiting for more output comparisons to be made. Examples 14-10 and 14-11 show how to generate a 1-second delay using the output compare flag to generate an interrupt. (See also Table 14-6.) The delay is achieved by waiting for 250 complete 4-ms delay times generated by the output compare. This is done by reading the TNCT register in *line 34* and then generating an interrupt every 4 ms after that. After 250 interrupts, the LED will toggle. A counter for this is initialized in *line 32*. After the C2F flag is cleared and interrupts are

TABLE 14-6 Pseudocode for 1-second Delay Using Output Compare Interrupts

Program Requirements: Generate a 1-second delay using the output compare interrupts. Flash LED1 at a 0.5-Hz rate to demonstrate the delay.

```
; Initialize I/O to make PA-0 output and turn the LED off
; Enable the timer
; Enable Output Compare channel 2
; Initialize a counter to count 250 interrupts occurring at 4 ms intervals
; Read the current value of the TCNT register
; Save it in TC2
; Clear OC2 interrupt flag
; Enable OC2 interrupts
; Unmask interrupts
; DO
;     WHILE the counter is not zero
;         Wait for an interrupt
;     END WHILE
;     Reinitialize the counter
;     Toggle the LED display
; FOREVER

: Output Compare Channel 2 Interrupt Service Routine
; Decrement the counter
; Get the current value of TC2
; Add the number of clock pulses for 4 ms
; Store in TC2
; Clear the OC2 interrupt flag
; Return from interrupt
```

Example 14-10 A 1-second Delay Using Output Compare Interrupts

```
Metrowerks HC12-Assembler
(c) COPYRIGHT METROWERKS 1987-2003
Rel. Loc    Obj. code Source line
---- ------ --------- -----------
    1                 ; This is a test program showing a
    2                 ; 1 second delay using output compare
    3                 ; and interrupts. It flashes LED1 on the
    4                 ; Student Learning Kit with a CSM-12C32
                        CPU.
    5                 ;*******************************
    6                 ; Define the entry point for the main
                        program
    7                         XDEF     Entry, main,OC2_isr
    8                         XREF     __SEG_END_SSTACK
    9                         INCLUDE  timer.inc   ; Timer defns
   10                         INCLUDE  base.inc    ; Reg base defn
   11                         INCLUDE  bits.inc    ; Bit defns
   12                         INCLUDE  porta.inc   ; Port A defns
   13                 ;*******************************
   14                 ; Constants
   15     0000 7D00 MS_4:   EQU   32000    ; # Clocks for 4 ms
   16     0000 00FA NTIMES: EQU   250      ; Number of interrupts
   17     0000 0001 LED1:   EQU   BIT0     ; LED1 on Port A - 0
   18                 ; Code Section
   19                 MyCode: SECTION
   20                 Entry:
   21                 main:
   22 000000 CFxx xx          lds   #__SEG_END_SSTACK
   23                 ; Initialize the I/O
   24 000003 4C02 01          bset  DDRA,LED1 ; Make PA-0 output
   25 000006 4C00 01          bset  PTA,LED1; Initial LED display
   26                 ; Enable the timer
   27 000009 4C46 80          bset  TSCR1,TEN
   28                 ; Enable Output Compare Channel 2
   29 00000C 4C40 04          bset  TIOS,IOS2
```

```
30                          ; Generate a 1 sec delay
31                          ; Need NTIMES interrupts
32 00000F 180B FAxx         movb  #NTIMES,counter
   000013 xx
33                          ;   Grab the value of the TCNT register
34 000014 DC44              ldd   TCNT
35 000016 5C54              std   TC2
36                          ; Now have 8 ms to set up the system
37                          ; Set up the interrupts
38 000018 8604              ldaa  #C2F
39 00001A 5A4E              staa  TFLG1  ; Clear C2F
40                          ; Enable OC2 interrupts
41 00001C 4C4C 04           bset  TIE,C2I
42                          ; Unmask global interrupts
43 00001F 10EF              cli
44                          ; Wait until the counter is zero
45                    spin:
46 000021 3E                wai         ; Wait for interrupt
47 000022 F7xx xx           tst   counter
48 000025 26FA              bne   spin
49                          ; When the counter is zero
50                          ; Reinitialize the counter
51 000027 180B FAxx         movb  #NTIMES,counter
   00002B xx
52                          ; and toggle the LED
53 00002C 8601              ldaa  #LED1
54 00002E 9800              eora  PTA
55 000030 5A00              staa  PTA
56                          ; Return to wait for next interrupt
57 000032 20ED              bra   spin
58                          ; FOREVER
59                          ;******************************
60                          ; Output Compare Channel 2 overflow ISR
61                    OC2_isr:
62                          ; Decrement the counter the main program
63                          ; uses to delay for 1 sec
```

```
64 000034 73xx xx          dec    counter
65                     ; Set up TC2 for the next interrupt
66 000037 DC54             ldd    TC2
67                     ; Add the clock pulses
68 000039 C37D 00          addd   #MS_4
69 00003C 5C54             std    TC2
70                     ; Clear the OC2 flag
71 00003E 8604             ldaa   #C2F
72 000040 5A4E             staa   TFLG1
73 000042 0B               rti
74                     ;******************************
75                 Data:   SECTION
76 000000          counter: DS.B   1
```

Example 14-11 A 1-second Delay in C

```c
/*****************************************************************
 * This test program shows a one second delay using output
 * compare and interrupts. It flashes LED1 on the Student
 * Learning Kit with a CSM-12C32 CPU
 *****************************************************************/
#include <mc9s12c32.h>       /* derivative information */
#define NTIMES   250         /* Number of interrupts for 1 second */
#define MS_4     32000       /* Number of clocks for 4 ms */
/*****************************************************************/
void interrupt 10 OC2_isr (void);
    int counter;            /* Global counter for interrupts */
/*****************************************************************/
void main(void) {
/*****************************************************************/
  /* Initialize I/O */
    DDRA_BIT0 = 1;          /* Make Port A-0 output */
    PORTA_BIT0 = 0;         /* Turn the LED1 on */
    TSCR1_TEN = 1;          /* Enable the timer */
    TIOS_IOS2 = 1;          /* Enable output compare channel 2 */
  /* Initialize the interrupt counter */
    counter = NTIMES;
  /* Set up TC2 */
    TC2 = TCNT;
```

```
/* Now have ~ 8 ms to set up the system without an
 * interrupt
 */
  TFLG1 = TFLG1_C2F_MASK;      /* Clear C2F flag */
  TIE_C2I = 1;                 /* Enable the interrupt */
  EnableInterrupts;
/* DO */
  for(;;) {
    /* Wait until the counter is zero */
      while (counter > 0) {
        asm (wai);             /* Wait for interrupt */
      }
    /* When the counter is zero reinitialize it */
      counter = NTIMES;
    /* and toggle the LED */
      if (PORTA_BIT0 == 0) PORTA_BIT0 = 1;
      else PORTA_BIT0 = 0;
  /* FOREVER */
  }
}
/****************************************************************
 * The interrupt service routine decrements the counter
 * sets up TC2 for the next interrupt and resets the flag.
 ****************************************************************/
void interrupt 10 OC2_isr ( void ){
/****************************************************************/
  /* Decrement the counter */
    --counter;
  /* Set up TC2 */
    TC2 += MS_4;
  /* Clear the flag */
    TFLG1 = TFLG1_C2F_MASK; /* Clear C2F flag */
}
```

enabled and unmasked (*lines 38–43*), the processor waits for the interrupt to occur (*line 46*). When it does, the counter is checked to see if it is zero. After the counter reaches zero, it is reinitialized to NTIMES and the LED is toggled. The interrupt service routine decrements the counter in *line 64*. With a bus clock of 8 MHz, each time an interrupt occurs MS_4 (32,000 bus clock cycles) is added to the TC2 register in *lines 66–69*. Finally, the flag is cleared in *lines 71* and *72*. (See also Example 14-12.)

> **Example 14-12**
>
> In Example 14-10 the programmer calculates the count for the next interrupt by adding 32,000 clock cycles to the current value of the TC2 register (*lines 68–71*). Why didn't the programmer first read the TCNT register and then add 32,000 to find the time for the next interrupt?
>
> **Solution:** Interrupts are required every 4 ms in this example. If the TCNT register is used every time to calculate the time for the next interrupt, the time interval will be longer than 4 ms because the TCNT register increments with every clock cycle.

OUTPUT COMPARE INTERRUPT VECTORS

Each of the eight timer channels can generate interrupts and each has its own interrupt vector as given in Table 14-7.

Output Compare Bit Operation

Look again at Figure 14-3 and see that the output compare flags pass through an OR gate to a block called *Output Bit Control* and then to port T. The port T pins are multipurpose and may be programmed to be simple I/O pins, as shown in Chapter 11, or for use by the output compare functions. Let us first look at how C0F to C7F can be used as output compare pins. The registers that control this function are the *Timer Control Register 1* and *Timer Control Register 2*. When a successful output comparison is made, one of four actions may occur at the output pin in port T. It can be either disconnected, toggled, cleared, or set. Two bits for each port T output, *OMn* and *OLn*, are programmed to make this selection. OM7:OL7–OM0:OL0 are in the timer control registers TCTL *1* and TCTL *2*. Example 14-13 shows how to set bits OM2:OL2 to toggle automatically the port T-2 when output compares occur every 2 ms to generate a 250-Hz square wave. (See also Example 14-14.) Example 14-15 shows a C version of this program.

If you wish the output bit to toggle each time an output compare is made (overflows), instead of setting OMn:OLn to %01, the TOVn bit in the *Timer Toggle on Overflow* (*TTOV*) can be set.

TABLE 14-7 Output Compare Interrupt Vectors

Priority	Vector Address	Interrupt Source	Local Enable Bit	See Register	HPRIO Value to Promote
15	$FFE0:FFE1	Timer Channel 7	C7I	TIE	$E0
14	$FFE2:FFE3	Timer Channel 6	C6I	TIE	$E2
13	$FFE4:FFE5	Timer Channel 5	C5I	TIE	$E4
12	$FFE6:FFE7	Timer Channel 4	C4I	TIE	$E6
11	$FFE8:FFE9	Timer Channel 3	C3I	TIE	$E8
10	$FFEA:FFEB	Timer Channel 2	C2I	TIE	$EA
9	$FFEC:FFED	Timer Channel 1	C1I	TIE	$EC
8	$FFEE:FFEF	Timer Channel 0	C0I	TIE	$EE

TCTL1—Base + $0048—Timer Control Register 1
TCTL2—Base + $0049—Timer Control Register 2

	Bit 7	6	5	4	3	2	1	0
TCTL1	OM7	OL7	OM6	OL6	OM5	OL5	OM4	OL4
Reset:	0	0	0	0	0	0	0	0

	Bit 7	6	5	4	3	2	1	0
TCTL2	OM3	OL3	OM2	OL2	OM1	OL1	OM0	OL0
Reset:	0	0	0	0	0	0	0	0

Read: Anytime. Write: Anytime.

OMn:OLn

Output Compare Bit Control

OMn = Output Mode.
OLn = Output Level.

These pairs of bits specify the output action to be taken as a result of a successful output compare. When either OMn or OLn is one, the pin associated with OCn becomes an output tied to OCn regardless of the state of the associated DDRT bit. (See Table 14-8.)

TABLE 14-8 Compare Result Output Action

OMn	OLn	Bit-n Action
0	0	Timer disconnected from output pin logic (default).
0	1	Toggle OCn output line.
1	0	Clear OCn output line to zero.
1	1	Set OCn output line to one.

Example 14-13 Output Compare Bit Operation

```
Metrowerks HC12-Assembler

(c) COPYRIGHT METROWERKS 1987-2003


Rel. Loc     Obj. code Source line

---- ------ --------- -----------

      1                 ; This is a test program showing a
      2                 ; 250 Hz square wave using output compare
```

```
 3                       ; bit control and interrupts
 4                       ;********************************
 5                       ; Define the entry point for the main program
 6                                XDEF     Entry, main,OC2_isr
 7                                XREF     __SEG_END_SSTACK
 8                                INCLUDE  timer.inc   ; Timer defns
 9                                INCLUDE  base.inc    ; Reg base defn
10                                INCLUDE  bits.inc    ; Bit defns
11                       ;********************************
12                       ; Constants
13      0000 3E80 HALF_P: EQU    16000              ; 1/2 period
14                       ; Code Section
15                       MyCode: SECTION
16                       Entry:
17                       main:
18 000000 CFxx xx                lds    #__SEG_END_SSTACK
19                       ; Initialize the I/O
20                       ; Enable the timer
21 000003 4C46 80                bset   TSCR1,TEN
22                       ; Enable Output Compare Channel 2
23 000006 4C40 04                bset   TIOS,IOS2
24                       ; Grab the value of the TCNT register
25 000009 DC44                   ldd    TCNT
26 00000B 5C54                   std    TC2
27                       ; Now have 8 ms to set up the system
28                       ; Set up the interrupts
29 00000D 8604                   ldaa   #C2F
30 00000F 5A4E                   staa   TFLG1  ; Clear C2F
31                       ; Enable OC2 interrupts
32 000011 4C4C 04                bset   TIE,C2I
33                       ; Set up output compare action to toggle
34                       ; bit-2 (DDRT does not need to be set)
35 000014 4D49 20                bclr   TCTL2,OM2
36 000017 4C49 10                bset   TCTL2,OL2
37                       ; Unmask global interrupts
38 00001A 10EF                   cli
39                       spin:
40 00001C 3E                     wai        ; Wait for interrupt
```

```
41 00001D 20FD              bra    spin
42                  ;*******************************
43                  ; Output Compare Channel 2 ISR
44                  ;*******************************
45                  OC2_isr:
46                  ; Set up TC2 for the next interrupt
47 00001F DC54              ldd    TC2
48                  ; Add the clock pulses
49 000021 C33E 80           addd   #HALF_P
50 000024 5C54              std    TC2
51                  ; Clear the OC2 flag
52 000026 8604              ldaa   #C2F
53 000028 5A4E              staa   TFLG1
54 00002A 0B               rti
55                  ;*******************************
```

Example 14-14

Write a short section of code to cause port T, bit 5 to clear the bit when an output comparison is made.

Solution: Port T, bit 5 is connected to Output Compare 5. Therefore use the code

```
; Set OM5, OL5 in TCTL1 to 1 0
      bset   TCTL1, %00001000
      bclr   TCTL1, %00000100
```

Example 14-15 Output Compare Bit Operation in C

```
/****************************************************************
 * This program outputs a 250 Hz square wave on
 * Port T, bit-0 using timer channel 0 interrupts and the
 * output compare bit operations.
 ****************************************************************/
#include <mc9s12c32.h>     /* derivative information */
#define  HALF_P  16000      /* Number clocks per 1/2 period */
```

```
/*****************************************************************/
void interrupt 8 OC0_isr (void);
/*****************************************************************/
void main(void) {
/*****************************************************************/
  /* Initialize I/O */
    TSCR1_TEN = 1;            /* Enable the timer */
    TIOS_IOS0 = 1;           /* Enable output compare channel 0 */
  /* Set up TC0 */
    TC0 = TCNT;
  /* Now have ~ 8 ms to set up the system without an
   * interrupt
   */
    TFLG1 = TFLG1_C0F_MASK;/* Clear C0F flag */
    TIE_C0I = 1;             /* Enable the interrupt */
  /* Set up the output compare action to toggle
   * Port T, bit-0. (DDRT does not need to be set.)
   */
    TCTL2_OM0 = 0;
    TCTL2_OL0 = 1;
    EnableInterrupts;       /* Unmask interrupts */
  /* DO */
    for(;;) {
        asm (wai);          /* Wait for interrupt */
  }
  /* FOREVER */
}
/*****************************************************************
 * The interrupt service routine simply sets up the TC0
 * for the next interrupt time.
 *****************************************************************/
void interrupt 8 OC0_isr ( void ){
/*****************************************************************/
  /* Set up TC0 */
    TC0 += HALF_P;
  /* Clear the flag */
    TFLG1 = TFLG1_C0F_MASK; /* Clear C2F flag */
}
```

TTOV—Base + $0047—Timer Toggle on Overflow Register

	Bit 7	6	5	4	3	2	1	0
Read:	TOV7	TOV6	TOV5	TOV4	TOV3	TOV2	TOV1	TOV0
Write:								
Reset:	0	0	0	0	0	0	0	0

Read: Anytime. Write: Anytime.

TOVn

Toggle on Overflow

0 = Toggle output compare pin on overflow disabled (default).
1 = Toggle output compare pin on overflow enabled.

When this bit is set and the associated output compare channel overflows, the output compare pin is toggled. This is the same as setting OMn:OLn = %01.

One Output Compare Controlling Up to Eight Outputs

Output Compare 7 can simultaneously switch up to eight
outputs.

The Output Compare 7 channel has special features that are controlled by the *OC7M* and *OC7D* registers. OC7M and OC7D work together to define the action taken on port T, bits 7–0. OC7D is a data register and its contents are transferred to the output bits on port T when a successful output comparison is made on channel 7. OC7M is a mask register and a 1 in a bit position in the mask means that the corresponding data bit in the data register, OC7D, is transferred to the output bit in port T. Thus up to 8 bits can be simultaneously changed by one output comparison. This is useful in applications where bit streams are controlling devices that must be changed in synchronism. Any change designated by a bit in OC7D overrides any action from the other output compare channels. See Examples 14-16 and 14-17.

Example 14-16 Control Four Bits with One Output Compare

Show an assembly language program to output the waveforms shown in Figure 14-4(a), where the output high times are 2, 1, 0.5, and 0.25 ms.

Solution:

```
Metrowerks HC12-Assembler
(c) COPYRIGHT METROWERKS 1987-2003

Rel. Loc  Obj. code Source line
---- ---  --------- ------ ----
       1              ; This example shows how to control
       2              ; four bits simultaneously using the
```

```
 3                       ; output compare channel 7.
 4                       ;*******************************
 5                       ; Define the entry point for the main program
 6                               XDEF      Entry, main
 7                               XREF      __SEG_END_SSTACK
 8                               INCLUDE   timer.inc  ; Timer defns
 9                               INCLUDE   base.inc   ; Reg base defn
10                               INCLUDE   bits.inc   ; Bit defns
11                       ;*******************************
12                       ; Constants
13      0000 00F0 D_BITS: EQU   %11110000          ; Initial
                                                     pattern
14      0000 00F0 O_BITS: EQU   %11110000          ; The bits to
                                                     output
15      0000 3E80 DELAY1: EQU   16000              ; 2 ms - Ch 7
16      0000 1F40 DELAY2: EQU   8000               ; 1 ms - Ch 6
17      0000 0FA0 DELAY3: EQU   4000               ; 0.5 ms - Ch 5
18      0000 07D0 DELAY4: EQU   2000               ; 0.25 ms - Ch 4
19                       ;*******************************
20                       ; Code Section
21                       MyCode: SECTION
22                       Entry:
23                       main:
24 000000 CFxx xx                lds    #__SEG_END_SSTACK
25                       ; Initialize the I/O
26                       ; Enable Output Compare Ch 7-4
27 000003 4C40 F0                bset   TIOS,O_BITS
28                       ; Initialize the mask register
29 000006 86F0                   ldaa   #O_BITS
30 000008 5A42                   staa   OC7M
31                       ; Initialize the data register
32 00000A 86F0                   ldaa   #D_BITS
33 00000C 5A43                   staa   OC7D
34                       ; Enable the timer
35 00000E 4C46 80                bset   TSCR1,TEN
36                       ; Wait until the C7F is set
37                       main_loop:
38                       ; DO
39                       ;   Wait until output compare 7 hits
```

```
40                    spin:
41 000011 4F4E 80FC        brclr TFLG1,C7F,spin
42                 ;    The data in OC7D is output to bits 7 - 4
43                 ;    Now set up for the next interval on
44                 ;    each channel
45 000015 DC5E             ldd   TC7
46 000017 C307 D0          addd  #DELAY4  ; 0.25 ms
47 00001A 5C58             std   TC4
48 00001C DC5E             ldd   TC7
49 00001E C30F A0          addd  #DELAY3  ; 0.5 ms
50 000021 5C5A             std   TC5
51 000023 DC5E             ldd   TC7
52 000025 C31F 40          addd   #DELAY2 ; 1.0 ms
53 000028 5C5C             std   TC6
54 00002A DC5E             ldd   TC7
55 00002C C33E 80          addd   #DELAY1 ; 2.0 ms
56 00002F 5C5E             std   TC7
57                 ;   Reset C7F
58 000031 86C0              ldaa  #C7F|C6F
59 000033 5A4E              staa  TFLG1
60                 ;   Toggle the bits in OC7D
61 000035 9643              ldaa  OC7D
62 000037 88F0              eora  #O_BITS
63 000039 5A43              staa  OC7D
64                 ; FOREVER
65                 ;   Return to spin
66 00003B 20D4              bra   spin
67                 ;******************************
```

Setting the comparison values in the TC6–TC0 registers shorter than what is in TC7 allows these output comparisons also to switch their outputs as shown in Example 14-16. When the channel 7 output compare matches, OC7F is set and the data bits in OC7D are transferred to the channels whose mask bits are set in OC7M. At this time the program falls out of the spin loop at *line 41* and reinitializes TC4 to 0.25 ms (*line 46*), TC5 to 0.5 ms (*line 49*), TC6 to 1 ms (*line 52*), and TC7 to 2 ms (*line 55*). This gives the waveform shown in Figure 14-4(a). If TC6–TC0 registers have comparison values equal to or greater than TC7, the TC7 comparison causes all output bits to switch at the same time. See Figure 14-4(b) and Example 14-18.

OC7M—Base + $0042—Output Compare 7 Mask Register

	Bit 7	6	5	4	3	2	1	0
Read:	OC7M7	OC7M6	OC7M5	OC7M4	OC7M3	OC7M2	OC7M1	OC7M0
Write:								
Reset:	0	0	0	0	0	0	0	0

Read: Anytime. Write: Anytime.

OC7Mn

Setting any bit in OC7M enables the corresponding bit in OC7D to be output to the corresponding pin in port T. Setting OC7Mn sets the port T bit to be an output regardless of the state of bits in DDRT.

OC7D—Base + $0043—Output Compare 7 Data Register

	Bit 7	6	5	4	3	2	1	0
Read:	OC7D7	OC7D6	OC7D5	OC7D4	OC7D3	OC7D2	OC7D1	OC7D0
Write:								
Reset:	0	0	0	0	0	0	0	0

Read: Anytime. Write: Anytime.

OC7Dn

When a bit in OC7M is set, the data value in the corresponding bit in OC7D is output to the corresponding pin in port T when the output compare in channel 7 is made.

EXPLANATION OF EXAMPLE 14-16

Example 14-16 illustrates how to control 4 bits simultaneously using one output compare. PT7–PT4, defined by O_BITS in *line 14,* are to be toggled. The initial data state for these bits is %1111xxxx as defined in *line 13* by D_BITS. Output compare channels 7 to 4 are initialized for use in *lines 27–35.* All four channels must be activated as output compare channels (IOSn = 1). The mask register, OC7M, is set up in *line 30* and the initial data is stored in OC7D in *line 33.* A spin loop is entered at *line 41* to wait for the output compare to be made on channel 7. When it does, the data in OC7D is transferred to the output pins automatically. *Lines 45–56* set up each channel for the next output compare time and reset the C7F flag. *Lines 61–63* then complement the data bits in the OC7D data register. Because each of the channels 4 to 6 have an output comparison time shorter than channel 7, their values will be switched to the value of the bit in OC7D when their comparisons are made at 0.25, 0.5, and 1.0 ms, respectively.

Example 14-17 Control Four Bits with One Output Compare in C

Show a C program to output the waveforms shown in Figure 14-4(a) where the output high times are 2, 1, 0.5, and 0.25 ms.

Solution:

```c
/***********************************************************
 * This example shows how to control four bits simultaneously
 * using the output compare channel 7 with each channel
 * switching at the same time.
 ***********************************************************/
#include <mc9s12c32.h>   /* derivative information */
#define DELAY1  16000    /* Number clocks for 2 ms */
#define DELAY2   8000    /* 1 ms */
#define DELAY3   4000    /* 0.5 ms */
#define DELAY4   2000    /* 0.25 ms */
#define O_BITS 0xF0      /* Bits to be output */
#define D_BITS 0xF0      /* Initial data value to output */
/***********************************************************/
void main(void) {
/***********************************************************/
  /* Initialize I/O */
    TIOS = O_BITS;          /* Enable output compare bits 7 - 4 */
    OC7M = O_BITS;          /* Initialize the mask register */
    OC7D = D_BITS;          /* Initialize the data register */
    TSCR1_TEN = 1;          /* Enable the timer */
  /* DO */
    for(;;) {
      /* Wait until output compare 7 hits */
        while (TFLG1_C7F == 0) {}
      /* The data in OC7D is output to bits 7 - 4 */
      /* Now set up each channel for the next bit change */
        TC4 = TC7+DELAY4;
        TC5 = TC7+DELAY3;
        TC6 = TC7+DELAY2;
        TC7 = TC7+DELAY1;
      /* Reset the output compare flag */
        TFLG1 = TFLG1_C7F_MASK;
      /* Toggle the bits in the data register */
        OC7D = OC7D ^ O_BITS;
    }
  /* FOREVER */
}
```

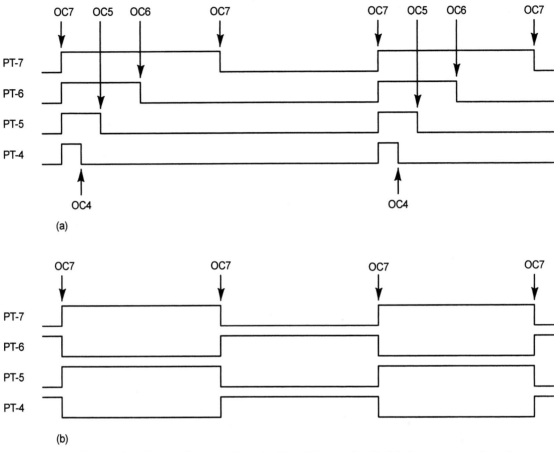

Figure 14-4 Output Compare 7 controlling 4 bits each with (a) shorter comparison times and (b) equal comparison times.

Example 14-18 Control Four Bits with One Output Compare in C (Equal Comparison Values)

Show a C program to output the waveforms shown in Figure 14-4(b) where the output high times are 2 ms.

Solution:

```
/****************************************************************
 * This example shows how to control four bits simultaneously
 * using the output compare channel 7 with each channel
 * switching at the same time.
 ****************************************************************/
#include <mc9s12c32.h>      /* derivative information */
#define DELAY1  16000       /* Number clocks for 2 ms */
#define O_BITS 0xF0         /* Bits to be output */
```

```
#define D_BITS 0xA0          /* Initial data value to output */
/*****************************************************************/
void main(void) {
/*****************************************************************/
  /* Initialize I/O */
    TIOS = O_BITS;        /* Enable output compare bits 7 - 4 */
    OC7M = O_BITS;        /* Initialize the mask register */
    OC7D = D_BITS;        /* Initialize the data register */
    TSCR1_TEN = 1;        /* Enable the timer */
  /* DO */
    for(;;) {
      /* Wait until output compare 7 hits */
        while (TFLG1_C7F == 0) {}
      /* The data in OC7D is output to bits 7 - 4 */
      /* Now set up each channel for the next bit change */
        TC4 = TC7;
        TC5 = TC7;
        TC6 = TC7;
      /* Reset the output compare flag */
        TFLG1 = TFLG1_C7F_MASK;
      /* Toggle the bits in the data register */
        OC7D = OC7D ^ O_BITS; //O_BITS;
    }
    /* FOREVER */
}
```

Very Short Duration Pulses

Pulses as short as one bus clock
period can be generated.

Output Compare 7 can be used with another Output Compare channel to produce very short duration pulses. In Example 14-19 a 2-µs pulse is generated using Output Compare 7 and 6. OC7 can control the normal OC6 output bit by using OC7M6 and OC7D6. In this example, OC7 is set to output a 1 on port T, bit 6. Sixteen clock cycles later OC6 will reset the bit to zero. (See Examples 14-19 and 14-20.)

Example 14-19 A Very Short Duration Pulse

```
Metrowerks HC12-Assembler
(c) COPYRIGHT METROWERKS 1987-2003

 Rel. Loc  Obj. code Source line
 --------  --------- -----------
    1                ; This is a test program showing how
```

```
 2                         ; to use output compare 6 and 7 to
 3                         ; generate a 2 µs pulse on port T,  bit-6.
 4                         ;*******************************
 5                         ; Define the entry point for the main program
 6                                 XDEF     Entry, main
 7                                 XREF     __SEG_END_SSTACK
 8                                 INCLUDE  timer.inc   ; Timer defns
 9                                 INCLUDE  base.inc    ; Reg base defn
10                                 INCLUDE  bits.inc    ; Bit defns
11                                 INCLUDE  portt.inc   ; Port T defns
12                         ;*******************************
13                         ; Constants
14      0000 0010 PULSE:  EQU    16               ; 2 µs
15                         ;*******************************
16                         ; Code Section
17                         MyCode: SECTION
18                         Entry:
19                         main:
20000000 CFxx xx                  lds    #__SEG_END_SSTACK
21                         ; Initialize the I/O
22                         ; Enable the timer
23000003 4C46 80                  bset   TSCR1,TEN
24                         ; Enable Output Compare Channel 6 and 7
25000006 4C40 C0                  bset   TIOS,IOS6|IOS7
26                         ;   Grab the value of the TCNT register
27000009 DC44                     ldd    TCNT
2800000B 5C5E                     std    TC7
29                         ; Set TC6 to compare PULSE cycles later
3000000D C300 10                  addd   #PULSE
31000010 5C5C                     std    TC6
32                         ; Reset the output compare flags
33000012 86C0                     ldaa   #C7F|C6F
34000014 5A4E                     staa   TFLG1
35                         ; Initialize the mask register
36000016 8640                     ldaa   #BIT6
37000018 5A42                     staa   OC7M
38                         ; Initialize the data register
3900001A 5A43                     staa   OC7D   ; Will set it high
40                         ; Set up OC6 to reset the bit
```

```
41                      ; Set OM6=1, OL6=0
4200001C 4C48 20              bset    TCTL1,OM6
4300001F 4D48 10              bclr    TCTL1,OL6
44                      ; Wait until C7F is set to set the output
45                      spin7:
46000022 4F4E 80FC            brclr TFLG1,C7F,spin7
47                      ; Wait until C6F is set to reset the output
48                      spin6:
49000026 4F4E 40FC            brclr TFLG1,C6F,spin6
50                      ; Reset the C7F and C6F
5100002A 86C0                 ldaa    #C7F|C6F
5200002C 5A4E                 staa    TFLG1
53                      ; Return to spin
5400002E 20F2                 bra     spin7
55                      ;*******************************
```

EXPLANATION OF EXAMPLE 14-19

To create a very short pulse we first set up output compare channel 7 to successfully compare at some time in the future (*line 28*) and then set output compare channel 6 to compare 2 µs later by initializing TC6 in *line 31*. OC7M and OC7D are set to output a one on port T6 when output compare 7 is made (*lines 36–39*) and output compare 6 resets the bit (*lines 42–43*). The two spin loops at *line 46* and *line 49* wait until OC7 and then OC6 are made.

Forced Output Compares

The final feature of the timer output comparison section is the *Forced Output Compare*. Figure 14-3 shows that the *CFORC* register bit (*FOCn*) is ORed with the output compare flag. Writing a one to the CFORC register forces a comparison action to occur at the output pins. This forced comparison does not set the output compare flag and therefore no interrupt will be generated.

Example 14-20

Example 14-19 shows a program to output a 2-µs pulse on port T, bit 6. What is the period of this pulse?

Solution: The output compare registers are not changed after their initialization. Therefore the period of the pulse is 8.192 ms.

CFORC—Base + $0041—Timer Compare Force Register

	Bit 7	6	5	4	3	2	1	0
Read:								
	FOC7	FOC6	FOC5	FOC4	FOC3	FOC2	FOC1	FOC0
Write:								
Reset:	0	0	0	0	0	0	0	0

Read: Anytime (but will always return $00). Write: Anytime.

FOC7:FOC0

Force Output Compare Action for Channels 7–0.

Writing a one to any of these bits causes the action programmed by OMn:OLn bits to occur. The action taken will be the same as if a successful comparison had just taken place except the interrupt flag is not set.

Output Compare Software Checklist

Here are the steps to check when writing your software for timer output compares:

1. Initialize the interrupt vector(s) for the timer channel(s) if interrupts are to be used.
2. Set bit 7 (TEN) in TSCR1 to enable the timer.
3. Set bits in TIOS to enable the Output Compare channels.
4. Initialize the OMn and OLn bits in TCTL1 and TCTL2 if the timer output pins are to be activated.
5. Set the bits in OC7M and OC7D if any timer channel 7 override is to be done.
6. Load the current value for TCNT and add to it the required delay.
7. Store the (TCNT + delay) value into the TCn register to be used.
8. Reset the flag(s) in TFLG1.
9. Enable any required interrupts in TIE.
10. Unmask interrupts by clearing the I bit in the condition code register.
11. Wait for the output compare to set the flag by either polling the flag or waiting for the interrupt.
12. After the output compare action occurs, reinitialize the TCn register with the new delay value by adding the delay to the current TCn value.
13. Reset the CnF flag bit to prepare for the next output compare.

14.5 Input Capture

Input Capture allows the TCNT register value to be latched when an external event occurs.

The input capture hardware is shown in Figure 14-5. Again, the 16-bit free-running TCNT register is the heart of the system, and the same eight, 16-bit *Timer Input Capture/Output Compare* registers, *TC7–TC0*, latch the value of the free-running counter in response to a program-selected, external signal coming from port T. For example, the period of a pulse train can be found by capturing the TCNT at the start of the period, signified by a rising or falling edge, and storing it. The next rising or falling edge will capture the count

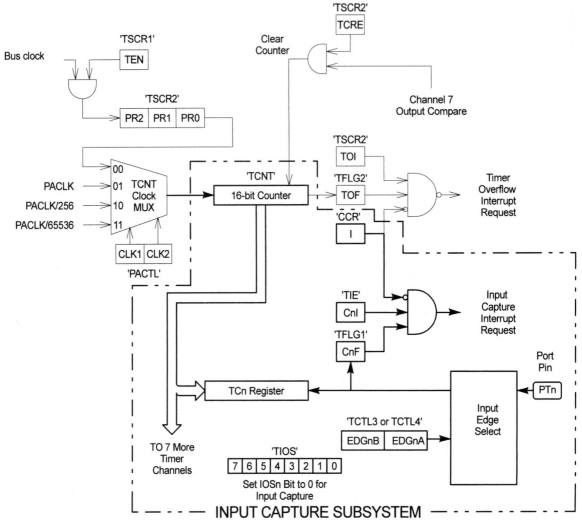

Figure 14-5 Input capture hardware.

at the end of the period. The difference in the two counts, taking into account timer overflows, will be the period in bus clock cycles. The length of the positive pulse can be measured by capturing the time at the rising edge and then again at the falling edge.

Two bits for each Input Capture channel, *EDGnB* and *EDGnA* in *Timer Control Registers 3* and *4*, control when the signal on port T causes the capture event to occur. You may select rising, falling, or both edges to be active.

The input capture interrupts operate just like output compare interrupts. The interrupt enable bits in TIE must be set. Then, when the flag in TFLG1 is set by the selected input capture edge, the interrupt request is forwarded to the CPU.

Example 14-21 shows a subroutine that measures the period of a waveform, providing it is less than 8.192 ms. If the period if longer than this, you can change the counter prescaler as shown in Example 14-9 or keep track of timer overflows. In this case, input capture 1 is to be

TCTL3—Base + $004A—Timer Control Register 3
TCTL4—Base + $004B—Timer Control Register 4

	Bit 7	6	5	4	3	2	1	0
TCTL3	EDG7B	EDG7A	EDG6B	EDG6A	EDG5B	EDG5A	EDG4B	EDG4A
Reset:	0	0	0	0	0	0	0	0

	Bit 7	6	5	4	3	2	1	0
TCTL4	EDG3B	EDG3A	EDG2B	EDG2A	EDG1B	EDG1A	EDG0B	EDG0A
Reset:	0	0	0	0	0	0	0	0

Read: Anytime. Write: Anytime.

OMn:OLn

Input Capture Edge Control

These pairs of bits configure the input capture edge selection. (See Table 14-9.)

TABLE 14-9 Input Capture Edge Selection

EDGnB	EDGnA	Edge to Capture TCNT
0	0	Capture disabled (default).
0	1	Capture on rising edges only.
1	0	Capture on falling edges only.
1	1	Capture on any edge (falling or rising).

used; it is enabled in *line 28*, and a rising edge is selected in *lines 31* and *32*. The IC1 flag is reset in *lines 34* and *35*. The program then waits until the first positive edge appears on Input Capture 1 in *line 38*. When this happens, the contents of the TCNT register are latched into the TC1 register and the program leaves the *line 38* spin loop. The first count is saved in a buffer in *line 41*, the IC1 flag reset, and the second spin loop (*line 47*) is entered. After the second rising edge, the duration of the pulse is calculated by subtracting the second TCNT value from the first. Note that modulo 2^{16} arithmetic will return the correct number of clock pulses even if the second TCNT value is less than the first. (See also Example 14-22.)

Example 14-21 A Subroutine to Measure the Period of a Waveform
Write a C-callable function in assembly language that will measure the period of a waveform on input capture channel 2.

Solution:

```
Metrowerks HC12-Assembler

(c) COPYRIGHT METROWERKS 1987-2003
```

```
Rel. Loc   Obj. code  Source line
---- ---    ---------  -----------
   1                   ;*******************************
   2                   ; Subroutine get_period
   3                   ;*******************************
   4                   ; C Callable Function
   5                   ; int get_period( void );
   6                   ;*******************************
   7                   ; Subroutine to measure the period of
   8                   ; a wave. It measures the time
   9                   ; in TCNT units between two successive
  10                   ; rising edges on Input Capture channel 1.
  11                   ; Limited to waveforms with frequency
  12                   ; higher than 122 Hz with an 8 MHz clock.
  13                   ; Input Parameters: None
  14                   ; Output Parameters: D register returns
  15                   ;     the period in clock cycles
  16                   ; Registers Modified: A, B, CCR
  17                   ; Stack Use: 2 bytes
  18                   ;*******************************
  19                   ; Define the entry point for the main program
  20                           XDEF     get_period
  21                           INCLUDE  timer.inc   ; Timer defns
  22                   ;*******************************
  23                   get_period:
  24                   ;*******************************
  25                   ; Enable the timer system
  26 000000 4C46 80        bset   TSCR1,TEN
  27                   ; Reset TIOS bit to enable input capture
  28 000003 4D40 02        bclr   TIOS,IOS1       ; Channel 1
  29                   ; Initialize IC1 for rising edge
  30                   ; EDG1B=0, EDG1A=1
  31 000006 4D4B 08        bclr   TCTL4,EDG1B
```

```
32 000009 4C4B 04              bset   TCTL4,EDG1A
33                     ; Reset C1F flag
34 00000C 8602               ldaa   #C1F
35 00000E 5A4E               staa   TFLG1
36                     ; Wait for the next rising edge
37                     spin1:
38 000010 4F4E 02FC          brclr TFLG1,C1F,spin1
39                     ; Now get the count that was latched
40 000014 DC52               ldd    TC1
41 000016 3B                 pshd        ; Save count on the stack
42                     ; Reset C1F flag
43 000017 8602               ldaa   #C1F
44 000019 5A4E               staa   TFLG1
45                     ; Wait for the next rising edge
46                     spin2:
47 00001B 4F4E 02FC          brclr TFLG1,C1F,spin2
48                     ; Get the ending count
49 00001F DC52               ldd    TC1
50                     ; Calculate the period
51 000021 A380               subd   0,SP ; Subtract the first count
52                     ; Return with the value in D
53 000023 1B82               leas   2,sp  ; Adjust SP
54 000025 3D                 rts
55                     ;******************************
```

Example 14-22

Show how the calculation in *line 51* of Example 14-21, which calculates the number of TCNT clock cycles between two successive rising edges, can return the correct answer if the second number captured in *line 49* is less than the first number captured in *line 40*.

Solution: Modulo 2^{16} arithmetic is being used on the 16-bit number. To see how this works, take a simple example with, say, a 3-bit number. Let the first count be 101_2 and the second number 100_2. For this to occur, seven clock cycles must have elapsed. The 3-bit arithmetic, $100_2 - 101_2 = 111_2$.

Input Capture Software Checklist

Here is a list of the steps to check when writing your software for input captures:

1. Initialize the interrupt vector(s) for the timer channel(s) if interrupts are to be used.
2. Set bit 7 (TEN) in TSCR to enable the timer.
3. Reset bits in TIOS to disable the Output Compare channels.
4. Initialize the EDGnB and EDGnA bits in TCTL3 and TCTL4 to select the active edge for the input capture trigger signal.
5. Reset the flag(s) in TFLG1.
6. Enable any required interrupts in TMSK1.
7. Unmask interrupts by clearing the I bit in the condition code register.
8. Wait for the input capture to set the flag by either polling the flag or waiting for the interrupt.
9. After the input capture event occurs, use the data in the TCn register.
10. Reset the CnF flag bit to prepare for the next input capture.

14.6 Pulse Accumulator

> The *pulse accumulator* can be used to count external events.

The pulse accumulator is a 16-bit counter that can operate as an *event counter*, counting external clock pulses, or a *gated time accumulator*. In gated time operation, the bus clock is divided by 64 and gated by the pulse accumulator input into the accumulator. These operating modes are shown in Figure 14-6. Timer channel 7 (port T7) is used as the input for the pulse accumulator. You may select the edge (positive or negative) for event counting or the level (high or low) for gated time accumulation. There are three registers to be programmed when using the pulse accumulator. The *PACTL* register enables the system, selects the mode of operation, and enables pulse accumulator interrupts. Bits in PACTL also select the clock source for the TCNT counter as shown in Figure 14-1. The *PAFLG* register contains the interrupt flags and the 16-bit *PACNT* register records the accumulated input pulses.

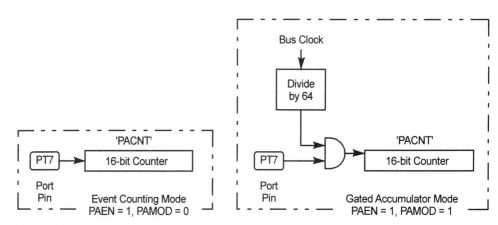

Figure 14-6 Pulse accumulator operating modes.

PACTL—Base + $0060—16-bit Pulse Accumulator Control Register

	Bit 7	6	5	4	3	2	1	0
Read:	0	PAEN	PAMOD	PEDGE	CLK1	CLK0	PAOVI	PAI
Write:								
Reset:	0	0	0	0	0	0	0	0

▓▓▓ = reserved, unimplemented, or cannot be written to.

Read: Anytime. Write: Anytime.

PAEN

Pulse Accumulator System Enable

0 = Pulse accumulator system disabled (default).

1 = Enabled.

PAEN is independent of the timer system enable bit TEN. The pulse accumulator can function (if PAEN = 1) when the timer is disabled (TEN = 0).

PAMOD

Pulse Accumulator Mode Select

0 = Event counter mode (default).

1 = Gated time accumulation mode.

PEDGE

Pulse Accumulator Edge Control

For PAMOD = 0 (event counter mode):

0 = Falling edges on PT7 cause the count to be increased (default).

1 = Rising edges are counted.

For PAMOD = 1 (gated time accumulation mode):

0 = PT7 going high enables the bus clock/64 to pulse the accumulator. The subsequent falling edge of PT7 sets the *Pulse Accumulator Input Edge* (*PAIF*) flag.

1 = PT7 going low enables the clock to the accumulator. The subsequent rising edge of PT7 sets the PAIF flag.

If the timer is not enabled (TEN = 0), there is no divided by 64 clock since this clock is generated by the timer prescaler.

CKL1:CLK0

Clock Select Bits

The clock select bits control a multiplexer that selects the clock used by the TCNT counter. (See Figure 14-1.) If the pulse accumulator is disabled (PAEN = 0), the prescaler clock is always used as an input clock to the timer counter. For more information, see Section 14.7. (See Table 14-10.)

TABLE 14-10 Timer Clock Selection

CLK1	CLK0	TCNT Clock	TCNT Clock Frequency
0	0	Use timer prescaler clock as TCNT counter clock.	Bus clock/prescaler (default)
0	1	Use PACLK as input to the TCNT counter clock.	Bus clock/64
1	0	Use PACLK/256 as TCNT counter clock.	Bus clock/16,384
1	1	Use PACLK/65536 as TCNT counter clock.	Bus clock/16,777,216

PAOVI

Pulse Accumulator Overflow Interrupt Enable

0 = Interrupt disabled (default).
1 = Interrupt requested if PAOVF is set.

PAIF

Pulse Accumulator Input Interrupt Enable

0 = Interrupt disabled (default).
1 = Interrupt requested if PAIF is set.

PACNT—Base + $0062—16-bit Pulse Accumulator Counter Register

Bit 15	14	13	12	11	10	9	8
PACNT15	PACNT14	PACNT13	PACNT12	PACNT11	PACNT10	PACNT9	PACNT8

Reset: 0   0   0   0   0   0   0   0

Bit 7	6	5	4	3	2	1	0
PACNT7	PACNT6	PACNT5	PACNT4	PACNT3	PACNT2	PACNT1	PACNT0

Reset: 0   0   0   0   0   0   0   0

Read: Anytime. Write: Anytime.

This 16-bit register holds the number of accumulated input pulses (active edges) since the last reset. You may write it at any time and you should use a 16-bit read to read simultaneously the high and low bytes. When PACNT overflows from $FFFF to $0000, the PAOVF overflow flag in the PAFLG register is set.

Pulse Accumulator Interrupts

The pulse accumulator can interrupt when the counter register overflows and/or when an input edge is detected.

Pulse accumulator interrupts operate like the other functions in the timer section. A flag is set by the hardware when the appropriate condition is true, and if an interrupt enable is set, the interrupt request is generated. There are two flags and two interrupts that can be generated; these are controlled by the *PAOVI* and *PAI* bits in the *PACTL* register. When the pulse accumulator overflows, the PAOVF flag bit in the PAFLG register is set. PAOVI enables this flag to generate an interrupt request. *PAI* is the *Pulse Accumulator Input Edge* interrupt enable. When the selected input edge occurs (chosen by PEDGE in PACTL), the *Pulse Accumulator Input Edge Flag* (*PAIF*) in the PAFLG is set. If PAI is set, an interrupt is generated. These interrupts are enabled by following the procedures outlined in the previous sections.

The pulse accumulator can interrupt the processor after a number of external events have occurred. Let us say that a sensor on a conveyor belt is detecting a product passing by and a crate is to be filled after 24 counts. Example 14-23 shows pseudocode to initialize the interrupt and to do the interrupt service routine.

PAFLG—Base + $0061—Pulse Accumulator Flag Register

	Bit 7	6	5	4	3	2	1	0
Read:	0	0	0	0	0	0	PAOVF	PAIF
Write:								
Reset:	0	0	0	0	0	0	0	0

 = reserved, unimplemented, or cannot be written to.

Read: Anytime. Write: Anytime.

PAOVF

Pulse Accumulator Overflow Flag

This bit is set when the 16-bit pulse accumulator overflows from $FFFF to $0000. The bit is cleared by writing a one to the bit (see Section 14.8).

PAIF

Pulse Accumulator Input Edge Flag

This bit is set when the selected edge (PEDGE) is detected at the IOC7 input pin. In event mode (PAMOD = 0) the event edge sets PAIF. In the gated time accumulation mode (PAMOD = 1) the trailing edge of the gate signal sets PAIF.

This bit is cleared by writing a one to the bit (see Section 14.8).

Example 14-23 Pulse Accumulator Overflow Interrupt Pseudocode Design

An interrupt service routine will be used. The pulse accumulator is initialized with −24 and is incremented with each external event. After 24 counts, the pulse accumulator overflows and generates an interrupt. The pseudocode design is

```
; ENABLE the pulse accumulator in event counter mode and select the
    correct input edge in PACTL.
; CLEAR the pulse accumulator overflow flag PAOVF in PAFLG.
; SET the PAOVI bit in PACTL to enable pulse accumulator overflow
    interrupts.
; INITIALIZE the pulse accumulator register (PACNT) to -24.
; CLEAR the interrupt mask in the condition code register.
; DO Foreground Job.
```

The interrupt service routine pseudocode is

```
; DO Whatever is needed at the 24th count.
; INITIALIZE the pulse accumulator register (PACNT) to -24.
; CLEAR the pulse accumulator overflow flag PAOVF in PAFLG.
; RETURN from interrupt.
```

TABLE 14-11 Pulse Accumulator Interrupt Vectors

Priority	Vector Address	Interrupt Source	Local Enable Bit	See Register	HPRIO Value to Promote
18	$FFDA:FFDB	Pulse Accumulator Input Edge	PAI	PACTL	$DA
17	$FFDC:FFDD	Pulse Accumulator Overflow	PAOV	PACTL	$DC

PULSE ACCUMULATOR INTERRUPT VECTORS

These vectors are summarized in Table 14-11.

14.7 Plain and Fancy Timing

The *Pulse Accumulator Control Register* (PACTL) has two clock select bits (CLK1:CLK0) that can select the clock source for the TCNT register. Normally, when the pulse accumulator is disabled (PAEN = 0), the bus clock is prescaled by the PR2:PR1:PR0 bits (in TSCR2), giving us a counter clock frequency ranging from 8 MHz to 250 kHz when the external oscillator is 16 MHz. However, as shown in Figure 14-7 and Table 14-12, when the pulse accumulator is enabled (PAEN = 1), the TCNT clock source can be derived from either the event signal on PT7 in event counting mode (PAMOD = 0) or the bus clock divided by 64 and then further divided by 1, 256, or 65,536. PAEN:CLK1:CLK0 are the select inputs for the TCNT clock multiplexer and PAEN and PAMOD select different sources for the clock. Table 14-12 gives all the various combinations. Special counting and timing requirements, such as counting long pulse streams or long periods can use these special features of the HCS12 timer.

14.8 Clearing Timer Flags

All timer flags are cleared by *writing a one* to the bit.

A common theme for all elements of the timer is the setting and resetting of various flags. The hardware, such as the timer overflow, sets the flag, and the software you write must reset it. When interrupts are enabled, the setting of the flag also generates the interrupt (if the I bit is clear). The flag must always be reset, either

TABLE 14-12 TCNT Counter Clock Selection

PAMOD	PAEN	CLK1	CLK0	Pulse Accumulator Mode	Clock Used for TCNT
X	0	x	x	Disabled	The normal prescaled Bus
x	1	0	0	Enabled, either mode	Normal prescaled bus clock
0	1	0	1	Event counter	Pulse accumulator input (PT7)
0	1	1	0	Event counter	PT7 divided by 256
0	1	1	1	Event counter	PT7 divided by 65,536
1	1	0	1	Gated accumulator	Bus clock divided by 64
1	1	1	0	Gated accumulator	Bus clock divided by (64×256)
1	1	1	1	Gated accumulator	Bus clock divided by $(64 \times 65,536)$

in the interrupt service routine or in the polling software. In all cases, the flag is reset by *writing a 1* to the flag. For example, resetting the timer channel 7 interrupt flag can be done with the following code sequences:

```
ldaa   #%10000000
staa   TFLG1
```

An alternative is

```
bclr   TFLG1,#%01111111
```

The bit clear (BCLR) instruction has a mask byte with ones in the bit positions at which zeros are to be written (bits cleared). The way this instruction works is as follows:

The data byte is read from TFLG1, say, %11000000 (timer channels 7 and 6 flags both set).

The mask byte is complemented %01111111 → %1000000.

The complemented mask byte is ANDed with the register value and this result is written back to the register; that is, TFLG1 is written with %10000000.

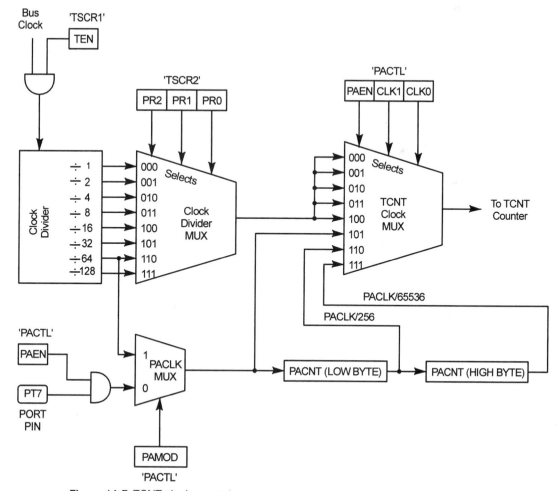

Figure 14-7 TCNT clock generator.

The bit set instruction (BSET) *will not work*. The instruction

```
bset   TFLG1,#%10000000
```

operates like this:

> The data byte is read from TFLG1, say, %11000000 (timer channels 7 and 6 flags are both set).
>
> The mask byte is ORed with the data byte and written back to TFLG1; that is, TFLG1 is written with %11000000! This resets *both* the channel 7 and 6 flags.

Clearing Timer Flags in C

The principles just presented must be followed when clearing flags in a C program, and we must avoid the `bset` and `bclr` instructions when resetting the interrupt flags. Example 14-24 shows a code segment to use the store instruction.

The CodeWarrior system gives us a header file that defines all registers and bits within registers. We should use unsigned char assignments to the registers to reset the flags as shown in Example 14-25.

Example 14-24 Resetting Flags in C

```
 1: /************************************************
 2: * Resetting flags in C
 3: ************************************************/
 4: volatile unsigned char TFLG1 @(0x0043); /* TFLG1 location */
 5: #define C0F 1    /* Timer flag channel 0 */
 6: #define C1F 2    /* Timer flag channel 1 */
 7: #define C2F 4    /* Timer flag channel 2 */
 8:                  /* etc */
 9:  void main(void) {
10:     /* . . . */
11:     /* Reset bit-0 inTFLG1 */
12:     TFLG1 = C0F;
0000 c601          LDAB   #1
0002 5b00          STAB   TFLG1
13:     /* Reset bit-0 and bit-2 in TFLG1 */
14:     TFLG1 = C0F|C2F;
0004 8605          LDAA   #5
0006 5a00          STAA   TFLG1
15:     /* . . . */
16:  }  0008 3d          RTS
```

Example 14-25 Resetting Flags with CodeWarrior C

```
 4:   /************************************************
 5:    * Resetting flags in CodeWarrior C
 6:    ************************************************/
 7:   #include <mc9s12c32.h>          /* derivative information */
 8:   void main(void) {
 9:   /************************************************/
10:     /* . . . */
11:     /* Reset bit-0 in PIFP */
12:     PIFP = PIFP_PIFP0_MASK;
0000 c601          LDAB   #1
0002 7b0000        STAB   _PIFP
13:     /* Reset bit-7 and bit-6 in PIFP */
14:     PIFP = PIFP_PIFP7_MASK | PIFP_PIFP6_MASK;
0005 86c0          LDAA   #192
0007 7a0000        STAA   _PIFP
15:     /* . . . */
16:   }
000a 3d            RTS
```

Fast Timer Flag Clearing

Each of the methods for clearing flags shown previously incurs some software overhead. The *Timer System Control Register 1* (*TSCR1*) has the *Timer Fast Flag Clear All* (*TFFCA*) bit that operates as follows when TFFCA = 1:

TFLG1 contains the flags for all eight timer channels; a read from an input capture or output compare channel (TC0–TC7) causes the corresponding channel flag, CnF in TFLG1, to be reset automatically.

TFLG2 has the timer overflow flag; any access to the TCNT register (e.g., reading it) clears the TOF in TFLG2.

PAFLG has the pulse accumulator overflow and input edge flags. Any access to the pulse accumulator count register (PACNT) clears both the PAOVF and PAIF in the PAFLG register.

This method eliminates software overhead associated with clearing the flag but the programmer must be wary of accidental flag clearing if an unintended access to the registers occurs.

14.9 Real-Time Interrupt

The *Real-Time Interrupt* (*RTI*) operates like the timer overflow interrupt except that the rate at which interrupts are generated can be selected. See Figure 14-8. The real-time interrupt is a component of the *Clocks and Reset Generator* (*CRG*) module. Chapter 21 describes the various clock sources produced by this module and used in the HCS12 in detail.

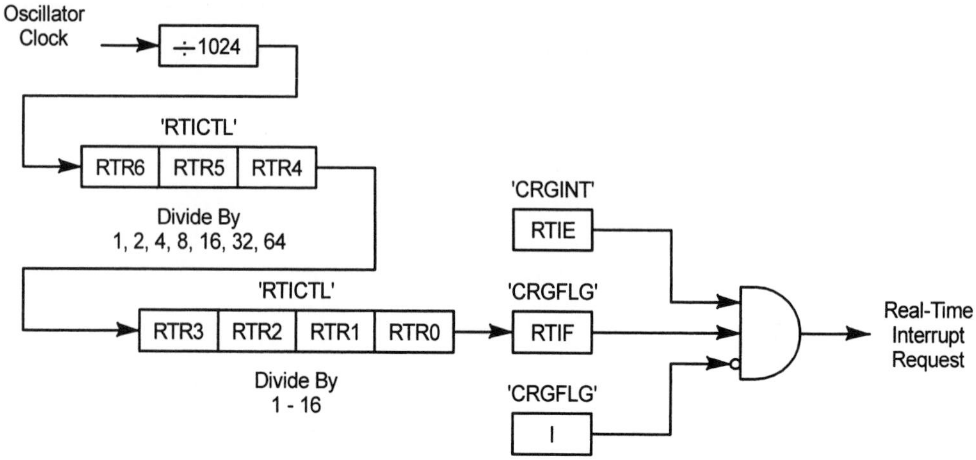

Figure 14-8 Real-time interrupt hardware.

The real-time interrupt rate is generated by a 10-bit counter that divides the system oscillator clock (twice the frequency of the bus clock) by 1024. The clock can be divided further by a programmable prescaler like the prescaler used for the TCNT register. The control bits for this are in the *Real-Time Interrupt Control Register* (*RTICTL*), and Table 14-13 shows the interrupt intervals assuming a 16-MHz oscillator clock.

TABLE 14-13 RTI Interrupt Rate

		RTR[6:4] Divisor							
		Off	2^{10}	2^{11}	2^{12}	2^{13}	2^{14}	2^{15}	2^{16}
		RTR[6:4]							
		000	001	010	011	100	101	110	111
RTR[3:0] Divisor		RTI Interrupt Interval—seconds (16-MHz Oscillator)							
1	0000	Off	6.40E − 05	1.28E − 04	2.56E − 04	5.12E − 04	1.02E − 03	2.05E − 03	4.10E − 03
2	0001	Off	1.28E − 04	2.56E − 04	5.12E − 04	1.02E − 03	2.05E − 03	4.10E − 03	8.19E − 03
3	0010	Off	1.92E − 04	3.84E − 04	7.68E − 04	1.54E − 03	3.07E − 03	6.14E − 03	1.23E − 02
4	0011	Off	2.56E − 04	5.12E − 04	1.02E − 03	2.05E − 03	4.10E − 03	8.19E − 03	1.64E − 02
5	0100	Off	3.20E − 04	6.40E − 04	1.28E − 03	2.56E − 03	5.12E − 03	1.02E − 02	2.05E − 02
6	0101	Off	3.84E − 04	7.68E − 04	1.54E − 03	3.07E − 03	6.14E − 03	1.23E − 02	2.46E − 02
7	0110	Off	4.48E − 04	8.96E − 04	1.79E − 03	3.58E − 03	7.17E − 03	1.43E − 02	2.87E − 02
8	0111	Off	5.12E − 04	1.02E − 03	2.05E − 03	4.10E − 03	8.19E − 03	1.64E − 02	3.28E − 02
9	1000	Off	5.76E − 04	1.15E − 03	2.30E − 03	4.61E − 03	9.22E − 03	1.84E − 02	3.69E − 02
10	1001	Off	6.40E − 04	1.28E − 03	2.56E − 03	5.12E − 03	1.02E − 02	2.05E − 02	4.10E − 02
11	1010	Off	7.04E − 04	1.41E − 03	2.82E − 03	5.63E − 03	1.13E − 02	2.25E − 02	4.51E − 02
12	1011	Off	7.68E − 04	1.54E − 03	3.07E − 03	6.14E − 03	1.23E − 02	2.46E − 02	4.92E − 02
13	1100	Off	8.32E − 04	1.66E − 03	3.33E − 03	6.66E − 03	1.33E − 02	2.66E − 02	5.32E − 02
14	1101	Off	8.96E − 04	1.79E − 03	3.58E − 03	7.17E − 03	1.43E − 02	2.87E − 02	5.73E − 02
15	1110	Off	9.60E − 04	1.92E − 03	3.84E − 03	7.68E − 03	1.54E − 02	3.07E − 02	6.14E − 02
16	1111	Off	1.02E − 03	2.05E − 03	4.10E − 03	8.19E − 03	1.64E − 02	3.28E − 02	6.55E − 02

RTICTL—Base + $003B—RTI Control Register

	Bit 7	6	5	4	3	2	1	0
Read:	0	RTR6	RTR5	RTR4	RTR3	RTR2	RTR1	RTR0
Write:								
Reset:	0	0	0	0	0	0	0	0

= reserved, unimplemented, or cannot be written to.

Read: Anytime. Write: Anytime.

RTR6:RTR0

RTR6:RTR4—Real-Time Interrupt Prescale Select Bits

These bits select the prescale rate for the RTI.

RTR4:RTR0—Real-Time Interrupt Modulus Counter Select Bits

These bits select the modulus counter value to provide additional granularity.

See Table 14-13 for all possible values of RTR[6:4] and RTR[4:0].

CRGINT—Base + $0038—CRG Interrupt Enable Register

	Bit 7	6	5	4	3	2	1	0
Read:	RTIE	0	0	LOCKIE	0	0	SCMIE	0
Write:								
Reset:	0	0	0	0	0	0	0	0

= reserved, unimplemented, or cannot be written to.

Read: Anytime. Write: Anytime.

RTIE

Real-Time Interrupt Enable Bit

0 = Interrupt requests from RTI are disabled (default).

1 = Enable the RTI system. Interrupts will be requested whenever RTIF is set.

LOCKIE, SCMIE

LOCKIE—Lock Interrupt Enable Bit

SCMIE—Self Clock Mode Interrupt Enable Bit

See Chapter 21 for the details of these bits.

Real-time interrupts are enabled by the RTIE enable bit in the *CRG Interrupt Enable Register (CRGINT)* register. The real-time interrupt flag, *RTIF*, is in the *CRG Flags Register (CRGFLG)*, and a bit that controls the RTI operation in wait mode is found in the *CRG Clock Select Register (CLKSEL)*. Table 14-14 gives the interrupt vector for the real-time interrupt. See Chapter 21 for more information about the RTI. (See Examples 14-26 and 14-27).

CRGFLG—Base + $0037—CRG Flag Register

	Bit 7	6	5	4	3	2	1	0
Read:	RTIF	PORF	LVRF	LOCKIE	LOCK	TRACK	SCMIF	SCM
Write:								
Reset:	0	—	—	0	0	0	0	0

▨ = reserved, unimplemented, or cannot be written to.

RTIF

Real-Time Interrupt Flag

0 = RTI time-out has not occurred (default).
1 = RTI time-out has occurred.

This bit is set at the end of an RTI period. If RTI interrupts are enabled (RTIE = 1), RTIF causes an interrupt request. The flag can be cleared only by writing a one to the bit (see Section 14.8).

PORF, LVRF, LOCKIE, LOCK, TRACK, SCMIF, SCM

PORF—Power on Reset Flag
LVRF—Low Voltage Reset Flag
LOCKIF—PLL Lock Interrupt Flag
LOCK—Lock Status Bit
TRACK—Track Status Bit
SCMIF—Self Clock Mode Interrupt Flag
SCM—Self Clock Mode Status Bit

See Chapter 21 for the details of these bits.

TABLE 14-14 Real-Time Interrupt Vectors

Priority	Vector Address	Interrupt Source	Local Enable Bit	See Register	HPRIO Value to Promote
7	$FFF0:FFF1	Real-Time Interrupt	RTIE	CRGINT	$F0

CLKSEL—Base + $0039—CRG Clock Select Register

	Bit 7	6	5	4	3	2	1	0
Read:	PLLSEL	PSTP	SYSWAI	ROAWAI	PLLWAI	CWAI	RTIWAI	COPWAI
Write:								
Reset:	0	0	0	0	0	0	0	0

RTIWAI

RTI Stops in Wait Mode

0 = RTI keeps running in wait mode (default).
1 = RTI stops and initializes the RTI dividers whenever the microcontroller goes into wait mode.

See Chapter 21.

PLLSEL, PSTP, SYSWAI, ROAWAI, PLLWAI, CWAI, COPWAI

PLLSEL—PLL Select Bit
PSTP—Pseudo Stop Bit
SYSWAI—System Clocks Stop in Wait Mode
ROAWAI—Reduced Oscillator Amplitude in Wait Mode
PLLWAI—PLL Stops in Wait Mode
COPWAI—COP Stops in Wait Mode

See Chapter 21 for the details of these bits.

Example 14-26 Real-Time Interrupt 1-kHz Square Wave

```
Metrowerks HC12-Assembler
(c) COPYRIGHT METROWERKS 1987-2003


Rel. Loc    Obj. code Source line
--- ---     --------- -----------
   1                  ;********************************
   2                  ; Real Time Interrupt Example
   3                  ; Generate ~1 kHz square wave on Port T,
                        bit-0
   4                  ;********************************
   5                  ; Define the entry point for the main
                        program
   6                          XDEF  Entry, main, RTI_isr
   7                          XREF  __SEG_END_SSTACK ; Note
                        double underbar
   8                  ;********************************
   9                  ; Include files
  10                          include portt.inc
  11                          include rti.inc
  12                          include bits.inc
  13                  ;********************************
  14                  ; Constant definitions
  15        0000 0040 RTI_RATE: EQU  %01000000    ; 5.12E-04 sec
                                                    intervals
```

```
16                        ;********************************
17                        ; Code Section
18                        MyCode: SECTION
19                        Entry:
20                        main:
21                        ;********************************
22                        ; Initialize stack pointer register
23 000000 CFxx xx              lds    #__SEG_END_SSTACK
24                        ;********************************
25                        ; Initialize Port T bit-0 for output
26 000003 1C02 4201            bset   DDRT,BIT0
27                        ; Set Port T bit-0 low
28 000007 1C02 4001            bset   PTT,BIT0
29                        ; Initialize RTICTL to select interrupt
30                        ; rate of 5.12E-04 sec
31 00000B 180B 4000            movb   #RTI_RATE,RTICTL
   00000F 3B
32                        ; Enable the RTI interrupt
33 000010 4C38 80              bset   CRGINT,RTIE
34                        ; Reset RTI flag
35 000013 8680                 ldaa   #RTIF
36 000015 5A37                 staa   CRGFLG
37                        ; Unmask interrupts
38 000017 10EF                 cli
39                        ; DO
40                        ;    Wait for the interrupt
41                        ; FOREVER
42                        ; DO
43                        main_loop:
44 000019 3E                   wai
45                        ; FOREVER
46 00001A 20FD                 bra    main_loop
47                        ;********************************
48                        ; Real Time Interrupt Service Routine
49                        ; Merely toggles Port T, bit-0
50                        ;********************************
51                        RTI_isr:
52                        ; Toggle Port T, bit-0
53                        ; IF Port T, bit-0 is 1
54 00001C 1F02 4001            brclr PTT,BIT0,set_bit
   000020 06
```

```
55                           ; THEN reset it
56 000021 1D02 4001                  bclr   PTT,BIT0
57 000025 2004                       bra    end_if
58                           ; ELSE set it
59                           set_bit:
60 000027 1C02 4001                  bset   PTT,BIT0
61                           ; ENDIF
62                           end_if:
63                           ; Reset the RTI flag
64 00002B 8680                       ldaa   #RTIF
65 00002D 5A37                       staa   CRGFLG
66                           ; Return
67 00002F 0B                         rti
```

Example 14-27 Real-Time Interrupt in C

```c
/*****************************************************************
 * Real Time Interrupt example
 * Output ~1 kHz square wave on Port T, bit-0
 *****************************************************************/
#include <mc9s12c32.h>              /* derivative information */
/*****************************************************************/
void main(void) {
/*****************************************************************/
  /* Initialize Port T bit-0 for output */
    DDRT = 0x01;
  /* Set Port T bit-0 low */
    PTT_PTT0 = 0;
  /* Initialize RTICTL to select interrupt rate of 5.12E-04 */
    RTICTL = 0x40;
  /* Enable the RTI interrupt */
    CRGINT_RTIE = 1;
  /* Reset RTI flag */
    CRGFLG = CRGFLG_RTIF_MASK;
  /* Unmask interrupts */
    EnableInterrupts;
  /* DO */
```

```
      /*  FOREVER */
  }
  /*************************************************************
   * RTI Interrupt Service Routine
   *************************************************************/
  void interrupt 7 RTI_isr( void ) {
    /* Toggle the Port T, bit-0 */
      if (PTT_PTT0 == 1) PTT_PTT0 = 0;
      else PTT_PTT0 = 1;
    /* Reset the RTI flag */
      CRGFLG = CRGFLG_RTIF_MASK;
  }
```

EXPLANATION OF EXAMPLES 14-26 AND 14-27

The assembler and C programs using the real-time interrupt are very similar to the interrupt software for other timer subsystems, as you can see by inspecting the pseudocode in Table 14-15. A major difference is that you must choose a real-time interrupt interval from Table 14-13 to suit your application. In this case, to generate a 1-kHz square wave, interrupts should be generated at $5.00E - 04$ second. The closest available is $5.12E - 04$ second, which results in a period of 1.24 ms. If this is not accurate enough for your application, you may have to use other timing features such as an output compare timer.

TABLE 14-15 Pseudocode for Real-Time Interrupt Program

Program Requirements: Use the real-time interrupt (RTI) module to generate approximately 1-kHz square wave on port T, bit 0.

```
; Initialize Port T bit-0 for output
; Set Port T bit-0 low
; Initialize RTICTL to select interrupt rate of 5.12E-04
; Enable the RTI interrupt
; Reset RTI flag
; Unmask interrupts
; DO
;    Wait for the interrupt
; FOREVER

; Real Time Interrupt Service Routine
; Toggle Port T, bit-0
; Reset the RTI flag
; Return from interrupt
```

14.10 Pulse-Width Modulator

The HCS12 can output *pulse-width modulated waveforms* continuously without causing program overhead.

The HCS12 has a *pulse-width modulation (PWM)* module giving up to six independent pulse-width modulated waveforms. After the PWM module has been initialized and enabled, PWM waveforms will be output automatically with no further action required by the program. This is very useful in applications such as controlling stepper motors.

Figure 14-9 shows a pulse-width modulated waveform. Two time intervals must be specified and controlled. These are the period (t_{PERIOD}) and the time the output is high (t_{DUTY}). A term used to describe a pulse-width modulated waveform is *duty cycle*. Duty cycle is defined as the ratio of t_{DUTY} to t_{PERIOD} and is usually given as a percent.

$$Duty\ Cycle = \frac{t_{DUTY}}{t_{PERIOD}} \times 100\%$$

A simplified block diagram of the pulse-width modulator in the HCS12 is shown in Figure 14-10. An 8-bit (or 16-bit) counter, *PWMCNTn*,[4] is clocked by a clock signal. This clock is derived by dividing the system bus clock by a prescaler and other division logic and there are two clock choices for each PWM channel. There are registers that control the period, *PWMPERn*, and the duty cycle, *PWMDTYn*. The system must be initialized with values in these two registers and a clock frequency must be selected. When PWMCNTn is reset, the pulse-width modulator output, *PWMn*, is set high (or low, depending on the *PPOLn* polarity control bit). PWMCNTn counts up and when it matches the value in PWMDTYn, the *8-bit Duty Cycle Comparator* causes the output to go low (or high). As PWMCNTn continues to count, it ultimately matches the value in the *8-bit Period Comparator* that sets the output high (or low) again and resets PWMCNTn to start the process over.

The hardware shown in Figure 14-10 gives a *left-aligned* pulse as shown in Figure 14-11(a). *Center-aligned* pulses are available as shown in Figure 14-11(b) as well. Slightly different hardware is used but the concept of registers defining the duty cycle and the period is the same.

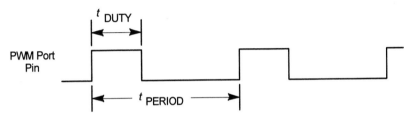

Figure 14-9 Pulse-width modulation waveform.

---

[4] n is 0–5. A 16-bit counter may be contrived by combining two, 8-bit PWMCNTn registers.

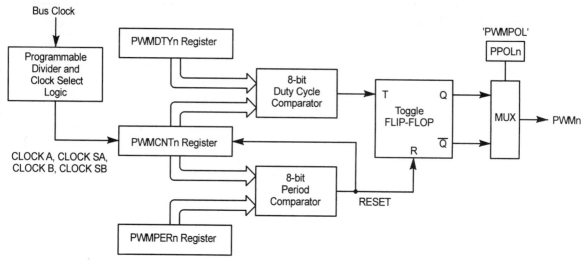

Figure 14-10 Simplified block diagram of the pulse-width modulator for left-aligned pulses.

Figure 14-12 shows three pulse-width modulated waves with 25%, 50%, and 75% duty cycle signals running simultaneously. Figure 14-12(a) illustrates left-aligned pulses and Figure 14-12(b) center-aligned. If it is not necessary for the signals to be aligned as shown in Figure 14-12(a), the center-aligned pulses will result in less system noise because all outputs are not switching at the same time.

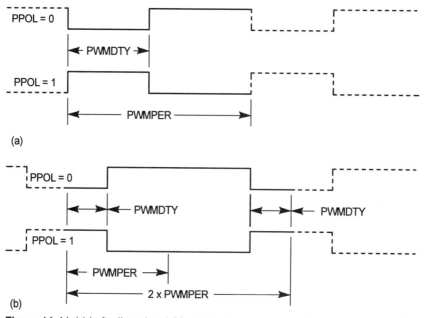

Figure 14-11 (a) Left-aligned and (b) center-aligned pulse-width modulator waveforms.

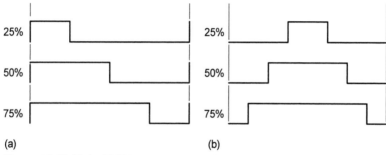

Figure 14-12 Three PWM waves with 25%, 50%, and 75% duty cycle pulses: (a) left-aligned pulses and (b) center-aligned pulses.

The pulse-width modulation registers and the counter may be concatenated in pairs to give 16-bit timing resolution. For example, the PWMCNT registers for channels 5 and 4, channels 2 and 3, and channels 1 and 0 can be concatenated. This gives a longer period and higher duty cycle resolution than can be achieved with the normal 8-bit operation. A control bit in one of the registers can be set to enable 16-bit operation. You may have six 8-bit, three 16-bit, or any combination of PWM registers.

Pulse-Width Modulator Clock Control

You must select a clock frequency and calculate values for PWMPER and PWMDTY registers when initializing the pulse-width modulator.

To initialize the PWM module you select the clock rate, the polarity of the output, left- or center-aligned pulses, and which of the PWM outputs are enabled.

There are four clock sources derived from the system bus clock. These are *Clock A*, *Clock B*, *Clock SA*, and *Clock SB*. Bits in the *PWM Prescale Clock Register* (*PWMPRCLK*) control the divider stages for Clock A and Clock B. These two independent clocks may be slower than the bus clock by a factor of 1, 2, 4, 8, 16, 32, 64, or 128.

PWMPRCLK—Base + $00E3—PWM Prescale Clock Select Register

	Bit 7	6	5	4	3	2	1	0
Read:	0	PCKB2	PCKB1	PCKB0	0	PCKA2	PCKA1	PCKA0
Write:								
Reset:	0	0	0	0	0	0	0	0

███ = reserved, unimplemented, or cannot be written to.

Read: Anytime. Write: Anytime. *If these bits are changed while a PWM signal is being generated, a truncated or stretched pulse may occur.*

PCKB2:PCKB0, PCKA2:PCKA0

Prescaler Bits for Clock A and Clock B

(See Table 14-16).

TABLE 14-16 PWM Clock B and Clock A Prescaler Select Bits

PCKB2 PCKA2	PCKB1 PCKA1	PCKB0 PCKA0	Value of Clock B Value of Clock A
0	0	0	Bus clock (default)
0	0	1	Bus clock/2
0	1	0	Bus clock/4
0	1	1	Bus clock/8
1	0	0	Bus clock/16
1	0	1	Bus clock/32
1	1	0	Bus clock/64
1	1	1	Bus clock/128

Clocks SA and SB are *scaled* versions of Clocks A and B and are produced by further dividing Clock A and Clock B by *twice* the value in *PWMSCLA* (for Clock SA) and *PWMSCLB* (for Clock SB). Clocks SA and SB may be anywhere from 1/2 to 1/512th the frequency of Clock A and Clock B. One of these four clocks must be selected by the clock select logic to be used by PWM channels 0 through 5. This is done by bits in the *PWM Clock Select Register* (*PWMCLK*) and each channel has a choice of two clocks as shown in Table 14-17.

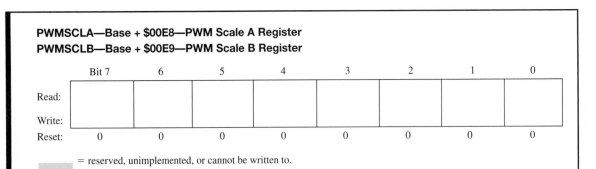

PWMSCLA—Base + $00E8—PWM Scale A Register
PWMSCLB—Base + $00E9—PWM Scale B Register

	Bit 7	6	5	4	3	2	1	0
Read:								
Write:								
Reset:	0	0	0	0	0	0	0	0

　= reserved, unimplemented, or cannot be written to.

Read: Anytime. Write: Anytime.

Clock A and Clock B Scaling Bits

These registers contain the programmable scaling value for Clock A and Clock B to produce Clocks SA and SB. Clock SA/SB is generated by taking Clock A/B and dividing by *twice* the value in PWMSCLA/PWMSCLB. When a register is $00, the scale value is 512.

PWMCLK—Base + $00E2—PWM Clock Select Register

	Bit 7	6	5	4	3	2	1	0
Read:	0	0	PCKL5	PCLK4	PCLK3	PCLK2	PCLK1	PCLK0
Write:								
Reset:	0	0	0	0	0	0	0	0

[shaded] = reserved, unimplemented, or cannot be written to.

Read: Anytime. Write: Anytime.

PCKL5:PCLK0

PWM Channel Clock Select Bits

(See Table 14-17.)

TABLE 14-17 Pulse Width Channel Clock Select

PWM Channel	PCLKn	Clock Source
5	0	Clock A (default)
	1	Clock SA
4	0	Clock A (default)
	1	Clock SA
3	0	Clock B (default)
	1	Clock SB
2	0	Clock B (default)
	1	Clock SB
1	0	Clock A (default)
	1	Clock SA
0	0	Clock A (default)
	1	Clock SA

Pulse-Width Modulator Control Registers

A number of registers control the pulse-width modulator. Table 14-18 summarizes these and the following sections provide detailed information for each.

TABLE 14-18 PWM Control Registers

Register	Name	Address (Base +)	Function
PWME	PWM Enable	$00E0	PWM Enable. Enables or disables each PWM channel.
PWMPOL	PWM Polarity	$00E1	Polarity select bits for each channel.
PWMCLK	PWM Clock Select	$00E2	Selects which of two clock sources is to be used.
PWMPRCLK	PWM Prescale Clock Select	$00E3	Sets prescaler value for Clock A and Clock B.
PWMCAE	PWM Center Align	$00E4	Selects left-aligned or center-aligned pulses.
PWMCTL	PWM Control	$00E5	Concatenation (16-bit) control and stop mode control.
PWMTST	PWM Test	$00E6	For factory testing only.
PWMPRSC	PWM Prescale Counter	$00E7	For factory testing only.

TABLE 14-18 (Continued)

Register	Name	Address (Base +)	Function
PWMSCLA	PWM Scale A	$00E8	PWM scale register for Clock SA.
PWMSCLB	PWM Scale B	$00E9	PWM scale register for Clock SB.
PWMSCNTA	PWM Scale A Counter	$00EA	For factory testing only.
PWMSCNTB	PWM Scale B Counter	$00EB	For factory testing only.
PWMCNT0	PWM Channel 0 Counter	$00EC	Eight-bit, up/down counters for channels 0–5.
PWMCNT1	PWM Channel 1 Counter	$00ED	
PWMCNT2	PWM Channel 2 Counter	$00EE	
PWMCNT3	PWM Channel 3 Counter	$00EF	
PWMCNT4	PWM Channel 4 Counter	$00F0	
PWMCNT5	PWM Channel 5 Counter	$00F1	
PWMPER0	PWM Channel 0 Period	$00F2	Eight-bit period definition registers for channels 0–5.
PWMPER1	PWM Channel 1 Period	$00F3	
PWMPER2	PWM Channel 2 Period	$00F4	
PWMPER3	PWM Channel 3 Period	$00F5	
PWMPER4	PWM Channel 4 Period	$00F6	
PWMPER5	PWM Channel 5 Period	$00F7	
PWMDTY0	PWM Channel 0 Duty	$00F8	Eight-bit duty cycle definition registers for channels 0–5.
PWMDTY1	PWM Channel 1 Duty	$00F9	
PWMDTY2	PWM Channel 2 Duty	$00FA	
PWMDTY3	PWM Channel 3 Duty	$00FB	
PWMDTY4	PWM Channel 4 Duty	$00FC	
PWMDTY5	PWM Channel 5 Duty	$00FD	

PULSE-WIDTH MODULATION ENABLE

PWM channel enable bits PWMEN5–PWME0 in the *PWME* register allow the selected clock signal to be gated to the PWM. If all channels are disabled, the clock prescaler shuts off to reduce power consumption.

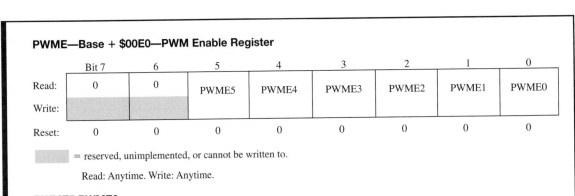

PWME—Base + $00E0—PWM Enable Register

	Bit 7	6	5	4	3	2	1	0
Read:	0	0	PWME5	PWME4	PWME3	PWME2	PWME1	PWME0
Write:								
Reset:	0	0	0	0	0	0	0	0

▢ = reserved, unimplemented, or cannot be written to.

Read: Anytime. Write: Anytime.

PWME5:PWME0

Pulse Width Modulation Channel Enable

When a PWM channel is enabled, the signal becomes available at the output bit when the clock source begins its next cycle. (See Table 14-19.)

TABLE 14-19 Pulse Width Channel Enable

PWM Channel	PWMEn	
5	0	PWM Channel 5 enabled.
	1	PWM Channel 5 disabled.
4	0	PWM Channel 4 enabled. If CON45 = 1, PWME4 has no effect and the channel is disabled.
	1	PWM Channel 4 disabled.
3	0	PWM Channel 3 enabled.
	1	PWM Channel 3 disabled.
2	0	PWM Channel 2 enabled. If CON23 = 1, PWME2 has no effect and the channel is disabled.
	1	PWM Channel 2 disabled.
1	0	PWM Channel 1 enabled.
	1	PWM Channel 1 disabled.
0	0	PWM Channel 0 enabled. If CON01 = 1, PWME1 has no effect and the channel is disabled.
	1	PWM Channel 0 disabled.

PWM POLARITY CONTROL

The starting polarity of each of the PWM channels is controlled by the PPOLn bits in the PWMPOL register. If the polarity bit is one, the output is high at the beginning of the cycle and then goes low at the end of the duty time. The opposite is true if the PPOLn bit is zero.

PWMPOL—Base + $00E1—PWM Polarity Select Register

	Bit 7	6	5	4	3	2	1	0
Read:	0	0	PPOL5	PPOL4	PPOL3	PPOL2	PPOL1	PPOL0
Write:								
Reset:	0	0	0	0	0	0	0	0

☐ = reserved, unimplemented, or cannot be written to.

Read: Anytime. Write: Anytime. If the polarity bit changes while a PWM signal is being generated, a truncated or stretched pulse may occur.

PPOL5:PPOL0

PWM Output Polarity

0 = The associated channel is low at the beginning of the period and goes high when the duty count is reached (default).

1 = The associated channel is high at the beginning of the period and goes low when the duty count is reached.

PWM CENTER-ALIGNMENT CONTROL

Left-aligned or center-aligned pulse-width modulated waveforms as shown in Figure 14-11 may be generated. Notice that with center-aligned pulses the period is *twice* the time given by the value in the PWMPER register. The alignment is controlled by the *PWMCAE* register.

PWMCAE—Base + $00E4—PWM Center Align Enable Register

	Bit 7	6	5	4	3	2	1	0
Read:	0	0	CAE5	CAE4	CAE3	CAE2	CAE1	CAE0
Write:								
Reset:	0	0	0	0	0	0	0	0

███ = reserved, unimplemented, or cannot be written to.

Read: Anytime. Write: Anytime but write only when the corresponding channel is disabled.

CAE5:CAE0

Center Aligned Output Enable

0 = Associated channel operates in left-aligned mode (default).
1 = Associated channel operates in center-aligned mode.

PWM CONCATENATE CONTROL REGISTER

Channels 4 and 5, 2 and 3, and 0 and 1 can be concatenated to create 16-bit PWMs. The high bytes, low bytes, and other control bits are specified in Table 14-20. A 16-bit PWM can have longer periods and greater time resolution than an 8-bit pulse modulator.

PWMCTL—Base + $00E5—PWM Control Register

	Bit 7	6	5	4	3	2	1	0
Read:	0	CON45	CON23	CON01	PSWAI	PFRZ	0	0
Write:								
Reset:	0	0	0	0	0	0	0	0

███ = reserved, unimplemented, or cannot be written to.

Read: Anytime. Write: Anytime.

CON45, CON23, CON01

Concatenate Control Bits

0 = Channels are separate 8-bit PWMs (default).
1 = Channels are combined to be 16-bit PWMs.

(See Table 14-20.)

TABLE 14-20 PWM Concatenate Controls

	High Byte	Low Byte	Output Pin	Clock Select	Polarity Bit	Enable Bit	Center Alignment Bit
CON45 = 1	Ch 4	Ch 5	PWM5	PCLK5	PPOL5	PWME5	CAE5
CON23 = 1	Ch 2	Ch 3	PWM3	PCLK3	PPOL3	PWME3	CAE3
CON01 = 1	Ch 0	Ch 1	PWM1	PCLK1	PPOL1	PWME1	CAE1

PSWAI

PWM Stops in Wait Mode

0 = Allow the clock to the prescaler to continue while in wait mode (default).

1 = Stop the clock when in wait mode. Enabling this allows for lower power consumption when in wait mode.

PFRZ

PWM Counters Stop in Freeze Mode

0 = Allow PWM to continue in freeze mode (default).

1 = Disable the clock to the prescaler when in freeze mode.

PWM CHANNEL COUNTER REGISTERS

Each channel has a dedicated 8-bit up/down counter that runs at the selected clock source rate. The counter can be read at any time without affecting the PWM waveform. In left-aligned mode, the counter starts at $00 and counts to the value in the period register minus one ($PWMPERn - 1$). In center-aligned mode, the counter counts from $00 up to the value in the period register and then back down to $00.

PWMCNT0—Base + $00EC—Counter Channel 0 Register
PWMCNT1—Base + $00ED—Counter Channel 1 Register
PWMCNT2—Base + $00EE—Counter Channel 2 Register
PWMCNT3—Base + $00EF—Counter Channel 3 Register
PWMCNT4—Base + $00F0—Counter Channel 4 Register
PWMCNT5—Base + $00F1—Counter Channel 5 Register

	Bit 7	6	5	4	3	2	1	0
Read:								
Write:	0	0	0	0	0	0	0	0
Reset:	0	0	0	0	0	0	0	0

■ = reserved, unimplemented, or cannot be written to.

Read: Anytime. Write: Anytime (any value written causes PWM counter to be reset to $00).

PWMCNT0—PWMCNT5

PWM Channel Counters

The channel counters contain the current count against which the channel period registers (PWMPERn) and the duty cycle registers (PWMDTYn) are compared.

A counter is reset to $00 and the direction is set to up anytime it is written. At this time the duty and period registers are loaded and the output is changed according to the polarity bit.

The counter is cleared at the end of a period.

To avoid a truncated period, you should write to the PWMCNTn register when the counter is disabled.

If a counter channel is disabled (PWMEn = 0), the PWMCNTn register does not count.

If you disable a PWM counter and then enable it again, the PWM channel counter starts at the count in the PWMCNTn register.

The counter may be read at any time without affecting the value of the counter.

PWM PERIOD AND DUTY REGISTERS

As you can see in Figure 14-10, the values in the PWMPERn and PWMDTYn registers control length of the period and the active high or low time of the pulse modulated waveform.

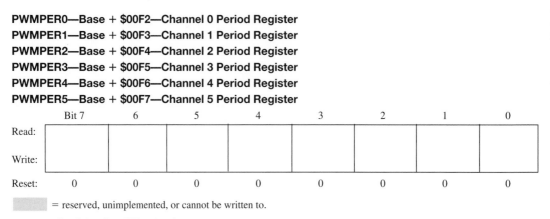

PWMPER0—Base + $00F2—Channel 0 Period Register
PWMPER1—Base + $00F3—Channel 1 Period Register
PWMPER2—Base + $00F4—Channel 2 Period Register
PWMPER3—Base + $00F5—Channel 3 Period Register
PWMPER4—Base + $00F6—Channel 4 Period Register
PWMPER5—Base + $00F7—Channel 5 Period Register

	Bit 7	6	5	4	3	2	1	0
Read:								
Write:								
Reset:	0	0	0	0	0	0	0	0

= reserved, unimplemented, or cannot be written to.

Read: Anytime. Write: Anytime.

PWMPER0—PWMPER5

PWM Channel Period Registers

The channel period registers contain the count specifying the end of the period.

If a PWMPERn register is written while the PWM is enabled, the new value will not take effect until the existing period terminates.

To start a new period immediately, write the PWMPERn register and then write to the PWMCNTn register to reset it to $00 or else disable and then enable the channel.

PWMPERn can be read at any time to find the most recent value written.

The PWM period is as follows:

Left-Aligned Waveform (CAEn = 0): Period = Channel_Clock_Period × PWMPERn

Center-Aligned Waveform (CAEn = 1): Period = Channel_Clock_Period × 2 × PWMPERn

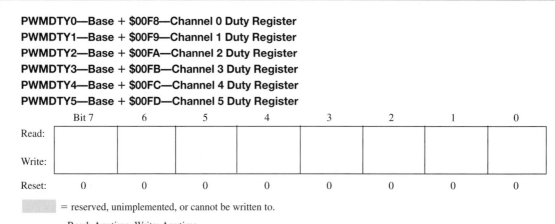

PWMDTY0—Base + $00F8—Channel 0 Duty Register
PWMDTY1—Base + $00F9—Channel 1 Duty Register
PWMDTY2—Base + $00FA—Channel 2 Duty Register
PWMDTY3—Base + $00FB—Channel 3 Duty Register
PWMDTY4—Base + $00FC—Channel 4 Duty Register
PWMDTY5—Base + $00FD—Channel 5 Duty Register

	Bit 7	6	5	4	3	2	1	0
Read:								
Write:								
Reset:	0	0	0	0	0	0	0	0

= reserved, unimplemented, or cannot be written to.

Read: Anytime. Write: Anytime.

PWMDTY0—PWMDTY5

PWM Channel Duty Registers

The channel duty registers contain the count specifying the end of the present high (or low) duty time.

If a PWMDTYn register is written while the PWM is enabled, the new value will not take effect until the existing period terminates, or the PWMCNTn register is written, or the channel is disabled.

If PWMDTYn is greater than or equal to the period register PWMPERn, there will be no PWM pulse.

If the polarity bit PPOLn = 1, PWMDTYn defines the time high. If PPOLn = 0, PWMDTYn defines the time low.

PWMPERn can be read at any time to find the most recent value written.

The duty cycle (high time as a percent) can be calculated as follows:

Left-Aligned Waveform:

Active High (PPOLn = 1): Duty Cycle = [(PWMDTYn)/(PWMPERn)] × 100%
Active Low (PPOLn = 0): Duty Cycle = [(PWMPERn-PWMDTYn)/(PWMPERn)] × 100%

Center-Aligned Waveform:

Active High (PPOLn = 1): Duty Cycle = [(PWMDTYn)/(PWMPERn)] × 100%
Active Low (PPOLn = 0): Duty Cycle = [(PWMPERn-PWMDTYn)/(PWMPERn)] × 100%

PULSE-WIDTH MODULATOR BOUNDARY CASES

Boundary cases are those cases where the register values are zero or otherwise do not make sense, such as PWMDTY being greater than PWMPER. See Table 14-21.

Using Port T Outputs for PWM Waveforms

On some versions of the MC9S12C microcontroller family, the port P, PWM pins may not be available for use. As we described in Chapter 11, you may switch the PWM outputs to bits on port T. Of course in these cases, the port T bit cannot be used for other timer functions. The register in the port integration module used to do this is the *Port T Module Routing Register* (*MODRR*). When using this you do not have to set the data direction register for *port T*.

TABLE 14-21 PWM Boundary Cases

PWMDTYn	PWMPERn	PPOLn	PWMn Output
$00 Indicates no duty	>$00	1	Always low
$00 Indicates no duty	>$00	0	Always high
XX	$00[a] Indicates no period	1	Always high
XX	$00 Indicates no period	0	Always low
>= PWMPER	XX	1	Always high
>= PWMPER	XX	0	Always low

[a] Counter = $00 and does not count.

MODRR—Base + $0247—Port T Module Routine Register

	Bit 7	6	5	4	3	2	1	0
Read:	0	0	0	MODRR4	MODRR3	MODRR2	MODRR1	MODRR0
Write:								
Reset:	0	0	0	0	0	0	0	0

◻ = reserved, unimplemented, or cannot be written to.

Read: Anytime. Write: Anytime.

MODRR4:MODRR0

Port T Module Selection Bits

0 = Associated pin is connected to the timer module (default).
1 = Associated pin is connected to the PWM module.

Choosing Pulse-Width Modulation Counter Prescaler and Scaler Values

There are several clock dividers and counter registers that must be initialized before using the pulse-width modulators. When PWMs drive motors or other mechanical systems, the PWM frequency (if too low) can cause objectionable audible noise. For this reason, it is important to be able to get the PWM frequency above 20 kHz (not easy or practical in software). If the frequency is too high, however, some components may not be able to follow the rate of change. Another common use of the PWM is to use it to drive an RC low-pass filter to get a crude digital-to-analog converter. In this case the PWM frequency needs to be high enough to be able to choose reasonable resistors and capacitors. Here is a strategy to pick appropriate values for the PWM registers based on the period and duty cycle of the pulse-width modulated waveform.

1. Choose PWMPERn = 255. This gives the best time resolution when choosing the duty cycle count for PWMDTYn.

2. Divide t_{PERIOD} by 255. This establishes the period of the PWMCNTn clock.

$$T_{PWMCNT} = \frac{t_{PERIOD}}{255}$$

3. Divide the PWMCNTn period by the bus clock period. This gives the total divisor needed.

$$Total\ Divisor = \frac{T_{PWMCNT}}{Bus\ Clock_{PERIOD}}$$

If the total divisor is greater than 65,536—[(*128 in the PWMPRCLK prescaler*) × (*256 in the PWMSCLn clock scale register*) × (*2*)]—then 16-bit concatenated registers are needed.

If the total divisor is not an integer, choose the next higher, even integer value,

$$T_{\text{PWMCNT}} = Total\ Divisor \times Bus\ Clock_{\text{PERIOD}}$$

and recalculate the PWMCNTn period.

$$PWMPER = \frac{t_{\text{PERIOD}}}{T_{\text{PWMCNT}}}$$

Recalculate PWMPERn.

4. Calculate the number of counts needed in the PWMDTYn register based on the PWMCNTn period.

$$PWMDTY = \frac{t_{\text{DUTY}}}{T_{\text{PWMCNT}}}$$

If PWMDTYn is not an integer, round up or down to the nearest integer value. Make sure PWMDTYn $\leq$ 255 (or 65,535 if 16-bit registers are to be used).

5. Select an appropriate divider combination to give the total divisor and a clock source to give PWMCNTn. If the total divisor is not a power of 2, you will have to use a combination of Clocks A and SA or Clocks B and SB and PWMSCLn to achieve an exact period and duty cycle. If the total divisor is not an integer, round to the nearest even integer and recalculate PWMPERn and PWMDTYn register values. Again, be sure PWMPERn does not exceed the maximum.

6. If you are using Clock A or Clock B and the total divisor is not a power of 2, choose the nearest value and recalculate PWMPERn and PWMDTYn register values.

(See Example 14-28.)

Another strategy to pick appropriate values for each register is based on the *duty cycle resolution* required by the application.

Example 14-28 Choosing PWM Register Values

Choose values for the PWMDTY1, PWMPER1, and PWSCLN registers, and the PCKA2:PCKA0, PCKB2:PCKB0, PPOL1, CAE1, and CON01 control bits to generate a pulse-width modulated waveform on PWM channel 1 with a frequency of 20 kHz and a high duty cycle time of 10 µs. Assume the bus clock is 8 MHz.

Solution:

1. The t_{PERIOD} is 50 µs.

2. Choose PWMPER1 = 255 and calculate the period of the PWMCNT1 clock:

$$T_{\text{PWMCNT1}} = t_{\text{PERIOD}}/255 = 50\ \mu s/255 = 0.196\ \mu s$$

3. Calculate the Total Divisor required:

$$Total\ Divisor = T_{\text{PWMCNT1}}/Bus\ Clock_{\text{PERIOD}} = 0.196\ \mu s/0.125\ \mu s = 1.57$$

This is not an even integer, so choose the next higher even number: Total Divisor = 2.

4. Recalculate T_{PWMCNT1}:

$$T_{\text{PWMCNT1}} = \textit{Total Divisor} \times \textit{Bus Clock}_{\text{PERIOD}} = 2 \times 0.125 \ \mu s = 0.25 \ \mu s$$

5. Recalculate PWMPER1:

$$PWMPER1 = t_{\text{PERIOD}}/T_{\text{PWMCNT1}} = 50 \ \mu s/0.25 \ \mu s = 200$$

6. Calculate the number of counts required in PWMDTY1 for 10 μs.

$$PWMDTY1 = t_{\text{DUTY}}/T_{\text{PWMCNT1}} = 10 \ \mu s/0.25 \ \mu s = 40$$

7. Choose a divisor combination to achieve a Total Divisor = 2. This can be done by choosing Clock A.

$$Clock\ A = \textit{Bus Clock}/\textit{Prescaler}$$

For a Total Divisor of 2, choose the Prescaler = 2.

The control bits are PCKA2:PCKA0 = %001, PWMSCLA = don't care or whatever might be required if Clock SA is used on another channel, PCKB2:PCKB0 are don't cares or whatever might be needed for Clock B, PCLK1 = 0, CON01 = 0, CAE1 = 0, and PPOL1 = 1. See Example 14-32 for a program example.

Duty cycle resolution is the smallest unit of time by which t_{DUTY} (Figure 14-9) may change. This may be given as the actual time or as a percentage of the full PWM period. Duty cycle resolution determines the minimum count value for the PWMPERn register.

$$\textit{Duty Cycle Resolution} = \frac{\Delta t_{\text{DUTY}}}{t_{\text{PERIOD}}} \times 100\% = \frac{1}{PWMPER} \times 100\%$$

Here is a method to choose PWMPER, PWMDTY, and the clock frequency.

1. Find the minimum value for the PWMPER register based on the duty cycle resolution. Choose any value greater than this but less than or equal to 255 (or 65,535 in a 16-bit system.)
2. Divide t_{PERIOD} by the value chosen for PWMPER. This establishes the period of the PWMCNT clock.

$$T_{\text{PWMCNT}} = \frac{T_{\text{PERIOD}}}{PWMPER}$$

3. Calculate the number of counts needed in the PWMDTY register based on the PWMCNT period.

$$PWMDTY = \frac{t_{\text{DUTY}}}{T_{\text{PWMCNT}}}$$

If PWMDTY is not an integer, choose the next higher value and recalculate the PWMCNT period and a new PWMPER value. Make sure PWMPER $\leq$ 256 (or 65,536).

4. Divide the PWMCNT period by the bus clock period. This gives the total divisor needed.

$$Total\ Divisor = \frac{T_{PWMCNT}}{Bus\ Clock_{PERIOD}}$$

If the total divisor is greater than 65,536 —[*(128 in the PWMPRCLK prescaler)* $\times$ *(256 in the PWMSCLn clock scale register)* $\times$ *(2)*]—then 16-bit concatenated registers are needed.

5. Select an appropriate divider combination to give the total divisor and a clock source to give PWMCNT. If the total divisor is not a power of 2, you will have to use a combination of Clocks A and SA and PWMSCLA to achieve an exact period and duty cycle. If the total divisor is not an integer, round down to the nearest integer and recalculate PWMPER and PWMDTY register values. Again, be sure PWMPER does not exceed the maximum.

6. If you are using Clock A or Clock B and the total divisor is not a power of 2, choose the nearest value and recalculate PWMPER and PWMDTY register values.

(See Examples 14-29 to 14-31.)

Example 14-29 Choosing PWM Register Values

Choose values for the PWMDTY1, PWMPER1, and PWSCLn registers, and the PCKA2:PCKA0, PCKB2:PCKB0, PPOL1, CAE1, and CON01 control bits to generate a pulse-width modulated waveform on PWM channel 1 with a period of 20 ms and a high duty cycle time of 7 ms. The duty cycle resolution is to be 0.5%. Assume the bus clock is 8 MHz.

Solution:

1. The required duty cycle resolution is 0.5%. This tells us that the ratio of Δt_{DUTY} to T_{PERIOD} is 1:200 and that an 8-bit PWMPER register should be sufficient. Choose a value of 200 for PWMPER1.

2. Calculate the period of the PWMCNT1 clock:

$$T_{PWMCNT1} = t_{PERIOD}/200 = 20\ ms/200 = 100\ \mu s$$

3. Calculate the number of counts required in PWMDTY1 for 7 ms.

$$PWMDTY1 = t_{DUTY}/T_{PWMCNT1} = 7\ ms/100\ \mu s = 70$$

4. Calculate the total divisor required.

$$Total\ Divisor = T_{PWMCNT1}/Bus\ Clock_{PERIOD} = 100\ \mu s/0.125\ \mu s = 800$$

5. Choose a divisor combination to achieve 800. This can be done by choosing Clock SA.

$$Clock\ SA = Bus\ Clock/(Prescaler \times PWSCLA \times 2)$$

For a Total Divisor of 800, choose the Prescaler = 16 and PWSCLA = 25 ($16 \times 25 \times 2 = 800$).

The registers and control bits are PCKA2:PCKA0 = %100, PWSCLA = 25, PCKB2:PCKB0 are don't cares or whatever might be needed for Clock B, PCLK1 = 1, CON01 = 0, CAE1 = 0, and PPOL1 = 1. See Example 14-34.

Example 14-30 Choosing PWM Register Values

Specify the divider value and PWMPER3 and PWMDTY3 register values for a pulse-width modulated waveform on PWM3 using Clock B. The period is to be 1.8 ms, $t_{DUTY} = 0.1$ ms, and the duty cycle resolution is to be less than (better than) 1%. Assume the bus clock is 8 MHz. State the final t_{DUTY}, t_{PERIOD}, and duty cycle resolution achieved by your design. Calculate the percent error in t_{DUTY} and t_{PERIOD} if the design does not achieve exact timing.

Solution:

1. Duty cycle resolution is 1%. Therefore PWMPER3 must be ≥ 100. Choose PWMPER3 = 200.
2. Calculate PWMCNT3 period.

$$T_{PWMCNT3} = t_{PERIOD}/200 = 1.8 \text{ ms}/200 = 9 \text{ } \mu s$$

3. $$PWMDTY3 = t_{DUTY}/T_{PWMCNT3} = 0.1 \text{ ms}/9 \text{ } \mu s = 11.1$$

PWMDTY3 is not an integer, so choose the next higher even integer PWMDTY3 = 12. To achieve t_{DUTY} when PWMDTY3 = 12,

$$T_{PWMCNT3} = t_{DUTY}/PWMDTYT3 = 0.1 \text{ ms}/12 = 8.33 \text{ } \mu s$$

$$\text{New } PWMPER3 = t_{PERIOD}/T_{PWMCNT3} = 1.8 \text{ ms}/8.33 \text{ } \mu s = 216$$

4. $$Total \ Divisor = 8.33 \text{ } \mu s/0.125 \text{ } \mu s = 66.64$$

The next smaller Clock B divisor is 64, so choose that.

5. Recalculate PWMCNT3 period and new values for PWMPER3 and PWMDTY3.

$$T_{PWMCNT3} = 0.125 \text{ } \mu s \times 64 = 8 \text{ } \mu s$$

$$PWMPER3 = 1.8 \text{ ms}/8 \text{ } \mu s = 225$$

$$PWMDTY3 = 0.1 \text{ ms}/8 \text{ } \mu s = 12.5. \ \text{ Choose } 12.$$

6. Calculate errors.

$$Final \ t_{PERIOD} = 225 \times 8 \text{ } \mu s = 1.8 \text{ ms}$$

$$Error = 0\%$$

$$Final \ t_{DUTY} = 12 \times 8 \text{ } \mu s = 0.096 \text{ ms}$$

$$Error = (0.1 - 0.096)/0.1 = 4 \%$$

$$Duty \ Cycle \ Resolution = 1/225 = 0.4\%$$

Example 14-31

What is the longest PWM period that can be obtained using a single 8-bit PWM register assuming an 8-MHz bus clock and left-aligned waveforms?

Solution: The longest period is achieved using an 8-bit PWMPER register and the slowest Clock S. The slowest Clock SA or SB is formed by dividing the bus clock by 128 to give Clock A or B and then by setting PWSCAL to $00, which divides Clock A or B by 512. The period is PWMPER times this clock period. Therefore the longest period is

$$P\text{-}clock_{PERIOD} \times 128 \times 512 \times 256 = 2.09 \text{ seconds}$$

Pulse-Width Modulation Software Checklist

Here are the steps to check when writing your software for timer output compares.

1. Set the module routing register (MODRR) to route the PWM output to port T if needed.
2. Choose which clock source is to be used, Clock A or Clock B, and set the total divisor in the PWMPRCLK register.
3. If a scaled clock, Clock SA or Clock SB, is to be used, set the scaler value in PWMSCLA or PWMSCLB.
4. Assign the clock source to the PWM channel being used in PWMCLK.
5. Set the polarity bit in PWMPOL.
6. Enable left-aligned or center-aligned pulses in PWMCAE.
7. Set the period register, PWMPERn.
8. Set the duty register, PWMDTYn.
9. Enable the channel in PWME.

PULSE-WIDTH MODULATION EXAMPLES

See Examples 14-32 to 14-34.

Example 14-32 PWM Waveform Using Clock A

```
Metrowerks HC12-Assembler
(c) COPYRIGHT METROWERKS 1987-2003


Rel. Loc  Obj. code Source line
---- ---  --------- -----------
    1                   ; This is a test program showing a
    2                   ; 20 kHz PWM waveform with 10 us high
    3                   ; duty time. The output is Port T-1.
    4                   ;******************************
    5                   ; Define the entry point for the main program
    6                           XDEF    Entry, main
```

```
 7                              XREF      __SEG_END_SSTACK
 8                              INCLUDE   pwm.inc    ; Timer defns
 9                              INCLUDE   base.inc   ; Reg base defn
10                              INCLUDE   bits.inc   ; Bit defns
11                      ;********************************
12                ; Constants
13     0000 00C8 PERIOD: EQU    200         ; PWMPER1
14     0000 0028 DUTY:   EQU    40          ; PWMDTY1
15     0000 0001 DIVISOR:EQU    %00000001    ; Clk A=BusClock/2
16                ; Code Section
17                MyCode: SECTION
18                Entry:
19                main:
20 000000 CFxx xx         lds   #__SEG_END_SSTACK
21                ; Initialize the I/O
22                ; Set the module routing register to
23                ; route PWM1 to Port T, bit-1
24 000003 1C02 4702      bset   MODRR,BIT1
25                ; Set Clock A prescaler to divide by 2
26 000007 180B 0100      movb   #DIVISOR,PWMPRCLK
   00000B E3
27                ; Set Clock SA scale register
28                ;    (Don't care, not used)
29                ; Set PWM Channel 1 clock to Clock A
30 00000C 4DE2 02        bclr   PWMCLK,PCLK1
31                ; Enable active high polarity
32 00000F 4CE1 02        bset   PWMPOL,PPOL1
33                ; Enable left-aligned pulse
34 000012 4DE4 02        bclr   PWMCAE,CAE1
35                ; Set the period register
36 000015 180B C800      movb   #PERIOD,PWMPER1
   000019 F3
37                ; Set the duty register
38 00001A 180B 2800      movb   #DUTY,PWMDTY1
   00001E F9
39                ; Enable Channel 1
40 00001F 4CE0 02        bset   PWME,PWME1
41                ; Main loop - do nothing
42                spin:
43 000022 20FE           bra    spin
44                ;********************************
```

EXPLANATION OF EXAMPLE 14-32

Example 14-32 is a program outputting a 20-kHz PWM waveform with a 10-μs high duty time. The MC9S12C32 processor used for this code does not have the PWM output pins so port T, bit 1 is to be used for the PWM channel 1 output. The MODRR register is initialized to make this switch in *line 24*. The Clock A prescaler, period, and duty times have been calculated as shown in Example 14-28. The Clock A prescaler is initialized in *line 26* and Clock A is assigned to PWM channel 1 in *line 30*. Active-high polarity and left-aligned pulses are selected in *lines 31–34*. The period and duty registers are initialized in *lines 35–38* and the PWM is enabled in *line 40*. Once the PWM is enabled, it starts to generate the waveform and the program enters a main loop, in this case doing nothing.

Example 14-33 PWM Waveform in C

```
/*******************************************************************
 * Pulse Width Modulation waveform.
 * This outputs a 20 kHz waveform with a high duty time that varies
 * from 10 us to 50 us. The output is Port T, bit-1
 *******************************************************************/
#include <mc9s12c32.h>                    /* derivative information */
#define PERIOD   200
#define DUTY     40
/*******************************************************************/
void main(void) {
  unsigned char duty;
/*******************************************************************/
  /* Set the module routing register to route PWM1
   * to Port T bit-1 */
   MODRR_MODRR1 = 1;
  /* Set Clock A prescaler to divide by 2 */
   PWMPRCLK = 1;
  /* Set Clock SA scale register */
  /* Not used in this example */
  /* Set PWM Channel 1 clock to Clock A */
   PWMCLK_PCLK1 = 0;
  /* Enable active high polarity */
   PWMPOL_PPOL1 = 1;
  /* Enable left-aligned pulse */
   PWMCAE_CAE1 = 0;
  /* Set the period register */
   PWMPER1 = PERIOD;
  /* Enable Channel 1 */
```

```
      PWME_PWME1 = 1;
   /* Scan the output high time from 10 us to 50 us */
    for(duty = DUTY;duty <= PERIOD;++duty) {
      PWMDTY1 = duty;
    } /* wait forever */
}
```

Example 14-34 PWM Waveform Using Clock SA

```
Metrowerks HC12-Assembler
(c) COPYRIGHT METROWERKS 1987-2003

Rel. Loc   Obj. code Source line
---- ---    -------- -----------
     1                  ; This is a test program showing a
     2                  ; PWM waveform with 20 ms period and
     3                  ; 7 ms high duty time. The output is on
     4                  ; Port P-5.
     5                  ;********************************
     6                  ; Define the entry point for the main program
     7                          XDEF     Entry, main
     8                          XREF     __SEG_END_SSTACK
     9                          INCLUDE  pwm.inc   ; Timer defns
    10                          INCLUDE  base.inc  ; Reg base defn
    11                          INCLUDE  bits.inc  ; Bit defns
    12                  ;********************************
    13                  ; Constants
    14     0000 00C8 PERIOD: EQU    200              ; PWMPER1
    15     0000 0046 DUTY:   EQU    70               ; PWMDTY1
    16     0000 0004 DIVISOR:EQU    %00000100  ; Clk A=BusClock/16
    17     0000 0019 SCALER: EQU    25               ; Clk SA scaler
    18                  ; Code Section
    19                  MyCode: SECTION
    20                  Entry:
    21                  main:
    22 000000 CFxx xx          lds    #__SEG_END_SSTACK
```

```
23                        ; Initialize the I/O
24                        ; Set Clock A prescaler to divide by 16
25 000003 180B 0400          movb   #DIVISOR,PWMPRCLK
   000007 E3
26                        ; Set Clock SA scale register
27 000008 180B 1900          movb   #SCALER,PWMSCLA
   00000C E8
28                        ; Set PWM Channel 5 clock to Clock SA
29 00000D 4CE2 20          bset   PWMCLK,PCLK5
30                        ; Enable active high polarity
31 000010 4CE1 20          bset   PWMPOL,PPOL5
32                        ; Enable left-aligned pulse
33 000013 4DE4 20          bclr   PWMCAE,CAE5
34                        ; Set the period register
35 000016 180B C800          movb   #PERIOD,PWMPER5
   00001A F7
36                        ; Set the duty register
37 00001B 180B 4600          movb   #DUTY,PWMDTY5
   00001F FD
38                        ; Enable Channel 1
39 000020 4CE0 20          bset   PWME,PWME5
40                        ; Main loop - do nothing
41                        spin:
42 000023 20FE                bra   spin
43                        ;*******************************
```

EXPLANATION OF EXAMPLE 14-34

Example 14-34 is a program outputting a PWM waveform with a 20-ms period and a 7-ms high duty time. The Clock A prescaler, Clock SA scaler, period, and duty times have been calculated as shown in Example 14-29. The Clock A prescaler is initialized in *line 28* and the Clock SA scaler in *line 25*. Clock SA is assigned to PWM channel 1 in *line 29*. Active-high polarity and left-aligned pulses are selected in *lines 30–33*. The period and duty registers are initialized in *lines 35–37* and the PWM is enabled in *line 39*. Once the PWM is enabled, it starts to generate the waveform and the program enters a main loop, in this case doing nothing.

14.11 Timer System Register Address Summary

This summary is given in Table 14-22.

TABLE 14-22 Timer System Registers

Name	Register	Address Base +
CRGINT	CRB Interrupt Enable Register	$0038
CLKSEL	CRG Clock Select Register	$0039
RTICTL	RTI Control Register	$003B
TIOS	Timer Input Capture/Output Compare Select	$0040
CFORC	Timer Compare Force Register	$0041
OC7M	Output Compare 7 Mask Register	$0042
OC7D	Output Compare 7 Data Register	$0043
TCNT	Timer Counter Register	$0044
TSCR1	Timer System Control Register 1	$0046
TTOV	Timer Toggle on Overflow Register	$0047
TCTL1	Timer Control Register 1	$0048
TCTL2	Timer Control Register 2	$0049
TCTL3	Timer Control Register 3	$004A
TCTL4	Timer Control Register 4	$004B
TIE	Timer Interrupt Enable Register	$004C
TSCR2	Timer System Control Register 2	$004D
TFLG1	Timer Interrupt Flag Register 1	$004E
TFLG2	Timer Interrupt Flag Register 2	$004F
TC0	Timer Input Capture/Output Compare Register 0	$0050
TC1	Timer Input Capture/Output Compare Register 1	$0052
TC2	Timer Input Capture/Output Compare Register 2	$0054
TC3	Timer Input Capture/Output Compare Register 3	$0056
TC4	Timer Input Capture/Output Compare Register 4	$0058
TC5	Timer Input Capture/Output Compare Register 5	$005A
TC6	Timer Input Capture/Output Compare Register 6	$005C
TC7	Timer Input Capture/Output Compare Register 7	$005E
PACTL	Pulse Accumulator Control Register	$0060
PACNT	Pulse Accumulator Counter Register	$0061
PAFLG	Pulse Accumulator Flag Register	$0061
PWME	PWM Enable Register	$00E0
PWMPOL	PWM Polarity Select Register	$00E1
PWMCLK	PWM Clock Select Register	$00E2
PWMPRCLK	PWM Prescale Clock Select Register	$00E3
PWMCAE	PWM Center Align Enable Register	$00E4
PWMCTL	PWM Control Register	$00E5
PWMCNT0	PWM Counter Channel 0 Register	$00EC
PWMCNT1	PWM Counter Channel 1 Register	$00ED
PWMCNT2	PWM Counter Channel 2 Register	$00EE
PWMCNT3	PWM Counter Channel 3 Register	$00EF
PWMCNT4	PWM Counter Channel 4 Register	$00F0
PWMCNT5	PWM Counter Channel 5 Register	$00F1
PWMPER0	PWM Channel 0 Period Register	$00F2
PWMPER1	PWM Channel 1 Period Register	$00F3
PWMPER2	PWM Channel 2 Period Register	$00F4
PWMPER3	PWM Channel 3 Period Register	$00F5
PWMPER4	PWM Channel 4 Period Register	$00F6
PWMPER5	PWM Channel 5 Period Register	$00F7
PWMDTY0	PWM Channel 0 Duty Register	$00F8
PWMDTY1	PWM Channel 1 Duty Register	$00F9
PWMDTY2	PWM Channel 2 Duty Register	$00FA
PWMDTY3	PWM Channel 3 Duty Register	$00FB
PWMDTY4	PWM Channel 4 Duty Register	$00FC
PWMDTY5	PWM Channel 5 Duty Register	$00FD
MODRR	Port T Module Routine Register	$0247

TABLE 14-23 Summary of HCS12 Timer Features

Timer Feature	Use	Polled	Interrupts	See Section
Timer Overflow	Crude time intervals (± 4.096 ms with 8-MHz clock).	Y	Y	14.2 14.3
Output Compare	High time resolution (1 clock cycle) timing; automatically generate output waveforms; output multiple waveforms simultaneously; output very short pulses.	Y	Y	14.4
Input Capture	Capture the internal TCNT register value with an external signal; measure pulse width and frequency.	Y	Y	14.5
Pulse Accumulator	Count external events; time external events with gated clock.	Y	Y	14.6
Real-Time Interrupt	Generate interrupts at intervals ranging from 6.40×10^{-5} to 6.55×10^{-2} seconds.	N	Y	14.9
Pulse-Width Modulator	Generate PWM waveforms.	N	N	14.10

14.12 Timer Features

A summary of timer features is given in Table 14-23.

14.13 Conclusion and Chapter Summary Points

The timer features of the HCS12 are useful in many applications. Although the programming and control of the elements seem complex, the operation of all functions is similar with similar control requirements. The common elements are as follows:

- Timing is derived from the bus clock.
- A common bus clock frequency is 8 MHz.
- The bus clock may be prescaled (divided) by 1, 2, 4, 8, 16, 32, 64, or 128 to generate clock frequencies in the range of 8 MHz to 62.5 kHz.
- The TEN bit in the TSCR1 register must be set to start the timer counting.
- A 16-bit free-running counter, TCNT, provides the basic counting functions in the system.
- TCNT generates a timer overflow every 65,536 clock cycles.
- Eight channels of output compare can set an output compare flag when the TCNT register is equal to the output compare register.
- The output compare can automatically change output bits when the comparison is made.
- Timer resolution can be as fine as one bus clock cycle.
- Eight channels of input capture can latch the TCNT on an input signal.
- A pulse accumulator can count external pulses and generate interrupts.
- All timer functions set a flag to indicate when their particular event has occurred.
- Each timer flag can be ANDed with an interrupt enable bit to generate an interrupt when the particular event has occurred.
- In all events, the flag is reset by software writing a one to the flag.

- The timer module has a real-time interrupt that can generate interrupts at a variety of rates.
- The HCS12 has a pulse-width modulator waveform generator.
- Pulse-width modulation waveforms are automatically generated by the hardware without burdening the software.

14.14 Bibliography and Further Reading

CPU12 Reference Manual, S12CPUV2.PDF, Freescale Semiconductor, Inc., July 2003.

Morales, A., *Using Real-Time Interrupt on HCS12 Microcontrollers*, Application Note AN2882, Rev 1, 5/2005, Freescale Semiconductor, Inc., 2005.

More, G. M., *Audio Reproduction HCS12 Microcontrollers*, Application Note AN2250, Rev 0, 1/2002, Freescale Semiconductor, Inc., 2002.

More, G. M. and M. Gallop, *PWM Generation Using HCS12 Timer Channels*, Application Note AN2612, Rev 0, 11/2003, Freescale Semiconductor, Inc., 2003.

PWM_8B6C Block User Guide, S12PWM8B6CV1/D, Freescale Semiconductor, Inc., March 2002.

TIM_16B8C Block User Guide, S12TIM16B8CV1/D.PDF, Freescale Semiconductor, Inc., October 2001.

14.15 Problems

Basic

14.1 What is wrong with the following code to get the 16-bit value of the TCNT register? **[b, k]**

```
ldaa $44   ; Get the high byte
ldab $45   ; Get the low byte
```

14.2 What is wrong with the following code to get the 16-bit value of the TCNT register? **[b, k]**

```
ldab $45   ; Get the low byte
ldaa $44   ; Get the high byte
```

14.3 How should you read the 16-bit TCNT value? **[k]**

14.4 How is the TCNT clock prescaler programmed? **[a]**

14.5 Give the name of the bit, the name of the register that it is in, the register's address, which bit, and the default or reset state of the bit for each of the following. **[k]**

   a. What bit indicates that the timer has overflowed?

   b. What bit enables the timer overflow interrupts?

   c. What bits are used to prescale the timer clock?

14.6 When is the timer overflow flag set? **[k]**

14.7 How is the timer overflow flag reset? **[k]**

14.8 What timing resolution can be achieved with the output compare? **[a]**

14.9 The TCNT register is receiving an 8-MHz clock. **[a, c, k]**

   a. How many clock cycles will constitute a delay of 5.8 ms?

   b. If you are using an output compare with interrupts to delay 5.8 ms, can this be done without multiple interrupts?

14.10 Give the name of the bit, the name of the register that it is in, the register's address, which bit, and the default or reset state of the bit for each of the following. **[k]**

   a. What bit indicates that a comparison has been made on Output Compare 2?

   b. What bit enables the Output Compare 2 interrupt?

   c. What bits are used to set the Output Compare 3 I/O pin high on a successful comparison?

14.11 Write a small section of code to set the Output Compare 2 I/O pin to toggle on every comparison. **[c]**

14.12 Write a small section of code to enable Output Compare 7 to set bits PT7, PT6, and PT5 to one on the next successful comparison. **[c]**

14.13 The HCS12 timer channel 3 interrupt flag (C3F) is bit 3 in the TFLG1 (Timer Interrupt Flag 1) register ($004E). Show a snippet of code that you would use to reset this flag. **[c]**

14.14 What two registers control which data bits are output when the Output Compare 7 flag is set? **[c]**

14.15 How does the programmer select the active edge for Input Capture 2? **[k]**

14.16 Write a short section of code demonstrating how to reset the Input Capture 2 flag, C2F. **[c, k]**

14.17 Write a short section of code demonstrating how to enable the Input Capture 1 interrupts. **[c, k]**

14.18 What bits in what registers must be set to enable the Pulse Accumulator Input Edge interrupt? **[k]**

14.19 Give the definition of duty cycle. **[a]**

14.20 Give the name of the bit, the name of the register that it is in, the register's address, which bit, and the default or reset state of the bit for each of the following. **[k]**

a. What bit enables PWM channel 0?

b. What bit must be set if you want an active-high pulse for the duty cycle time on PWM channel 0?

c. What bit must be set to combine PWM channels 0 and 1 to create a 16-bit PWM channel?

d. What bit must be set to switch the PWM output from channel 3 to port T, bit 3?

Intermediate

14.21 Write a short section of code demonstrating how to enable the real-time interrupt and to set the nominal rate to 16.384 ms assuming an 8-MHz oscillator clock. **[c, k]**

14.22 Write a short section of code demonstrating how to enable the Pulse Accumulator as an event counter counting rising edges. **[c, k]**

14.23 Write a short section of code demonstrating how to enable the Pulse Accumulator as a gated time accumulator with a high level enable accumulation. **[c, k]**

14.24 The infamous 68FC12 processor has a 12-bit timer subsystem similar to the HCS12's. It has a 12-bit TCNT register with one 12-bit output compare register and similar flags and controls. Assume that it has a 1 MHz clock. **[c, k]**

a. What is the interval between timer overflows?

b. Assuming the present value of the TCNT register is $D18, what value should be loaded into the output compare register to create a delay of (i) 100 microseconds and (ii) 1 millisecond?

Advanced

14.25 Choose values for the PWMDTY0, PWMPER0, and PWSCLN registers, and the PCKA2:PCKA0, PCKB2:PCKB0, PPOL0, CAE0, and CON01 control bits to generate a PWM waveform on PWM channel 0 with a period of 20 ms and a left-aligned high duty cycle time of 5 ms. Assume the bus clock is 8 MHz. **[k]**

14.26 The exact total divisor (470.58) is not achievable in Problem 14.25. With the total divisor = 472, PWMPER0 = 254, and PWMDTY0 = 85, what is the actual period and duty high time? **[k]**

14.27 Design and implement, in either assembly language or C, a real-time clock using the MC9S12C32 timer. Assume an 8-MHz bus clock. The program is to maintain a 24-hour clock, 00:00–23:59 as BCD digits in memory locations tens_hours, ones_hours, tens_minutes, ones_minutes. The clock is to be updated every minute. **[c]**

14.28 Design a traffic light controller. **[c]**

Imagine an intersection with North/South and East/West streets. There are to be six traffic light signals:

RedE_W, YellowE_W, GreenE_W

RedN_S, YellowN_S, GreenN_S

Assume the time elements in the following figure are 10 seconds.

a. Write an assembly language program to create a timer that generates an interrupt every 10 seconds for the traffic light controller.

b. Write an interrupt service routine that implements the traffic light controller to output signals to the traffic lights as shown in the figure.

14.29 Repeat Problem 14.28 in C. **[c]**

14.30 Write an assembly language function that measures the duty cycle of a PWM waveform on port T, bit 0. The routine is to return the duty cycle as a 16-bit fraction in the D register. **[c]**

14.31 Repeat Problem 14.30 in C. **[c]**

14.32 Write an assembly language program using interrupts that output a 30-kHz square wave on port T bit 0. Assume an 8-MHz bus clock. **[c]**

14.33 Repeat Problem 14.32 in C. **[c]**

14.34 Write an assembly language program using interrupts that output a 30-kHz waveform with a 25% duty cycle on port T bit 0. **[c]**

14.35 Repeat Problem 14.34 in C. **[c]**

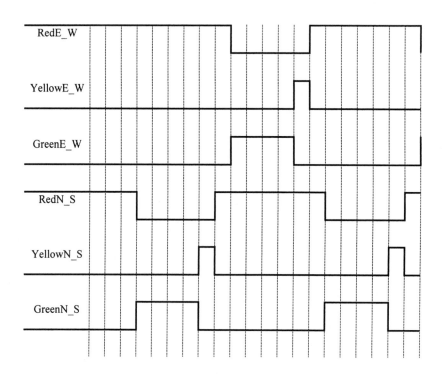

15 HCS12 Serial I/O—SCI and SPI

OBJECTIVES

This chapter discusses the HCS12 serial I/O capabilities. The Asynchronous Serial Communications Interface, SCI, and the Synchronous Serial Peripheral Interface, SPI, are covered. The Freescale Scalable Controller Area Network, MSCAN, is covered in the next chapter. Interfacing the SCI to external RS-232 devices is covered in Chapter 18.

15.1 Introduction

The HCS12 can contain a variety of serial interfaces. Among these are the following:

- *Asynchronous Serial Communications Interface—SCI.* This is a universal asynchronous receiver–transmitter (UART) designed for serial communications to a terminal or other asynchronous serial devices such as personal computers.
- *Synchronous Serial Peripheral Interface—SPI.* This is a high-speed, synchronous serial interface used between an HCS12 and a serial peripheral such as a Maxim 512 Serial 10-bit A/D converter or between two HCS12s.
- *Motorola Scalable Controller Area Network—MSCAN.* This module is a serial communications controller implementing the CAN 2.0 A/B protocol widely used in automotive applications.

Figure 15-1(a) shows the *asynchronous SCI.* It is frequently used to communicate with other computers, terminals, and modems. In the personal computer world, this interface is called the *COM* port. Usually an RS-232-C interface standard is adopted with voltage levels that require a CMOS-to-RS-232-C translation chip. The RS-232-C standard also defines handshaking signals such as *Request to Send (RTS)* and *Clear to Send (CTS).* These signals have not been implemented in the SCI but port S, bits 2 and 3 may be used for this purpose. Chapter 18 shows how to interface the SCI to RS-232-C devices.

Figure 15-1(b) shows the *MSCAN.* This interface is widely used in the automotive industry and is a two-wire, carrier-sense, multiple access bus. A CAN transceiver is required to create

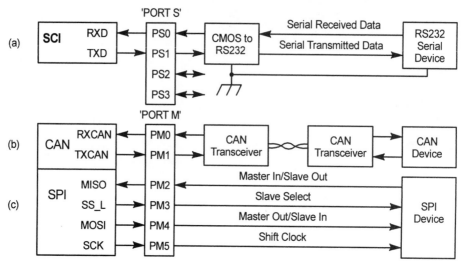

Figure 15-1 HCS12 (a) asynchronous serial communications interface (SCI), (b) controller area network (MSCAN) interface, and (c) synchronous serial peripheral (SPI) interface.

a bus structure supporting multiple CAN devices. Chapter 16 shows how to use the MSCAN interface.

Figure 15-1(c) is the synchronous *SPI*. This interface allows a much higher data transfer between peripheral SPI devices and the microcontroller than does the SCI. Section 15.3 shows how to use the SPI and Chapter 18 gives interfacing examples.

15.2 Asynchronous Serial Communications Interface (SCI)

The SCI is a *Universal Asynchronous Receiver–Transmitter (UART).*

The SCI is a full duplex, asynchronous, serial interface. It has an on-chip, independent, baud rate generator that can produce standard serial communication rates from normal HCS12 bus clock frequencies. The receiver and transmitter are double buffered and although they operate independently, they use the same baud rate and data format. The SCI can send and receive 8- or 9-bit data, has a variety of interrupts, and is fully programmable.

Using and programming each SCI can be broken into three parts: (1) initialization of the device's data rate, word length, parity, and interrupting capabilities; (2) writing to the SCI data register, taking care not to exceed the data transmission rate, and (3) reading data from the SCI data register making sure to read the incoming data before the next serial data arrives.

The input and output transmitted data pins are in port S. Two of the four bidirectional port S pins are shared with the serial communications interface (*SCI*). If the SCI is enabled, bit 0 is assigned to received-data (*RXD*) and is configured as an input. Bit 1 is assigned to transmitted-data (*TXD*) and configured as an output. Bit 2 and bit 3 are not assigned and can be used to implement flow control in the serial interface.

PTS—Base + $0248—Port S I/O Register

	Bit 7	6	5	4	3	2	1	0
Read:	0	0	0	0	PTS3	PTS2	PTS1	PTS0
Write:								
Reset:	0	0	0	0	0	0	0	0
SCI	—	—	—	—	—	—	TXD	RXD

▨ = reserved, unimplemented, or cannot be written to.

SCI Data

> Serial data are read from and
> written to the SCIDR register.

Two data registers, *SCIDRH* and *SCIDRL*, shown in Figure 15-2, contain the 8- or 9-bit serial data received and to be transmitted. The SCIDRL register is two separate registers occupying the same memory address. Data to be transmitted serially are written to and serial data received are read from these registers.

SCIDRH and SCIDRL form a 9-bit data register when sending and receiving 9-bit data. The ninth transmit bit (T8) does not have to be changed each time new serial data are sent. The same value will be transmitted until T8 is changed. If 9-bit data are to be used, the SCIDRH bit should be written before SCIDRL to ensure the correct data are transferred to the SCI transmit data register when SCIDRL is written. If an 8-bit data format is being used, only SCIDRL needs to be read or written.

SCIDRL contains the 8-bit serial transmitted and received data. *Reading* from SCIDRL reads the last *received* data and *writing* to it sends (transmits) the serial data. Writing to SCIDRL does not overwrite or destroy the contents of the data that has been received.

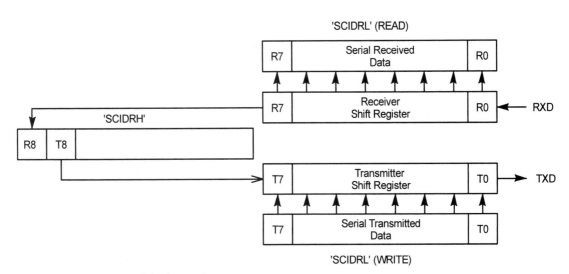

Figure 15-2 SCI data registers.

SCIDRH—Base + $00CE—SCI Data Register High

	Bit 7	6	5	4	3	2	1	0
Read:	R8	T8	0	0	0	0	0	0
Write:								
Reset:	0	0	0	0	0	0	0	0

▨ = reserved, unimplemented, or cannot be written to.

Read: Anytime; reading accesses SCI receive data register. Write: Anytime; write accesses SCI transmit data register; writing to R8 has no effect.

SCIDRL—Base + $00CF—SCI Data Register Low

	Bit 7	6	5	4	3	2	1	0
Read:	R7	R6	R5	R4	R3	R2	R1	R0
Write:	T7	T6	T5	T4	T3	T2	T1	T0
Reset:	0	0	0	0	0	0	0	0

Read: Anytime; reading accesses SCI receive data register. Write: Anytime; write accesses SCI transmit data register.

R8:R0, T8:T0

Serial Receive and Transmit Data

R8 and T8 are the ninth data bit received and transmitted when the SCI is in 9-bit data format (M = 1).

R7–R0 and T7–T0 are the received and transmitted 8-bit or 9-bit data formats.

When transmitting in 9-bit data format and using 8-bit write instructions (STAA, etc.), write the SCIDRH register and then SCIDRL.

If the value of T8 is the same as the previous transmission, T8 does not have to be rewritten.

In 8-bit data format, only SCIDRL has to be written and read.

SCI Initialization

RECEIVER AND TRANSMITTER ENABLE

As with any programmable device, the SCI must be initialized before use and there are a variety of registers and bits to be programmed. First, the transmitter and receiver are enabled by setting *TE* and *RE* in the *SCI Control Register 2* (*SCICR2*). SCICR2 contains bits to enable various interrupts, the SCI wake-up control, and a break generation bit. We will discuss these in a later section.

SCICR2—Base + $00CB—SCI Control Register 2

	Bit 7	6	5	4	3	2	1	0
Read:	TIE	TCIE	RIE	ILIE	TE	RE	RWU	SBK
Write:								
Reset:	0	0	0	0	0	0	0	0

Read: Anytime. Write: Anytime.

TIE

Transmitter Interrupt Enable

0 = Transmitter Data Register Empty (TDRE) flag interrupt requests disabled (default).

1 = Transmit Data Register Empty (TDRE) flag interrupt requests enabled.

TCIE

Transmission Complete Interrupt Enable

0 = Transmission complete (TC) flag interrupt requests disabled (default).

1 = Transmission complete (TC) flag interrupt requests enabled.

RIE

Receiver Full Interrupt Enable

0 = Receive data register full (RDRF) flag interrupt requests disabled (default).

1 = Receive data register full (RDRF) flag interrupt requests enabled.

ILIE

Idle Line Interrupt Enable

0 = Idle line (IDLE) flag interrupt requests disabled (default).

1 = Idle line (IDLE) flag interrupt requests enabled.

TE

Transmitter Enable

0 = SCI transmitter disabled (default).

1 = Transmitter enabled.

RE

Receiver Enable

0 = SCI receiver disabled (default).

1 = Receiver enabled.

RWU

Receiver Wake-up

0 = SCI normal operation (default).

1 = Enables the wake-up function and disables further receiver interrupt requests.

SBK

Send Break

0 = No break characters (default).

1 = Transmit break characters.

SCI MODE CONTROL

In addition to enabling the receiver/transmitter, the SCI operation mode must be initialized. This is done in the *SCI Control Register 1 (SCICR1)*.

SCICR1—Base + $00CA—SCI Control Register 1

	Bit 7	6	5	4	3	2	1	0
Read:	LOOPS	SCIWAI	RSRC	M	WAKE	ILT	PE	PT
Write:								
Reset:	0	0	0	0	0	0	0	0

Read: Anytime. Write: Anytime.

LOOPS

Loop Select

0 = Normal operation (default).
1 = Loop operation enabled.

In loop mode, the RXD pin is disconnected from its normal Port S input pin and connected to the transmitter as defined by the RSRC bit. Both transmitter and receiver must be enabled (TE = 1, RE = 1).

SCIWAI

SCI Stop in Wait Mode

0 = SCI enabled in wait mode (default).
1 = SCI disabled in wait mode.

RSRC

Receiver Source

0 = Receiver input internally connected to the transmitter output (default).
1 = Receiver input externally connected to the transmitter output.
(See Table 15-1.)

TABLE 15-1 Loop Functions

LOOPS	RSRC	Function
0	X	Normal operation.
1	0	Loop mode with RXD input internally connected to transmitter output.
1	1	Single-wire mode with RXD input connected externally to TXD output.

M

Data Format Select

0 = One start, eight data, one stop bit (default).
1 = One start, eight data, plus ninth data, one stop bit.

WAKE

Wake-up Condition

0 = Idle line wake-up (default).

1 = Address line wake-up.

ILT

Idle Line Type

0 = Idle character bit count starts after start bit (default).

1 = Idle character bit count starts after stop bit.

An idle character contains all ones with no start or stop bit. ILT determines when the bit count starts for detecting an idle line. The M bit determines the number of bits in an idle character.

PE

Parity Enable

0 = Parity is disabled (default).

1 = Parity is enabled.

When parity is enabled, PT determines the type of parity. The parity bit is inserted in the data word in the most significant bit position.

PT

Parity Type

0 = Even parity (default).

1 = Odd parity.

SCI Loop Mode The HCS12 offers a loop mode useful for testing serial I/O software when the external serial device may not be available. Loop mode is enabled by setting *LOOPS* to one. In the loop mode, the receiver is connected directly to the transmitter and thus anything you send out comes straight back in. You have a choice of how the receiver is connected to the transmitter. If the *Receiver Source (RSRC)* bit is zero, the receiver is connected internally to the transmitter. If RSRC is one, the SCI operates in single-wire mode with the receiver externally connected to the transmitter. See Table 15-1.

SCI Single-Wire Mode When LOOPS = 1 and RSRC = 1, the *single-wire mode* is selected. This allows a half-duplex operation between two SCI devices. The normal RXD pins (port S, bit 0) is disconnected from the SCI and can be used for general purpose I/O. The TXD output is enabled or disabled by controlling *TXDIR* in the *SCISR2* register. Thus in half-duplex or single-wire mode, the transmitter can be active (TXDIR = 1) when it is sending data and inactive (TXDIR = 0) when it is receiving. Table 15-1 gives a summary of the TXD activity for these modes.

Eight- or Nine-Bit Data The number of data bits to be sent and received is controlled by the *M* bit in SCICR1. M = 0 for one start, eight data, and one stop bit; M = 1 for one start, nine data, and one stop bit.

Idle Line An *idle* line is one in which the receive line is in a mark[1] condition (logic high) for more than one character time. Idle line detect may be used in half-duplex, or single-wire systems, in which the line needs to be "turned around" when the remote transmitter is done transmitting. The idle condition is detected when the RXD input remains in the mark (1) condition for a full character time. In long idle detect mode, the idle detect logic does not start counting logic 1-bit times until after a stop bit. In short idle detect mode, the idle detect logic starts counting logic ones after a start bit so the stop bit time and any contiguous ones at the end of a serial character can cause an idle line to be detected earlier than expected.

Parity The HCS12 can generate a parity bit in hardware. Parity generation for transmitting and checking for receiving is enabled by setting the *PE* bit. The type of parity, even or odd, is selected by the *PT* bit. The chosen parity bit is inserted in the most significant bit position.

SCISR2—Base + $00CD—SCI Status Register 2

	Bit 7	6	5	4	3	2	1	0
Read:	0	0	0	0	0	BK13	TXDIR	RAF
Write:								
Reset:	0	0	0	0	0	0	0	0

= reserved, unimplemented, or cannot be written to.

Read: Anytime. Write: Anytime.

BK13

Break Transmit Character Length

0 = Break character is 10 or 11 bits long (default).
1 = Break character is 13 or 14 bits long.

TXDIR

Transmit Data Direction in Single-Wire Mode

0 = TXD pin is used as an input (default).
1 = TXD pin is used as an output.

RAF

Receiver Active Flag

0 = No reception in progress (default).
1 = Reception in progress.

This bit is set when the receiver detects a start bit and is cleared when an idle character is detected.

---

[1] Mark and space are terms used in asynchronous serial communications to denote a logic one (mark) and zero (space).

SCI BAUD RATE SELECTION

The rate at which serial data bits are sent is called the *baud rate*. Both the receiver and transmitter use the same rate and a 13-bit divider derives standard baud rates from the bus clock. The bus clock is normally one-half the external oscillator frequency. Table 15-2 shows how the bits in the *SCI Baud Rate Control Registers* select standard rates. Nonstandard baud rates can be chosen by initializing the baud rate (BR) registers with other BR values. The value to choose is given by the following relationship and shown in Table 15-2 for an 8-MHz bus clock.

$$SCI\ Baud\ Rate\ =\ \frac{Bus\ Clock}{(16 \times SCIBR[12{:}0])}$$

SCIBDH—Base + $00C8—SCI Baud Rate Register High

	Bit 7	6	5	4	3	2	1	0
Read:	0	0	0	SBR12	SBR11	SBR10	SBR9	SBR8
Write:								
Reset:	0	0	0	0	0	0	0	0

SCIBDL—Base + $00C9—SCI Baud Rate Register Low

	Bit 7	6	5	4	3	2	1	0
Read:	SBR7	SBR6	SBR5	SBR4	SBR3	SBR2	SBR1	SBR0
Write:								
Reset:	0	0	0	0	0	0	0	0

▨ = reserved, unimplemented, or cannot be written to.

Read: Anytime. Write: Anytime.

SBR12:SBR0

SCI Baud Rate Select

(See Table 15-2.)

TABLE 15-2 Baud Rates (8-MHz Bus Clock)

Baud Rate	SBR12:SBR0$_{10}$
38,400	13
19,200	26
9,600	52
4,800	104
2,400	208
1,200	417
600	833
300	1,667
150	3,333
110	4,545

SCI Status Flags

The SCI system has several status flags and interrupts to inform you of its progress and of error conditions that may occur. Your program may poll the flags or make use of the interrupts. The status flags are in the *SCI Status Registers 1* and *2* (*SCISR1, SCISR2*) and the interrupt enable bits are in the *SCI Control Register 2* (*SCICR2*).

SCISR1—Base + $00CC—SCI Status Register 1

	Bit 7	6	5	4	3	2	1	0
Read:	TDRE	TC	RDRF	IDLE	OR	NF	FE	PF
Write:								
Reset:	1	1	0	0	0	0	0	0

= reserved, unimplemented, or cannot be written to.

Read: Anytime. Write: Anytime.

TDRE

Transmitter Data Register Empty Flag

0 = No byte transferred to the transmit shift register.

1 = Byte has been transferred to the transmit shift register and the transmit data register is empty (default).

This flag is set when the last character written to the SCI data register (SCIDRL) has been transferred to the output shift register. Normally the program should check this bit before writing the next character to the SCIDRL. The flag is reset (cleared) by reading the SCISR1 register with TDRE set and then writing the next byte to the SCIDRL.

TC

Transmit Complete Flag

0 = Transmission in progress.

1 = No transmission in progress (default).

This bit is different than the TDRE bit. It shows when the last character has been completely sent from the output shift register. The flag is reset (cleared) by reading the SCISR1 register with TC set and then writing the next byte to the SCIDRL.

RDRF

Receive Data Register Full Flag

0 = Data not available in the SCI data register (default).

1 = Received data is available.

The flag is reset (cleared) by reading the SCISR1 register with RDRF set and then reading SCIDRL.

IDLE

Idle Line Flag

0 = The receive line is either active now or has never been active since IDLE was last reset (default).

1 = The receive line has become idle.

The IDLE flag is cleared by reading SCISR1 (while IDLE = 1) and then reading SCIDRL. After the IDLE flag has been reset, it will not be set again until the receive line becomes active and then idle again. When the receiver wake-up bit (RWU) is set, an idle line does not set the IDLE flag.

OR

Receiver Overrun Flag

0 = No overrun error (default).

1 = An overrun has occurred.

Overrun occurs if a new character has been received before the old data have been read by the program. The new data are lost and the old data preserved. The flag is cleared by reading SCISR1 (with OR = 1) and then reading the SCIDRL.

NF

Noise Flag

0 = No noise was detected during the last character (default).

1 = Noise was detected.

The hardware takes three samples of the received signal near the middle of each data bit and the stop bit. Seven samples are taken during the start bit. If the samples in each bit do not agree, the noise flag is set. The flag may be reset by reading SCISR1 (with NF = 1) followed by a read of SCIDRL.

FE

Framing Error Flag

0 = No framing error occurred (default).

1 = Framing error occurred.

A framing error occurs if the receiver detects a space during the stop bit time instead of a mark. This kind of error can occur if the receiver misses the start bit or if the sending and receiving data rates are not equal. It is possible to have bad framing or mismatched baud rates but not get a framing error indication if a mark is detected when a stop bit was expected. The flag is reset by reading SCISR1 (with FE T 1) and then reading the SCIDRL.

PF

Parity Error Flag

0 = Parity on the last received data is correct (default).

1 = Parity incorrect.

PE is reset by reading SCISR1 with PF set and then reading SCIDRL.

SCI Flag Clearing

All SCI flags are cleared in a two-step operation. First, the status register is read followed by a read or write to the SCI data register. If the flag has been set, it will be reset by this operation. The two instructions may have other instructions between them but the order of operations must be maintained.

SCI Interrupts

The SCI interrupts are enabled by setting bits in the *SCI Control Register 2 (SCICR2)*. The interrupts are serviced by an interrupt service routine whose vector is shown in Table 15-3. There are five potential sources of interrupts shown in Table 15-3 with one interrupt vector. When an SCI system interrupt occurs, the service routine must test the SCISR1 status register

TABLE 15-3 SCI Serial System Interrupts

Priority	Vector Address	Interrupt Source	Local Enable (SCICR2)	Local Flag (SCISR1)	SCIDRL to Reset	HPRIO Value to Promote
20	$FFD6:FFD7	Transmit Data Register Empty	TIE	TDRE	Write	$D6
		Transmission Complete	TCIE	TC	Write	
		Receive Data Register Full	RIE	RDRF	Read	
		Receiver Overrun	RIE	OR	Read	
		Idle Line Detected	ILIE	IDLE	Read	

to find out which condition caused the interrupt. As in the case of all other interrupting sources, the flag that caused the interrupt must be cleared in the interrupt service routine. This is done by reading the SCISR1 register and then either reading or writing the next data byte to the SCIDRL.

SCI Wake-up

The SCI features a sleep and wake-up mode. This may be used in multireceiver applications in which one HCS12 is broadcasting data to many serial receivers in a network. Software in each receiver puts it to sleep (by setting the *RWU* bit in the *SCICR2* register) until the programmed wake-up sequence is received. At the start of a broadcast, each receiver automatically wakes up and software in all receivers decodes for whom the message is intended. Only the addressed station stays awake to receive the message. Each of the others is put back to sleep until the start of the next broadcast. Receivers that are asleep do not respond to received data. However, only the SCI receiver is asleep and the CPU can continue to operate and do other chores. The CPU can wake up the SCI by resetting the RWU bit to zero although the automatic hardware mechanism normally wakes up sleeping receivers. The wake-up mode and receiver wake-up enable are controlled by bits in SCICR1 and SCICR2.

When the program puts a receiver to sleep by writing a one to the RWU bit, all receiver interrupts are disabled until the receiver is awakened by one of two wake-up methods. If WAKE = 0, a full character of idle line (a mark) wakes up the receiver. If WAKE = 1, any byte with a one in the most significant bit wakes it up.

SCI Break Character

The SCI can send a break character, 10 or 11 zeros (depending on the M bit), by the program writing a one into the SBK bit in the SCICR2 register. Break characters are used in some systems to wake up the receiving end.

Port S SCI I/O

Port S provides the serial I/O data and control bits. For any serial function not enabled, the bits may be used for general purpose I/O. The data direction register bits do not have to be initialized for bit 0 and bit 1 when these bits are in service as RXD and TXD (when TE and RE are set).

SCI Programming Examples

SCI I/O DRIVERS IN ASSEMBLY LANGUAGE

Examples 15-1 to 15-5 show subroutines for initializing the SCI to operate at 9600 baud, one start and stop bit, and eight data bits with no parity. The examples show initialization, data output, receiver status check, data input, and an example of data output subroutines. These routines are written so that they may be used as I/O drivers for C programs. See Examples 15-6 and 15-7.

Example 15-1 SCI Initialization

```
Metrowerks HC12-Assembler
(c) COPYRIGHT METROWERKS 1987-2003

Rel. Loc    Obj. code Source line
---- -----  --------- -----------
    1                 ;***************************************
    2                 ; HCS12 SCI I/O Example
    3                 ; Initialize SCI 9600,8,N,1
    4                 ; Assembler routines:
    5                 ;   init_sci, initsci
    6                 ; C Routines
    7                 ;   void initsci( void);
    8                 ;***************************************
    9                 ; Initialize SCI to 1 start, 8 data and 1
                      stop
   10                 ; bit, no parity and 9600 Baud.
   11                 ; Inputs:   None
   12                 ; Outputs:  None
   13                 ; Reg Mod:  CCR
   14                 ;***************************************
   15                 ; SCI port definitions
   16                         INCLUDE "sci.inc"
   17                         XDEF   init_sci, initsci
   18                 ;***************************************
   19                 init_sci:
   20                 initsci:
   21000000 3B                pshd       ; Save D reg
   22                 ; Set 1 start, 8 data and 1 stop bit
   23000001 4DCA 10          bclr   SCICR1,MODE
   24                 ; Disable parity for no parity
   25000004 4DCA 02          bclr   SCICR1,PE
```

```
26                          ; Enable transmitter and receiver
27000007 4CCB 0C                    bset    SCICR2,TE|RE
28                          ; Set Baud rate
2900000A CC00 34                    ldd    #B9600
3000000D 5CC8                       std    SCIBDH
3100000F A7                         nop
32000010 5CC8                       std    SCIBDH
33000012 3A                         puld      ; Restore D
34000013 3D                         rts
35                          ;****************************************
```

Example 15-2 SCI Put Character

Metrowerks HC12-Assembler
(c) COPYRIGHT METROWERKS 1987-2003

```
Rel. Loc   Obj. code Source line
---- ----- --------- -----------
   1                  ;****************************************
   2                  ; HCS12 SCI I/O Example
   3                  ; Put char to SCI 0
   4                  ; Assembler routines:
   5                  ;   put_sci_0, putchar
   6                  ; C Routines
   7                  ;   void putchar( char );
   8                  ;****************************************
   9                  ; Send SCI data
  10                  ; Inputs:   B register = data to send
  11                  ; Outputs:  None
  12                  ; Reg Mod:  CCR
  13                  ;****************************************
  14                  ; SCI port definitions
  15                          INCLUDE "sci.inc"
  16                          XDEF  put_sci, putchar
  17                  ;****************************************
  18                  put_sci:
  19                  putchar:
  20                  ; Wait until the transmit data reg is empty
  21                  spin_put:
22000000 4FCC 80FC           brclr SCISR1,TDRE,spin_put
  23                  ; Output the data and reset TDRE
```

```
24000004 5BCF                    stab   SCIDRL
25000006 3D                      rts
26                      ;****************************************
```

Example 15-3 SCI Get Character

```
Metrowerks HC12-Assembler
(c) COPYRIGHT METROWERKS 1987-2003


Rel. Loc   Obj. code Source line
---- ----- --------- -----------
   1                      ;****************************************
   2                      ; HCS12 SCI I/O Example
   3                      ; Get char from SCI
   4                      ; Assembler routines:
   5                      ;   get_sci, getchar
   6                      ; C Routines
   7                      ;   char getchar( void );
   8                      ;****************************************
   9                      ; Get a character from SCI
  10                      ; Waits until character has been received
  11                      ; Inputs:   None
  12                      ; Outputs:  B = character
  13                      ; Reg Mod:  B, CCR
  14                      ;****************************************
  15                      ; SCI port definitions
  16                              INCLUDE "sci.inc"
  17                              XDEF   get_sci, getchar
  18                              XREF   sci_char_ready
  19                      ;****************************************
  20                      get_sci:
  21                      getchar:
  22000000 36                     psha
  23                      wait:
  24                      ; IF a char is there waiting
  25000001 16xx xx               jsr   sci_char_ready
  26000004 24FB                  bcc   wait   ; ELSE wait
  27                      ; THEN get it and return
  28000006 D6CF                  ldab  SCIDRL
  29000008 32                    pula
  30000009 3D                    rts
  31                      ;****************************************
```

Example 15-4 SCI Character Ready

```
Metrowerks HC12-Assembler
(c) COPYRIGHT METROWERKS 1987-2003

Rel. Loc   Obj. code Source line
---- ----- --------- -----------
   1                 ;****************************************
   2                 ; HCS12 SCI I/O Example
   3                 ; Check for char in SCI receiver
   4                 ; Assembler routines:
   5                 ;   sci_char_ready, scichar_ready
   6                 ; C Routines
   7                 ;   int scichar_ready(void);
   8                 ;****************************************
   9                 ; Check the RDRF flag
  10                 ; If a character is ready, returns with C=1
  11                 ; the character in the B register, and the
  12                 ; status information in the A register.
  13                 ; Otherwise, C=0 and the A and B regs 0
  14                 ; Inputs:   None
  15                 ; Outputs:  B = character, Carry bit T or F
  16                 ;           A = status information
  17                 ; Reg Mod:  A,B CCR
  18                 ;****************************************
  19                 ; SCI port definitions
  20                         INCLUDE "sci.inc"
  21                         XDEF   sci_char_ready
  22                         XDEF   scichar_ready
  23                 ;****************************************
  24                 sci_char_ready:
  25                 scichar_ready:
  26000000 10FE          clc          ; Clear carry
  27000002 87            clra         ; Clear A
  28000003 C7            clrb         ; Clear B
  29                 ; IF RDRF is set
  30000004 4FCC 2006     brclr SCISR1,RDRF,exit_ready
  31                 ; THEN the character is there
  32000008 D6CF          ldab SCIDRL  ; Get the data
  3300000A 96CC          ldaa SCISR1  ; Get the status
  3400000C 1401          sec          ; Set the carry
  35                 ; ENDIF
  36                 exit_ready:
  3700000E 3D            rts
  38                 ;****************************************
```

Example 15-5 SCI Put String

```
Metrowerks HC12-Assembler
(c) COPYRIGHT METROWERKS 1987-2003

Rel. Loc  Obj. code Source line
---- ---- --------- -----------
   1                ;***************************************
   2                ; HCS12 SCI I/O Example
   3                ; Output null terminated string
   4                ; Assembler routines:
   5                ;   put_str, putstr
   6                ; C Routines
   7                ;   int putstr( char *);
   8                ;***************************************
   9                ; Put null terminated string to sci
  10                ; Inputs:   D = starting address of string
  11                ; Outputs:  D = number of characters printed
  12                ; Reg Mod:  D, CCR
  13                ;***************************************
  14                ; SCI port definitions
  15                        INCLUDE "sci.inc"
  16                        XDEF    put_str, putstr
  17                        XREF    putchar
  18                ;***************************************
  19                put_str:
  20                putstr:
  21000000 34           pshx
  22000001 35           pshy
  23000002 B745         tfr   d,x      ; Initialize pointer
                                           to string
  24                ; Use the Y register as counter of chars
                    printed
  25000004 CD00 00       ldy   #0       ; Initialize counter
  26                ; WHILE character to print is not a null
  27                while_do:
  28000007 E600         ldab  0,x
  29000009 2707         beq   done
  30                ; DO
  31                ;   Print the character
  3200000B 16xx xx       jsr   putchar
  33                ;   Increment the counter and pointer
  3400000E 02           iny
  3500000F 08           inx
  36000010 20F5         bra   while_do
```

```
37                      done:
38 000012 31                    puly
39 000013 30                    pulx
40 000014 3D                    rts
41                      ;****************************************
```

SCI PROGRAMMING IN C

Example 15-6 shows a C program that uses the SCI I/O drivers given earlier. Often I/O drivers such as these are written in assembly language and then used as library functions for the higher level C programs. Example 15-6 prints a string on the screen and then, after a character is typed, echoes the character back to the screen. It also displays the binary ASCII code for the character on eight LEDs.

Example 15-7 illustrates an interrupt driven character handler. There are a variety of interrupts that can be generated as shown in the previous *SCI Interrupts* section. The *Receive Data Register Full* interrupt is generated when the SCI receives a character. We can use this to interrupt whatever else is going on so that we will not miss a character. The RDRF flag in the SCISR1 register generates the interrupt request and it is reset by reading the character. You do not have to do a separate flag resetting operation like some of the other interrupting devices such as the timer.

Example 15-6 SCI I/O in C

```
/************************************************************
 * This SCI programming example prints a message to the SCI and
 * then waits for a character to be returned. The ASCII code for
 * that character is displayed on 8 LEDs.
 * This program uses the SCI drivers shown in the previous
   examples.
 ***********************************************************/
#include <mc9s12c32.h>              /* derivative information */
/***********************************************************/
/* Function prototypes */
void initsci( void );
void putchar( char );
char getchar( void );
int scichar_ready( void );
int putstr( char * );
/***********************************************************/
void main(void) {
char input_char;
/***********************************************************/
```

```
      /* Initialize Port T for 8 bits output */
        DDRT = 0xFF;
     /* Set all the Port T bits low */
        PTT = 0;
     /* Initialize the SCI 9600, 8, N, 1 */
        initsci();
     /* DO */
          for(;;) {
          putstr("Type a character -> ");
          /* Get the character the user types */
            input_char = getchar();
          /* Echo the character to the screen */
            putchar( input_char );
          /* Display the ASCII code on the LEDs */
            PTT = input_char;
          /* Go to the next line */
            putstr("\n\n\r");
        }
     /* FOREVER */}
}
```

Example 15-7 SCI Interrupts in C

```
/****************************************************************
 * This SCI programming example shows an interrupt driven character
 * receiver. Each time a character is received an interrupt is
 * generated and the interrupt handler echoes the character back
 * to the terminal.
 * This program uses the SCI drivers shown in the previous
 examples.
 ****************************************************************/
#include <mc9s12c32.h>                /* derivative information */
/****************************************************************/
/* Function prototypes */
void initsci( void );
void putchar( char );
char getchar( void );
int scichar_ready( void );
int putstr( char * );
/****************************************************************/
```

```
void main(void) {
/*****************************************************/
  /* Initialize the SCI 9600, 8, N, 1 */
    initsci();
  /* Enable the SCI Receiver Data Register Full Interrupt */
    SCICR2_RIE = 1;
  /* Unmask interrupts */
  EnableInterrupts;
  /* DO */
      for(;;) {        }
  /* FOREVER */
}
/******************************************************
 * SCI Receiver Data Register Full Interrupt Handler
 * Get the character and echo it back to the terminal
 *****************************************************/
void interrupt 20 SCI_isr( void ) {
char input_char;
  /* An interrupt has occurred
   * Reset the RDRF by reading the data */
  input_char = getchar();
  /* Echo the character to the terminal */
   putchar( input_char );
}
```

SCI Interfacing

Please see Chapter 18 for more information about the SCI and to see how to connect to other serial interface devices.

15.3 Synchronous Serial Peripheral Interface (SPI)

The SPI is designed to send high-speed serial data to peripherals and other SPI equipped MCUs and digital signal processors. Normally the SPI is used for short distance communication links for chip-to-chip signaling on a single circuit board. The SCI and the CAN, described in Chapter 16, is used for longer distances and for board-to-board or system-to-system communications.

Interprocessor Serial Communications

Figure 15-3 shows a typical application of the SPI. Two HCS12s are connected in a master/slave arrangement. The 8-bit shift registers in the master and slave make a circular

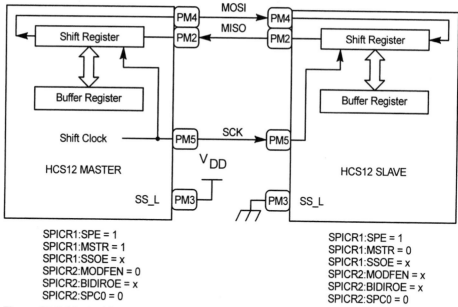

SPICR1:SPE = 1
SPICR1:MSTR = 1
SPICR1:SSOE = x
SPICR2:MODFEN = 0
SPICR2:BIDIROE = x
SPICR2:SPC0 = 0

SPICR1:SPE = 1
SPICR1:MSTR = 0
SPICR1:SSOE = x
SPICR2:MODFEN = x
SPICR2:BIDIROE = x
SPICR2:SPC0 = 0

Figure 15-3 Master/slave serial peripheral interface.

> The SPI is a synchronous serial interface because the master provides a clock to shift data in and out.

16-bit register. When data are to be transmitted from the master to the slave, a clock signal, *SCK* (port M, bit 5), is generated by the master device to *synchronize* the transfer of each bit. Data are transferred out of each shift register simultaneously so that the master receives what was in the slave. The transmitted data are single buffered; this means that the program must wait until the last transmitted data bit is shifted out before writing new data to the register. An *SPI Transfer Complete Flag* (*SPIF*) is available for polling or interrupts. Received data, on the other hand, are buffered and so the program has one character time to read the data before the next data overwrites it. The slave select signal, SS_L, must be low to select a HCS12 as a slave and high for the master.

SPI Data Register

The *SPI Data Register* (*SPIDR*) is similar to the SCI Data Register in that two registers occupy one memory location, $00DD. Data to be transmitted serially are written to this register and serial data received are read from it.

SPIDR—Base + $00DD—SPI Data Register

	Bit 7	6	5	4	3	2	1	0
Read:	Bit 7	6	5	4	3	2	1	0
Write:								
Reset:	0	0	0	0	0	0	0	0

Read: Anytime; normally only after SPIF is set. Write: Anytime.

SPI Initialization

The SPI is initialized by the *SPI Control Registers SPICR1* and *SPICR2*.

SPICR1—Base + $00D8—SPI Control Register 1

	Bit 7	6	5	4	3	2	1	0
Read:	SPIE	SPE	SPTIE	MSTR	CPOL	CPHA	SSOE	LSBFE
Write:								
Reset:	0	0	0	0	0	1	0	0

Read: Anytime. Write: Anytime.

SPIE

SPI Interrupt Enable

0 = SPI interrupts are disabled (default).
1 = Interrupts are enabled.
See SPI Status Register and SPI Interrupts sections.

SPE

SPI System Enable

0 = SPI system is in a low-power, disabled state (default).
1 = SPI is enabled and port M pins are dedicated to SPI functions.

SPTIE

SPI Transmit Interrupt Enable

0 = SPTEF interrupt request disabled (default).
1 = SPTEF interrupt enabled.
See SPI Status Register and SPI Interrupts sections.

MSTR

SPI Master/Slave Mode Select

0 = Slave mode (default).
1 = Master mode.

CPOL, CPHA

SPI Clock Polarity and Clock Phase
See SPI Data Rate and Clock Formats section.

SSOE

Slave Select Output Enable
The slave select output feature allows the SS_L pin to act as an input or output when in master mode. The SSOE, MODFEN
in SPICR2. and MSTR bits determine the operation as shown in Table 15-4. See SPI Mode Faults section.

TABLE 15-4 SS_L Input/Output Selection

MODFEN	SSOE	Master Mode (MSTR = 1)	Slave Mode (MSTR = 0)
0	0	SS_L not used by SPI	SS_L input
0	1	SS_L not used by SPI	SS_L input
1	0	SS_L input with MODF feature	SS_L input
1	1	SS_L is slave select output	SS_L input

LSBFE

SPI Least Significant Bit First

0 = Data are transferred most significant bit first (default).

1 = Date are transferred least significant bit first.

SPICR2—Base + $00D9—SPI Control Register 2

	Bit 7	6	5	4	3	2	1	0
Read:	0	0	0	MODFEN	BIDIROE	0	SPISWAI	SPC0
Write:								
Reset:	0	0	0	0	0	0	0	0

☐ = reserved, unimplemented, or cannot be written to.

Read: Anytime. Write: Anytime.

MODFEN

Mode Fault Enable

0 = SS_L port pin is not used by the SPI (default).

1 = SS_L port pin is used with the mode fault feature.

See Table 15-4 and SPI Mode Faults section.

BIDIROE

Output Enable in Bidirectional Mode

0 = Output buffer is disabled (default).

1 = Output buffer is enabled.

See SPI Master and Slave Modes section.

SPISWAI

SPI Stop in Wait Mode

0 = SPI clock operates normally in wait mode (default).

1 = Stop SPI clocks when in wait mode.

SPC0

Serial Pin Control Bit 0

See SPI Master and Slave Modes section.

SPI Master and Slave Modes

An SPI system consists of a master unit, which controls all timing and data transfer, and one or more slave units. The *MSTR* bit in the *SPICR1* register chooses the master mode (MSTR = 1) or the slave mode (MSTR = 0). In addition to this control bit, the Slave Select (SS_L) pin (port M, bit 5) is controlled by the hardware configuration. Normally, in a system with one master and one slave, the SS_L is pulled high on the master and tied low on the slave as shown in Figure 15-3.

MASTER MODE

The SPI operates in master mode when the MSTR bit is set. Only a master can transmit data from one SPI to another. The transmission begins when a data byte is written to the SCIDR register. The byte is shifted out the *Master Out/Slave In* (*MOSI*) pin at the bit rate set by the serial clock—*SCK*. The serial clock rate is set as discussed later. The baud rate is set by the master device, and as shown in Figure 15-3, the slave device uses the clock to shift the data into its shift register. The Master Out/Slave In (MOSI) and Master In/Slave Out (MISO) pin functions are controlled by the Serial Pin Control (SPC0) and Bidirectional Output Enable (BIDIROE) pins as shown in Table 15-5.

The slave select pin, SS_L, has a variety of functions and attributes controlled by Mode Fault Enable (MODFEN) and Slave Select Output Enable (SSOE). When MODFEN is high and SSOE is low, SS_L is configured as an input pin to be used by the master to detect a mode fault. The pin is driven by an external device and while normally high for the master device to operate as a master, when it is pulled low, it indicates another master is attempting to send data. This is a mode fault—there cannot be more than one master in the SPI system. In this case, the SPI switches to slave mode by clearing the MSTR bit. All outputs are disabled and SCK, MOSI, and MISO are inputs. An interrupt can be generated if the SPI interrupts are enabled (SPIE = 1).

SLAVE MODE

The SPI operates in slave mode when MSTR = 0. SCK is an input that clocks the data into the shift register. The MISO and MOSI pins are controlled by SPC0 and BIDIROE as shown in Table 15-5.

The slave select pin, SS_L, must be low before and during a data transmission. If it is high, the SPI is in an idle mode.

TABLE 15-5 Bidirectional Pin Configurations

Pin Mode	SPC0	BIDIROE	MISO	MOSI
Master Mode (MSTR = 1)				
Normal	0	X	Master In	Master Out
Bidirectional	1	0	MISO not used by SPI	Master In
	1	1		Master In/Out
Slave Mode (MSTR = 0)				
Normal	0	X	Slave Out	Slave In
Bidirectional	1	0	Slave In	MOSI not used by SPI
	1	1	Slave In/Out	

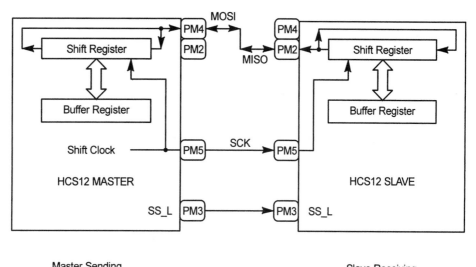

Master Sending
SPICR1:SPE = 1
SPICR1:MSTR = 1
SPICR1:SSOE = 0
SPICR2:MODFEN = 1
SPICR2:BIDIROE = 1
SPICR2:SPC0 = 1

Slave Receiving
SPICR1:SPE = 1
SPICR1:MSTR = 0
SPICR1:SSOE = x
SPICR2:MODFEN = x
SPICR2:BIDIROE = 0
SPICR2:SPC0 = 1

Master Receiving
SPICR1:SPE = 1
SPICR1:MSTR = 1
SPICR1:SSOE = 0
SPICR2:MODFEN = 1
SPICR2:BIDIROE = 0
SPICR2:SPC0 = 1

Slave Sending
SPICR1:SPE = 1
SPICR1:MSTR = 0
SPICR1:SSOE = x
SPICR2:MODFEN = x
SPICR2:BIDIROE = 1
SPICR2:SPC0 = 1

Figure 15-4 SPI bidirectional mode.

SPI BIDIRECTIONAL MODE

The *SPC0* bit in the *SPICR2* register can select either a normal, two-wire mode or a single-wire, bidirectional mode. Normally a two-wire mode is used in which two wires,[2] *MOSI* and *MISO*, are needed to transfer the data. In the single-wire, bidirectional mode, MISO (in the slave) and MOSI (in the master) must be controlled by a program in each unit when transmitting and receiving. The bidirectional modes allow the unused SPI bits to be used as general purpose I/O. Table 15-5 shows the SPI pin names and their various functions under different operating modes. Figure 15-4 show the connections and bit values for the bidirectional mode.

MULTIPLE SPI SLAVES

Figure 15-5 shows a design for one master connected to multiple slaves. Only one slave may be selected at a time, which is accomplished by outputting a slave select signal from port T. When a slave is not selected (SSN_L = 1), its output driver on the MISO line is high impedance.

---

[2] Actually, it is three wires because there needs to be a ground reference for the signals.

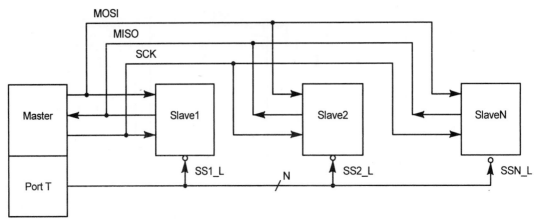

Figure 15-5 Master and multiple slaves.

SPI MODE FAULTS

A mode fault occurs if SS_L is pulled low while the SPI is in master mode. This indicates that some other SPI device is trying to act as a master and a data collision may occur. An interrupt may be generated if SPI interrupts are enabled (SPIE = 1).

SPI Data Rate and Clock Formats

The rate at which data are transferred and the format of the shifting clock are controlled by the *SPI Baud Rate Register* (*SPIBR*) and the *CPOL* and *CPHA* bits in the *SPICR1* register. The bus clock (see Chapter 21) controls the basic clocking rate for the data in the SPI. It can be further divided by 6 bits, *SPPR2, SPPR1, SPPR0*, and *SPR2, SPR1*, and *SPR0* as given by the following equations:

$$Baud\ Rate\ Divisor = (SPPR + 1) \times 2^{(SPR+1)}$$

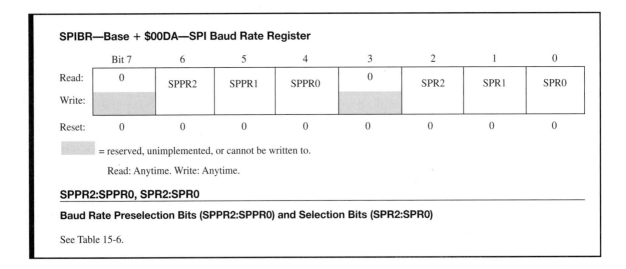

and

$$Baud\ Rate = Bus\ Clock/Baud\ Rate\ Divisor$$

Even though 6 bits are used to form the Baud Rate Divisor, there are only 36 unique divisors as shown in Table 15-6.

The SPI clock output from the master, *SCK* (port M, bit 4), controls the baud rate at which data are transferred as given in Table 15-6. In addition, both the master and the slave modes can be programmed to sample the data according to four modes controlled by the CPOL and CPHA bits in the SPICR1 register. The four choices shown in Table 15-7 and Figure 15-6 allow a variety of SPI devices to be used.

TABLE 15-6 SPI Baud Rate Selection (8-MHz Bus Clock)

SPPR2	SPPR1	SPPR0	SPR2	SPR1	SPR0	SPIBR Value$_{10}$	Baud Rate Divisor	Baud Rate
0	0	0	0	0	0	0	2	4.00 MHz
0	0	0	0	0	1	1	4	2.00 MHz
0	1	0	0	0	0	32	6	1.33 MHz
0	0	0	0	1	0	2	8	1.00 MHz
1	0	0	0	0	0	64	10	800.00 kHz
0	1	0	0	0	1	33	12	666.67 kHz
1	1	0	0	0	0	96	14	571.43 kHz
0	0	0	0	1	1	3	16	500.00 kHz
1	0	0	0	0	1	65	20	400.00 kHz
0	1	0	0	1	0	34	24	333.33 kHz
1	1	0	0	0	1	97	28	285.71 kHz
0	0	0	1	0	0	4	32	250.00 kHz
1	0	0	0	1	0	66	40	200.00 kHz
0	1	0	0	1	1	35	48	166.67 kHz
1	1	0	0	1	0	98	56	142.86 kHz
0	0	0	1	0	1	5	64	125.00 kHz
1	0	0	0	1	1	67	80	100.00 kHz
0	1	0	1	0	0	36	96	83.33 kHz
1	1	0	0	1	1	99	112	71.43 kHz
0	0	0	1	1	0	6	128	62.50 kHz
1	0	0	1	0	0	68	160	50.00 kHz
0	1	0	1	0	1	37	192	41.67 kHz
1	1	0	1	0	0	100	224	35.71 kHz
0	0	0	1	1	1	7	256	31.25 kHz
1	0	0	1	0	1	69	320	25.00 kHz
0	1	0	1	1	0	38	384	20.83 kHz
1	1	0	1	0	1	101	448	17.86 kHz
0	0	1	1	1	1	23	512	15.63 kHz
1	0	0	1	1	0	70	640	12.50 kHz
0	1	0	1	1	1	39	768	10.42 kHz
1	1	0	1	1	0	102	896	8.93 kHz
0	1	1	1	1	1	55	1024	7.81 kHz
1	0	0	1	1	1	71	1280	6.25 kHz
1	0	1	1	1	1	87	1536	5.21 kHz
1	1	0	1	1	1	103	1792	4.46 kHz
1	0	1	1	1	1	119	2048	3.91 kHz

TABLE 15-7 SCK Polarity and Phase

CPOL	CPHA	Clock Polarity	Sample Time	Sample Edge	SCK Idle State
0	0	Active high	Odd edges	Rising	Low
0	1	Active low	Odd edges	Falling	Low
1	0	Active high	Even edges	Rising	High
1	1	Active low	Even edges	Falling	High

SPI Status Register

The *SPI Status Register* (*SPISR*) contains status and error bits. You may poll them or use them to generate interrupts.

SPI Interrupts

There are three bits that can generate an interrupt. These are the *SPI Transfer Complete Flag* (*SPIF*), *SPI Transmit Empty Interrupt Flag* (*SPTEF*), and the *Mode Fault Error Flag* (*MODF*). The *SPI Interrupt Enable* (*SPIE*) (in SPICR1) must be set and when an interrupt occurs, SPIF, SPTEF, and MODF must be polled to determine the source of the interrupt. The interrupt vector and available interrupt sources are shown in Table 15-8.

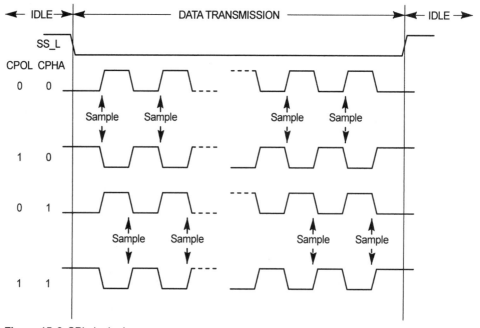

Figure 15-6 SPI clock phases.

SPISR—Base + $00DB—SPI Status Register

	Bit 7	6	5	4	3	2	1	0
Read:	SPIF	0	SPTEF	MODF	0	0	0	0
Write:								
Reset:	0	0	1	0	0	0	0	0

⬜ = reserved, unimplemented, or cannot be written to.

Read: Anytime. Write: Has no effect.

SPIF

SPI Interrupt Flag

0 = Data transfer not complete (default).
1 = New data copied to SPIDR.

This flag is set at the end of an SPI transfer. It is cleared by reading the SPISR with SPIF set, followed by an access (reading or writing) of the SPI data register, SPIDR. If the SPI interrupt enable bit (SPIE) is set, an SPI interrupt is generated.

SPTEF

SPI Transmit Empty Interrupt Flag

0 = SPI data register is not empty.
1 = SPI data register is empty (default).

This flag is set to indicate the transmit data register is empty. It is cleared by reading the SPISR with SPTEF set, followed by writing to the SPI data register, SPIDR. Any write to the SPIDR without first reading SPIDR with SPTEF set will be ignored. If the SPI interrupt enable bit (SPIE) is set, an SPI interrupt is generated.

MODF

Mode Fault Flag

0 = Mode fault has not occurred (default).
1 = Mode fault has occurred.

This flag is set if the SS_L input becomes low while the SPI is configured as a master mode and mode fault detection is enabled (MODFEN = 1 in SPICR2). The flag is cleared by reading the SPISR with MODF set followed by a write to SPICR1.

TABLE 15-8 SPI Serial System Interrupts

Priority	Vector Address	Interrupt Source	Local Enable Bit (SPICR1)	Local Flag (SPISR1)	HPRIO Value to Promote
19	$FFD8:FFD9	SPI New Data in SPIDR	SPIE	SPIF	$D8
		Mode Fault	SPIE	MODF	
		SPI Transmit Empty	SPTIE	SPTEF	

TABLE 15-9 SCI and SPI System Registers

Name	Register	Address Base +
SCIBDH	SCI Baud Rate Register High	$00C8
SCIBDL	SCI Baud Rate Register Low	$00C9
SCICR1	SCI Control Register 1	$00CA
SCICR2	SCI Control Register 2	$00CB
SCISR1	SCI Status Register 1	$00CC
SCISR2	SCI Status Register 2	$00CD
SCIDRH	SCI Data Register High	$00CE
SCIDRL	SCI Data Register Low	$00CF
SPICR1	SPI Control Register 1	$00D8
SPICR2	SPI Control Register 2	$00D9
SPIBR	SPI Baud Rate Register	$00DA
SPISR	SPI Status Register	$00DB
SPIDR	SPI Data Register	$00DD
PTS	Port S I/O Register	$0248

SPI Programming Examples

Chapter 18 gives examples showing how to use the SPI in a variety of applications including a digital-to-analog converter and an LCD display and gives programming examples.

15.4 SCI and SPI Register Address Summary

This summary is presented in Table 15-9.

15.5 Conclusion and Chapter Summary Points

The serial interfaces in the HCS12 support asynchronous data transfer between "normal" serial devices such as terminals, printers, and other computers. It also has a high speed, synchronous data transfer mode for communications with other SPI equipped devices.

- The SCI gives the HCS12 UART capabilities.
- The SCI can send and receive 8- or 9-bit data.
- There can be hardware parity generation.
- The SCI has its own programmable baud rate generator.
- The SCI status registers provide the following bits:
  TDRE—Transmit Data Register Empty
  TC—Transmission Complete
  RDRF—Receive Data Register Full
  IDLE—Idle Line Detect
  OR—Receiver Overrun Error
  NF—Noise Detected During Last Character
  FE—Framing Error
  PF—Parity Incorrect Flag
  RAF—Receiver Active Flag

- The SCI can generate interrupts for the following conditions:

  Transmit Data Register Empty

  Transmission Complete

  Receiver Data Register Full and Receiver Overrun Error

  Idle Line Detected

- The software must poll the status register to see which of the receiver interrupts has occurred.
- The SPI is a high-speed synchronous serial peripheral interface.
- The SPI can transfer serial data at up to 4 Mb/s.
- The SPI status register provides the following bits:

  SPIF—SPI Transfer Complete Flag

  WCOL—Write Collision Error Flag

  MODF—Mode Fault Error Flag

- The SPI can generate interrupts for the following conditions:

  SPI Transfer Complete and Mode Fault Error

- The SPI status register must be checked to see which of the two interrupting sources has occurred.

15.6 Bibliography and Further Reading

Cady, F. M., *Microcontrollers and Microprocessors*, Oxford University Press, New York, 1997.

CPU12 Reference Manual, S12CPUV2.PDF, Freescale Semiconductor, Inc., Austin, TX., July 2003.

HCS12 Serial Communications Interface (SCI) Block Guide, S12SCIV2/D, Freescale Semiconductor, Inc., Austin, TX, April 2004.

HCS12 Microcontrollers MC9S12C Family, Rev 1.15, Freescale Semiconductor, Inc., Austin, TX, July 2005.

MC9S12C Family Device User Guide V01.05, 9S12C128DGV1/D, Freescale Semiconductor, Inc., Austin, TX, 2004.

SPI Block Guide V03.06, S12SPIV3/D, Freescale Semiconductor, Inc., Austin, TX, February 2003.

15.7 Problems

Basic

15.1 How does an asynchronous serial port achieve synchronization of the bits it is sending or receiving? [a]

15.2 An SCI is transmitting data at the following baud rates. The format is eight data bits, no parity, one stop bit. For each case, what is the maximum number of characters per second that can be transmitted? [a]

    a. 56 kbaud.

    b. 9600 baud.

15.3 For the SCI, give the name of the bit, the name of the register it is in, the register's address, which bit, and the default or reset state of the bit for each of the following. [k]

    a. What bit enables the SCI transmitter?

    b. What bit enables the SCI receiver?

    c. What bit determines how many data bits are sent?

    d. What bit can the user test to see if the last character has cleared the transmit data buffer?

e. What bit can the user test to see if a new character has been received?

f. What bit is used to indicate the software is not reading data from the SCIDRL fast enough?

g. What bit is an indication that the communication channel is noisy?

h. What bit is an indication that the sending and receiving baud rates may not be identical?

15.4 For the SCI, give the name of the bit, the name of the register it is in, the register's address, which bit, and the default or reset state of the bit for each of the following. **[k]**

a. What bit enables an interrupt when the transmit buffer is empty?

b. What bit enables an interrupt when the transmitter has completely emptied its serial shift register?

c. What bit enables interrupts by the SCI receiver?

15.5 What SCI receiver conditions can generate an interrupt? **[k]**

15.6 What different status information do the SCI status bits TDRE and TC give? **[k]**

15.7 Give the meanings of the following mnemonics. **[a]**

TDRE, TC, RDRF, OR, FE, NF

15.8 What is the HCS12 I/O address for the SCIBDH register? **[a]**

15.9 What is the value used to initialize the SCI for 4800 baud assuming a Bus Clock of 8.0 MHz? **[k]**

15.10 On which port and which bits are the serial communications interface (SCI) transmitted and received data? **[a]**

15.11 What is the SCIDRH register used for? **[a]**

15.12 For the SPI, give the name of the bit, the name of the register it is in, the register's address, which bit, and the default or reset state of the bit for each of the following. **[k]**

a. What bit enables the SPI?

b. What bit selects the master or slave mode?

c. What bits select the data transfer rate?

d. What bit is the Master Output/Slave Input?

e. What bit is the Master Input/Slave Output?

15.13 For the SPI, give the name of the bit, the name of the register it is in, the register's address, which bit, and the default or reset state of the bit for each of the following. **[k]**

a. What bit indicates the SPI has completely sent the last data?

b. What bit is set to enable SPI interrupts?

15.14 What do the following mnemonics mean in the operation of the SPI?

SS_L, SCK, MOSI, MISO, SISO, SPIE, SPE, MSTR

Intermediate

15.15 An SCI is transmitting data at 19.2 kbaud. The format is seven data bits, even parity, one stop bit. How long does it take to send a document that is 1 megabyte long? **[a]**

15.16 How does the SPI differ from the SCI? **[g]**

15.17 How does a slave station SPI send data to the master station? **[g]**

Advanced

15.18 Implement an asynchronous serial transmitter that does not use the SCI. This should be done as a function or a subroutine that accepts a byte and then transmits it at 1200 baud using port T, bit 7. When the transmitter completes sending the data, bit 7 in a memory location ASY_SR1 is to be set to simulate the TDRE flag. **[c]**

15.19 Assuming an 8-MHz bus clock, what is the highest data rate (bits/s) that your solution to problem 15.18 can provide? **[a]**

15.20 Using the interrupting capabilities of port P, bit 5, implement an asynchronous serial receiver that does not use the SCI. An interrupt is to be generated by port P, bit 5 when it receives a start bit. The interrupt service routine is to shift in the rest of the bits and store the received data in a memory location RX_DATA. When the data have been received, bit 5 in a memory location ASY_SR1 is to be set to simulate the RDRF flag in the SCI. Assume the data rate is 1200 baud. **[c]**

15.21 Assuming an 8-MHz bus clock, what is the highest baud rate at which your solution to Problem 15.20 can receive characters without errors? **[c]**

15.22 Implement a synchronous serial transmitter that outputs 8 data bits using port T, bit 0 as the master out, port T, bit 1 as the serial clock, and port T, bit 2 as SS_L. This should be done as a function or subroutine that accepts a byte and then transmits it at 50 kHz. **[c]**

16 HCS12 Serial I/O—MSCAN

OBJECTIVES

This chapter discusses the HCS12 serial I/O capabilities using the Motorola Scalable Controller Area Network (MSCAN) module.

16.1 Introduction

The HCS12 contains another type of serial interface, in addition to the SCI and SPI, called the *Motorola Scalable Controller Area Network*, or *MSCAN*. This module is a serial communication controller implementing the CAN 2.0 A/B protocol widely used in automotive applications.

16.2 Motorola Scalable Controller Area Network Interface (MSCAN)

Robert Bosch introduced the *Controller Area Network*, or *CAN*, serial bus at the Society of Automotive Engineers congress in February 1986. The CAN bus can reliably handle short messages (up to 8 bytes) with multiple-master access. Although originally developed for automotive markets, it is finding uses in many other applications.

This section describes the various registers and tools to allow you to develop a Controller Area Network. We will not describe fully many of the details of the CAN 2.0A/B specification. Please refer to the references given in Section 16.11 for more information.

CAN Definitions

The CAN serial interface has its own jargon and terms. Here are a few definitions to help you understand some of the following CAN descriptions.

- **Acceptance Filter**: A digital key word that incoming messages must match before the receiver accepts them.
- **Basic CAN**: Basic CAN devices implement in hardware only the basic functions of the protocol, such as generation and checking of the bit stream. All message management, such as accepting the message, must be done in software. See Full CAN.

- **Baud Rate**: Number of bits per second for data transmitted on the CAN bus. See Time Quantum.
- **CRC—Cyclic Redundancy Check**: A 15-bit error checking word used to detect bit errors in the preceding data and in itself.
- **CSMA/CD—Carrier Sense, Multiple Access with Collision Detection**: A method for avoiding or resolving errors when multiple devices try to send messages at the same time.
- **Data Frame**: There are two kinds of frames in CAN—data frames and remote frames. Data frames are used when the node wants to send data. Remote frames are a request for information. A frame with the RTR (Remote Transmission Request) bit set means that the transmitting node is requesting information of the type specified by the identifier.
- **Dominant Level**: A logic low level.
- **EOF—End-of-Frame**: A recessive (logic high) bit, similar to a stop bit in an asynchronous serial interface, that signifies the end of the current message buffer.
- **Extended Frame**: A data frame defined by CAN 2.0B with a 29-bit identifier.
- **Frame**: A message consisting of the start-of-frame (SOF), arbitration, control, data, CRC, acknowledge (ACK), and end-of-frame (EOF) fields.
- **Full CAN**: A Full CAN device implements the whole bus protocol in hardware including acceptance filtering and the message management. The Freescale MSCAN module is a full CAN device. See Basic CAN.
- **Idle Bus**: A bus in the recessive mode (logic high) for more than three bit times.
- **Initialization Mode**: A mode where the MSCAN is disconnected from the CAN bus, allowing system initialization to be done.
- **Recessive Level**: A logic high level.
- **Remote Frame**: See Data Frame.
- **SOF—Start-of-Frame**: A dominant (logic low) bit used like the start bit in an asynchronous serial communications system.
- **Standard Frame**: A data frame defined by CAN 2.0A with an 11-bit identifier.
- **Synchronization Jump**: An increment of time quanta used to synchronize a receiver's bit sampling time with the incoming data.
- **Time Quantum**: A time interval less than the bit time. There may be 8–25 time quanta per bit.

CAN Serial Communications

The CAN bus is a serial bus system with each of the CAN devices, called *nodes*, connected to the bus capable of being a *master*. A master device can initiate data transmission to any of the other nodes on the bus, unlike the SPI, which allows only one master at a time with multiple slaves. The bus uses a single wire (actually two) to reduce the amount of wiring needed in its applications. The bus provides clock synchronization based on the data stream. These concepts require the clever design that Robert Bosch introduced in 1986.

CAN SERIAL BUS BASICS

Recessive bits are logic high and *dominant* bits are logic low.

Figure 16-1 shows a CAN bus. It may have two or more nodes. Because these nodes can be widely separated, by as much as 1000 meters, each has no knowledge of the other nodes. This can lead to a collision of data bits if two or more

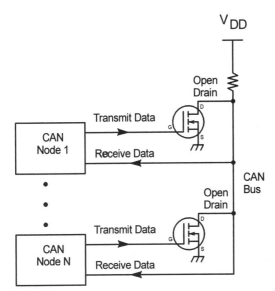

Figure 16-1 Basic CAN bus.

start to transmit at the same time. The CAN design overcomes this problem by defining the electrical characteristics of the bus to be a wired-AND type as shown in Figure 16-1. Each of the CAN nodes has a transmitter and a receiver. The transmitter uses an open-drain connection to the CAN bus. A pull-up resistor establishes the logic levels on the bus. A logic high is called a *recessive* bit. The high is active because none of the nodes are pulling the bus low. (This is why it is called a wired-AND.) A logic low is called a *dominant* bit because one node can dominate all other nodes that are sourcing a recessive bit.

CAN SERIAL BUS COLLISION DETECTION AND ARBITRATION

The problem of two or more nodes starting to transmit at the same time is solved in the following way:

- Each of the nodes continuously monitors the bus with its received data line.
- Each bus transmission is started with a dominant (low) bit, called the *start-of-frame* (SOF), and proceeds with a multiple bit *identifier* (11 or 29 bits long) that defines the type of message data that is to follow.
- If two nodes are transmitting at the same time, eventually one of the identifiers will be different with a low, dominant bit in place of a high, recessive bit.
- Because the low is dominant, and because each of the nodes is monitoring the bus while it is transmitting, the node transmitting the recessive bit will recognize there is another node out there transmitting.
- The node with the recessive bit stops transmitting and allows the other node to continue.
- The node that stops waits until bus activity ceases and tries to send its message again.

This scheme is called *Carrier Sense, Multiple Access with Collision Detection (CSMA/CD)* because nodes are able to detect other transmitters. A prioritization scheme is in effect because the node with the lower binary number for its identifier wins control of the bus. The message that follows contains up to 8 bytes of data and a 15-bit error checking code. The system is able to detect a variety of errors, and it provides an acknowledgment bit so the receiving node can let the transmitting node know the message was received without errors. As you might expect, the protocol to manage this consists of many more details and we will discuss those in the following sections of this chapter.

CAN SERIAL BUS INTERFACE

While the single-ended bus shown in Figure 16-1 explains the concept of the dominant and recessive bits, often in practice a differential, twisted-pair bus is used. The twisted-pair cable provides a transmission line with well behaved characteristic impedance. This allows it to be terminated with a resistance to reduce reflections. It also has noise reduction properties to preserve data quality in noisy industrial environments.

A CAN bus transceiver, such as a Linear Technology LT1796, is used normally to convert the CAN node's single-ended transmit and received data lines to the balanced, differential CAN system signals CAN_H and CAN_L as shown in Figure 16-2. The performance of the system in noise is greatly enhanced by the common mode rejection of the differential receivers. The bus may be twisted-pair wires either unshielded or shielded for additional noise rejection. The data bits are sent with start and stop bits, similar to the SCI described in Chapter 15, with additional characters to define the data frame.

Non-Return-to-Zero (NRZ) signaling encodes the data bits. Each node has its own clock and synchronizes it with the incoming data by detecting bit transitions. To assist clock synchronization, a *bit-stuffing* scheme is used. The receiver may lose bit synchronization if a number of consecutive bits of the same polarity are transmitted. To combat this, the transmitter will insert an additional bit of the opposite polarity into the bit stream after five consecutive ones or zeros. The receiver automatically detects the stuffed bit and removes it from the data.

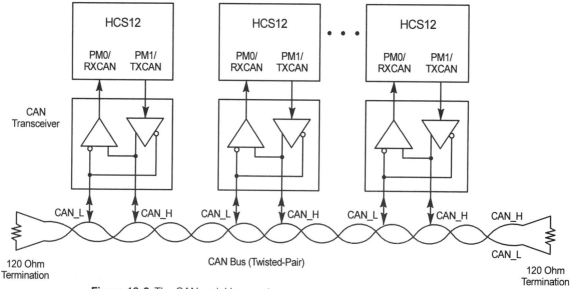

Figure 16-2 The CAN serial bus system.

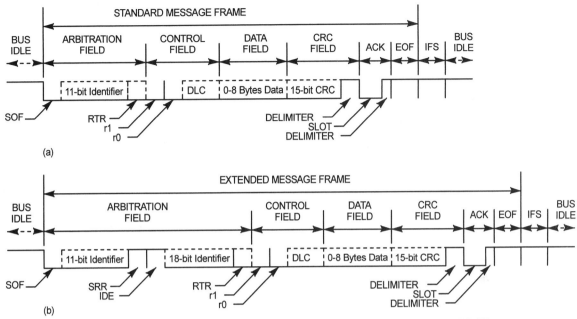

Figure 16-3 CAN message frames: (a) standard, V2.0A and (b) extended, V2.0B.

THE CAN MESSAGE

The CAN bus protocol includes address information, called the identifier, in the message frame. This is not a node address. Instead, it identifies the type of information being transmitted. For example, in an automotive application nodes on the CAN bus may be sending engine RPM, coolant temperature, fuel level, and so on. Each node that needs a particular piece of information has a matching identifier filter and can pick off messages that are relevant to its job and discard others.

The format for standard and extended CAN frames is shown in Figure 16-3 and Table 16-1.

CAN DATA TRANSMISSION

Each node on the system is responsible for *broadcasting* information to the system about its sensors. This means the transmitter does not necessarily know which node is to be the receiver. When a node has information to send, it checks the CAN bus and if the bus is idle (recessive) it starts sending a data packet by asserting the start-of-frame bit. The 11- or 29-bit identifier identifies the type of information to come. The DLC is a code for the number of data bytes in the message, zero to eight, and this is followed by the data and a 15-bit cyclic redundancy check (CRC) error detection word. We will describe the purpose of the other bits in the data transmission later.

CSMA/CD is used to resolve bus contention.

If more than one CAN starts to transmit data at the same time, a Carrier Sense, Multiple Access with Collision Detect (CSMA/CD) protocol allows an arbitration to occur. This results in one of the controllers receiving the highest priority so that data can be sent without collision errors. If a bus node wants to transmit a message, it first checks that the bus is in the idle, or recessive, state. This is the *carrier sense* part, and if no other nodes are transmitting, it

TABLE 16-1 CAN Message Frame Bits

Bit	Name	Function	Register
Bus Idle	—	Bus Idle is a recessive state (logic high).	—
SOF	Start-of-Frame	Acts as a start bit for the frame; inserted by the CAN hardware.	—
11-Bit Identifier	Identifier Bits	Standard or extended message frame identifier bits.	IDR0, IDR1
RTR	Remote Transmission Request	Status of the Remote Transmission Request in the CAN frame.	IDR1-4 (Standard) IDR3-0 (Extended)
SRR	Substitute Remote Request	Used only in extended format. SRR = 1 in transmit buffers and as received in receive buffers.	IDR1-4
r1, r0	—	Two dominant bits reserved for future use.	—
IDE	ID Extended	Identifies extended (=1) or standard (= 0) format.	IDR1-3
18-Bit Identifier	Identifier Bits	Extended format Identifier Bits.	IDR1, IDR2, IDR3
DLC	Data Length Code	Defines the data length (0–8 bytes).	DLR
Data Field	Data Bytes	Up to 8 bytes of data.	DSR0–DSR7
15-Bit CRC	Cyclic Redundancy Check	Error checking generated by the CAN hardware.	—
ACK	Acknowledge	Inserted by the CAN hardware.	—
EOF	End-of-Frame	Acts as stop bit for the frame; inserted by the CAN hardware.	—
IFS	Interframe Space	At least three recessive bit times are required after the frame and before the next frame to allow internal processing in the nodes.	—

becomes the master and sends its message. The first transmitted bit is the dominant *Start-of-Frame* (*SOF*) and during this time all other nodes switch to the receive mode. Each node checks the message identifier and if it matches its own identifier, it stores the message and if not, discards it. Each node that accepts a message sends an acknowledgment (ACK) that it has received a message correctly (internode handshaking).

If two or more nodes start to transmit at the same time (*multiple access*), bit-wise checking (*collision detection*) with nondestructive arbitration is used to resolve the problem. Each node starts by sending the identifier bits of the message and, during each bit time, monitors the bus. If another node sends a dominant bit, a node sending a recessive bit loses the bus (*lost arbitration*) and switches back to receive mode to wait for the bus to return to the idle state. Because the dominant bit is logic low, identifiers with a lower binary number will win the arbitration and thus have higher priority. This scheme means that multiple nodes cannot use identical identifiers. When the bus is idle, all nodes that lost arbitration will try again.

The identifiers used are either 11 bits, called the *standard CAN frame*, or 29 bits, called the *extended CAN frame*. The standard CAN is defined in the specification V2.0A and the extended CAN in V2.0B. Nodes operating as CAN V2.0B devices can transmit and receive both standard and extended frames but V2.0A devices can use only standard frames.

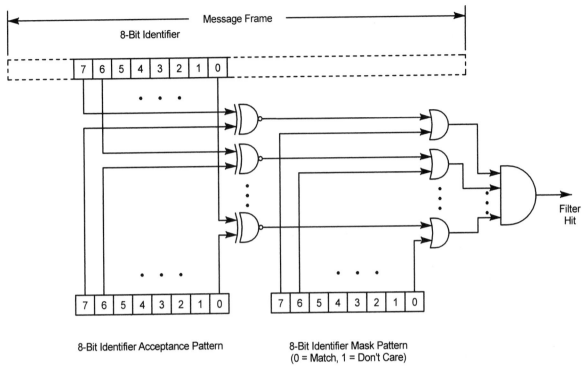

Figure 16-4 Receiver acceptance filtering.

CAN MESSAGE RECEIVING

The CAN receiver detects the start-of-frame bit and starts to clock message bits similar to the SCI receiver described in Chapter 15. The identifier bits in the message are used by the receiver to determine if it is a message for itself or if it should be ignored. Figure 16-4 shows how the receiver accepts or ignores CAN messages.

Although there are 11 or 29 identifier bits in the arbitration field of the message, the receiver may use 8-, 16-, or 32-bit patterns to identify its messages. As the message arrives it is shifted into a message buffer and the identifier (shown here as 8 bits) is compared with an *Identifier Acceptance Pattern*. When they match, all outputs of the exclusive-NOR gates in Figure 16-4 and the *Filter Hit* signal will be asserted. For added flexibility in identifying messages, an *Identifier Mask Pattern* can allow some of the bits in the identifier to be don't cares. A one in the identifier mask pattern sets the acceptance bit to be a don't care. When the filter hit signal is asserted, the rest of the message is accepted by the receiver.

The receiver checks the 15-bit CRC code to make sure there were no errors in the transmission and then, at the correct time in the message (see Figure 16-3), asserts the *Acknowledge (ACK)* bit to let the transmitter know the message was received correctly.

CAN MESSAGE TRANSMISSION SPEEDS AND DISTANCES

The transmission bit rates and distances are interrelated and depend on the physical layer implementation. Table 16-2 shows bit rates and distances for typical CAN bus applications using terminated, twisted-pair bus wires.

TABLE 16-2 CAN Bus Length Versus Data Rate

Bus Length	Bit Rate
40 meters	1 Mbit/s
40 to 300 meters	500 kbits/s
300 to 600 meters	100 kbits/s
600 to 1000 meters	50 kbits/s

16.3 MSCAN Initialization and Control Registers

As you might expect, you must initialize a number of registers and control bits to operate the MSCAN system.

Initialization Mode

When in *initialization mode*, the MSCAN is disconnected from the CAN bus to avoid creating errors on the bus.

All MSCAN initialization is done while the module is in *initialization mode*. In this mode the MSCAN is disconnected from the CAN bus to avoid creating any problem for other nodes on the bus. If your MSCAN is active and you wish to initialize it, the recommended procedure is to place it into sleep mode by setting the sleep request bit, SLPRQ = 1 in the CANCTL0 register, and waiting until the sleep acknowledge bit, SLPAK in CANCTL1, is asserted. At reset, MSCAN is inactive and can be initialized directly without this step. The MSCAN is initialized by setting the INITRQ bit in the CANCTL0 register and waiting until the acknowledge bit, INITAK in CANCTL1, is asserted. MSCAN registers enter their hard reset state and default values are restored. Both INITRQ and INITAK must be set before control bits are written into various registers to set up the operation of the MSCAN module. Table 16-3 summarizes the bits that may be initialized only in initialization mode. Other bits controlling the operation of the

TABLE 16-3 Bits to be Set in Initialization Mode

Control Bit	Name	Register
CANE	MSCAN Enable	CANCTL1
CLKSRC	Clock Source	CANCTL1
LOOPB	Loop Back Mode	CANCTL1
LISTEN	Listen Only Mode	CANCTL1
WUPM	Wake-up Mode	CANCTL1
SLPAK	Sleep Mode Acknowledge Flag	CANCTL1
INITAK	Initialization Mode Acknowledge Flag	CANCTL1
SJW1:SJW0	Synchronization Jump Width	CANBTR0
BRP5:BRP0	Baud Rate Prescaler	CANBTR0
SAMP	Sampling	CANBTR1
TSEG22:TSEG20	Time Segment 2	CANBTR1
TSEG13:TSEG10	Time Segment 1	CANBTR1
IDAM1:IDAM0	Identifier Acceptance Mode	CANIDAC

TABLE 16-4 Bits that May Be Set During MSCAN Operation

Control Bits	Name	Register
CSWAI	CAN Stops in Wait Mode	CANCTL0
TIME	Timer Enable	CANCTL0
WUPE	Wake-up Enable	CANCTL0
SLPRQ	Sleep Mode Request	CANCTL0
INITRQ	Initialization Mode Request	CANCTL0
WUPIE	Wake-up Interrupt Enable	CANRIER
CSCIE	CAN Status Change Interrupt Enable	CANRIER
RSTATE1:RSTATE0	Receiver Status Change Enable	CANRIER
TSTATE1:TSTATE0	Transmitter Status Change Enable	CANRIER
OVRIE	Overrun Interrupt Enable	CANRIER
RXFIE	Receiver Full Interrupt Enable	CANRIER
IDAMN1:IDAM0	Identifier Acceptance Mode	CANIDAC
AC7:AC0	Identifier Acceptance Bits	CANIDAR0–CANIDAR7
AM7:AM0	Identifier Mask Bits	CANIDMR0–CANIDMR7
ID28:ID0	Identifier Bits	CANTIDR0–CANTIDR3
DB7:DB0	Data Bits	CANTDSR0–CANTSDR7
DLC3:DLC0	Data Length Code	CANTDLR
PRIO7:PRIO0	Transmitter Priority Code	CANTBPR

MSCAN may be set or reset while in an operating mode. These are shown in Table 16-4. The following sections give full programming details for all control bits in the MSCAN.

After writing all initialization mode bits, leave the initialization mode by resetting INITRQ and waiting for INITAK to be reset. You may then either leave sleep mode or remain there as desired.

CANCTL0—Base + $0140—MSCAN Control 0 Register

	Bit 7	6	5	4	3	2	1	0
Read:	RXFRM	RXACT	CSWAI	SYNCH	TIME	WUPE	SLPRQ	INITRQ
Write:								
Reset:	0	0	0	0	0	0	0	1

[shaded] = reserved, unimplemented, or cannot be written to.

Read: Anytime. Write: After initialization mode is left (INITRQ = 0 and INITAK = 0). Exceptions are the read-only bits (RXACT and SYNCH) and RXFRM, which is set by the module when a frame has arrived, and INITRQ, which is writable in initialization mode.

RXFRM

Received Frame Flag

0 = No valid message was received since the last time the flag was cleared (default).

1 = A valid message was received.

This bit may be read and cleared by the program. When a frame has been received, the bit is set and may be reset by writing a 1 to the bit. The bit is not valid in loop back mode.

RXACT

Receiver Active Status

0 = MSCAN is transmitting or idle (default).

1 = MSCAN is receiving a message.

RXACT is a read-only bit and indicates that the MSCAN is receiving a message. The bit is not valid in loop back mode.

CSWAI

CAN Stops in Wait Mode

0 = The module continues to run in wait mode (default).

1 = The module is not clocked in wait mode, allowing reduced power consumption.

SYNCH

Synchronized Status

0 = MSCAN is not synchronized to the CAN bus (default).

1 = MSCAN is synchronized to the CAN bus and can take part in CAN bus communications.

TIME

Timer Enable

0 = Internal timer is disabled (default).

1 = Internal timer is enabled.

The MSCAN module has a 16-bit free-running counter clocked by the bit clock. When the timer is enabled, a 16-bit "time stamp" (not a real time) is assigned to each transmitted or received message. As soon as the message is acknowledged on the CAN bus, the time stamp will be written to the highest bytes of the appropriate transmit or receive message buffer.

WUPE

Wake-up Enable

0 = Wake-up is disabled. Traffic on the CAN bus is ignored (default).

1 = Wake-up is enabled and the MSCAN may restart.

When this bit is set, the MSCAN can restart from sleep mode when traffic on the CAN bus is detected. You must make sure that the WUPE bit and MSCAN Receiver Interrupt Enable Register (CANRIER) are initialized if you wish to wake up from HCS12 CPU stop or wait modes.

SLPRQ

Sleep Mode Request

0 = MSCAN is running normally (default).

1 = MSCAN can enter sleep mode when the CAN bus is idle.

This bit allows the MSCAN to enter a sleep mode, saving power, when the CAN bus is idle. When the module enters sleep mode it sets the SLPAK bit in the MSCAN Control 1 Register (CANCTL1). Sleep mode will be active until cleared by the CPU or when activity is detected on the CAN bus and at which time SLPRQ is cleared.

INITRQ

Initialization Mode Request

0 = Normal operation (default).

1 = MSCAN is in initialization mode.

Setting this bit causes the MSCAN to go into initialization mode. MSCAN registers enter their hard reset state and default values are restored. When the bit is cleared, MSCAN tries to synchronize itself with the CAN bus. See INITAK in CANCTL1.

CANCTL1—Base + $0141—MSCAN Control Register 1

	Bit 7	6	5	4	3	2	1	0
Read:	CANE	CLKSRC	LOOPB	LISTEN	0	WUPM	SLPAK	INITAK
Write:								
Reset:	0	0	0	1	0	0	0	1

⬜ = reserved, unimplemented, or cannot be written to.

Read: Anytime. Write: Anytime when in initialization mode (INITRQ = 1 and INITAK = 1).

CANE

MSCAN Enable

0 = MSCAN is disabled (default).
1 = MSCAN is enabled.

May be written once in normal mode or anytime in special modes when MSCAN is in initialization mode (INITRQ = 1 and INITAK = 1).

CLKSRC

MSCAN Clock Source

0 = MSCAN clock is the oscillator clock (default).
1 = Clock is the bus clock (one-half oscillator clock).

LOOPB

Loop Back Self Test Mode

0 = Loop back mode is disabled (default).
1 = Loop back mode is enabled.

When in loop back mode the transmitter output is connected internally to the receiver. RXCAN input pin is ignored and the transmitter goes to the recessive state.

LISTEN

Listen Only Mode

0 = Normal operation.
1 = Listen only mode activated (default).

This configures the MSCAN as a bus monitor. No acknowledgment or error frames are sent and error counters are frozen.

WUPM

Wake-up Mode

0 = MSCAN wakes-up the CPU when it detects a recessive to dominant edge on the CAN bus and when WUPE in CANCTL0 is set (default).
1 = MSCAN wakes up the CPU only when it detects a dominant pulse on the bus and WUPE in CANCTL0 is set.

SLPAK

Sleep Mode Acknowledge Flag

0 = Running; MSCAN is operating normally (default).
1 = Sleep mode is active.

This flag indicates that the MSCAN module has entered the sleep mode. It may be used as a handshake flag for the SLPRQ sleep mode request. When SLPRQ = 1 and SLPAK = 1, sleep mode is active. If WUPE = 1, MSCAN is able to restart from sleep mode.

INITAK

Initialization Mode Acknowledge Flag

0 = Running; MSCAN is operating normally (default).

1 = MSCAN has entered initialization mode.

This flag is a handshaking flag for the INITRQ initialization mode request.

CAN Bus Timing

Between 8 and 25 *Time Quanta* clocks define a bit time.

Figure 16-5 shows the clock source for the MSCAN module. Either the bus clock or the oscillator clock is the timing source. The CAN bus timing specification has a rather tight oscillator tolerance so one should choose a stable oscillator source.

The bus clock normally used in HCS12 systems is the external oscillator. In addition, when the data rate is 1 Mbits/s, the duty cycle of the clock should be in the range of 45–55%. After the basic oscillator is divided by the prescaler (bits BRP5:BRP0 in CANBTR0), the *Time Quanta* (T_Q) clock is produced. Between 8 and 25 T_Q periods define one bit time as shown in Figure 16-6.

Each bit time shown in Figure 16-6 is comprised of several time quanta. The CAN 2.0 specification defines the following elements in each bit time:

- **Synchronization Segment (SYNC_SEG)**: This segment starts at the time of transmission. It is one time quantum long and all nodes expect a bit transition during this segment.
- **Propagation Segment (PROP_SEG)**: This segment is one to eight time quanta long and compensates for physical delays in the network. It should be twice the sum of the signal's propagation time on the bus plus the input comparator delay plus the output driver delay. In MSCAN, this time is included with PHASE_SEG1 and specified by TSEG13:TSEG10 in CANBTR1.
- **Phase Segment 1 (PHASE_SEG1) and Phase Segment 2 (PHASE_SEG2)**: These segments compensate for phase edge errors and may be lengthened or shorted by the synchronization jump width. PHASE_SEG1 is one to eight time quanta.
- **Information Processing Time**: This is the time starting with the sample point reserved for calculating the actual bit level. The Information Processing Time is less than or equal to two time quanta.

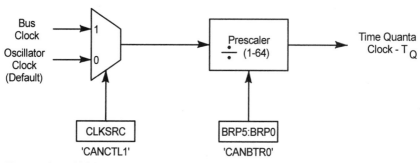

Figure 16-5 MSCAN clock source.

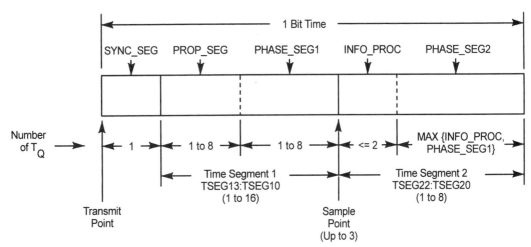

Figure 16-6 CAN bus bit time segments.

- **Phase Segment 2 (PHASE_SEG2)**: The maximum of the Information Processing Time or PHASE_SEG1. In MSCAN, this time is specified by TSEG22:TSEG20 in CANBTR1.
- **Sample Point**: The data bit is sampled at the end of PHASE_SEG1. Typically, the sample point should be at about 60–70% of the bit time. You may choose one or three sample points.

SYNCHRONIZATION AND RESYNCHRONIZATION

The MSCAN hardware is able to synchronize its bit sampling time with the incoming data stream. This allows a wide variety of devices with differing clocks and clock stabilities to coexist on the CAN bus. There are two kinds of synchronization processes to consider—*hard synchronization* at the start of a frame and *resynchronization* within a frame.

Hard synchronization takes place when a frame starts with a recessive to dominant bit change. The MSCAN hardware synchronizes its timing so that the bit change lies within the SYNC_SEG time quantum.

If the oscillators in two nodes are slightly different and the bits in the frame continue to arrive, the bit changes relative to the sample point may drift around. If this occurs, and the hardware detects a bit change outside the SYNC_SEG, it can adjust the sample point by instituting a jump in the number of time quanta in PHASE_SEG1 or PHASE_SEG2. This is called the *synchronization jump width* and is controlled by SJW1:SJW0 in the CANBTR0 register.

TIME SEGMENT PROGRAMMING

The baud rate prescaler, the number of time quanta in Time Segment 1 and Time Segment 2, and the synchronization jump width must be initialized during initialization mode. The following are some requirements for programming the time segments:

- PROP_SEG + PHASE_SEG1 ≥ INFO_PROC + PHASE_SEG2.
- Time Segment 1 ≥ Time Segment 2.
- Synchronization Jump Width < Time Segment 2.

CANBTR0—Base + $0142—MSCAN Bus Timing Register 0

	Bit 7	6	5	4	3	2	1	0
Read:	SJW1	SJW0	BRP5	BRP4	BRP3	BRP2	BRP1	BRP0
Write:								
Reset:	0	0	0	0	0	0	0	0

　　　= reserved, unimplemented, or cannot be written to.

Read: Anytime. Write: Anytime in initialization mode (INITRQ = 1 and INITAK = 1).

SJW1:SJW0

Synchronization Jump Width

The synchronization jump width defines the maximum amount of time in time quanta (T_Q) clock cycles a bit may be shortened or lengthened to resynchronize to data transitions on the bus. (See Table 16-5.)

TABLE 16-5 Synchronization Jump Width

SJW1	SJW0	Jump Width (T_Q Clock Cycles)
0	0	1 (default)
0	1	2
1	0	3
1	1	4

BRP5:BRP0

Baud Rate Prescaler

These bits determine the time quanta (T_Q) clock used to set the individual bit timing. (See Table 16-6.)

TABLE 16-6 Baud Rate Prescaler

BRP5	BRP4	BRP3	BRP2	BRP1	BRP0	Prescaler
0	0	0	0	0	0	1 (default)
0	0	0	0	0	1	2
0	0	0	0	1	0	3
—	—	—	—	—	—	—
1	1	1	1	1	0	63
1	1	1	1	1	1	64

　　The bit time for the CAN bus is determined by the oscillator frequency, Bus Clock, the prescaler value given by BRP5:BRP0, and the number of time quanta given by the TSEG2 and TSEG1 values. (See Example 16-1.)

$$Bit\ Time = \frac{PRESCALER\ VALUE}{f_{CANCLK}} \times [TSEG2 + TSEG1 + 1]$$

CANBTR1—Base + $0143—MSCAN Bus Timing Register 1

	Bit 7	6	5	4	3	2	1	0
Read:	SAMP	TSEG22	TSEG21	TSEG20	TSEG13	TSEG12	TSEG11	TSEG10
Write:								
Reset:	0	0	0	0	0	0	0	0

⬛ = reserved, unimplemented, or cannot be written to.

Read: Anytime. Write: Anytime when INITRQ = 1 and INITAK = 1.

SAMP

Bit Sampling Control

0 = One sample per bit (default).

1 = Three samples per bit.

SAMP controls the number of samples of the CAN bus taken per bit time. If SAMP = 1, three samples are taken and in this case PHASE_SEG1 must be at least two time quanta. When three samples are taken, one is at the sample point shown in Figure 16-6 with two proceeding samples. A majority rule determines the bit level.

TSEG22:TSEG20

Time Segment 2

These bits set the number of T_Q clock cycles in Time Segment 2 (see Table 16-7). See Figure 16-6.

TABLE 16-7 Time Segment 2 Values

TSEG2	TSEG1	TSEG0	Time Segment 2 (T_Q Clock Cycles)
0	0	0	Not valid
0	0	1	2
0	1	0	3
—	—	—	—
1	1	0	7
1	1	1	8

TSEG13:TSEG10

Time Segment 1

These bits set the number of T_Q clock cycles in Time Segment 1 (see Table 16-8). See Figure 16-6.

TABLE 16-8 Time Segment 1 Values

TSEG13	TSEG12	TSEG11	TSEG10	Time Segment 1 (T_Q Clock Cycles)
0	0	0	0	Not Valid
0	0	0	1	2
0	0	1	0	3
—	—	—	—	—
1	1	1	0	15
1	1	1	1	16

> **Example 16-1 CANBTR0 and CANBTR1 Initialization**
>
> Calculate the prescaler divider, the Time Segment 2 length needed, and the synchronization jump width assuming the oscillator clock = 8 MHz, Time Segment 1 = 8 T_Q, and the desired bit rate is 250 kHz. Give the values for the CANBTR0 and CANBTR1 registers. Assume Time Segment 2 = Time Segment 1 = 8 T_Q.
>
> **Solution:** The bit time is 1/250 kHz = 4 μs. Assume Time Segment 2 = Time Segment 1 = 8 T_Q; this gives 17 time quanta (1 + 8 + 8) per bit. Thus T_Q = 4 μs/17 = 235.3 ns. The Bus Clock period is 1/8 MHz = 125 ns. Thus the prescaler divider is 235.3/125 = 1.9. The closest integer value is 2, giving T_Q = 250 ns. To achieve the desired bit rate of 250 kHz, the number of T_Q should be adjusted. The total allowed is 4 μs/250 ns = 16, so set Time Segment 2 to 7. The synchronization jump width must always be less than Time Segment 2. In this case the SJW value in CANBTR0 may be any value from 1 to 4.
>
> CANBTR0 = %00000001,   CANBTR1 = %01100111

MSCAN Identifiers

Each message transmitted on the CAN bus by various nodes has the format shown in Figure 16-3. The arbitration field consists of either 11 identifier bits in the standard frame or 29 bits in the extended format message. The transmitting node sets these bits to identify the type of information. For example, a sensor measuring the engine RPM may send that information to an engine control module while a fuel sensor may send the state of the gas tank to an instrument panel module. Both messages will have an identifier that allows the appropriate receiving node to accept its messages and ignore those destined for somewhere else. Furthermore, the way bus collisions are resolved, an identifier with a lower binary number will have higher priority in the event of a bus collision. In this case, we might expect the engine RPM message to be higher priority than fuel level information.

> CAN identifiers allow messages to be captured by the receiver for which the information is intended.

The MSCAN receiver has a filtering mechanism that allows it to load the receiver buffers with acceptable messages and discard all others. There are eight *Identifier Acceptance Registers (CANIDAR0–CANIDAR7)* and eight *Identifier Mask Registers (CANIDMR0–CANIDMR7)* that work together to form up to eight different filters, which when matched, allow the present message to be entered into the next available receiver message buffer. These allow four modes of filtering and a selection of filters to allow nodes to accept multiple messages. The IDAM1:IDAM0 bits in the CANIDAC register select the filter mode. Table 16-9 shows the four modes available, which bits are checked, and the resultant filter hit when an acceptable bit pattern provides a match to the data.

Figures 16-7 to 16-9 show how the bits in the Identifier Acceptance and Mask Registers match up with the transmitted message in both standard and extended formats for 32-bit, 16-bit, and 8-bit filters.

32-BIT IDENTIFIERS

Figure 16-7 shows two 32-bit acceptance filters. These filters can be used only in a CAN 2.0B system. The message is received in the receiver identifier register CANRIDR3:CANRIDR0 buffer. The receiver initializes two 32-bit acceptance filters CANIDAR3:CANIDAR0 and

TABLE 16-9 Identifier Acceptance Filtering

IDAM1	IDAM0	Mode	Identifier Acceptance Registers	Standard Frame Bits Checked	Extended Frame Bits Checked	Filter Hit
0	0	Two 32-bit filters	CANIDAR0:CANIDAR3	ID10–ID0 plus	ID28–ID0 plus	0
			CANIDAR4:CANIDAR7	RTR and IDE	RTR, IDE, and SRR	1
0	1	Four 16-bit filters	CANIDAR0:CANIDAR1	ID10–ID0 plus	ID28–ID15 plus	0
			CANIDAR2:CANIDAR3	RTR and IDE	IDE and SRR	1
			CANIDAR4:CANIDAR5			2
			CANIDAR6:CANIDAR7			3
1	0	Eight 8-bit filters	CANIDAR0	ID10–ID3	ID28–ID15	0
			CANIDAR1			1
			CANIDAR2			2
			CANIDAR3			3
			CANIDAR4			4
			CANIDAR5			5
			CANIDAR6			6
			CANIDAR7			7
1	1	Filter closed				

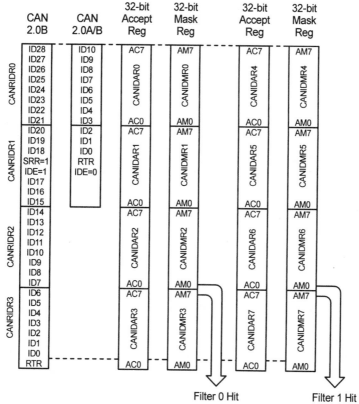

Figure 16-7 Two 32-bit acceptance filters.

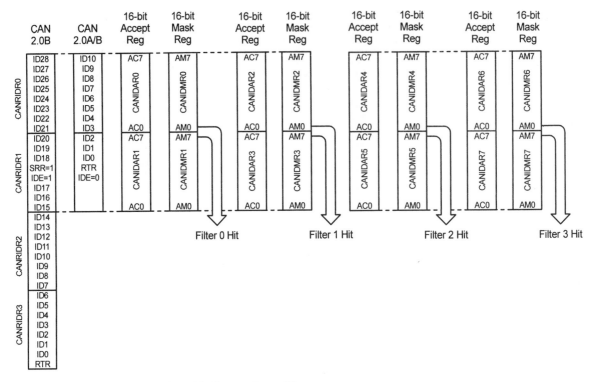

Figure 16-8 Four 16-bit acceptance filters.

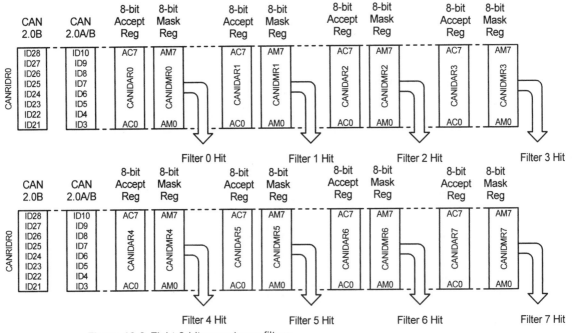

Figure 16-9 Eight 8-bit acceptance filters.

CANIDAR7:CANIDAR4 and two 32-bit mask registers CANIDMR3:CANIDMR0 and CANIDMR7:CANIDMR4. When the acceptance filter conditioned by the mask register agrees with the bits received in the receiver identifier register, Filter 0 Hit or Filter 1 Hit is asserted.

16-BIT IDENTIFIERS

Figure 16-8 shows that a receiver may use four 16-bit identifiers. These can be used only in a CAN 2.0B system. The 16 bits of the identifier are in the receiver buffer CANRIDR1: CANRIDR0 (ID15–ID28, and include SRR = 1 and IDE = 1). The four acceptance and mask register combinations are as shown in Table 16-9.

8-BIT IDENTIFIERS

Eight 8-bit identifier acceptance filters can be configured as shown in Figure 16-9. Bits ID21:ID28 in a CAN 2.0B system and ID3:ID10 in a standard 2.0A system are received in CANRIDR0. Eight filter hit signals are asserted when the acceptance filter/mask register combinations agree with the received identifier.

RECEIVER ACCEPTANCE FILTERING

Figure 16-10 shows how the MSCAN accepts a message. In this example an 8-bit identifier is used. The received identifier is written into the CANIDAR0 register and is compared with the

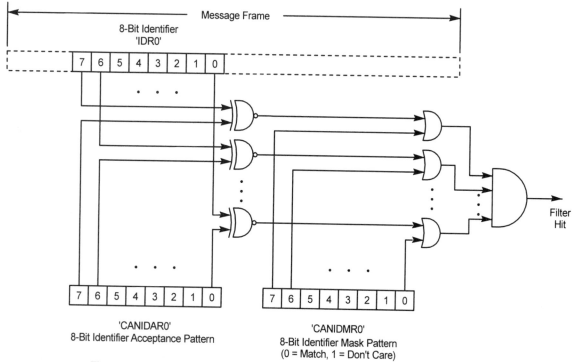

Figure 16-10 Receiver acceptance filtering.

8-bit identifier in the message. Each matched bit produces a logic one at the exclusive-NOR gates. The CANIDMR0 register is the mask register. Zeros in this register mean that the corresponding bits in the message must match the identifier bits in the CANIDAR0 register. A one means that the message bit does not have to match the identifier and is thus a don't care. When all bits match or are don't cares, the Filter Hit signal is asserted and the message is loaded into one of the receiver message buffers.

IDENTIFIER REGISTERS

CANIDAC—Base + $014B—MSCAN Identifier Acceptance Control Register

	Bit 7	6	5	4	3	2	1	0
Read:	0	0	IDAM1	IDAM0	0	IDHIT2	IDHIT1	IDHIT0
Write:								
Reset:	0	0	0	0	0	0	0	0

▨ = reserved, unimplemented, or cannot be written to.

Read: Anytime. Write: Anytime in initialization mode except for IDHITn bits, which are read only.

IDAM1:IDAM0

Identifier Acceptance Mode Bits

The application program sets these bits to define the identifier acceptance filter organization. (See Table 16-10.)

TABLE 16-10 Identifier Acceptance Mode Settings

IDAM1	IDAM0	Identifier Acceptance Mode
0	0	Two 32-bit Acceptance Filters
0	1	Four 16-bit Acceptance Filters
1	0	Eight 8-bit Acceptance Filters
1	1	Filter Closed

IDHIT2:IDHIT0

Identifier Acceptance Hit Indicator

The MSCAN sets these flags to indicate an identifier acceptance hit. (See Table 16-11.)

TABLE 16-11 Identifier Acceptance Hit Indication

IDHIT2	IDHIT1	IDHIT0	Identifier Acceptance Hit
0	0	0	Filter 0 Hit
0	0	1	Filter 1 Hit
0	1	0	Filter 2 Hit
0	1	1	Filter 3 Hit
1	0	0	Filter 4 Hit
1	0	1	Filter 5 Hit
1	1	0	Filter 6 Hit
1	1	1	Filter 7 Hit

CANIDAR0—Base + $0150—MSCAN Identifier Acceptance Register 0
CANIDAR1—Base + $0151—MSCAN Identifier Acceptance Register 1
CANIDAR2—Base + $0152—MSCAN Identifier Acceptance Register 2
CANIDAR3—Base + $0153—MSCAN Identifier Acceptance Register 3
CANIDAR4—Base + $0158—MSCAN Identifier Acceptance Register 4
CANIDAR5—Base + $0159—MSCAN Identifier Acceptance Register 5
CANIDAR6—Base + $015A—MSCAN Identifier Acceptance Register 6
CANIDAR7—Base + $015B—MSCAN Identifier Acceptance Register 7

	Bit 7	6	5	4	3	2	1	0
Read:	AC7	AC6	AC5	AC4	AC3	AC2	AC1	AC0
Write:								
Reset:	0	0	0	0	0	0	0	0

    = reserved, unimplemented, or cannot be written to.

Read: Anytime. Write: Anytime in initialization mode.

AC7–AC0

Acceptance Code Bits

These bits form a user defined sequence of bits with which the corresponding bits of the related identifier register (IDRn) of the receive message buffer are compared. The result of this comparison is then masked with the corresponding identifier mask register.

CANIDMR0—Base + $0154—MSCAN Identifier Mask Register 0
CANIDMR1—Base + $0155—MSCAN Identifier Mask Register 1
CANIDMR2—Base + $0156—MSCAN Identifier Mask Register 2
CANIDMR3—Base + $0157—MSCAN Identifier Mask Register 3
CANIDMR4—Base + $015C—MSCAN Identifier Mask Register 4
CANIDMR5—Base + $015D—MSCAN Identifier Mask Register 5
CANIDMR6—Base + $015E—MSCAN Identifier Mask Register 6
CANIDMR7—Base + $015F—MSCAN Identifier Mask Register 7

	Bit 7	6	5	4	3	2	1	0
Read:	AM7	AM6	AM5	AM4	AM3	AM2	AM1	AM0
Write:								
Reset:	0	0	0	0	0	0	0	0

    = reserved, unimplemented, or cannot be written to.

Read: Anytime. Write: Anytime in initialization mode.

AM7–AM0

Acceptance Mask Bits

0 = Match corresponding acceptance code register and identifier bits (default).
1 = Ignore corresponding acceptance code register bit.

If a bit in this register is zero, the corresponding bit in the identifier acceptance register (CANIDARn) must be the same as its identifier bit before a match is detected. If all such bits match, the message is accepted.

If a bit is set, the state of the corresponding bit in the identifier acceptance register does not affect whether or not the message is accepted.

MSCAN IDENTIFIER OPERATION EXAMPLES

These are presented in Examples 16-2 to 16-5.

Example 16-2

A receiver is to use a single 8-bit identifier equal to $11. It is to accept only messages with this identifier. Specify the registers to be used and how they should be initialized.

Solution: Set IDAM1:IDAM0 in CANIDAC = %10 to select 8-bit acceptance filters; set CANIDAR0 = $11 and CANIDMR0 = $00 to match all identifier bits.

Example 16-3

A receiver is to receive messages from eight different nodes that send messages with identifiers $50–$57. Specify the registers used and how they should be initialized.

Solution: Set IDAM1:IDAM0 in CANIDAC = %10 to select 8-bit acceptance filters. CANIDAR0–CANIDAR7 are initialized to $50–$57. CANIDMR0–CANIDMR0 = $00 to match all identifier bits.

Example 16-4

Explain how the receiver in Example 16-3 can determine which message has been accepted.

Solution: After receiving the message the IDHIT2:IDHIT0 bits in CANIDAC are checked to see which filter has been hit.

Example 16-5

Another receiver in the system described in Example 16-3 needs to receive all messages with identifiers $50–$57. Specify the registers used and how they should be initialized.

Solution: Set IDAM1:IDAM0 in CANIDAC = %10 to select 8-bit acceptance filters. Only one of the 8-bit filters is needed. CANIDAR0 is initialized to $50 and CANIDMR0 initialized to $07 to match identifier bits 7–2 and make identifier bits 2–0 don't cares.

TABLE 16-12 Receive Message Buffer Organization

Receive Address Base +	Register Name
$0160	Identifier Register 0—CANRIDR0
$0161	Identifier Register 1—CANRIDR1
$0162	Identifier Register 2—CANRIDR2
$0163	Identifier Register 3—CANRIDR3
$0164	Data Segment Register 0—CANRDSR0
$0165	Data Segment Register 1—CANRDSR1
$0166	Data Segment Register 2—CANRDSR2
$0167	Data Segment Register 3—CANRDSR3
$0168	Data Segment Register 4—CANRDSR4
$0169	Data Segment Register 5—CANRDSR5
$016A	Data Segment Register 6—CANRDSR6
$016B	Data Segment Register 7—CANRDSR7
$016C	Data Length Register—CANRDLR
$016D	Not Used in the Receiver
$016E	Time Stamp Register (High Byte)—CANRTSRH[a]
$016F	Time Stamp Register (Low Byte)—CANRTSRL[a]

[a] Read-only register.

Programmer's Model of Message Storage

All transmitted and received messages are organized in message buffers with associated control registers. Receive and transmit message registers have the same format, and each message buffer allocates 16 bytes of memory. Tables 16-12 and 16-13 show the general organization of both message buffers.

TABLE 16-13 Transmit Message Buffer Organization

Transmit Address Base +	Register Name
$0170	Identifier Register 0—CANTIDR0
$0171	Identifier Register 1—CANTIDR1
$0172	Identifier Register 2—CANTIDR2
$0173	Identifier Register 3—CANTIDR3
$0174	Data Segment Register 0—CANTDSR0
$0175	Data Segment Register 1—CANTDSR1
$0176	Data Segment Register 2—CANTDSR2
$0177	Data Segment Register 3—CANTDSR3
$0178	Data Segment Register 4—CANTDSR4
$0179	Data Segment Register 5—CANTDSR5
$017A	Data Segment Register 6—CANTDSR6
$017B	Data Segment Register 7—CANTDSR7
$017C	Data Length Register—CANTDLR
$017D	Transmit Buffer Priority Register
$017E	Time Stamp Register (High Byte)—CANTTSRH[a]
$017F	Time Stamp Register (Low Byte)—CANTTSRL[a]

[a] Read-only register.

IDENTIFIER REGISTERS

The CANTIDR0–CANTIDR3 identifier registers are set by the transmitter to identify the message. These bits are received in the CANRIDR0–CANRIDR3 registers in the receiver.

Identifier Registers CANRIDR0–CANRIDR3 (Receive) and CANTIDR0–CANTIDR3 (Transmit)

The identifier registers may use an extended format using 32 bits or standard format using 13 bits. Table 16-14 gives the addresses for these registers.

TABLE 16-14 MSCAN Identifier Registers

CANx		Bit 7	6	5	4	3	2	1	0
IDR0	Extended:	ID28	ID27	ID26	ID25	ID24	ID23	ID22	ID21
	Standard:	ID10	ID9	ID8	ID7	ID6	ID5	ID4	ID3
IDR1	Extended:	ID20	ID19	ID18	SRR=1	IDE=1	ID17	ID16	ID15
	Standard:	ID2	ID1	ID0	RTR	IDE=0			
IDR2	Extended:	ID14	ID13	ID12	ID11	ID10	ID9	ID8	ID7
	Standard:								
IDR3	Extended:	ID6	ID5	ID4	ID3	ID2	ID1	ID0	RTR
	Standard:								
	Reset:	x	x	x	x	x	x	x	x

▓ = reserved, unimplemented, or cannot be written to. x = don't care (unknown at reset).

Read: Anytime for transmit buffers; only when RXF flag is set for receive buffers. Write: Anytime for transmit buffers when the Transmitter Empty (TXEn) flag (in CANTFLG) is set and the corresponding transmit buffer is selected in CANTBSEL.

ID28:ID0

Extended Format Identifier

The identifier is 29 bits (ID28:ID0). IDE = 1. ID28 is the most significant bit and is transmitted first on the bus during the arbitration procedure. The priority of an identifier is the highest for the smallest binary number.

ID10:ID0

Standard Format Identifier

The identifier is 11 bits (ID10:ID0). IDE = 0. ID10 is the most significant bit and is transmitted first on the bus during the arbitration procedure. The priority of an identifier is the highest for the smallest binary number.

SRR

Substitute Remote Request

This is a fixed recessive bit and is used only in extended format. It must be set by the user for transmit buffers and is stored as received on the CAN bus for receive buffers.

IDE

ID Extended Flag

0 = Standard format (11 bits).

1 = Extended format (29 bits).

The IDE flag indicates if this buffer uses the extended or the standard format. For a receive buffer, the flag is stored as received and indicates how to process the buffer identifier registers. In the case of a transmit buffer, the flag indicates to the MSCAN what type of identifier to send.

RTR

Remote Transmission Request

0 = Data frame.
1 = Remote frame.

This flag indicates the status of the Remote Transmission Request bit in the CAN frame. In the case of a receive buffer, it indicates the status of the received frame and supports the transmission of an answering frame in software. In the case of a transmit buffer, the flag defines the setting of the RTR bit to be sent. (See Table 16-15.)

TABLE 16-15 Receive and Transmit Identifier Registers

Receive Register	Address Base +	Transmit Register	Address Base +
CANRIDR0	$0160	CANTIDR0	$0170
CANRIDR1	$0161	CANTIDR1	$0171
CANRIDR2	$0162	CANTIDR2	$0172
CANRIDR3	$0163	CANTIDR3	$0173

DATA SEGMENT REGISTERS

The transmitter places the data to be sent in the CANTDSR0–CANTDSR7 registers and the receiver retrieves the data by reading CANRDSR0–CANRDSR7.

CANRDSR0–CANRDSR7 (Receive) and CANTDSR0–CANTDSR7 (Transmit) Data Segment Registers

There are eight data segment registers for the transmitted data and eight for the received data. The number of bytes to be sent or that are received is found in the data length code in the corresponding CANTDLR or CANRDLR register. See Table 16-16.

TABLE 16-16 Receive and Transmit Data Segment Registers

Receive Register	Address Base +	Transmit Register	Address Base +
CANRDSR0	$0164	CANTDSR0	$0174
CANRDSR1	$0165	CANTDSR1	$0175
CANRDSR2	$0166	CANTDSR2	$0176
CANRDSR3	$0167	CANTDSR3	$0177
CANRDSR4	$0168	CANTDSR4	$0178
CANRDSR5	$0169	CANTDSR5	$0179
CANRDSR6	$016A	CANTDSR6	$017A
CANRDSR7	$016B	CANTDSR7	$017B

DATA LENGTH REGISTERS

The transmitter initializes the CANTDLR register with the number of bytes in the data message. The receiver can check the CANRDLR register to see how many bytes have been received.

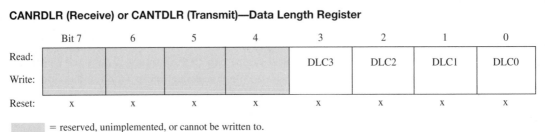

CANRDLR (Receive) or CANTDLR (Transmit)—Data Length Register

	Bit 7	6	5	4	3	2	1	0
Read:					DLC3	DLC2	DLC1	DLC0
Write:								
Reset:	x	x	x	x	x	x	x	x

= reserved, unimplemented, or cannot be written to.

Read: Anytime. Write: Anytime.

DLC3:DLC0

Data Length Code

The data length code contains the number of bytes of the respective (transmitted or received) message. During transmission of a remote frame, the data length code is transmitted as programmed while the number of transmitted bytes is always zero. The data byte count ranges from 0 to 8 for a data frame. (See Table 16-17.)

TABLE 16-17 Data Length Codes

DLC3	DLC2	DCL1	DLC0	Data Byte Count
0	0	0	0	0
0	0	0	1	1
0	0	1	0	2
0	0	1	1	3
0	1	0	0	4
0	1	0	1	5
0	1	1	0	6
0	1	1	1	7
1	0	0	0	8

TABLE 16-18 Receive and Transmit Data Length Registers

Receive Register	Address Base +	Transmit Register	Address Base +
CANRDLR	$016C	CANTDLR	$017C

TRANSMIT BUFFER PRIORITY REGISTER

The MSCAN transmitter described in Section 16.4 has three transmitter buffers. Two or more messages may be waiting in the buffers for transmission if, for example, the bus is busy with other traffic. The transmit buffer priority register CANTBPR can be set to prioritize the order in which the queued messages are sent. A lower binary number has priority over a higher number.

CANTBPR—Base + $017D—MSCAN Transmit Buffer Priority Register

	Bit 7	6	5	4	3	2	1	0
Read:	PRIO7	PRIO6	PRIO5	PRIO4	PRIO3	PRIO2	PRIO1	PRIO0
Write:								
Reset:	x	x	x	x	x	x	x	x

░ = reserved, unimplemented, or cannot be written to.

Read: Anytime. Write: Anytime.

PRIO7:PRIO0

Transmit Buffer Priority

This register contains the local priority of the associated transmit message buffer. The local priority is used for the internal prioritization process of the MSCAN and is the highest for the smallest binary number. The MSCAN implements the following internal prioritization mechanisms:

- All transmission buffers with a cleared TXEn flag participate in the prioritization immediately before the SOF (start-of-frame) is sent.
- The transmit buffer with the lowest local priority field wins the prioritization.

TIME STAMP REGISTERS

If the TIME bit in CANCTL0 is set, a value from an internal free-running CAN bit clock will be written to these registers when a message has been acknowledged on the bus. The CPU can only read these registers.

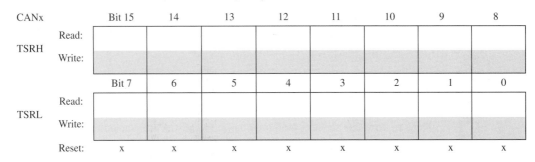

CANRTSRH—MSCAN Receiver Time Stamp Register High
CANRTSRL—MSCAN Receiver Time Stamp Register Low
CANTTSRH—MSCAN Transmit Time Stamp Register High
CANTTSRL—MSCAN Transmit Time Stamp Register Low

CANx		Bit 15	14	13	12	11	10	9	8
TSRH	Read:								
	Write:								
		Bit 7	6	5	4	3	2	1	0
TSRL	Read:								
	Write:								
	Reset:	x	x	x	x	x	x	x	x

░ = reserved, unimplemented, or cannot be written to.

Read: Anytime. Write: Unimplemented.

If the TIME bit in CANCTL0 is set, the MSCAN will write a time stamp derived from a free-running internal counter in the MSCAN as soon as a message has been acknowledged on the CAN bus. (See Table 16-19.)

TABLE 16-19 Receive and Transmit Time Stamp Registers

Receive Register	Address Base +	Transmit Register	Address Base +
CANRTSRH	$016E	CANTTSRH	$017E
CANRTSRL	$016F	CANTTSRL	$017F

MSCAN Initialization

Here are the steps, bits, and control registers to be initialized when using the MSCAN. (See Examples 16-6 and 16-7.)

1. Set INITRQ in CANCTL0 to put MSCAN into initialization mode.
2. Wait for INITAK in CANCTL1 to be set acknowledging initialization mode has been entered.
3. Initialize CANCTL1.
   a. CANE = 1 to enable MSCAN.
   b. CLKSRC = 0 to select the oscillator clock (default).
   c. LOOPB = 0 (default) or 1 for loop back mode testing.
   d. LISTEN = 0 for normal operation (default).
   e. WUPM = 0 for normal wake-up mode (default).
4. Initialize CANBTR0. The synchronization jump width bits, SJW1:SJW0, and baud rate prescaler bits, BRP5:BRP0, may be calculated as shown in Example 16-1.
5. Initialize CANBTR1. Choose SAMP = 0 (default) or 1 depending on the noise in the system. SAMP = 1 (three samples per bit) gives some noise filtering. Calculate values for TSEG2:TSEG0 and TSEG13:TSEG10 as shown in Example 16-1.
6. Initialize CANIDAC IDAM1:IDAM0 to choose the identifier acceptance filter organization.
7. Initialize identifier acceptance registers CANIDAR0–CANIDAR7.
8. Initialize identifier mask registers CANIDMR0–CANIDMR7.
9. Clear INITRQ in CANCTL0 to leave initialization mode.
10. Wait for INITAK in CANCTL1 to be reset acknowledging initialization mode has been left.

Example 16-6 MSCAN Initialization

Show a short code segment used to initialize the MSCAN. It should operate in loop back mode with a data rate of 250 kHz with a 16-MHz oscillator clock. Eight 8-bit filters with identifier numbers $00–$07 are to be used. None of the identifier bits are don't cares.

Solution:

```
;*******************************
; INITIALIZATION SECTION
;(1) Put MSCAN into initialization mode.
```

```
;      Set INITRQ in CANCTL0
        bset   CANCTL0,INITRQ
;   Wait until INITAK is set
wait_for_initak_set:
        brclr CANCTL1,INITAK,wait_for_initak_set
;(2) Initialize CANCTL1
;      CANE = 1 (enable MSCAN)
;      LOOPB = 1 (loop back mode for testing)
        bset   CANCTL1,CANE|LOOPB
;      LISTEN = 0 (normal operation)
        bclr   CANCTL1,LISTEN
;(3) Initialize CANBTR0
;   Osc clock = 16 MHz, data rate = 250 kHz.
;      Prescaler = 4, SJW = 1
        movb  #%00000011,CANBTR0
;    Initialize CANBTR1
;    SAMP = 0 one per bit, TSEG2 = 7,TSEG1 = 8,
        movb  #%01100111,CANBTR1
;(4) Initialize CANIDAC
;    IDAM1:IDAM0 = 10 = 8, 8-bit filters
        movb  #%00100000,CANIDAC
;(5) Initialize Identifier Acceptance Registers
        movb   #0,CANIDAR0
        movb   #1,CANIDAR1
        movb   #2,CANIDAR2
        movb   #3,CANIDAR3
        movb   #4,CANIDAR4
        movb   #5,CANIDAR5
        movb   #6,CANIDAR6
        movb   #7,CANIDAR7
;    Initialize Identifier Mask Registers
        movb   #0,CANIDMR0
        movb   #0,CANIDMR1
        movb   #0,CANIDMR2
        movb   #0,CANIDMR3
        movb   #0,CANIDMR4
        movb   #0,CANIDMR5
        movb   #0,CANIDMR6
        movb   #0,CANIDMR7
;(6) Leave Initialization mode
        bclr   CANCTL0,INITRQ
```

```
;   Wait until INITAK is reset
wait_for_initak_reset:
        brset CANCTL1,INITAK,wait_for_initak_reset
;*******************************
```

See Example 16-13 for a complete MSCAN program.

Example 16-7 MSCAN Initialization in C

Show how to do the initialization for Example 16-6 in C but using an interrupt for the receiver.

Solution:

```
/**************************************************************
 * INITIALIZATION SECTION
 * Data rate = 250 kHz with 16 MHz clock.
 * 1 sample/bit, 8, 8-bit acceptance filters
 **************************************************************/
void initCAN( void ){
/**************************************************************/
  /* Enter initialization mode */
    CANCTL0_INITRQ = 1;
    while (CANCTL1_INITAK == 0);  /* Wait until acknowledged */
  /* Initialize CANCTL1
   * Enable MSCAN (CANE=1), Loopback mode (LOOPB=1),
   * Normal Ops (LISTEN=0) */
    CANCTL1 = CANCTL1_CANE_MASK | CANCTL1_LOOPB_MASK;
  /* Initialize CANBTR0
   * For oscillator clock = 16 MHz and 250 kHz data rate
   * prescaler = 1 and set SJW = 1 */
    CANBTR0 = 0b00000011;
  /* Initialize CANBTR1
   * One sample/bit (SAMP = 0), TSEG2 = 7, TSEG1 = 8 */
    CANBTR1 = 0b01100111;
  /* Initialize CANIDAC
   * 8, 8-bit filters (IDAM = 0b10) */
    CANIDAC = 0b00100000;
  /* Initialize 8 8-bit identifier acceptance registers
   * with the 8 message IDs */
    CANIDAR0 = ID0;
    CANIDAR1 = ID1;
    CANIDAR2 = ID2;
    CANIDAR3 = ID3;
```

```
        CANIDAR4 = ID4;
        CANIDAR5 = ID5;
        CANIDAR6 = ID6;
        CANIDAR7 = ID7;
   /* Initialize the identifier mask registers. */
        CANIDMR0 = MASK;
        CANIDMR1 = MASK;
        CANIDMR2 = MASK;
        CANIDMR3 = MASK;
        CANIDMR4 = MASK;
        CANIDMR5 = MASK;
        CANIDMR6 = MASK;
        CANIDMR7 = MASK;
   /* Leave initialization mode */
        CANCTL0_INITRQ = 0;
        while (CANCTL1_INITAK == 1);
   /* Initialize the receiver interrupt after leaving init mode */
        CANRIER_RXFIE = 1;
}
```

See Example 16-14 for a complete MSCAN C program.

16.4 MSCAN Data Transmitter

CAN Transmitter Buffers

Figure 16-11 shows the general organization of the CAN transmitter message buffers. The application program interface is through a set of three 16-byte buffers signified by TX0BG–TX2BG. Table 16-13 shows the buffer organization. One of these three buffers is mapped to the *CAN Transmit Foreground Buffer* (*CANTXFG*) that provides the application program's access to all three background buffers. To send a message, the application program must determine which of the three background buffers is available to accept a message. When a buffer becomes empty, it sets its *Transmitter Empty Flag* (*TXEn*) in the CANTFLG register. The TXn bits in the CANTBSEL register allow you to determine which of the three transmit buffers is available for a message to be sent. A bit set in TXn signifies an available register. The application software to find which register is available uses a two-step process. Assume that TX2 and TX1 are available and that TX0 is busy. A read from CANTFLG returns the binary %00000110. If this value is written back to CANTBSEL and then read again, %00000010 will be returned. The one returned in bit 1 signifies the TX1 register is to be used next. After initializing the transmit buffer, resetting this flag in CANTFLG enables this message to be sent. See Example 16-8.

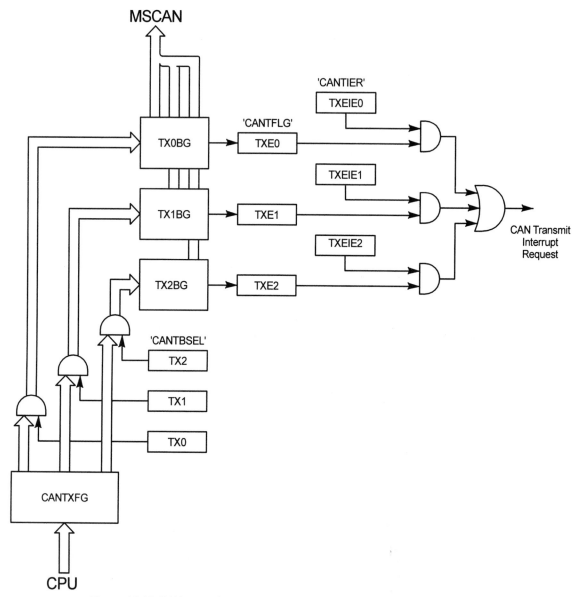

Figure 16-11 CAN transmit message buffers.

CANTFLG—Base + $0146—MSCAN Transmitter Flag Register

	Bit 7	6	5	4	3	2	1	0
Read:	0	0	0	0	0	TXE2	TXE1	TXE0
Write:								
Reset:	0	0	0	0	0	1	1	1

= reserved, unimplemented, or cannot be written to.

Read: Anytime. Write: Anytime when not in initialization mode.

TXE2, TXE1, TXE0

Transmit Buffer Empty Flags

0 = The associated message buffer is full (loaded with a message due for transmission).
1 = The associated message buffer is empty (not scheduled for transmission) (default).

This flag, when set, indicates that the associated transmit message buffer is empty. To send the message after it is set up in the buffer and due for transmission, the application program must clear this flag (by writing a one to it).

The MSCAN sets the flag after the message is sent successfully or when the transmission request is successfully aborted due to a pending abort request.

Clearing a TXEn flag also clears the corresponding abort acknowledge flag (ABTAKn) in the CANTAAK register.

When a TXEn flag is set, the corresponding abort request bit ABTRQn in the CANTARQ register is cleared.

When listen mode is active (LISTEN in CANCTL1 is set), the TXEn flags cannot be cleared and no transmission is started.

Read and write accesses to the associated transmit buffer will be blocked if the corresponding TXEn bit is cleared and the buffer is scheduled for transmission.

CANTBSEL—Base + $014A—MSCAN Transmit Buffer Selection Register

	Bit 7	6	5	4	3	2	1	0
Read:	0	0	0	0	0	TX2	TX1	TX0
Write:								
Reset:	0	0	0	0	0	0	0	0

  = reserved, unimplemented, or cannot be written to.

Read: Anytime. Write: Anytime when not in initialization mode.

TX2, TX1, TX0

Transmit Buffer Select

0 = The associated message buffer is deselected (default).
1 = The associated message buffer is selected, if it is the lowest numbered bit.

The lowest numbered bit that is set defines which of the transmit buffers is available for the next message to be sent.

Example 16-8
Show a code segment to determine which transmit buffer should be used next. Assume TX2 and TX0 are free and TX1 is busy. Show what is returned each time CANTBSEL is read.

Solution:

```
ldab    CANTFLG     ; B <- %00000101
stab    CANTBSEL    ; Value written is %00000101
ldab    CANTBSEL    ; Value read is %00000001
stab    CANTFLG     ; Reset the flag to send the message
```

The TXn bits in the CANTBSEL register allow you to determine which of the three transmit buffers is available for a message to be sent. An available register is signified by a one in TXn. A two-step process is used by the application software to find which register is available. See Example 16-8.

Transmission Aborts

> The application program can stop a message from being transmitted by asserting an *Abort Request Flag*.

If for some reason the application software wants to stop a message before it is transmitted, say, new data are available, it can request that the transmitter abort a scheduled message. When one of the Abort Request (ABTRQn) bits is set, the MSCAN will delete the message from the transmission queue if it has not already started transmission. The Abort Acknowledge (ABTAK) flag in CANTAAK and the associated TXEn flag in CANTFLG are set when the message is aborted.

CANTARQ—Base + $0148—MSCAN Transmitter Message Abort Control Register

	Bit 7	6	5	4	3	2	1	0
Read:	0	0	0	0	0	ABTRQ2	ABTRQ1	ABTRQ0
Write:								
Reset:	0	0	0	0	0	0	0	0

▨ = reserved, unimplemented, or cannot be written to.

Read: Anytime. Write: Anytime when not in initialization mode.

ABTRQ2, ABTRQ1, ABTRQ0

Abort Request

0 = No abort request (default).

1 = Abort request pending.

The CPU may set these bits to request that a scheduled message buffer (TXEn = 0) be aborted. The MSCAN grants the request if the message has not already started transmission or if the transmission is not successful (lost arbitration or error). When a message is aborted, the associated TXEn flag in CANTFLG and Abort Acknowledge Flag (ABTAK in CANTAAK) are set. A CAN transmitter interrupt may occur if it is enabled (TXEIEn = 1 in CANTIER). ABTRQn is reset whenever the associated TXEn flag in CANTFLG is set.

CANTAAK—Base + $0149—MSCAN Transmitter Message Abort Acknowledge Register

	Bit 7	6	5	4	3	2	1	0
Read:	0	0	0	0	0	ABTAK2	ABTAK1	ABTAK0
Write:								
Reset:	0	0	0	0	0	0	0	0

▨ = reserved, unimplemented, or cannot be written to.

Read: Anytime. Write: Unimplemented.

ABTAK2, ABTAK1, ABTAK0

Abort Acknowledge Flags

0 = The message was not aborted (default).

1 = The message was aborted.

These flags acknowledge that a message was aborted due to a pending abort request. After a particular message buffer is flagged empty, the application software can identify whether the message was aborted successfully or was sent anyway. The ABTAKn flag is cleared whenever the corresponding TXEn flag in CANTFLG is cleared

Remote Transmission Request and Substitute Remote Request

A *Remote Transmission Request* is sent from a destination to a source of information to request data be sent.

There are two kinds of frames in CAN—data frames and remote frames. Data frames are used when the node wants to send data. Remote frames are a request for information. A frame with the RTR (Remote Transmission Request) bit set means that the transmitting node is requesting information of the type specified by the identifier.

The standard and extended CAN frames shown in Figure 16-3 and Table 16-14 show the RTR bit for the standard frame occupies bit 4 of CANIDR1 and for the extended frame bit 0 of CANIRD3. When RTR = 0 (a dominant bit), the frame is a data frame; when RTR = 1 (a recessive bit), a destination node is requesting a transmission from a source node. In an extended frame, CANIDR1 bit 4 is called Substitute Remote Request (SRR) and is a recessive bit transmitted in place of the standard frame RTR bit. Because it is recessive, a collision between a standard data frame with RTR = 0 and an extended frame with SRR = 1 will be resolved in favor of the standard frame. The RTR bit for extended frames is CANIDR3 bit 0.

MSCAN Transmitting

Here are the steps to follow when transmitting data (see Examples 16-9 and 16-10):

1. Wait for a transmit buffer to be available by waiting for a nonzero value in CANTFLG.
2. Select the next available buffer by the following:
   a. Read the value in CANTFLG.
   b. Write it to CANTBSEL.
   c. Read CANTBSEL to find the next available buffer. Save this value.
3. Initialize the identifier registers CANTIDR0–CANTIDR3 with the identifier used for the message.
4. Put up to eight message data bytes into the transmit data segment registers CANTDSR0–CANTDSR7.
5. Initialize the data length register CANTDLR with the number of bytes to be sent.
6. Initialize the transmit buffer priority register CANTBPR if necessary.
7. Send the data by clearing the TXE flag. This is done by writing the byte saved in step 2.

Example 16-9 MSCAN Transmitting

Show a segment of code to send a single byte of data using an identifier of $01.

Solution:

```
;********************************
; TRANSMIT SECTION
; Send a message with identifier $01 (CANIDAR1)
;(1) Find the next available transmit buffer
;    Wait until one is available.
wait_for_txe:
        ldab  CANTFLG        ; Find out what is empty
        beq   wait_for_txe   ; If all bits zero, wait
        stab  CANTBSEL       ; Select the next avail
        ldab  CANTBSEL       ; B = buffer to be used
;(2) Initialize CANTIDR0 - CANTIDR1
        movb  CANIDAR1,CANTIDR0   ; Use CANIDAR1
        bclr  CANTIDR1,RTR_STAND  ; Not remote request
        bclr  CANTIDR1,IDE        ; Standard frame
;(3) CANTIDR2 and CANTIDR3 are don't cares
;(4) Put a byte in CANTDSR0
        movb  CAN_TEST,CANTDSR0
;(5) Initialize data length register
        movb  #1,CANTDLR
;(6) Initialize transmit priority register if needed
;(7) Send the data by clearing the TXE flag
        stab  CANTFLG
;********************************
```

See Example 16-13 for a complete MSCAN program.

Example 16-10 MSCAN Transmitter in C

Show a C function to transmit up to 8 bytes of data.

Solution:

```
/***************************************************************
 * TRANSMIT SECTION
 * Send CAN Data
 * Use an 8-bit identifier, up to 8-bytes of data.
 * If length > 8, the data stream is truncated.
 ***************************************************************/
```

```
void putCAN( unsigned char id, unsigned char length, unsigned char
data[] ){
  unsigned char i,max_length, *dataseg_reg;
  volatile unsigned char tx_buf;
/*****************************************************************/
 /* Check to see if there is an available transmit buffer and wait
  * until one is available */
  while( CANTFLG == 0 );
   tx_buf = CANTFLG;              /* Find the lowest number buffer */
   CANTBSEL = tx_buf;
   tx_buf = CANTBSEL;
 /* Initialize CANTIDRD0 with the 8-bit message ID */
   CANTXIDR0 = id;
 /* Initialize CANTIDR1: RTR = 0 and IDE = 0 */
   CANTXIDR1 = 0;
 /* Transfer the data */
 /* Check to make sure not more than 8 bytes of data */
   if (length > 8) max_length = 8; /* Truncate the data stream at
   8 bytes */
   else max_length = length;
   dataseg_reg = &CANTXDSR0;
   for (i = 0; i < max_length; ++i){
     dataseg_reg[i] = data[i];
   }
 /* Set the data length byte */
   CANTXDLR = max_length;
 /* Send the data */
   CANTFLG = tx_buf;
}
```

See Example 16-14 for a complete MSCAN C program.

16.5 MSCAN Data Receiver

The MSCAN receiver *filters* all incoming messages and accepts only those that pass the filter.

The MSCAN receiver is shown in Figure 16-12. As message bits are transmitted on the CAN bus, the receiver tests each of the identification bits against the acceptance filters set up by CANIDAC, CANIDAR0–CANIDAR7, and CANIDMR0–CANIDMR7 (see Table 16-9 and Figures 16-7 to 16-9). The Filter Hit signal is asserted when a match is found and the message is transferred to the next available Receiver Background Buffer (RXBG). If the Receiver Full Flag (RXF) is not set, indicating the Receiver Foreground Buffer (RXFG) is not full, the message is shifted into the RXFG and the RXF flag is set. If the receiver interrupts are enabled, an interrupt will be generated when the RXF flag is set. The application program receive handler must read the message and reset the RXF to acknowledge the interrupt

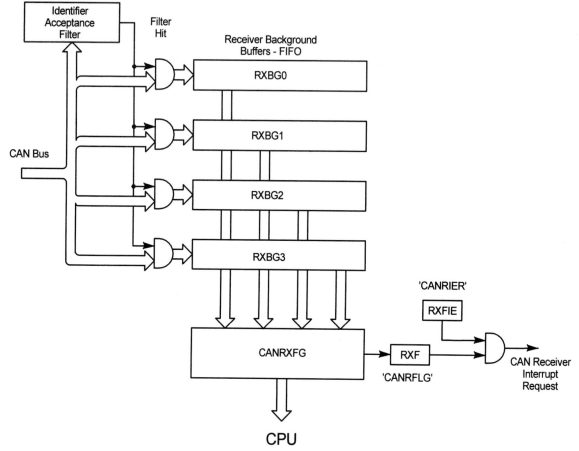

Figure 16-12 CAN receiver message buffers.

and release the foreground buffer. Subsequent messages arriving before the application program releases the RXFG will be stored in next available RXBG buffer until they are full. An overrun error occurs when all RXBG buffers are full and another acceptable message arrives. In this case, the latter message is discarded and a receiver overrun error interrupt is generated. The organization of the receive foreground message buffer is the same as the transmit message buffer and is shown in Table 16-13.

MSCAN Receiving

Here are the steps to follow when receiving messages on the CAN bus (see Example 16-11):

1. Either wait for data to be ready by polling the RXF flag in CANRFLG to be set or generate an interrupt when the flag is set and read the data in the interrupt service routine. (See Example 16-12.)
2. Find the number of bytes in the message by reading CANRDLR.
3. Read the data in CANRDSR0–CANRDSR7.
4. Reset RXF flag in CANRFLG by writing a one to the bit.

Example 16-11 MSCAN Receiving

Show a segment of code that polls the RXF flag until a message is received and accepts a 1-byte message with identifier $01.

Solution:

```
;*****************************
; RECEIVER SECTION
;(1) Wait for receiver to receive the byte
can_rx_wait_1:
        brclr CANRFLG,RXF,can_rx_wait_1
;(2) Read the contents of the receive buffer
        ldaa  CANRDSR0
;(3) Reset the RXF to release buffer for next message
        movb  #RXF,CANRFLG
;*****************************
```

See Example 16-13 for a complete MSCAN program.

Example 16-12 MSCAN Receiving in C

Repeat Example 16-11 in C using an interrupt handler.

Solution:

```
 * RECEIVE SECTION
 * Receive CAN data.
 * This interrupt handler reads the buffer and
 * resets the receive buffer full flag RXF.
 ***************************************************************/
void interrupt 38 getCAN_isr( void ){
  unsigned char length, i;
  unsigned char *CAN_rx_buf_ptr;
/***************************************************************/
  CAN_rx_buf_ptr = &CANRXDSR0;
  length = CANRXDLR;
  if (length > 8) length = 8; /* Make sure max length not exceeded */
  /* Transfer the data to the external data buffer */
    for (i = 0; i < length; ++i ){
      CANdata[i] = *CAN_rx_buf_ptr++;
```

```
        }
    /* Reset the RXF flag */
        CANRFLG = CANRFLG_RXF_MASK;
    }
```

See Example 16-14 for a complete MSCAN C program.

16.6 MSCAN Interrupts

The CAN interface is an autonomous, asynchronous message-handling device. The application program's interface is through the transmit and receive buffer registers shown in Table 16-13, and handshaking flags, which may generate an interrupt request, are provided to allow the application program to synchronize with the asynchronous data flow. (See Table 16-20.)

MSCAN Sleep Mode and Wake-up

The MSCAN can go to sleep when there is no activity on the bus to reduce power consumption.

The application program asserting the *Sleep Request* (*SLPRQ*) bit in the CANCTL0 register places the MSCAN module into a power-saving sleep mode. MSCAN then goes into sleep mode after sending all messages scheduled for transmission or completing the reception of any incoming messages. The SLPAK acknowledge bit in CANCTL1 is asserted when sleep mode is entered. There are two ways to exit sleep mode. The application program may reset SLPRQ (only if SLPAK is set) or if wake-up mode is active and bus activity occurs.

The MSCAN can automatically wake up from sleep mode when activity occurs on the bus by enabling the wake-up mode. Waking up can also generate an interrupt request.

Wake-up mode is entered by setting the *Wake-Up Enable* (*WPUE*) bit in the CANCTL0 register. When this bit is set, any activity on the bus will wake up the MSCAN module and allow it to receive a message. An additional bit, *Wake-Up Mode* (*WPUM*) in CANCTL1 allows you to choose if the module wakes up when any recessive to dominant edge is detected (WPUM = 0) or if a dominant pulse of length T_{WUP} is detected.

TABLE 16-20 MSCAN Interrupts

Interrupt Source	Flag	Flag Register	Enable	Enable Register	Vector	Priority	HPRIO Value to Promote
CAN Wake-up	WUPIF	CANRFLG	WUPIE	CANRIER	$FFB6:FFB7	30	$B6
CAN Status Change	CSCIF	CANRFLG	CSCIE	CANRIER	$FFB4:FFB5	31	$B4
CAN Receiver Overrun	OVRIF	CANRFLG	OVRIE	CANRIER	$FFB4:FFB5	31	$B4
Receive Buffer Full	RXF	CANRFLG	RXFIE	CANRIER	$FFB2:FFB3	32	$B2
CAN Transmit Buffer 2 Empty	TXE2	CANTFLG	TXEIE2	CANTIER	$FFB0:FFB1	33	$B0
CAN Transmit Buffer 1 Empty	TXE1	CANTFLG	TXEIE1	CANTIER	$FFB0:FFB1	33	$B0
CAN Transmit Buffer 0 Empty	TXE0	CANTFLG	TXEIE0	CANTIER	$FFB0:FFB1	33	$B0

CANRFLG—Base + $0144—MSCAN Receiver Flag Register

	Bit 7	6	5	4	3	2	1	0
Read:	WUPIF	CSIF	RSTAT1	RSTAT0	TSTAT1	TSTAT0	OVRIF	RXF
Write:								
Reset:	0	0	0	0	0	0	0	0

▨ = reserved, unimplemented, or cannot be written to.

Read: Anytime. Write: Anytime when out of initialization mode, except RSTATn and TSTATn, which are read only. Writing a 1 to the flag clears it.

WUPIF

Wake-up Interrupt Flag

0 = No wake-up activity while in sleep mode (default).
1 = Activity detected on the CAN bus and wake-up is requested.

WUPIE is set if the MSCAN module detects bus activity while in sleep mode and WUPE in CANTL0 is set. If WUPIE in CANRIER is enabled, an interrupt request will be generated.

CSIF

CAN Status Change Interrupt Flag

0 = No change in bus status since the last interrupt (default).
1 = MSCAN changed current bus status.

CSIF is set when the current bus status is changed due to the actual values in the Transmit Error Counter (CANTXERR) and the Receiver Error Counter (CANRXERR). RSTAT[1:0] and TSTAT[1:0] inform the system of the actual bus status as shown later. If CSIE in CANRIER is set, a CAN error interrupt is enabled and pending while CSIF is set. When CSIF is set, it blocks any further changes in RSTAT and TSTAT even if the error counters change. RSTAT and TSTAT will not be updated until CSIF is cleared.

RSTAT1:RSTAT0

Receiver Status Bits

The bus status is controlled by the values of the receiver and transmitter error counters. When these change, setting CSIF, the RSTAT bits indicate the receiver related bus status. (See Table 16-21.)

TABLE 16-21 Receiver Status Bits

RSTAT1	RSTAT0	Status	Error Counts
0	0	Receiver OK	$0 \leq$ Receiver Error Counter ≤ 96
0	1	Receiver Warning	$96 <$ Receiver Error Counter ≤ 127
1	0	Receiver Error	$127 <$ Receiver Error Counter
1	1	Bus-Off	Transmitter Error Counter > 255

TSTAT1:TSTAT0

Transmitter Status Bits

The bus status is controlled by the values of the receiver and transmitter error counters. When these change, setting CSIF, the TSTAT bits indicate the transmitter related bus status. (See Table 16-22.)

TABLE 16-22 Transmitter Status Bits

TSTAT1	TSTAT0	Status	Error Counts
0	0	Transmitter OK	$0 \leq$ Transmitter Error Counter ≤ 96
0	1	Transmitter Warning	$96 <$ Transmitter Error Counter ≤ 127
1	0	Transmitter Error	$127 <$ Transmitter Error Counter ≤ 255
1	1	Bus-Off	Transmitter Error Counter > 255

OVRIF

Overrun Interrupt Flag

0 = No data overrun condition (default).
1 = Data overrun detected.

OVRIF is set if a data overrun is detected. If OVRIE in CANRIER is set, a CAN error interrupt request is generated.

RXF

Receive Buffer Full Flag

0 = No new message available (default). Do not read the receive buffer register when RXF = 0.
1 = The receiver FIFO is not empty and a new message is available in the RXFG.

This flag is set when a new, correctly received message is shifted into the receiver. After the CPU has read the message from the RXFG buffer, the RXF flag must be cleared to release the buffer for the next message. The flag is cleared by writing a one to it. If RXFIE in CANRIER is set, a CAN receive interrupt will be generated.

CANRIER—Base + $0145—MSCAN Receiver Interrupt Enable Register

	Bit 7	6	5	4	3	2	1	0
Read: Write:	WUPIE	CSCIE	RSTATE1	RSTATE0	TSTATE1	TSTATE0	OVRIE	RXFIE
Reset:	0	0	0	0	0	0	0	0

= reserved, unimplemented, or cannot be written to.

Read: Anytime. Write: Anytime when out of initialization mode.

WUPIE

Wake-up Interrupt Enable

0 = No interrupt request generated by this event (default).
1 = Enables the wake-up event to generate a CAN wake-up interrupt request.

CSCIE

CAN Status Change Interrupt Enable

0 = No interrupt request generated by this event (default).
1 = A CAN status change event causes a CAN error interrupt request.

RSTATE1:RSTATE0

Receiver Status Change Enable

These bits control the sensitivity level for which receiver state changes cause CSCIF interrupts. The RSTAT flags in CANRFLG are independent of this level and still show the actual receiver state and are updated only if a CSCIF interrupt is not pending. See Table 16-23.

TABLE 16-23 Receiver Status Change Enable

RSTATE1	RSTATE0	CSCIF
0	0	Do not generate any CSCIF on all state changes.
0	1	Generate CSCIF interrupt only if the receiver enters or leaves the Bus-Off state. Discard other receiver state changes for generating CSCIF interrupt.
1	0	Generate CSCIF interrupt only if the receiver enters or leaves Receiver Error or Bus-Off state. Discard other receiver state changes for generating CSCIF interrupt.
1	1	Generate CSCIF interrupt on all state changes.

TSTATE1:TSTATE0

Transmitter Status Change Enable

These bits control the sensitivity level for which transmitter state changes cause CSCIF interrupts. The TSTAT flags in CANRFLG are independent of this level and still show the actual transmitter state and are updated only if a CSCIF interrupt is not pending. See Table 16-24.

TABLE 16-24 Transmitter Status Change Enable

TSTATE1	TSTATE0	CSCIF
0	0	Do not generate any CSCIF on all state changes.
0	1	Generate CSCIF interrupt only if the transmitter enters or leaves the Bus-Off state. Discard other transmitter state changes for generating CSCIF interrupt.
1	0	Generate CSCIF interrupt only if the transmitter enters or leaves Transmit Error or Bus-Off state. Discard other transmitter state changes for generating CSCIF interrupt.
1	1	Generate CSCIF interrupt on all state changes.

OVRIE

Overrun Interrupt Enable

0 = No interrupt is generated (default).

1 = An overrun event may cause a CAN error interrupt request.

RXFIE

Receiver Full Interrupt Enable

0 = Receiver full interrupts are disabled (default).

1 = A receiver buffer full (successful message reception) event causes a receiver interrupt request.

CANTIER—Base + $0147—MSCAN Transmitter Interrupt Enable Register

	Bit 7	6	5	4	3	2	1	0
Read:	0	0	0	0	0	TXEIE2	TXEIE1	TXEIE0
Write:								
Reset:	0	0	0	0	0	0	0	0

▨ = reserved, unimplemented, or cannot be written to.

Read: Anytime. Write: Anytime when not in initialization mode.

TXEIE2, TXEIE1, TXEIE0

Transmitter Empty Interrupt Enable

0 = No interrupt request is generated (default).

1 = A transmitter buffer empty causes a transmitter empty interrupt request.

16.7 MSCAN Errors

The CAN standard defines a variety of errors that can be detected with signaling and error indicators to make the CAN bus very reliable. Error-prone nodes are taken off the bus so that they do not disturb properly operating nodes.

Error Detection

There are five different error types.

- **Bit Errors**: While the transmitter is sending bits it is also monitoring the bus. A *bit error* is detected if the monitored bit is different from the transmitted bit.
- **Bit-Stuffing Errors**: Bits are stuffed into a stream of identical valued bits to ensure no more than five consecutive ones or zeros are sent. A receiver that detects six consecutive bits will detect a *stuff error*.
- **CRC Error**: The transmitter calculates the CRC sum based on all bits in the message and then adds it to the end. The receiver also calculates the CRC as it receives the message and a *CRC error* is detected if its calculation differs from what it receives.
- **Form Error**: The message frames shown in Figure 16-3 contain predefined bits, such as RTR, in predefined locations. If a receiver detects one or more invalid bits in these frame positions, a *form error* has been found.
- **Acknowledgment Error**: The ACK bit shown in Figure 16-3 is a dominant bit. A recessive bit in this position detected by a transmitter indicates an *acknowledgment error*.

Error Signaling

When an error is detected, the CAN device sends an *error flag* to prevent other nodes from receiving the message. An error flag can be one of two types—six consecutive dominant bits

TABLE 16-25 MSCAN Error Indicators

Register	Bits	Name	Purpose
CANRFLG	CSCIF	CAN Status Change Interrupt Flag	Set when the MSCAN changes its current bus status as indicated by changes in the Receiver and Transmit Error Counters.
CANRFLG	RSTAT1:RSTAT0	Receiver Status Bits	Indicate the level of receive errors detected when the CAN Status Change Interrupt Flag is set.
CANRFLG	TSTAT1:TSTAT0	Transmitter Status Bits	Indicate the level of transmit errors detected when the CAN Status Change Interrupt Flag is set.
CANRIER	CSCIE	CAN Status Change Interrupt Flag Enable	Enables the CSCIF to generate a CAN error interrupt request.
CANRIER	RSTATE1:RSTATE0	Receiver Status Change Enable	Controls the level at which receiver state changes are causing CSCIF interrupts.
CANRIER	TSTATE1:TSTATE0	Transmitter Status Change Enable	Controls the level at which transmitter state changes are causing CSCIF interrupts.
CANRXERR	RXERR7:RXERR0	Error Counter Bits	Counter for errors detected by the receiver.
CANTXERR	TXERR7:TXERR0	Error Counter Bits	Counter for errors detected by the transmitter.

or six consecutive recessive bits. The transmission of six dominant bits violates the bit-stuffing rule and so all other nodes detect this as an error and discard the message frame. Dominant bits are sent when the node is in *active error mode*. A node that detects more than 127 transmit or receive errors may be at fault itself and so it switches to a *passive error mode* and instead of sending dominant bits sends six recessive bits. This means it does not affect the bus and other nodes that do not detect errors (their receiver is all right) will continue to receive the message. If a transmitter error count exceeds 255, the CAN controller will remove the device from the bus, assuming it has become faulty.

MSCAN Error Indicators

The error detection and processing described previously happen automatically in the CAN hardware. The application software can determine the state of the error detection through several registers and bits in the MSCAN. See Table 16-25.

CANRXERR—Base + $014E—MSCAN Receive Error Counter Register

	Bit 7	6	5	4	3	2	1	0
Read:	RXERR7	RXERR6	RXERR5	RXERR4	RXERR3	RXERR2	RXERR1	RXERR0
Write:								
Reset:	0	0	0	0	0	0	0	0

▨ = reserved, unimplemented, or cannot be written to.

Read: Only when in sleep mode (SLPRQ = 1 and SLPAK = 1) or initialization mode. Write: Not implemented.

RXERR7:RXERR1

This register reflects the status of the MSCAN receive error counter.

CANTXERR—Base + $014F—MSCAN Transmit Error Counter Register

	Bit 7	6	5	4	3	2	1	0
Read:	TXERR7	TXERR6	TXERR5	TXERR4	TXERR3	TXERR2	TXERR1	TXERR0
Write:								
Reset:	0	0	0	0	0	0	0	0

     = reserved, unimplemented, or cannot be written to.

Read: Only when in sleep mode (SLPRQ = 1 and SLPAK = 1) or initialization mode. Write: Not implemented.

TXERR7:TXERR1

This register reflects the status of the MSCAN transmit error counter.

16.8 MSCAN Programming Examples

Full use of the features of the MSCAN to communicate using the CAN bus will require several layers of software appropriate to the final application. The example given here is to help properly configure the module in a basic mode. Example 16-13 covers these basics in assembly language. Example 16-14 gives a C program using a CAN receiver interrupt handler.

Example 16-13 Programming the MSCAN

```
Metrowerks HC12-Assembler
(c) COPYRIGHT METROWERKS 1987-2003

Rel. Loc  Obj. code Source line
---- ---  --------- -----------

    1                        ;*******************************
    2                        ; Test program for the CAN
    3                        ; This simply sends a byte in loop back
    4                        ; mode. The data value is incremented each
    5                        ; time through the loop.
    6                        ;*******************************
    7                        ; Define the entry point for the main program
    8                                XDEF   Entry, main
    9                                XREF   __SEG_END_SSTACK ; Note double
                                       underbar
   10                        ;*******************************
   11                        ; Include files
   12                                INCLUDE  mscan.inc
```

```
13                    ;********************************
14                    ; Code Section
15                    MyCode: SECTION
16                    Entry:
17                    main:
18                    ;********************************
19                    ; Initialize stack pointer register
20 000000 CFxx xx          lds    #__SEG_END_SSTACK
21                    ;********************************
22                    ; INITIALIZATION SECTION
23                    ;(1) Put MSCAN into initialization mode.
24                    ;     Set INITRQ in CANCTL0
25 000003 1C01 4001       bset   CANCTL0,INITRQ
26                    ;   Wait until INITAK is set
27                    wait_for_initak_set:
28 000007 1F01 4101    brclr CANCTL1,INITAK,wait_for_initak_set
   00000B FB
29                    ;(2) Initialize CANCTL1
30                    ;       CANE = 1 (enable MSCAN)
31                    ;       LOOPB = 1 (loop back mode for testing)
32 00000C 1C01 41A0       bset   CANCTL1,CANE|LOOPB
33                    ;   LISTEN = 0 (normal operation)
34 000010 1D01 4110       bclr   CANCTL1,LISTEN
35                    ;(3) Initialize CANBTR0
36                    ; Osc clock = 16 MHz, data rate = 250 kHz.
37                    ;       Prescaler = 4, SJW = 1
38 000014 180B 0301       movb   #%00000011,CANBTR0
   000018 42
39                    ;   Initialize CANBTR1
40                    ;   SAMP = 0 one per bit, TSEG2 = 7,TSEG1 = 8,
41 000019 180B 6701       movb   #%01100111,CANBTR1
   00001D 43
42                    ;(4) Initialize CANIDAC
43                    ;     IDAM1:IDAM0 = 10 = 8, 8-bit filters
44 00001E 180B 2001       movb   #%00100000,CANIDAC
   000022 4B
45                    ;(5) Initialize Identifier Acceptance
                      Registers
46 000023 180B 0001       movb   #0,CANIDAR0
   000027 50
```

```
47 000028 180B 0101        movb  #1,CANIDAR1
   00002C 51
48 00002D 180B 0201        movb  #2,CANIDAR2
   000031 52
49 000032 180B 0301        movb  #3,CANIDAR3
   000036 53
50 000037 180B 0401        movb  #4,CANIDAR4
   00003B 58
51 00003C 180B 0501        movb  #5,CANIDAR5
   000040 59
52 000041 180B 0601        movb  #6,CANIDAR6
   000045 5A
53 000046 180B 0701        movb  #7,CANIDAR7
   00004A 5B
54                  ;    Initialize Identifier Mask Registers
55 00004B 180B 0001        movb  #0,CANIDMR0
   00004F 54
56 000050 180B 0001        movb  #0,CANIDMR1
   000054 55
57 000055 180B 0001        movb  #0,CANIDMR2
   000059 56
58 00005A 180B 0001        movb  #0,CANIDMR3
   00005E 57
59 00005F 180B 0001        movb  #0,CANIDMR4
   000063 5C
60 000064 180B 0001        movb  #0,CANIDMR5
   000068 5D
61 000069 180B 0001        movb  #0,CANIDMR6
   00006D 5E
62 00006E 180B 0001        movb  #0,CANIDMR7
   000072 5F
63                  ;(6) Leave Initialization mode
64 000073 1D01 4001        bclr  CANCTL0,INITRQ
65                  ;    Wait until INITAK is reset
66                       wait_for_initak_reset:
67 000077 1E01 4101        brset CANCTL1,INITAK,wait_for_
                           initak_reset
   00007B FB
68                  ;(7) Enable  interrupts  here  if  you  are
                         using them.
69                  ;********************************
```

```
70                    ; Initialize the test data
71 00007C 180B 00xx        movb  #0,CAN_TEST
   000080 xx
72                    ;********************************
73                    main_loop:
74                    ; DO
75                    ;********************************
76                    ; TRANSMIT SECTION
77                    ; Send  a  message  with  identifier  $01
                         (CANIDAR1)
78                    ;(1) Find the next available transmit buffer
79                    ;    Wait until one is available.
80                    wait_for_txe:
81 000081 F601 46          ldab  CANTFLG      ; Find out what
                                                 is empty
82 000084 27FB             beq  wait_for_txe  ; If all bits
                                                 zero, wait
83 000086 7B01 4A          stab  CANTBSEL     ; Select the next
                                                 avail
84 000089 F601 4A          ldab  CANTBSEL     ; B = buffer  to
                                                 be used
85                    ;(2) Initialize CANTIDR0 - CANTIDR1
86 00008C 180C 0151    movb  CANIDAR1,CANTIDR0 ; Use CANIDAR1
   000090 0170
87 000092 1D01 7110    bclr CANTIDR1,RTR_STAND ; Not remote
                                                 request
88 000096 1D01 7108    bclr  CANTIDR1,IDE     ; Standard
                                                 frame
89                    ;(3) CANTIDR2 and CANTIDR3 are don't cares
90                    ;(4) Put a byte in CANTDSR0
91 00009A 180C xxxx     movb  CAN_TEST,CANTDSR0
   00009E 0174
92                    ;(5) Initialize data length register
93 0000A0 180B 0101    movb  #1,CANTDLR
   0000A4 7C
94                    ;(6) Initialize transmit priority register if
                         needed
95                    ;(7) Send the data by clearing the TXE flag
96 0000A5 7B01 46       stab  CANTFLG
97                    ;********************************
98                    ;********************************
99                    ; RECEIVER SECTION
```

```
100                          ;(1) Wait for receiver to receive the byte
101                          can_rx_wait_1:
102 0000A8 1F01 4401             brclr CANRFLG,RXF,can_rx_wait_1
    0000AC FB
103                          ;(2) Read the contents of the receive buffer
104 0000AD B601 64              ldaa  CANRDSR0
105                          ;(3) Reset the RXF to release buffer for next
                                  message
106 0000B0 180B 0101            movb  #RXF,CANRFLG
    0000B4 44
107                          ;*******************************
108                          ; TEST SECTION
109                          ; Put a breakpoint here and inspect A reg and
110                          ; $0160 - $017F to see receive and transmit
                                  buffers
111 0000B5 A7                    nop
112                          ; Change the data for the next time
113 0000B6 72xx xx             inc   CAN_TEST
114
115                          ; FOREVER
116 0000B9 20C6                 bra   main_loop
117                          ;*******************************
118                          MyData: SECTION
119 000000                   CAN_TEST: DS.B  1
```

Explanation of Example 16-13

INITIALIZATION SECTION

The initialization procedures for the MSCAN are shown in *lines 22–67*.

1. Put the MSCAN into initialization mode. The MSCAN must be inactive and disconnected from the CAN bus while it is initializing so it will not disturb other communications that may be taking place. If the MSCAN is currently active, a safe procedure is to place it into sleep mode by setting the SLPRQ bit and waiting for SLPAK to be set. Then MSCAN can be initialized by setting INITRQ and waiting for INITAK as shown in *lines 22–28*.

2. Initialize control register 1. CANCTL1 contains bits to control the operating mode. In this example, the MSCAN is enabled (CANE = 1), loop back mode is selected (LOOPB = 1), and normal operating mode is enabled (LISTEN = 0). The clock source remains with the default oscillator clock (CLKSRC = 0). See *lines 29–34*.

3. Initialize CANTBR0 and CANTBR1. The oscillator clock is 16 MHz and the data rate is to be 250 kHz. The bit time is $1/250$ kHz $= 4$ μs. Assume Time Segment 2 = Time

Segment 1 = 8 time quanta (T_Q). This gives 17 T_Q per bit (1 + 8 + 8, see Figure 16-6). Thus each T_Q = 4 μs/17 = 235.3 ns. The oscillator clock period is 1/16 MHz = 62.5 ns. Thus the prescaler divider is 235.3/62.5 = 3.76. The closest integer is 4, giving T_Q = 250 ns. To achieve the desired bit rate of 250 kHz, the number of T_Q per bit must be adjusted. The total allowed is 4 μs/250 ns = 16 so set Time Segment 2 to 7 and Time Segment 1 to 8 (Time Segment 1 $\geq$ Time Segment 2). The synchronization jump width must always be less than Time Segment 2. In this case the SJW value in CANBTR0 may be any value from 1 to 4. *Lines 35–41* initialize CANTBR0 and CANTBR1.

4. Initialize CANIDAC. The CANIDAC register chooses the identifier acceptance mode. For this example, eight 8-bit identifiers are chosen in *lines 42–44*.

5. Initialize identifier acceptance and mask registers. *Lines 45–62* allow the receiver to be programmed with the identifiers for which it is to accept messages.

6. Leave initialization mode. In *lines 63–67* the INITRQ bit is reset and a wait loop is entered to wait until INITAK is cleared.

7. Enable interrupts. If you are using interrupts, they are enabled after leaving the initialization mode.

TRANSMIT SECTION

The transmitter foreground buffer is to be loaded with the message identifier, the data, the data length, and a transmit buffer prioritization byte in preparation for sending the data.

1. Select a transmit buffer. There are three transmit buffers and one must be chosen for the current message. First, wait for a buffer to become available as shown in *lines 80–82*. When CANTFLG returns a nonzero value, at least one transmit buffer is available. The two-step sequence in *lines 83* and *84* accomplishes this. (See Example 16-8.)

2. Initialize CANTIDR0 and CANTIDR1. In this case, only an 8-bit identifier is to be used and CANTIDR0 is initialized with this number in *line 86*. CANTIDR1 contains the RTR and IDE bits that are reset in this example in *lines 87* and *88*. If 16-bit or 32-bit identifiers are used, CANTIDR1 must contain this information.

3. Initialize CANTIDR1 and CANTIDR2. In this example, these registers are not used.

4. Initialize the transmit data segment registers. You may send up to 8 bytes in one message. In this case, only one is sent and CANTDSR0 is written with this byte in *line 91*.

5. Initialize the data length register as shown in *line 93*.

6. Initialize the transmit buffer priority register for this message if needed. This is not done in this example.

7. Send the data by clearing the selected TXE flag as shown in *line 96*.

RECEIVE SECTION

An interrupt service routine can be written to receive data when it becomes available or, as shown here, the RXF flag can be polled until new data are ready.

1. Wait for data to be ready. The RXF flag in CANRFLG is polled in *line 102*.

2. Read the data (*line 104*).

3. Reset the RXF flag to release the buffer for the next message (*line 106*).

TEST SECTION

To test that the MSCAN has been initialized properly, a simple test is performed. The data message byte is incremented each time it is sent and a breakpoint can be set at *line 111* to inspect register A for the current received data value.

Example 16-14 Programming the MSCAN in C

```
/****************************************************************
 * C programming example for the MSCAN module
 * Initializes it for 250 kHz data rate and
 * sends a four byte message in loop back mode.
 * A CAN receiver interrupt handler receives the data.
 ****************************************************************/
#include <mc9s12c32.h>              /* derivative information */
/****************************************************************/
/* Define the acceptable message IDs and masks */
#define ID0 0
#define ID1 1
#define ID2 2
#define ID3 3
#define ID4 4
#define ID5 5
#define ID6 6
#define ID7 7
#define MASK 0   /* All masks are all 0's */
/* Declare external CAN data input buffer */
unsigned char CANdata[8];
/****************************************************************
 * INITIALIZATION SECTION
 * Data rate = 250 kHz with 16 MHz clock.
 * 1 sample/bit, 8, 8-bit acceptance filters
 ****************************************************************/
void initCAN( void ){
/****************************************************************/
  /* Enter initialization mode */
    CANCTL0_INITRQ = 1;
    while (CANCTL1_INITAK == 0);   /* Wait until acknowledged */
  /* Initialize CANCTL1
   * Enable MSCAN (CANE=1), Loopback mode (LOOPB=1),
```

```
 * Normal Ops (LISTEN=0) */
  CANCTL1 = CANCTL1_CANE_MASK | CANCTL1_LOOPB_MASK;
/* Initialize CANBTR0
 * For oscillator clock = 16 MHz and 250 kHz data rate
 * prescaler = 1 and set SJW = 1 */
  CANBTR0 = 0b00000011;
/* Initialize CANBTR1
 * One sample/bit (SAMP = 0), TSEG2 = 7, TSEG1 = 8 */
  CANBTR1 = 0b01100111;
/* Initialize CANIDAC
 * 8, 8-bit filters (IDAM = 0b10) */
  CANIDAC = 0b00100000;
/* Initialize 8 8-bit identifier acceptance registers
 * with the 8 message IDs */
  CANIDAR0 = ID0;
  CANIDAR1 = ID1;
  CANIDAR2 = ID2;
  CANIDAR3 = ID3;
  CANIDAR4 = ID4;
  CANIDAR5 = ID5;
  CANIDAR6 = ID6;
  CANIDAR7 = ID7;
/* Initialize the identifier mask registers. */
  CANIDMR0 = MASK;
  CANIDMR1 = MASK;
  CANIDMR2 = MASK;
  CANIDMR3 = MASK;
  CANIDMR4 = MASK;
  CANIDMR5 = MASK;
  CANIDMR6 = MASK;
  CANIDMR7 = MASK;
/* Leave initialization mode */
  CANCTL0_INITRQ = 0;
  while (CANCTL1_INITAK == 1);
/* Initialize the receiver interrupt after leaving init mode */
  CANRIER_RXFIE = 1;
}
/*****************************************************************
 * TRANSMIT SECTION
 * Send CAN Data
```

```
 * Use an 8-bit identifier, up to 8-bytes of data.
 * If length > 8, the data stream is truncated.
 ********************************************************************/
void putCAN( unsigned char id, unsigned char length, unsigned char
data[] ){
  unsigned char i,max_length, *dataseg_reg;
  volatile unsigned char tx_buf;
/********************************************************************/
  /* Check to see if there is an available transmit buffer and wait
   * until one is available */
    while( CANTFLG == 0 );
    tx_buf = CANTFLG;            /* Find the lowest number buffer */
    CANTBSEL = tx_buf;
    tx_buf = CANTBSEL;
  /* Initialize CANTIDRD0 with the 8-bit message ID */
    CANTXIDR0 = id;
  /* Initialize CANTIDR1: RTR = 0 and IDE = 0 */
    CANTXIDR1 = 0;
  /* Transfer the data */
  /* Check to make sure not more than 8 bytes of data */
    if (length > 8) max_length = 8; /* Truncate the data stream at
    8 bytes */
    else max_length = length;
    dataseg_reg = &CANTXDSR0;
    for (i = 0; i < max_length; ++i){
      dataseg_reg[i] = data[i];
    }
  /* Set the data length byte */
    CANTXDLR = max_length;
  /* Send the data */
    CANTFLG = tx_buf;
}
/********************************************************************
 * RECEIVE SECTION
 * Receive CAN data.
 * This interrupt handler reads the buffer and
 * resets the receive buffer full flag RXF.
 ********************************************************************/
void interrupt 38 getCAN_isr( void ){
```

```
  unsigned char length, i;
  unsigned char *CAN_rx_buf_ptr;
/****************************************************************/
  CAN_rx_buf_ptr = &CANRXDSR0;
  length = CANRXDLR;
  if (length > 8) length = 8; /* Make sure max length not exceeded */
  /* Transfer the data to the external data buffer */
    for (i = 0; i < length; ++i ){
      CANdata[i] = *CAN_rx_buf_ptr++;
    }
  /* Reset the RXF flag */
    CANRFLG = CANRFLG_RXF_MASK;
}
/****************************************************************/
void main(void) {
  unsigned char data[4] = {1,2,3,4},i;
/****************************************************************/
  initCAN();                     /* Initialize MSCAN */
  EnableInterrupts;
  for (;;) {
    putCAN( ID1, 4, data );          /* Send four bytes with ID1 */
    /* Set up next data message */
    for (i = 0; i < 4; ++i){
      data[i]+=1;
    }
  } /* wait forever */
}
```

Explanation of Example 16-14

Example 16-14 shows three C functions to use the MSCAN.

void initCAN(void): Each of the initialization steps needed to configure the MSCAN operation are in this function. If you are using an interrupt driven receiver, the receiver interrupt is enabled at the end. Other interrupts would be enabled here also.

void putCAN(unsigned char id, unsigned char length, unsigned char data[]): This is the transmitter section to send up to 8 bytes. It is useful for 8-bit identifiers. For a system with 16- or 32-bit identifiers, `unsigned char id` should be changed to `int` or `long` and the message identifier initialization changed appropriately.

TABLE 16-26 MSCAN Registers

Name	Register	Address
CANCTL0	Control Register 0	$0140
CANCTL1	Control Register 1	$0141
CANBTR0	Bus Timing Register 0	$0142
CANBTR1	Bus Timing Register 1	$0143
CANRFLG	Receiver Flag Register	$0144
CANRIER	Receiver Interrupt Enable Register	$0145
CANTFLG	Transmitter Flag Register	$0146
CANTIER	Transmitter Interrupt Enable Register	$0147
CANTARQ	Transmitter Message Abort Control Register	$0148
CANTAAK	Transmitter Message Abort Acknowledge Register	$0149
CANTBSEL	Transmit Buffer Selection Register	$014A
CANIDAC	Identifier Acceptance Control Register	$014B
CANRXERR	Receive Error Counter Register	$014E
CANTXERR	Transmit Error Counter Register	$014F
CANIDAR0–CANIDAR7	Identifier Acceptance Registers	$0150–$015B
CANIDMR0–CANIDMR7	Identifier Mask Registers	$0154–$015F
CANRIDR0–CANRIDR3	Receiver Identifier Registers	$0160–$0163
CANRDSR0–CANRDSR7	Receive Data Segment Registers	$0164–$016B
CANRDLR	Receive Data Length Register	$016C
CANRTSRH:CANRTSRL	Receiver Time Stamp Register	$016E:016F
CANTIDR0–CANTIDR3	Transmit Identifier Registers	$0170–$0173
CANTDSR0–CANTDSR7	Transmit Data Segment Registers	$0174–$017B
CANTDLR	Transmit Data Length Register	$017C
CANTBPR	Transmit Buffer Priority Register	$017D
CANTTSRH:CANTTSRL	Transmit Time Stamp Register	$017E:017F

void interrupt 38 getCAN_isr(void): The receiver is an interrupt driven receiver. When a message is received the RXF flag is set, generating the interrupt. The flag is reset by reading the data and then writing a 1 to RXF.

16.9 MSCAN Register Address Summary

This summary is presented in Table 16-26.

16.10 Conclusion and Chapter Summary Points

Members of the HCS12 family that have a MSCAN interface module offer significant hardware capabilities for automatically transmitting and receiving messages on a Controller Area Network. The common elements are as follows:

- Timing is derived from the bus clock.
- Each bit time is comprised of several time elements called Time Quanta.
- Between 8 and 25 time quanta comprise a bit.
- Bits may be dominant (low level) or recessive (high level).

- Dominant bits can control the bus where recessive bits cannot.
- The application program can send a message by initializing an identifier and adding up to 8 data bytes to the message.
- All elements of transmitting the message on the CAN bus, including multiple access detection and collision avoidance, are handled in the hardware.
- The application program can initialize up to eight message identifier filters to allow it, in the CAN hardware, to select only those messages relevant to the application.

16.11 Bibliography and Further Reading

CPU12 Reference Manual, S12CPUV2.PDF, Freescale Semiconductor, Inc., Austin, TX, July 2003.

MSCAN Block Guide V02.14, Freescale Semiconductor, Inc., Austin, TX, September 2002.

MSCAN Low-Power Applications, Application Note AN2255/D, Freescale Semiconductor, Inc., Austin, TX, February 2002.

Using the Motorola MSCAN Filter Configuration Tool, Application Note AN2010/D, Freescale Semiconductor, Inc., Austin, TX, 2000.

Using the MSCAN on the HCS12 Family, Application Note AN3034, Freescale Semiconductor, Inc., Austin, TX, September 2005.

16.12 Problems

Basic

16.1 Define the following terms. **[g, k]**

a. Acceptance filter.
b. CRC.
c. CSMA/CD.
d. Dominant level.
e. Recessive level.
f. Synchronization jump.
g. CAN bus.

16.2 Give the register name and address and the bit name for the following. **[k]**

a. The bit that enables the MSCAN.
b. The bit that sets the loop back mode.
c. The bit that puts MSCAN into initialization mode.
d. The bit that acknowledges initialization mode has been entered.

16.3 What is the default clock used in the MSCAN? **[k]**

16.4 Give definitions for the following MSCAN errors. **[g, k]**

a. Bit errors.
b. Bit-stuffing errors.
c. CRC errors.
d. Form errors.
e. Acknowledgment error.

Intermediate

16.5 Calculate the initialization values for CANBR0, CANBR1, and the Time Segment 2 for a data rate of 100 kHz, assuming the oscillator clock is 16 MHz. **[k]**

16.6 The CANTFLG register contains %00000110. **[k]**

a. Which transmit register is to be used for the next message?
b. Show a short segment of code to find the register and transmit the message.

Advanced

16.7 Describe how collision detection is accomplished in the CAN. **[g, k]**

17 HCS12 Analog Input

OBJECTIVES

In this chapter we learn how to initialize and use the HCS12 A/D converter system.

17.1 Introduction

An *analog-to-digital converter* converts a real-world analog signal to a digital representation to be used in our microcontroller system.

Analog-to-digital converters are also called *A/Ds, ATDs,* and *ADCs.*

The analog-to-digital converter performs a vital job in embedded systems. Our "real" world is an analog world—signals vary continuously from a minimum value in the noise of the system to some maximum determined by the particular device generating the signal. Often we wish to take action in our microcontroller system based on some analog value. To do that, we must convert the analog signal to a digital one.

There are a variety of analog-to-digital types. The one chosen depends on the application and on the performance required. Analog-to-digital converters are also called A/Ds, ATDs, and ADCs.

The A/D converter in the HCS12 is a successive approximation converter, as shown in Figure 17-1. Each bit in the successive approximation register is tested, starting at the most significant bit and working toward the least significant bit. As each bit is set, the output of the digital-to-analog (D/A) converter is compared with the input. If the D/A output is lower than the input signal, the bit remains set and the next bit is tried. Bits that make the D/A output higher than the analog input are reset. N bit-times are required to set and test each bit in the successive approximation register.

17.2 HCS12 A/D Converter

The HCS12 contains an *eight-channel, multiplexed, 8-bit or 10-bit, successive approximation* A/D converter with ± 1 *LSB accuracy.* Its *charge redistribution* input circuit eliminates the need for an external sample-and-hold. The A/D is *linear to* ± 1 *LSB* over its full temperature range and there are no missing codes. Both the *conversion time* and the sample-and-hold *aperture time* are *programmable.*

The A/D conversion is started by writing to a control register or by an external trigger signal (when that feature is enabled). With a 2-MHz A/D clock, an 8-bit single sample conversion is completed in 6 μs and 10 bits in 7 μs.

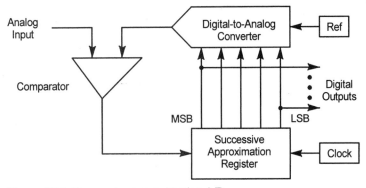

Figure 17-1 Successive approximation A/D.

Figure 17-2 shows a block diagram of the HCS12 analog-to-digital converter. Port AD bits AN0/PAD0–AN7/PAD7 are the analog or digital input pins. You may use any of these as general purpose digital inputs or outputs when the corresponding A/D channel is not in use, although you should not try to read a digital value while an analog conversion is in progress. An input multiplexer is controlled by the *A/D Control Register 5 (ATDCTL5)*. A switch and a single capacitor, C_{SAMP}, model the input sampling mechanism as shown in Figure 17-2. When a conversion is started, this capacitor is charged to the input signal voltage and then held constant for the duration of the A/D conversion. The 8-bit or 10-bit successive approximation A/D uses two voltages, V_{RH} and V_{RL}, to optimize resolution over the input signal range. Normally V_{RH} is set to the signal maximum and V_{RL} to the minimum. However, V_{RH} should not be higher than V_{DDA} and V_{RL} should not be less than V_{SSA}. V_{DDA} and V_{SSA} are the power supply and ground pins for the A/D converter. They should always be connected even if the A/D is not being used. The outputs from A/D conversions are placed into eight, 16-bit data registers ATDDR0–ATDDR7. You must initialize several control registers before operating the A/D. (See Example 17-1.)

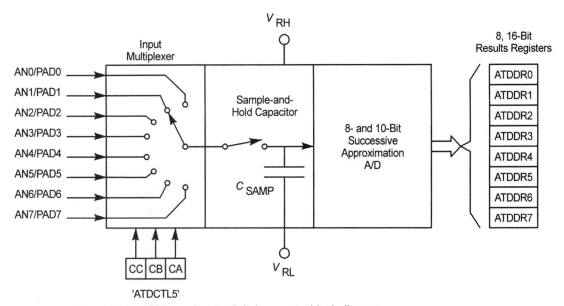

Figure 17-2 HCS12 analog-to-digital converter block diagram.

Example 17-1 A/D Resolution

The input signal to the A/D is unipolar and varies between 0 and 3 volts. The system hardware designer has set $V_{RH} = 5$ V and $V_{RL} = 0$. What is the resolution of the conversion assuming an 8-bit converter? A 10-bit converter?

Solution: The 8-bit resolution is

$$(V_{RH} - V_{RL})/256 = 19.5 \text{ mV}$$

The 10-bit resolution is

$$(V_{RH} - V_{RL})/1024 = 4.88 \text{ mV}$$

After reading the data book on the HCS12 A/D, the system designer realizes that V_{RH} should be set to the maximum signal. After this is done, what is the resolution of the measurement?

Solution: The 8-bit resolution now is $3/256 = 11.7$ mV, and the 10-bit resolution is $3/1024 = 2.9$ mV.

A/D Power Up

The A/D must be powered up before it is used. The *ATD power up bit—ADPU*—is in the *ATDCTL2* register. A short delay, of about 20 μs, must be observed before using the A/D after it is powered up. You may choose to go about other tasks or enter a short delay program.[1]

ATDCTL2—Base + $0082—ATD Control Register 2

	Bit 7	6	5	4	3	2	1	0
Read:	ADPU	AFFC	AWAI	ETRIGLE	ETRIGP	ETRIGE	ASCIE	ASCIF
Write:								
Reset:	0	0	0	0	0	0	0	0

■ = reserved, unimplemented, or cannot be written to.

Read: Anytime. Write: Anytime.

ADPU

ATD Power Up

0 = Power down A/D (default).

1 = Normal A/D function.

This bit is the power on/off control for the ATD. When the analog electronics are powered down, power consumption is reduced. You must wait at least 20 μs after setting ADPU before using the A/D.

---

[1] When the A/D is powered up, it draws 2 mA at 5 V. In power-sensitive applications you may choose to power down the A/D when it is not converting. When powering it up again you must wait the 20 μs.

AFFC

ATD Fast Flag Clear

0 = A/D conversion complete flag clearing operates normally. These flags are reset by reading the status register ATDSTAT1 before reading the result (default).

1 = Enables fast flag clear operation. Any access to a result register will clear the associated flag.

AWAI

ATD Power Down in Wait Mode

0 = A/D continues to run in wait mode (default).

1 = Halt conversion and power down the A/D in wait mode.

When entering wait mode (by executing the WAI instruction), the A/D may be powered down to reduce power consumption by setting this bit. The 20-μs delay described previously should be honored after the interrupt that causes the processor to exit wait mode before using the A/D.

ETRIGLE

External Trigger/Level Control

The A/D channel 7 input may be used as an external trigger signal. This bit allows the trigger to be a rising or falling edge or a high or low level.

See Table 17-1.

ETRIGP

External Trigger Polarity

This bit controls the polarity of the external trigger.
See Table 17-1.

ETRIGE

External Trigger Mode Enable

0 = Disable external trigger mode (default).

1 = Enable external trigger mode.

This bit enables the external trigger on A/D channel 7. The conversion results for external trigger channel 7 have no meaning when the external trigger mode is enabled.

TABLE 17-1 External Trigger Configurations

ETRIGLE	ETRIGP	External Trigger Sensitivity
0	0	Falling edge (default)
0	1	Rising edge
1	0	Low level
1	1	High level

ASCIE

ATD Sequence Complete Interrupt Enable

0 = A/D sequence complete interrupts are disabled (default).

1 = A/D sequence complete interrupts are enabled.

ASCIF

ATD Sequence Complete Interrupt Flag

If ASCIE = 1, the ASCIF flag is the same as the SCF flag.

A/D Conversion Sequence—ATDCTL3

As Figure 17-2 shows, there are up to eight analog inputs that may be sampled and converted to digital values. Each time an analog-to-digital conversion is initiated (by writing to ATD Control Register 5 or when an external trigger is asserted), a sequence of analog-to-digital conversions is done. You may choose the number of conversions done in a sequence and which of the A/D inputs are converted by initializing bits in three A/D control registers—ATDCTL3, ATDCTL4, and ATDCTL5.

ATDCTL3—Base + $0083—ATD Control Register 3

	Bit 7	6	5	4	3	2	1	0
Read:	0	S8C	S4C	S2C	S1C	FIFO	FRZ1	FRZ0
Write:								
Reset:	0	0	1	0	0	0	0	0

▒ = reserved, unimplemented, or cannot be written to.

Read: Anytime. Write: Anytime.

S8C:S1C

Conversion Sequence Length

These bits control the number of conversions per sequence. At reset, S4C is set to 1 (sequence length is four) to maintain software continuity with earlier members of the HC12 family. (See Table 17-2.)

TABLE 17-2 Conversion Sequence Length Coding

S8C	S4C	S2C	S1C	Number of Conversions per Sequence
0	0	0	0	8
0	0	0	1	1
0	0	1	0	2
0	0	1	1	3
0	1	0	0	4 (default)
0	1	0	1	5
0	1	1	0	6
0	1	1	1	7
1	X	X	X	8

FIFO

Result Register First In–First Out (FIFO) Mode

0 = Conversion results are placed in consecutive result registers up to the selected sequence length (default).

1 = Conversion results are placed in consecutive result registers with wrap around at the end.

When this bit is zero, non-FIFO mode is active. The A/D conversion results are placed into the result registers based on the conversion sequence. The result of the first conversion appears in the first result register (ATDDR0), the second in the second, and so on. The result location repeats during the next conversion sequence.

If this bit is one (FIFO mode), the conversion counter is not reset at the beginning or end of a conversion sequence and conversion results are placed in consecutive registers. The result register counter wraps around when it reaches the end of the result register file. The conversion counter value in the ATDSTAT0 register can be used to determine where in the result registers the current conversion result will be placed.

The conversion complete flags in ATDSTAT1 allow you to track which registers hold valid data.
See Table 17-4 and Example 17-2.

FRZ1:FRZ0

Background Debug Freeze Enable
These bits determine how the A/D will respond when a breakpoint (Freeze Mode) is encountered. (See Table 17-3.)

TABLE 17-3 A/D Behavior in Breakpoint Freeze Mode

FRZ1	FRZ0	Behavior in Freeze Mode (Breakpoint)
0	0	Continue conversion (default).
0	1	Reserved.
1	0	Finish current conversion, and then freeze.
1	1	Freeze immediately.

TABLE 17-4 Placement of Conversion Results for a Four Conversion Sequence

Non-FIFO Mode (FIFO = 0)				FIFO Mode (FIFO = 1)		
ATDDR0H	1:1[a]	2:1	3:1	ATDDR0H	1:1	3:1
ATDDR1H	1:2	2:2	3:2	ATDDR1H	1:2	3:2
ATDDR2H	1:3	2:3	3:3	ATDDR2H	1:3	3:3
ATDDR3H	1:4	2:4	3:4	ATDDR3H	1:4	3:4
ATDDR4H				ATDDR4H	2:1	
ATDDR5H				ATDDR5H	2:2	
ATDDR6H				ATDDR6H	2:3	
ATDDR7H				ATDDR7H	2:4	

[a] 1:1 = first conversion sequence:first channel.

Example 17-2 Placement of Results
The analog signals input at AN02, AN03, and AN04 are to be converted in a single conversion sequence. For non-FIFO and FIFO mode, show what registers are expected to receive the results of four successive conversions.

Solution:

Non-FIFO and FIFO Mode Result Registers

	Non-FIFO Mode (FIFO = 0)		FIFO Mode (FIFO = 1)	
Sequence	**Input**	**Result Register**	**Input**	**Result Register**
First	AN02	ATDDR0	AN02	ATDDR0
	AN03	ATDDR1	AN03	ATDDR1
	AN04	ATDDR2	AN04	ATDDR2

Non-FIFO and FIFO Mode Result Registers (Continued)

	Non-FIFO Mode (FIFO = 0)		FIFO Mode (FIFO = 1)	
Sequence	Input	Result Register	Input	Result Register
Second	AN02	ATDDR0	AN02	ATDDR3
	AN03	ATDDR1	AN03	ATDDR4
	AN04	ATDDR2	AN04	ATDDR5
Third	AN02	ATDDR0	AN02	ATDDR6
	AN03	ATDDR1	AN03	ATDDR7
	AN04	ATDDR2	AN04	ATDDR0
Fourth	AN02	ATDDR0	AN02	ATDDR1
	AN03	ATDDR1	AN03	ATDDR2
	AN04	ATDDR2	AN04	ATDDR3

A/D Resolution, Sampling Time, and Clock Selection—ATDCTL4

The resolution, sampling time, and conversion time for the HCS12 A/D are programmable. The resolution can be 8 or 10 bits, and the total sample time can range from 4 to 18 A/D clock periods. The *ATD Control Register 4* controls these features. The following successive approximation conversion takes either eight or ten clocks to complete the conversion.

The A/D derives its clock from the CPU's bus clock, which is the basic system clock (see Chapter 21) and is normally one-half the external oscillator's frequency. Figure 17-3 shows how the ATDCTL4 prescaler select bits (PRS4–PRS0) and the sample time select bits (SMP1–SMP0) affect the A/D conversion time.

The maximum and minimum conversion frequencies for the HCS12 A/D are 2 MHz and 500 kHz, respectively, and so the bus clock must be scaled so that the A/D receives a clock frequency within this range. The prescaler bits, PRS4–PRS0, are chosen as shown in Table 17-8 to generate an appropriate A/D clock frequency. See Example 17-3.

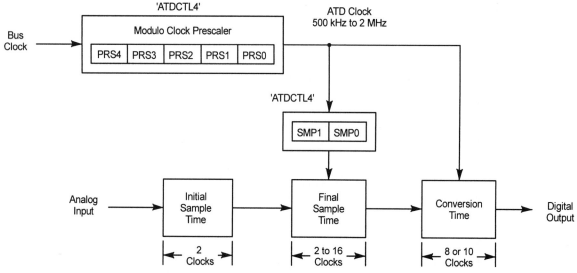

Figure 17-3 A/D sample and conversion timing.

Example 17-3 Selecting A/D Clock Prescaler Bits

The bus clock frequency is 8 MHz. Choose A/D clock prescaler bits PRS4–PRS0 so the A/D receives the highest possible clock frequency.

Solution: Referring to Table 17-8, for a bus clock of 8 MHz we would choose the total divisor of 4 to achieve an A/D clock of 2 MHz. PRS4:PRS0 = 00011.

Figure 17-3 shows that the A/D sampling process is divided into two subparts. The first is the *initial sample*, which occupies two A/D clocks. During this time, the input is connected through a unity gain buffer amplifier to the sample capacitor. In the second part, the input pin is connected directly to the sampling capacitor, which is then charged to the final input value. The optimal choice for the final sample time depends on the output impedance of the analog source. For high impedance sources choose a longer final sample time. Finally, the successive approximation A/D converter then requires eight or ten A/D clocks to complete the 8- or 10-bit conversion.

Table 17-5 shows how the final sample time selection affects the total conversion time and thus the maximum analog input frequencies. The *Nyquist* frequency is the maximum frequency that can be sampled without aliasing and results from requiring two samples per period of the input signal. (See Example 17-4.) This is called the sampling criterion.

APERTURE TIME

A significant error in a digitizing system is due to signal variation during the time the signal is being sampled. This is called the *aperture time* and it limits the maximum frequency that can be sampled. This error is shown in Figure 17-4, where the signal is changing when the aperture is open. A good design will attempt to have the uncertainty, ΔV, be less than one least significant bit (LSB). We can derive a design equation for the aperture time, t_{AP}, in terms of the maximum signal frequency, f_{MAX}, and the number of bits in the A/D converter, n, by observing the following:

$$v(t) = V_{MAX} \sin 2\pi f_{MAX} t$$

$$\Delta v = 2\pi f_{MAX} V_{MAX} \cos(2\pi f_{MAX} t) \, \Delta t = 2\pi f_{MAX} V_{MAX} \cos(2\pi f_{MAX} t) t_{AP}$$

TABLE 17-5 Final Sample Time Selection

SMP1	SMP0	Number Clocks Final Sample Time	Total Conversion Number A/D Clock Periods		Conversion Time (μs) for 2-MHz Clock		Nyquist Frequency (kHz) for 2-MHz Clock	
			8-bit	10-bit	8-bit	10-bit	8-bit	10-bit
0	0	2	12	14	6	7	83	71
0	1	4	14	16	7	8	71	63
1	0	8	18	20	9	10	57	50
1	1	16	26	28	13	14	38	36

Example 17-4 Nyquist Sampling Frequency
Calculate the Nyquist frequency for a 1-MHz A/D clock, a final sample time of four A/D clock periods, and an 8-bit conversion.

Solution:
The total conversion time is $(2 + 4 + 8) \times 1$ μs $= 14$ μs.
The maximum conversion frequency is $1/(14$ μs$) = 71.4$ kHz.
The Nyquist frequency is 71.4 kHz$/2 = 35.7$ kHz.

for $t = 0$ (worst case slope)

$$\Delta v = 2\pi f_{MAX} V_{MAX} t_{AP}$$

and for Δv to be less than 1 LSB,

$$\frac{\Delta v}{V_{MAX}} = \frac{1}{2^n} = 2\pi f_{MAX} t_{AP}$$

Solving for the aperture time,

$$t_{AP} = \frac{1}{2\pi f_{MAX} 2^n}$$

and the maximum frequency that can be converted with aperture errors less than $\pm 1/2$ LSB is given by

$$f_{MAX} = \frac{1}{2\pi t_{AP} 2^n}$$

Table 17-6 shows this effect.

This is a worst case scenario. The actual error due to aperture time is a function of the source impedance and the charging time of the sampling capacitor. For a more complete analysis, see Freescale Application Note AN2429 *Interfacing to the HCS12ATD Module.*

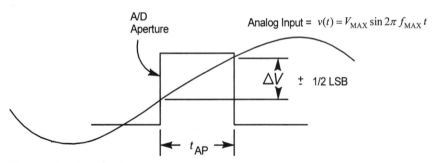

A/D
Aperture

Analog Input $= v(t) = V_{MAX} \sin 2\pi f_{MAX} t$

ΔV $\pm$ 1/2 LSB

t_{AP}

Figure 17-4 Aperture time error.

TABLE 17-6 Aperture Time Effect on Maximum Frequency

SMP1	SMP0	Phase 2 Clocks	Aperture Time (μs) with 2-MHz A/D Clock	f_{MAX} (Hz) 8-bit	f_{MAX} (Hz) 10-bit
0	0	2	1	621	155
0	1	4	2	310	78
1	0	8	4	155	39
1	1	16	8	78	19

ATDCTL4

ATDCTL4 selects the resolution of the A/D conversion (8 or 10 bits), the length of the second phase of the sample time, and the conversion clock frequency. Writes to this register will abort the current conversion sequence but will not start a new sequence.

ATDCTL4—Base + $0084—ATD Control Register 4

	Bit 7	6	5	4	3	2	1	0
Read:	SRES8	SMP1	SMP0	PRS4	PRS3	PRS2	PRS1	PRS0
Write:								
Reset:	0	0	0	0	0	1	0	1

Read: Anytime. Write: Anytime (aborts current conversion).

SRES8

A/D Resolution Select

0 = 10-bit resolution (default).

1 = 8-bit resolution.

SMP1:SMP0

Sample Time Select

These two bits select the length of the second phase of the sample time in units of A/D conversion clock cycles. The A/D clock period is a function of the prescaler bits (PRS4–PRS0) and the sample time consists of two phases. The first is two A/D conversion clock cycles long and during this time the sample is transferred quickly onto the A/D's storage capacitor via a buffer amplifier. The second phase attaches the external analog signal directly to the storage capacitor for final charging. Your conversion will be more accurate when longer times are chosen by SMP1:SMP0. (See Table 17-7.)

TABLE 17-7 Sample Time Select

SMP1	SMP0	Length of Phase Two Sample Time (A/D conversion clock periods)
0	0	2 (default)
0	1	4
1	0	8
1	1	16

PRS4:PRS0

ATD Clock Prescaler

These bits set the A/D clock prescaler. The prescaler divides the bus clock by the value given in Table 17-8. The final ATD clock frequency must be between 500 kHz and 2 MHz.

A/D Input Multiplexer and Input Scanning—ATDCTL5

The A/D has eight input channels selected by bits in the *ATD Control Register 5*. In addition to selecting which input signals are converted, you may control the output data format, and how and when conversions are done.

TABLE 17-8 Clock Prescaler Values

Prescaler n	Prescaler Bits PRS4–PRS0	Total Divisor $2 \times (n + 1)$	Max Bus Clock[a] (MHz)	Min Bus Clock[b] (MHz)
0	00000	2	4	1
1	00001	4	8	2
2	00010	6	12[5]	3
3	00011	8	16	4
4	00100	10	20	5
5	00101	12	24	6
6	00110	14	28	7
7	00111	16	32	8
8	01000	18	36	9
9	01001	20	40	10
10	01010	22	44	11
11	01011	24	48	12[c]
12	01100	26	52	13
13	01101	28	56	14
14	01110	30	60	15
15	01111	32	64	16
16	10000	34	68	17
17	10001	36	72	18
18	10010	38	76	19
19	10011	40	80	20
20	10100	42	84	21
21	10101	44	88	22
22	10110	46	92	23
23	10111	48	96	24
24	11000	50	100	25
25	11001	52	104	26
26	11010	54	108	27
27	11011	56	112	28
28	11100	58	116	29
29	11101	60	120	30
30	11110	62	124	31
31	11111	64	128	32

[a] For ATD clock = 2 MHz.
[b] For ATD clock = 500 kHz.
[c] Current (2006) versions of the MC9S12C-family processors have a maximum bus clock of 12 MHz.

The A/D converter is started by writing to *ATDCTL5* unless the ETRIGE = 1 in ATDCTL2. The A/D always completes a *sequence* of one to eight conversions, and the number of conversions in the sequence is selected by *S8C:S1C* in ATDCTL3. The *SCAN* bit controls whether the A/D converts only one of these sequences or starts a continuous stream of conversion sequences. The *MULT* bit determines if the conversion sequences are done on a single channel or on sequential channels. Finally, the channel select bits, *CC–CA*, choose which channel or channels are converted as shown in Tables 17-9 and 17-13.

Single-Channel Operation: Single-channel mode can be selected by setting MULT = 0. When this is done, that channel is converted the number of times given by S8C, S4C, S2C, and S1C and placed in successive registers starting at result register 0. See Tables 17-13 and 17-4.

Multiple-Channel Operation: In multiple-channel operation, MULT = 1 and the A/D result registers contain the conversions from channels selected as shown in Tables 17-13 and 17-4.

A/D Operation

The A/D conversion is started by writing to the ATDCTL5 register or asserting the external trigger signal; the A/D then does one to eight consecutive conversions. Each conversion requires the number of A/D clock cycles shown in Table 17-5. After the conversions are done, the A/D either waits for the program to write to the ATDCTL register again (when SCAN = 0) or starts another conversion cycle immediately (for SCAN = 1). See Examples 17-5 to 17-9.

ATDCTL5—Base + $0085—ATD Control Register 5

	Bit 7	6	5	4	3	2	1	0
Read:	DJM	DSGN	SCAN	MULT	0	CC	CB	CA
Write:								
Reset:	0	0	0	0	0	0	0	0

░ = reserved, unimplemented, or cannot be written to.

Read: Anytime. Write: Anytime.

DJM

Result Register Data Justification

0 = Left justified data in the result registers (default).
1 = Right justified data in the result registers.
This bit controls justification of the conversion data in the result registers. See Tables 17-10 and 17-11.

DSGN

Result Register Data Signed or Unsigned

0 = Unsigned data in the result registers (default).
1 = Signed data in the result registers.
DSGN selects either signed (two's complement) or unsigned data representation in the data registers. Signed data format is not available in right justification. See Tables 17-10 and 17-11.

SCAN

Continuous Conversion Sequence Mode

0 = Single conversion sequence (default).

1 = Continuous conversion sequences (scan mode).

This bit selects whether conversion sequences are performed continuously or only once.

MULT

Multichannel Sample Mode

0 = Sample only one channel (default).

1 = Sample across several channels (scan mode).

When MULT is zero, the A/D samples only from the specified analog input channel for an entire conversion sequence. The channel is selected by the channel selection code in bits CC, CB, and CA.

When MULT is one, the A/D samples across channels. The number of channels sampled is determined by the sequence length in bits S8C, S4C, S2C, and S1C in ATDCTL3. The first channel sampled is determined by CC, CB, and CA; subsequent channels in the sequence are determined by incrementing the channel selection code.

CC, CB, CA

Analog Input Channel Select Code

These bits select the analog input channel(s) that are sampled and converted to digital codes. When MULT = 0 and a single channel is to be converted, this selection code specifies the channel to be selected. When MULT = 1, the selection code represents the first channel to be converted in the conversion sequence. Subsequent channels are determined by incrementing the channel selection code to the maximum channel number when the code wraps to the minimum value.

TABLE 17-9 Analog Input Channel Select Coding

CC	CB	CA	Analog Input Channel
0	0	0	AN0
0	0	1	AN1
0	1	0	AN2
0	1	1	AN3
1	0	0	AN4
1	0	1	AN5
1	1	0	AN6
1	1	1	AN7

TABLE 17-10 Available Result Data Formats

SRES8	DJM	DSGN	Number of Bits	Justification	Data Format	Result Register Bit Mapping
1	0	0	8 bits	Left justified	Unsigned	ATDDRxH[15:8]
1	0	1	8 bits	Left justified	Signed	ATDDRxH[15:8]
1	1	x	8 bits	Right justified	Unsigned[a]	ATDDRxL[7:0]
0	0	0	10 bits	Left justified	Unsigned	ATDDRxH[15:8], ATDDRxL[7:6]
0	0	1	10 bits	Left justified	Signed	ATDDRxH[15:8], ATDDRxL[7:6]
0	1	x	10 bits	Right justified	Unsigned	ATDDRxH[1:0], ATDDRxL[7:0]

[a] Signed data not available in right justified format.

TABLE 17-11 Left Justified, Signed and Unsigned Output Codes

Input Signal V_{RL} = 0 volts V_{RH} = 5.0 volts	Signed 8-bit Codes	Unsigned 8-bit Codes	Signed 10-bit Codes	Unsigned 10-bit Codes
4.995	7F00	FF00	7FC0	FFC0
4.990	7F00	FF00	7F80	FF80
4.985	7F00	FF00	7F40	FF40
4.980	7F00	FF00	7F00	FF00
4.976	7F00	FF00	7EC0	FEC0
4.971	7F00	FF00	7E80	FE80
4.966	7E00	FE00	7E40	FE40
4.961	7E00	FE00	FE00	FE00
—	—	—	—	—
2.520	0100	8100	0100	8100
2.515	0100	8100	00C0	81C0
2.510	0100	8100	0080	8080
2.505	0000	8000	0040	8040
2.500	0000	8000	0000	8000
2.495	0000	8000	FFC0	7FC0
2.490	0000	8000	FF80	7F80
2.485	FF00	7F00	FF40	7F40
2.480	FF00	7F00	FF00	7F00
2.476	FF00	7F00	FEC0	7EC0
2.471	FF00	7F00	FE80	7E80
2.466	FE00	7E00	FE40	7E40
2.461	FE00	7E00	FE00	7E00
—	—	—	—	—
0.020	8100	0100	8100	0100
0.015	8100	0100	80C0	00C0
0.010	8100	0100	8080	0080
0.005	8000	0000	8040	0040
0.000	8000	0000	8000	0000

Example 17-5 Four-Conversion Sequence

The A/D is to be programmed to convert continuously a four-conversion sequence of channel 3 in single-channel mode. How must S8C, S4C, S2C, S1C, SCAN, MULT, and the CC–CA bits be initialized?

Solution: S8C, S4C, S2C, S1C = 0100; SCAN = 1; MULT = 0; and CC, CB, CA = 011.

Example 17-6 Aliasing Frequency

After the A/D has been programmed as specified in Example 17-5, and assuming the A/D clock is 2 MHz, the resolution is 10 bits, and the final sample time is 2 A/D clocks, what is the maximum frequency that can be digitized without aliasing?

Solution: Fourteen A/D clock cycles are required to complete one conversion (two for initial sample and two for final sample plus 10 for the 10-bit A/D) and the next can start immediately in continuous scan, single-channel mode. Therefore the conversion time is 7 μs and the Nyquist frequency is 71.4 kHz.

Example 17-7

When the A/D clock is 2 MHz, the resolution is 10 bits, and the final sample time is 2 A/D clocks, what is the maximum frequency that can be digitized without aperture errors?

Solution: The aperture time is 2 μs so the maximum frequency is $1/(2 \times \Pi \times 2 \times 10^{-6} \times 1024) = 77$ Hz.

Example 17-8 Four Channel Conversion

The A/D is to be programmed to convert continuously four channels (0–3). How must S8C, S4C, S2C, S1C, SCAN, MULT, and the CC–CA bits be initialized?

Solution: S8C, S4C, S2C, S1C = 0100; SCAN = 1; MULT = 1; CC, CB, CA = 000.

Example 17-9 Eight Channel Conversion

How should you initialize ATDCTL3, ATDCTL4, and ATDCTL5 to convert all eight inputs in a single conversion sequence with 10-bit resolution, left justified, unsigned results? The bus clock is 8 MHz.

Solution:

ATDCTL3: S8C, S4C, S2C, S1C = 0000; FIFO = 0; FRZ1, FRZ0 = 00.
ATDCTL4: SRES8 = 0; SMP1, SMP0 = 00; PRS4, PRS3, PRS2, PRS1 = 00001.
ATDCTL5: DJM = 0; DSGN = 0; SCAN = 0; MULT = 1; CC, CB, CA = 000.

Digital Results from the A/D

The HCS12 A/D converter has eight, *16-bit, result registers*. These registers contain the *right justified* or *left justified, signed* or *unsigned* results from the A/D conversion. The channel from which the result is obtained is chosen by the mode and channel select bits in ATDCTL5 as described previously. Table 17-12 shows the result register addresses. Note that the first conversion in a sequence in non-FIFO mode will appear in the ATDDR0 register regardless of which channel is being converted because the input channels do not map to the result registers. (See also Table 17-13.)

LEFT JUSTIFIED RESULTS REGISTERS

ATDDRxH—Base + $00xx—ATD Result Register, High Byte, Left Justified

	Bit 15	14	13	12	11	10	9	8
10-Bit	Bit 9 MSB	Bit 8	Bit 7	Bit 6	Bit 5	Bit 4	Bit 3	Bit 2
8-Bit	Bit 7 MSB	Bit 6	Bit 5	Bit 4	Bit 3	Bit 2	Bit 1	Bit 0
Reset:	0	0	0	0	0	0	0	0

Read only: Anytime.

Bit 9–Bit 2 (10-bit), Bit 7–Bit 0 (8-bit)

ATD Conversion High Byte Results

The high byte of the 16-bit A/D results registers contains bits 9–2 of a 10-bit conversion and bits 7–0 of an 8-bit conversion (controlled by SRES8 in ATDCTL4) when left justified format (DJM = 0 in ATDCTL5) is used.

ATDDRxL—Base + $00xx—ATD Result Register, Low Byte, Left Justified

	Bit 7	6	5	4	3	2	1	0
10-Bit	Bit 1	Bit 0 LSB	0	0	0	0	0	0
8-Bit	U	U	0	0	0	0	0	0
Reset:	0	0	0	0	0	0	0	0

Read only: Anytime.

Bit 1–Bit 0

ATD Conversion Low Byte Results

The low byte of the 16-bit A/D results registers contains bits 1–0 of a 10-bit conversion (controlled by SRES8 in ATDCTL4) when left justified format (DJM = 0 in ATDCTL5) is used.

RIGHT JUSTIFIED RESULTS REGISTERS

ATDDRxH—Base + $00xx—ATD Result Register, High Byte, Right Justified

	Bit 15	14	13	12	11	10	9	8
10-Bit	0	0	0	0	0	0	Bit 9 MSB	Bit 8
8-Bit	0	0	0	0	0	0	0	0
Reset:	0	0	0	0	0	0	0	0

Read only: Anytime.

Bit 9–Bit 8 (10-bit)

ATD Conversion High Byte Results

The high byte of the 16-bit A/D results registers contains bits 9–8 of a 10-bit conversion (controlled by SRES8 in ATDCTL4) when right justified format (DJM = 1 in ATDCTL5) is used.

ATDDRxL—Base + $00xx—ATD Result Register, Low Byte, Right Justified

	Bit 7	6	5	4	3	2	1	0
10-Bit	Bit 7	Bit 6	Bit 5	Bit 4	Bit 3	Bit 2	Bit 1	Bit 0 LSB
8-Bit	Bit 7 MSB	Bit 6	Bit 5	Bit 4	Bit 3	Bit 2	Bit 1	Bit 0 LSB
Reset:	0	0	0	0	0	0	0	0

Read only: Anytime.

Bit 7–Bit 0 (10-bit and 8-bit)

ATD Conversion Low Byte Results

The low byte of the 16-bit A/D results registers contains bits 7–0 of both 8- and 10-bit conversions (controlled by SRES8 in ATDCTL4) when right justified format (DJM = 1 in ATDCTL5) is used.

TABLE 17-12 ATD Result Register Addresses

A/D Result Register	Address Base +
ATDDR0H	$0090
ATDDR0L	$0091
ATDDR1H	$0092
ATDDR1L	$0093
ATDDR2H	$0094
ATDDR2L	$0095
ATDDR3H	$0096
ATDDR3L	$0097
ATDDR4H	$0098
ATDDR4L	$0099
ATDDR5H	$009A
ATDDR5L	$009B
ATDDR6H	$009C
ATDDR6L	$009D
ATDDR7H	$009E
ATDDR7L	$009F

TABLE 17-13 Examples of Resolution, Justification, Data Format, and Multichannel Mode Selections

SRES8	DJM	DSGN	SCAN	MULT	CC	CB	CA	Channel(s) Converted	Result Registers	Comments
For S8C:S4C:S2C:S1C = 0001 One Conversion per Sequence										
0	0	0	0	0	0	0	0	AN0	ATDDR0H	8-bit, left justified, unsigned, single conversion
1	0	0	0	0	0	0	1	AN1	ATDDR0H: ATDDR0L	10-bit, left justified, unsigned, single conversion
0	1	0	0	0	0	1	1	AN3	ATDDR0L	8-bit, right justified, unsigned, single conversion

TABLE 17-13 Continued

SRES8	DJM	DSGN	SCAN	MULT	CC	CB	CA	Channel(s) Converted	Result Registers	Comments
For S8C:S4C:S2C:S1C = 0100 Four Conversions per Sequence (Default)										
1	0	1	0	1	1	0	0	AN4–AN7	ATDDR0H: ATDDR0L– ATDDR3H: ATDDR3L	10-bit, left justified, 2's complement, single conversion, four channels

17.3 A/D Input Synchronization

Polling A/D Conversion Complete

The HCS12 A/D can generate interrupts when the conversion is complete or the user may poll the conversion completion flag. There is a *Sequence Complete Flag (SCF)* in ATDSTAT0 that is set when the conversion sequence is completed, and there are eight *Conversion Complete Flags (CCF7–CCF0)* that are associated with each A/D result register. These flags are in the *ATDSTAT1* register and are set when the current conversion writes into the associated result register.

The *ATDSTAT0* register contains bits to indicate error conditions. *External Trigger Overrun Flag (ETORF)* is set if the external trigger is asserted (in edge triggered mode only) before the previous conversion sequence is complete. A similar flag, *FIFO Overrun Flag (FIFOR)*, is set when the result registers are out of sequence with the input channel or if the result registers are not read before new results are written to them. The CC2, CC1, and CC0 bits give the binary value of the conversion counter and show which register is to receive the result of the current conversion.

ATDSTAT0—Base + $0086—ATD Status Register 0

	Bit 7	6	5	4	3	2	1	0
Read:	SCF	0	ETORF	FIFOR	0	CC2	CC1	CC0
Write:								
Reset:	0	0	0	0	0	0	0	0

▨ = reserved, unimplemented, or cannot be written to.

Read: Anytime. Write: Anytime, no effect on CC2, CC1, or CC0.

SCF

Sequence Complete Flag

0 = Conversion sequence has not been completed (default).
1 = Conversion sequence completed.

This flag is set when the A/D conversion sequence is completed. If conversion sequences are continuously performed (SCAN = 1), the flag is set after each sequence is completed. SCF is cleared when one of the following occurs:

1. Write "1" to SCF.
2. Write to ATDCTL5 to start a new conversion sequence.
3. If AFFC in ATDCTL2 is "1" and a result register is read.

ETORF

External Trigger Overrun Flag

0 = No external trigger overrun error has occurred (default).
1 = Overrun error has occurred.
When you are using the external edge trigger mode to start an A/D conversion (ETRIGLE = 0 and ETRIGE = 1 in ATDCTL2), the overrun flag is set if additional edges are detected while a conversion sequence is in progress. The flag is cleared when one of the following occurs:

1. Write "1" to ETORF.
2. Write to ATDCTL2, ATDCTL3, or ATDCTL4 to abort a conversion sequence.
3. Write to ATDCTL5 to start a new conversion sequence.

FIFOR

FIFO Overrun Flag

0 = No overrun has occurred (default).
1 = FIFO overrun has occurred.
This bit indicates that a result register has been written to before its associated conversion complete flag (CCF in ATDSTAT1) has been cleared. This flag is useful when using FIFO mode (FIFO = 1 in ATDCTL3) because it indicates that the result registers are potentially out of synchronism with the input channels. It can be used in non-FIFO modes where it indicates that a result register has been overwritten before it has been read and old data has been lost. The flag is cleared when one of the following occurs:

1. Write "1" to FIFOR.
2. Write to ATDCTL5 to start a new conversion sequence.

CC2, CC1, CC0

Conversion Counter

These bits are read only and give the binary value of the conversion counter. This points to the result register that is to receive the result of the current conversion. If in non-FIFO mode (FIFO = 0 in ATDCTL3), the conversion counter is initialized to "000" at the beginning and end of the conversion sequence. In FIFO mode, the counter is not initialized at the beginning of each conversion sequence and wraps around when its maximum value is reached.

ATDSTAT1—Base + $008B—ATD Status Register 1

	Bit 7	6	5	4	3	2	1	0
Read:	CCF7	CCF6	CCF5	CCF4	CCF3	CCF2	CCF1	CCF0
Write:								
Reset:	0	0	0	0	0	0	0	0

   = reserved, unimplemented, or cannot be written to.

Read: Anytime. Write: Anytime, no effect.

CCF7:CCF0

Conversion Complete Flags

0 = Conversion number x is not completed (default).

1 = Conversion number x is complete and the result is in ATDDRx.

A conversion complete flag is set at the end of each conversion in a conversion sequence. The CCF number is associated with the number of the conversion in a sequence and the result register in which the conversion is found. That is, CCF0 is set when the first conversion is complete and its result will be in ATDDR0H; CCF1 is set when the second conversion is complete with its result in ATDDR1H; and so on. The flag is cleared when one of the following occurs:

1. A write to ATDCTL5 starting a new conversion sequence.

2. If AFFC = 0 in ATDCTL2 (fast flag clearing mode disabled) and ATDSTAT1 is read followed by reading of the result register ATDDRx.

3. If AFFC = 1 in ATDCTL2 (fast flag clearing enabled) and result register ATDDRx is read.

Clearing Status Flags

The *ATD Fast Flag Clear All (AFFC)* bit in the *ATDCTL2* register controls how the status flags are reset.

ATDCTL2—Base + $0082—ATD Control Register 2

	Bit 7	6	5	4	3	2	1	0
Read:	ADPU	AFFC	AWAI	ETRIGLE	ETRIGP	ETRIGE	ASCIE	ASCIF
Write:								
Reset:	0	0	0	0	0	0	0	0

▓ = These bits are described fully elsewhere.

Read: Anytime. Write: Anytime.

AFFC

ATD Fast Flag Clear

0 = A/D flag clearing operates normally. The flag is reset by reading the status register ATDSTAT1 before reading the result register to clear the associated flag (default).

1 = Enables fast flag clear operation. Any access to a result register will clear the associated flag.

17.4 A/D Interrupts

The HCS12 can generate an interrupt when the current conversion sequence is completed. The *ATD Sequence Complete Interrupt Enable (ASCIE)* and the *ATD Sequence Complete Interrupt Flag (ASCIF)* in the ATDCTL2 register are used to enable and generate the interrupt. All normal interrupt service routine procedures (see Chapter 12) must be followed. The interrupt vector for the A/D is at $FFD2:FFD3.

ATDCTL2—Base + $0082—ATD Control Register 2

	Bit 7	6	5	4	3	2	1	0
Read:	ADPU	AFFC	AWAI	ETRIGLE	ETRIGP	ETRIGE	ASCIE	ASCIF
Write:								
Reset:	0	0	0	0	0	0	0	0

▨ = These bits are described fully elsewhere.

Read: Anytime. Write: Anytime.

ASCIE

ATD Sequence Complete Interrupt Enable

0 = A/D sequence complete interrupts are disabled (default).
1 = A/D sequence complete interrupts are enabled.

ASCIF

ATD Sequence Complete Interrupt Flag

If ASCIE = 1, the ASCIF flag is the same as the SCF flag.

17.5 A/D Low Power Modes

The A/D can be configured for lower power consumption in three ways:

1. Stop Mode: This halts A/D conversion. When exiting from Stop mode (with an interrupt), A/D conversion will continue. Because the A/D needs some time to recover (power up), this conversion should be ignored.
2. Wait Mode: When AWAI = 1 in ATDCTL2, A/D operation is halted when Wait mode is entered. Exit from Wait mode is made when an interrupt occurs but, again, due to the recovery time of the A/D this result should be ignored.
3. When ADPU = 0 in ATDCTL2, the A/D is powered down and any A/D conversion in process is aborted.

17.6 Digital Input to the A/D

The A/D channel bits PAD0–PAD7 may be used for general purpose digital inputs or outputs. Port AD pins not being used for analog input may be used for digital inputs (although port AD digital reads are not recommended during the sample period). These digital inputs (but not outputs) are through the *PORTAD Data Input Register*. In addition, as shown in Chapter 11, the Port Integration Module (PTAD) allows digital input and output from and to the PAD0–PAD7 pins.

Figure 17-5 shows the initialization needed to read digital data from a PADn pin in registers PORTAD, PTAD, or PTIAD. The *ATD Digital Input Enable Register* (*ATDDIEN*) must be set and the *Data Direction Register AD* (*DDRAD*) must be cleared for each bit that is to be used as an input. The input data can be read from any of the three registers.

Figure 17-5 shows that to output from PTAD to a PADn pin, the data direction register bit associated with the output bit must be set. Only PTAD may be used to output data.

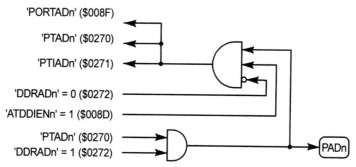

Figure 17-5 PORTAD and PTAD digital input and output.

ATDDIEN—Base + $008D—ATD Digital Input Enable Register

	Bit 7	6	5	4	3	2	1	0
Read:	IEN7	IEN6	IEN5	IEN4	IEN3	IEN2	IEN1	IEN0
Write:								
Reset:	0	0	0	0	0	0	0	0

Read: Anytime. Write: Anytime.

IEN7–IEN0

ATD Digital Input Enable

0 = Disable digital input buffer (default).

1 = Enable digital input buffer.

These bits control the digital input buffer from the microcontroller's analog input pin to the PORTAD and PTAD data registers. Setting this bit will enable the corresponding digital input buffer continuously. If this bit is set while simultaneously using it as an analog input port, there is a potential of increased power consumption because the input voltage may put the input electronics into a linear mode.

PORTAD—Base + $008F—Port AD Data Register

	Bit 7	6	5	4	3	2	1	0
Read:	PTAD7	PTAD6	PTAD5	PTAD4	PTAD3	PTAD2	PTAD1	PTAD0
Write:								
Reset:	1	1	1	1	1	1	1	1

░░░ = reserved, unimplemented, or cannot be written to.

Read: Anytime. Write: Anytime has no effect.

PTAD7–PTAD0

A/D Port Digital Input

If the digital input buffer is enabled (INEx = 1 in ATDDIEN), a read returns the logic level on that pin (signals not meeting V_{IL} or V_{IH} will have an indeterminate value.)

If the digital input buffer is disabled (INEx = 0) a read returns a "1."

PTAD—Base + $0270—Port AD I/O Register

	Bit 7	6	5	4	3	2	1	0
Read:	PTAD7	PTAD6	PTAD5	PTAD4	PTAD3	PTAD2	PTAD1	PTAD0
Write:								
Reset:	0	0	0	0	0	0	0	0

17.7 Miscellaneous A/D Registers

Three registers in the memory map of the A/D are reserved and not available for general use. These are ATDCTL0 (Base + $0080), ATDCTL1 (Base + $0081), and ATDTEST0 (Base + $0088).

ATDTEST1 contains a bit to enable special conversion values to be input to the A/D for testing. You may connect V_{RH}, V_{RL}, or $(V_{RH} + V_{RL})/2$ to the A/D.

ATDTEST1—Base + $0009—ATD Test Register 1

	Bit 7	6	5	4	3	2	1	0
Read:	U	U	0	0	0	0	0	SC
Write:	0	0	0	0	0	0	0	
Reset:	0	0	0	0	0	0	0	0

Read: Anytime, returns unpredictable values for bits 7 and 6. Write: Anytime.

SC

Special Channel Conversion

0 = Special channel conversion disabled (default).

1 = Special channel conversion enabled.

When this bit is set, the special channel conversion can be selected by the CC, CB, and CA bits in ATDCTL5 as shown in Table 17-14. When writing to SC, you should always write the remaining bits to zero. Otherwise, the A/D behavior might be unpredictable.

TABLE 17-14 Special Channel Select Coding

SC	CC	CB	CA	Analog Input Channel
0	—	—	—	Selected by CC, CB, CA
1	0	X	X	Reserved
1	1	0	0	V_{RH}
1	1	0	1	V_{RL}
1	1	1	0	$(V_{RH} + V_{RL})/2$
1	1	1	1	Reserved

17.8 A/D Programming Summary

Although there are many registers that must be initialized to operate the HCS12 A/D, the programming is straightforward. (See Table 17-15.)

1. Enable and power up the A/D by setting the ADPU bit in ATDCTL2.
2. Wait for 20 μs to allow the analog circuitry to settle before attempting to use the A/D.
3. Initialize the following bits in the ATDCTL2:

   AFFC to choose a flag clearing mode.

   AWAI to choose what happens to the A/D when in Wait mode.

   If you are using an external trigger on PAD7, set the ETRIGLE, ETRIGP, and ETRIGE bits.

   Enable ASCIE if sequence complete interrupts are desired.
4. Initialize ATDCTL3 by choosing and writing the following bits:

   S8C, S4C, S2C, S1C to choose the number of conversions per sequence.

   FIFO to choose the first in–first out mode wanted.

   FRZ1 and FRZ0 to choose what happens when a breakpoint is encountered.
5. Initialize ATDCTL4 by choosing and writing the following bits:

   SRES8 to choose 8- or 10-bit conversions.

   PRS[4:0] to keep the A/D clock in the range of 500 kHz to 2 MHz.

   SMP[1:0] to select the final sample time.
6. Choose the DJM, DSGN, SCAN, MULT, and CC–CA bits in ATDCTL5.
7. Write ATDCTL5 to start the conversion.
8. Wait for the conversion sequence to complete by polling the SCF bit in ATDSTAT or by using the sequence complete interrupt if it has been enabled.
9. Read the result in the ADR0H–ADR7H result registers.

17.9 Example ATD Programs

Examples 17-10 and 17-11 present ATD assembler and C programs.

TABLE 17-15 Examples of ATD Control Registers

Register	Value	Results
ATDCTL2	$80	Power up the A/D; normal flag clearing, no external triggering, no interrupts.
ATDCTL2	$8C	Power up the A/D; enable external, rising edge trigger, sequence complete interrupts enabled.
ATDCTL3	$00	Eight conversions/sequence; non-FIFO mode, continue conversion when in a breakpoint.
ATDCTL3	$08	One conversion/sequence; non-FIFO mode, continue conversion when in a breakpoint.
ATDCTL4	$61	10-bit conversion; 16 A/D clocks/sample; clock divider prescaler = 4.
ATDCTL4	$83	8-bit conversion, 2 A/D clocks/sample; clock divider prescaler = 8.
ATDCTL5	$00	Left justified, unsigned data; single conversion sequence; single channel; analog channel 0.
ATDCTL5	$94	Right justified, unsigned data; single conversion sequence; multiple channels, starting at analog channel 4.

Example 17-10 ATD Assembler Program Example

```
Metrowerks HC12-Assembler
(c) COPYRIGHT METROWERKS 1987-2003

Rel. Loc   Obj. code Source line

---- ----   --------- -----------
   1                  ;*******************************
   2                  ; ATD Test program
   3                  ;    This is a test program showing the use
   4                  ;    of the A/D converter.
   5                  ;    It converts the analog input on Ch 1.
   6                  ; Define the entry point for the main program
   7                          XDEF   Entry, main
   8                          XREF   __SEG_END_SSTACK ; Note double
                             underbar
   9                  ;*******************************
  10                  ; Register definitions
  11                  ;*******************************
  12                  ; Include files
  13                          INCLUDE  portt.inc
  14                          INCLUDE  portad.inc
  15                          INCLUDE  timer.inc
  16                          INCLUDE  atd.inc
  17                  ;*******************************
  18                  ; Constant Equates
  19                  ; AD Mode for ATDCTL5:
  20                  ;          DJM=1  Right justified
  21                  ;          DSGN=0 Unsigned data
  22                  ;          SCAN=0 Single conversion
  23                  ;          MULT=0 1 conversion on single
                             channel
  24     _            ;  CC,CB, CA=001  Analog channel 1
  25        0000 0081 ADMODE:    EQU    %10000001
  26                  ;*******************************
  27                  ; Code Section
  28                  MyCode: SECTION
  29                  Entry:
  30                  main:
  31                  ;*******************************
  32                  ; Initialize stack pointer register
  33 000000 CFxx xx           lds    #__SEG_END_SSTACK
```

```
34                        ;*******************************
35                        ; Initialize I/O
36                        ;*******************************
37                        ; Power up the A/D
38 000003 4C82 80                 bset  ATDCTL2,ADPU
39                        ; Generate a "short" delay > 20 microsec
40 000006 8614                    ldaa  #20        ; 20 loops for
41                 delay:                          ; 80 clock cycles
42 000008 A7                      nop
43 000009 0430 FC                 dbne  a,delay
44                        ; Set up the A/D
45                        ; Initialize ATDCTL2
46                        ;  Normal flag clearing, run in WAIT mode,
47                        ;  no external trigger, no interrupts
48 00000C 4D82 66                 bclr  ATDCTL2,AFFC|AWAI|ETRIGE|ASCIE
49                        ;  Initialize ATDCTL3
50                        ;  Set 1 conversion per sequence - S8C:S1C
                                  = 0001
51                        ;  Set non-FIFO mode - FIFO = 0
52                        ;  Continue in freeze mode - FRZ1:FRZ0 = 00
53 00000F 4D83 77                 bclr ATDCTL3,S8C|S4C|S2C|FIFO|FRZ1|FRZ0
54 000012 4C83 08                 bset  ATDCTL3,S1C
55                        ; Initialize ATDCTL4
56                        ;  Select 8-bit resolution - SRES = 1
57                        ;  Select A/D conversion clock periods
58                        ;   SMP1:SMP0 = 11 = 16 clocks
59                        ;  Set prescaler to divide by 8 -
                                  PRS4:PRS0 = 00011
60 000015 4D84 1C                 bclr  ATDCTL4,PRS4|PRS3|PRS2
61 000018 4C84 E3                 bset  ATDCTL4,SRES8|PRS1|PRS0|SMP1|SMP0
62                        ;*******************************
63                        ; Enable the POT on the Student Learning
                            Kit board
64 00001B 1C02 4208               bset  DDRT,BIT3  ; Make it an output
65 00001F 1C02 4008               bset  PTT,BIT3   ; Enable pot on AN01
66                        ;*******************************
67                 main_loop:
68                        ; DO
69                        ;  Start the conversion by writing the
                            scan select
70                        ;  information to ATDCTL5
71 000023 8681                    ldaa  #ADMODE
72 000025 5A85                    staa  ATDCTL5
```

```
73                         ;    Wait for the conversion to be complete
74                         spin:
75 000027 4F86 80FC              brclr ATDSTAT0,SCF,spin
76                         ;    Get the data
77 00002B D691                   ldab   ATDDR0L
78 00002D A7                     nop
79                         ;    Now do what you want with the digital
                                 data
80                         ; FOREVER
81 00002E 20F3                   bra    main_loop
82                         ;******************************
```

Example 17-11 ATD C Program Example

```c
/*****************************************************************
 * ATD Test program
 *    This is a test program showing the use
 *    of the A/D converter.
 *    It converts the analog input on Ch 1 four times
 *    and calculates the average.
 *****************************************************************/
#include <mc9s12c32.h>                /* derivative information */
/*****************************************************************
 * Define the value for ATDCTL2
 *    Power up ATD
 *    Normal flag clearing, run in wait mode,
 *    no external trigger, no interrupts */
#define ATDCTL2_VAL 0x80;
/*****************************************************************
 * Define the value for ATDCTL3
 *    4 conversions per sequence - S8C:S1C = 0100
 *    Non-FIFO mode - FIFO = 0
 *    Continue in freeze mode - FRZ1:FRZ0 = 00 */
#define ATDCTL3_VAL 0x20
/*****************************************************************
 * Define the value for ATDCTL4
 *    8-bit resolution - SRES8 = 1
```

```
   *    A/D conversion clock periods = 16 - SMP1:SMP0 = 11
   *    Set prescaler to divide by 8 - PRS4:PRS0 = 00011 */
#define ATDCTL4_VAL 0xE3
/***************************************************************
 * Define the AD mode for ATDCTL5
 *    DJM = 1    Right Justified Data
 *    DSGN = 0   Unsigned Data
 *    SCAN = 0   Single conversion
 *    MULT = 0   1 conversion on a single channel
 *    CC, CB, CA = 001   Analog channel 1 */
#define ATDCTL5_VAL   0x81
/***************************************************************/
void main(void) {
  char i;
  volatile unsigned int atd_value;
/***************************************************************/
  /* Initialize I/O */
  /*  Power up the ATD and initialize ATDCTL2 */
  ATDCTL2 = ATDCTL2_VAL;
  /* Generate a short delay > 20 µs */
  for ( i = 0; i < 20; ++i );
  /* Initialize ATDCTL3 */
  ATDCTL3 = ATDCTL3_VAL;
  /* Initialize ATDCTL4 */
  ATDCTL4 = ATDCTL4_VAL;
  /* Enable the POT on the Student Learning Kit attached to AN01 */
  DDRT_DDRT3 = 1;
  PTT_PTT3 = 1;
/***************************************************************/
  /* Main Loop */
  for(;;) {
    /* Start the conversion by writing the scan select information
     * to ATDCTL5 */
    ATDCTL5 = ATDCTL5_VAL;
    /* Wait for the conversion to be complete */
    while (ATDSTAT0_SCF == 0);
    atd_value = (ATDDR0L + ATDDR1L + ATDDR2L + ATDDR3L)/4;
  } /* wait forever */
}
```

17.10 Remaining Questions

- Where can I find out more about the electronics of the A/D? *See the* ATD_10B8C Block User Guide, *and Application Notes AN2428/D*, An Overview of the HCS12 ATD Module, *AN2429/D*, Interfacing to the HCS12 ATD Module, *and AN2438/D*, ADC Definitions and Specifications.

17.11 Conclusion and Chapter Summary Points

In this chapter we discussed the operation of the analog-to-digital converter.

- The HCS12 is an 8-bit or 10-bit successive approximation converter.
- The A/D must be powered up by writing a one to the ADPU bit in the ATDCTL2 register.
- A 20-µs delay must be observed after powering up the A/D before using it.
- There are eight input channels selected by an input multiplexer.
- The conversion time is programmable and can range from 12 to 28 A/D clock periods.
- The total sample-and-hold aperture time ranges from 4 to 18 A/D clock periods.
- One to eight channels may be converted in sequence with the results appearing in the eight A/D results registers.
- Analog input synchronization may be done by polling the Sequence Complete Flag or any of the eight Conversion Complete Flags in the ATDSTAT registers.
- The A/D can generate a Sequence Complete interrupt.

17.12 Bibliography and Further Reading

ATD_10B8C Block User Guide, S12ATD10B8CV2/D, Freescale Semiconductor, Inc., Austin, TX, February 2003.

Cady, F. M., *Microcontrollers and Microprocessors*, Oxford University Press, New York, 1997.

Fedeeler, J. and B. Lucas, *ADC Definitions and Specifications*, Application Note AN2438/D, Freescale Semiconductor, Inc., Austin, TX, 2003.

Gallop, M., *An Overview of the HCS12 ATD Module*, Application Note AN2428/D, Freescale Semiconductor, Inc., Austin, TX, 2003.

Gallop, M., *Interfacing to the HCS12 ATD Module*, Application Note AN2429/D, Freescale Semiconductor, Inc., Austin, TX, 2003.

HCS12 Microcontrollers MC9S12C Family, Rev 1.15, Freescale Semiconductor, Inc., Austin, TX, July 2005.

MC9S12C Family Device User Guide V01.05, 9S12C128DGV1/D, Freescale Semiconductor, Inc., Austin, TX, 2004.

17.13 Problems

Basic

17.1 Define conversion time. **[a]**

17.2 What is a sample-and-hold? **[a]**

17.3 How do you calculate the required aperture time for a sample-and-hold? **[a, k]**

17.4 How is the HCS12 A/D powered up? **[a, k]**

17.5 How long must the program delay before using the A/D after powering it up? **[a, k]**

Intermediate

17.6 An 8-bit successive approximation, general purpose (not HCS12) A/D converter has a 1-MHz clock. What is the maximum frequency it can convert without aliasing? **[a]**

17.7 An A/D converter is required to digitize a 5-kHz sinusoidal waveform. What is the maximum allowable conversion time for the A/D? (Assume a sample-and-hold circuit is being used.) **[a, c]**

17.8 An A/D converter is required to digitize a -5 to $+5$-volt analog signal to a resolution of 10 mV. How many bits are required? **[a, c]**

17.9 The A/D is programmed to convert a sequence of four channels in 8-bit continuous conversion mode. What is the maximum frequency signal on PAD0 that can be converted without aliasing (ignore aperture time effects; assume the final sample time is two ATD clocks and the ATD clock is 2 MHz)? **[b, k]**

17.10 The analog input ranges from 1 volt to 4 volts. **[a, k]**

   a. What should V_{RH} and V_{RL} be?

   b. What is the resolution for an 8-bit conversion?

   c. The analog result register shows $56. What is the analog voltage?

17.11 The analog input is 0–5 volts and $V_{RH} = 5$, $V_{RL} = 0$. The A/D reading is $24. What is the analog input voltage for a 10-bit conversion? **[b, k]**

Advanced

17.12 A 10-bit successive approximation A/D converter has the following specs: Minimum conversion time: 10^{-4}s; input voltage: -5 to $+5$ volts. **[a, c, k]**

   a. What is the maximum frequency that can be sampled without aliasing?

   b. For this frequency, what is the aperture time required so that errors in sampling are less than plus or minus ½ least significant bit?

   c. What is the resolution of this A/D in volts?

17.13 What sample rate must be used to sample a signal with a maximum frequency of 4-kHz? **[b, k]**

17.14 For Problem 17.13, assuming an 8-bit A/D converter, what is the maximum aperture time allowed? For a 10-bit converter? **[b, k]**

17.15 If the A/D is set up to measure 0–5 volts, state the binary value returned in the ATDDRxH:ATDDRxL registers assuming left justified, 8-bit, unsigned output code for the following input voltages. **[a, c]**

   a. 2.5 volts.

   b. 2.75 volts.

   c. 1.8 volts.

17.16 If the A/D is set up to measure 0–5 volts, state the binary value returned in the ATDDRxH:ATDDRxL registers assuming right justified, 8-bit, unsigned output code for the following input voltages. **[a, c]**

   a. 2.5 volts.

   b. 2.75 volts.

   c. 1.8 volts.

17.17 If the A/D is set up to measure 0–5 volts, state the binary value returned in the ATDDRxH:ATDDRxL registers assuming left justified, 8-bit, signed output code for the following input voltages. **[a, c]**

   a. 2.5 volts.

   b. 2.75 volts.

   c. 1.8 volts.

17.18 If the A/D is set up to measure 0–5 volts, state the binary value returned in the ATDDRxH:ATDDRxL registers assuming left justified, 10-bit, signed output code for the following input voltages. **[a, c]**

   a. 2.5 volts.

   b. 2.75 volts.

   c. 1.8 volts.

OBJECTIVES

This chapter describes hardware and software techniques to interface a variety of I/O devices to a single-chip microcontroller. Included are simple input devices (switches) and output devices (LEDs) and parallel I/O expansion to increase the number of I/O signals. The RS-232 interface used in SCI serial communications is described as is the SPI serial peripheral interface. A digital-to-analog converter using the SPI is shown as well as a liquid crystal display. I/O synchronization software is described.

18.1 The Single-Chip Microcontroller

A single-chip microcontroller, such as the MC9S12C32, used in an embedded system must be connected to the outside world to be useful. We have covered a variety of the I/O capabilities such as the analog-to-digital converter and serial and parallel I/O in previous chapters. This chapter gives examples showing how to interface external devices, such as keypads, LCD displays, LEDs, and serial interfaces such as the Serial Communications Interface (SCI) and Serial Peripheral Interface (SPI) devices to your microcontroller.

18.2 Simple Input Devices

Input Switches

The switch is the most basic of all binary input devices. Figure 18-1(a) shows a single-pole, single-throw (SPST) switch and a pull-up resistor. The switch output is high or low depending on the switch position. Figure 18-1(b) shows a multiple-pole, rotary switch. Pull-up resistors are necessary for each of these switches to provide a high logic level when the switch is open. The input ports in the MC9S12C32 and other microcontrollers often have a pull-up resistor on all inputs, and if so, the external resistors are not needed, but you may need to enable them on some ports. Check with Chapter 11 for details of the port you are using.

The *switch bounce* problem must be solved when using mechanical switches.

A problem with all switches is *switch bounce*. When a switch makes contact, its mechanical springiness will cause the contact to bounce, or make and break, for a few milliseconds as shown in Figure 18-1(c). In some cases, switch bounce

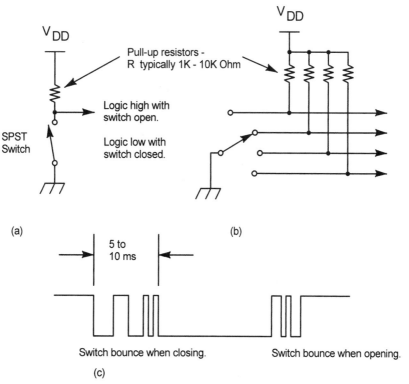

Figure 18-1 Switches used for binary input: (a) single-pole, single-throw (SPST) switch,
(b) multiple-pole switch, and (c) switch bounce.

may be observed when the switch is opened. If a program is counting switch closures and the
software is fairly fast, it may count several bounces and return more counts than are real. Thus,
depending on the application, switch debouncing may be necessary. There are several ways to
debounce switches, including both software and hardware methods.[1]

SOFTWARE DEBOUNCING

Here are two strategies for debouncing a switch in software. The first may be called *wait and
see*. Switch bouncing usually lasts only 5–10 milliseconds. If the software detects a low logic
level, indicating the switch has closed, it can simply wait for longer than the switch bounce
duration, say, 20–100 ms. Another approach is an integrating debouncer, which debounces
both switch closing and switch opening. We initialize a counter with a value of 10 and, after
the first logic low level is detected, poll the switch every millisecond. If the switch output is
low, decrement the counter. If the switch output is high, increment the counter. When the
counter reaches zero, we know that the switch output has been low for at least 10 milliseconds.
If, on the other hand, the counter reaches 20, we know that the switch has been open for at
least 10 milliseconds. The initial value of the counter should be adjusted to be longer than the
expected bounce time. Example 18-1 shows the pseudocode for this algorithm and a C program is shown in Example 18-2. (See also Example 18-3.)

---

[1] For a good review of switch bouncing and hardware and software methods for debouncing, see *A Guide to Debouncing*, by
Jack G. Ganssle, http://www.ganssle.com/debouncing.pdf, 2004.

Example 18-1 Pseudocode Design for Integrating Switch Debouncer

```
INITIALIZE Count = 10
WHILE ((Count > 0) and (Count < 20))
   DO
      Delay 1 millisecond
      Get Switch Input
      IF Switch Closed
         THEN Decrement Count
         ELSE Increment Count
      ENDIF Switch Closed
   ENDO
ENDWHILE ((Count > 0) and (Count < 20))
IF Count = 0
   THEN Switch is closed
   ELSE Switch is open
ENDIF Count = 0
```

Example 18-2 C Program Debouncer

Write a C program that implements the debouncer algorithm shown in Example 18-1.

Solution:

```
/*****************************************************************
 * Debounce routine using integrating debouncer.
 * Calls delay_X_ms which uses the timer channel 7.
 * Checks the switch on Port P, bit-5
 * Calling:
 * unsigned char debounce( unsigned int length );
 * where
 *    length is the time in milliseconds for switch bounce to last.
 *    The return is the final value of the switch (0 or 1).

 *****************************************************************/
#include <mc9s12c32.h>                    /* derivative information */
/*****************************************************************/
void delay_X_ms( unsigned int X );   /* Variable delay function */
/*****************************************************************/
unsigned char debounce(unsigned int bounce_length) {
/*****************************************************************/
unsigned int count;
```

```
unsigned char switch_val;
  /* initialize count */
  count = bounce_length/2;
  while (( count > 0 ) && ( count < bounce_length )){
    /* Delay a millisecond */
    delay_X_ms(1);
    /* If switch == 0, decrement count else increment it */
    if (PTP_PTP5 == 0)
      --count;
    else ++count;
  }
  if (count == 0) switch_val = 0;
  else switch_val = 1;
  return( switch_val );
}
```

Example 18-3 X Millisecond Delay Program
Write a C program that implements a variable delay program in steps of 1 millisecond.

Solution:

```
/********************************************************************
 * Approximately 1 millisecond delay using the HCS12
 * timer. The delay is only approximate
 * because not all clock cycles are accounted for.
 * This turns on the timer/counter (if it is not all
 * ready on), enables timer channel 7 for use as an output
 * compare channel and then waits for a millisecond.
 * The delay is repeated X times.
 * Calling:
 * void delay_X_ms( unsigned int X);
 * *****************************************************************/
#include <mc9s12c32.h>        /* derivative information */
/********************************************************************/
#define BUS_CLOCK   8000000   /* Assume 8 MHz bus clock */
#define DELAY       1         /* Number of milliseconds to delay */
#define DELAY_COUNT BUS_CLOCK*DELAY/1000 /* Clocks per ms */
/********************************************************************/
void delay_X_ms( unsigned int X ) {
  unsigned int i;
```

```
/**************************************************************/
/* Check to see if timer is on and turn it on if it is not */
   if ( TSCR1_TEN == 0 ) TSCR1_TEN = 1; /* Enable the timer */
/* Check to see if timer channel 7 is enabled as
   output compare */
   if ( TIOS_IOS7 == 0 ) TIOS_IOS7 = 1; /* Enable the channel */
/* Do the delay once */
   TC7 = TCNT + DELAY_COUNT;    /* Set up for the 1 ms delay */
   TFLG1 = TFLG1_C7F_MASK;      /* Reset the OC7 flag */
   while ( TFLG1_C7F == 0 );    /* Wait for the OC7 Flag to
                                   be set */
/* Now do the delay X-1 more times */
   for (i = 1; i < X; ++i){
     TC7 = TC7 + DELAY_COUNT;    /* Set TC7 register with the
                                    final count */

      TFLG1 = TFLG1_C7F_MASK;    /* Reset the OC7 flag */
     while ( TFLG1_C7F == 0 );   /* Wait for the OC7 Flag to
                                    be set */

   }
}
```

HARDWARE DEBOUNCING

Figure 18-2 shows two hardware debouncing schemes. In Figure 18-2(a), a single-pole, double-throw (SPDT) switch is debounced with a NAND latch. NOR gates can be used also for the latch. A disadvantage of this debouncer is that it requires a single-pole, double-throw switch. These switches are more expensive and bulky than the single-pole, single-throw (SPST) shown in Figure 18-1(a). An integrating debouncer is shown in Figure 18-2(b) with a Schmitt trigger gate. Before the switch closes the capacitor is charged and the output of the gate is low. When the switch closes the capacitor discharges and the gate output switches high. Because of the gate's built-in hysteresis it will not switch low again until the input voltage exceeds a threshold V_{T+}. The *RC* time constant of the circuit should be designed so the gate's input voltage does not exceed the threshold while the switch is bouncing. See Example 18-4.

Arrays of Switches

Switches can be organized as linear or matrix arrays; a linear array is shown in Figure 18-3. A variety of switches can be found including dual in-line package (DIP) switch arrays. The switch bounce problem may need to be solved and the array of switches must be scanned to find out which are closed or open. The output of the switch array could be interfaced directly to an 8-bit input port (at point A). To save some I/O lines, a 74HC151 eight-input multiplexer can be used.

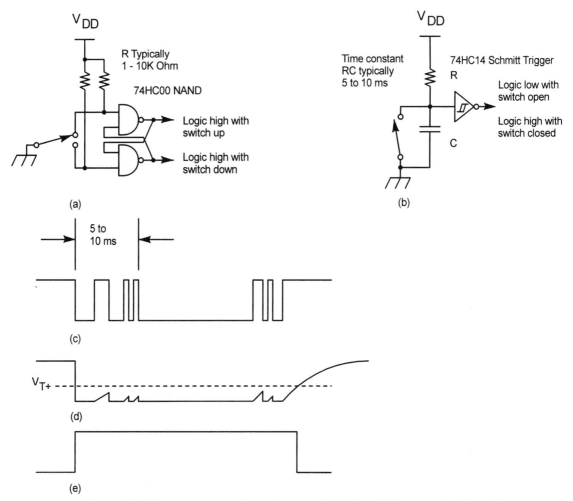

Figure 18-2 Hardware debouncing methods: (a) NAND latch debouncer, (b) Schmitt trigger, (c) switch bounce waveform, (d) Schmitt input voltage (V_C), and (e) Schmitt output.

Example 18-4 Schmitt Trigger Debouncer

For the Schmitt trigger debouncer shown in Figure 18-2, assume the following: $R = 10 \text{ k}\Omega$, $V_{T+} = 1.7$ V (positive-going threshold for the Schmitt trigger when $V_{DD} = 4.5$ V), and switch bounce shown in Figure 18-1 lasts no more than 10 ms. Calculate a value for the capacitor C so that the Schmitt trigger output will not switch during the switch bounce period.

Solution: The Schmitt trigger input voltage is given by

$$V_i = V_{DD}(1 - e^{-t/RC})$$

Setting $V_i = 1.7$ V and letting $t = 10$ ms gives $C = 2$ μF.

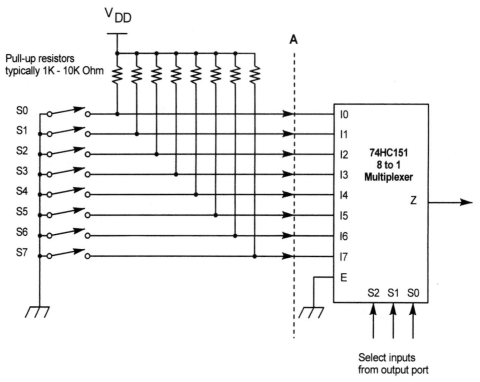

Figure 18-3 Linear array of switches.

Software is required to scan the array shown in Figure 18-3. As the software outputs a 3-bit sequence from 000 to 111, the multiplexer selects each of the switch inputs. The software scanner then reads one bit at an input port. See Examples 18-5 and 18-6.

Figure 18-3 shows pull-up resistors connected to the SPST switches. You may not have to use these when the switches are connected to your microcontroller if it has internal pull-ups enabled. If you use an 8-to-1 multiplexer you should use these resistors.

Example 18-5

For the linear array of switches in Figure 18-3, give a truth table showing which switch is read for each scan code output by the processor.

Solution:

Scan Code	Switch	Scan Code	Switch
000	S0	100	S4
001	S1	101	S5
010	S2	110	S6
011	S3	111	S7

Example 18-6 C Code for Scanning Switches

```c
/**************************************************************
 * unsigned char get_switches( void );
 * This is an example of a scanner to multiplex eight
 * switch inputs to one input port.
 * Port T, bits 2, 1, and 0 are used to switch the
 * multiplexer. Port A, bit 0 is the input.
 * All eight switches are read by this function and returned
 * as an unsigned char
 **************************************************************/
#include <mc9s12c32.h>               /* derivative information */
/**************************************************************/
typedef unsigned int bitfield;
/* Define a union to be able to set and clear bit
 * Port T bits 2, 1, 0 */
typedef union {
  unsigned char Byte;
  struct {
    bitfield BIT0 : 1;
    bitfield BIT1 : 1;
    bitfield BIT2 : 1;
    bitfield BIT3 : 1;
    bitfield BIT4 : 1;
    bitfield BIT5 : 1;
    bitfield BIT6 : 1;
    bitfield BIT7 : 1;
  } Bits;
}ByteBits;
/**************************************************************/
unsigned char get_switches( void ) {
  unsigned char switch_data;
  ByteBits mux_sel;
  unsigned char ddrt;
/**************************************************************/
  switch_data = 0;        /* Initialize the return data */
  /* Set Port T, 2, 1, and 0 as outputs if not already done */
  ddrt = DDRT & 0x07;        /* Get the current value of DDRT2-0 */
  if ( ddrt != 0x07 ) {
    /* Then the data direction register has not been set */
    DDRT_DDRT2 = 1;      /* Make DDRT2-0 output bits */
```

```
        DDRT_DDRT1 = 1;
        DDRT_DDRT0 = 1;
    }
    /* Now scan the multiplexer and read Port A bit-0 */
    for ( mux_sel.Byte = 0; mux_sel.Byte < 8; ++mux_sel.Byte ) {
        /* Output the mux scan code on Port T bits 5-7 without
           changing
         * other Port T bits */
        PTT_PTT2 = mux_sel.Bits.BIT2;
        PTT_PTT1 = mux_sel.Bits.BIT1;
        PTT_PTT0 = mux_sel.Bits.BIT0;
        /* Now read the bit selected on Port A, bit-0 and shift
         * it into the return value switch_data */
        switch_data = (2*switch_data) | PORTA_BIT0;
    }
    return( switch_data );
}
```

EXPLANATION OF EXAMPLE 18-6

The union of an unsigned char (*Byte*) with eight bit fields (*Bits*) allows us to output the scan code 000–111 (0–7) as individual bits on port T by bit addressing statements like `PTT_PTT2 = mux_sel.Bits.BIT2`. This lets us change the scan code bits without affecting the other bits on the port. An alternative is to directly output the scan code to the port. The can be done providing other output bits that may be in use are not altered. See Example 18-7 to see how this can be done.

Example 18-7 Alternative C Code for Scanning Switches

Show how to output a byte with a 3-bit scan code to port T bits 2–0 without disturbing bits 7–4.

Solution: Read the current contents of port T and clear bits 2–0 and then add (or OR) the scan code.

```
/************************************************************
 * Example code segment showing how to output 3-bit data
 * on bits 2 - 0 without changing bits 7 - 3.
 ************************************************************/
#include <mc9s12c32.h>                /* derivative information */
/************************************************************/
unsigned char get_switches (void) {
  unsigned char switch_data;
```

```
      unsigned char mux_sel,port_data;
      unsigned char ddrt;
/**************************************************************/
    switch_data = 0;
    /* Set Port T, 2, 1 and 0 as outputs if not already done */
    ddrt = DDRT & 0x07;   /* Get the current value of the DDRT2-0 */
    if ( ddrt != 0x07 ) {
      /* THEN the data direction register has not been set */
      DDRT_DDRT2 = 1;     /* Make 2-0 output bits */
      DDRT_DDRT1 = 1;
      DDRT_DDRT0 = 1;
    }
    /* Now scan the multiplexer and read Port A bit-0 */
    for (mux_sel = 0; mux_sel < 8; ++mux_sel) {
      /* Output the mux scan code on Port T bits 5-7 */
      /* Get the current value on Port T and reset bits 2-0 */
      port_data = PTT & 0xf8;
      port_data = port_data + mux_sel;  /* Add in the scan code */
      PTT = port_data;                  /* Output it to the port */
      /* Now read the bit selected on Port A bit-0 and shift
       * it into switch_data */
      switch_data = (2*switch_data) | PORTA_BIT0;
    }
    return( switch_data );
}
```

16-KEY KEYPAD

A keypad is an array of switches arranged in a two-dimensional matrix as shown in Figure 18-4. A switch and a diode connect each intersection of the vertical and horizontal lines as shown by the blow-up view, and closing the switch connects the horizontal line to the vertical. A 4×4 keypad can be interfaced directly to four output and four input bits. Port AD bits 3–0 are outputs and bits 7–4 are inputs. Software can scan the keyboard by outputting the 4-bit "ring" counter code as shown in Table 18-1 and then, for each of these codes, reading the values on input bits PAD7–4. The combination of the 4-bit output and input scan codes identifies which switch is closed. A lookup table can then convert the 8-bit code to a more convenient code, such as the ASCII character code, for the hexadecimal keypad. See Example 18-6.

A problem that occurs when a keypad user hits more than one key at once, or rapidly rolling the finger from one key to another, is called *n-key rollover*. Providing for two-key rollover is commonly done. The keyboard hardware and software store the rapidly depressed keys in a first-in, first-out (FIFO) buffer for later readout. An alternative strategy is *n-key lockout*, where

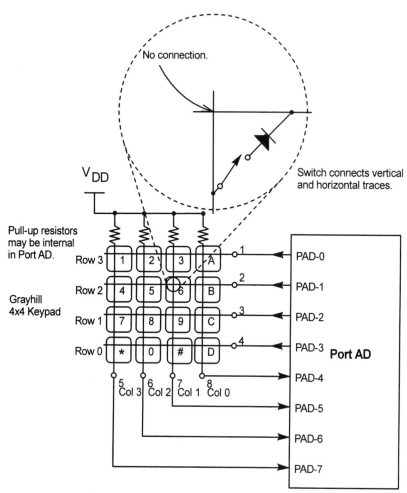

Figure 18-4 16-key keypad.

only the first or last of the sequence of keys depressed within some short period is recorded. Keyboard encoder chips incorporating all the scanning, debouncing, diodes, n-key rollover, and interrupt generation are available. Typical chips are the National Semiconductor 74C922 and the Intel i8279. Using these chips can eliminate scanning software and hardware and provide a much easier-to-implement keyboard interface at the expense of the additional hardware.

TABLE 18-1 Keyboard Scanning Codes

Output Scan Values on P3 P2 P1 P0	Values Input on P7 P6 P5 P4				
	1 1 1 1	0 1 1 1	1 0 1 1	1 1 0 1	1 1 1 0
			Key Pressed		
1 1 1 0	None	1	2	3	A
1 1 0 1	None	4	5	6	B
1 0 1 1	None	7	8	9	C
0 1 1 1	None	*	0	#	D

Example 18-8 shows a keypad scanning routine written in assembly language and in Example 18-9 we show a C module that calls the keypad scanner. Because the code in Example 18-8 does not debounce the switch, the C program enters a debounce routine in case the key was just pressed and is still bouncing.

```
Example 18-8 Keyboard Scanning
Metrowerks HC12-Assembler
(C) COPYRIGHT METROWERKS 1987-2003


Abs. Loc  Obj. code Source line
---- ---  --------- -----------
    1                 ; Hex keypad scanning module
    2                 ; This module scans a 16-key keypad
    3                 ; attached to Port AD.
    4                 ; Port AD bits:
    5                 ;   PAD-3 - PAD-0: Output: Scan row
                             scan codes
    6                 ;   PAD-7 - PAD-4: Input: Column code
    7                 ;              |    Col3 Col2 Col1 Col0
    8                 ;      Row   | Col Code
    9                 ; Row Code  |1111 0111 1011 1101 1110
   10                 ; ----------|-------------------------
   11                 ; 3  1110  |None   1    2    3    A  Key
   12                 ; 2  1101  |None   4    5    6    B  Pressed
   13                 ; 1  1011  |None   7    8    9    C
   14                 ; 0  0111  |None   *    0    #    D
   15                 ; ----------|-------------------------
   16                 ; Assembly calling:
   17                 ;   jsr  hex_key_scan
   18                 ; C calling
   19                 ;   unsigned char hex_key_scan( void );
   20                 ; Registers input:
   21                 ;   None
   22                 ; Registers output:
   23                 ;   B = ASCII Key code or user defined
   24                 ;       keycode given in Key_Codes table
   25                 ;     = $00 if no key pressed
   26                 ;       or multiple keys pressed
   27                 ;   Carry = 1 if key pressed, otherwise = 0
   28                 ; Registers modified: B, CCR
```

```
29                    ; Stack requirements: 8 bytes (including
30                    ;                          return address)
31                    ; ********************************************
32                    ; Define Grayhill Series 96 4x4 keypad
33      0000 0004 NUM_ROWS: EQU 4                ; Number of rows
34      0000 0010 NUM_KEYS: EQU 16               ; Number of keys
35                    ; PTAD Bit   Grayhill Keypad Pin
36                    ;   0          1
37                    ;   1          2
38                    ;   2          3
39                    ;   3          4
40                    ;   5          6
41                    ;   6          7
42                    ;   7          8
43                    ; ********************************************
44                    ; This module returns the first key pressed
45                    ; when scanning. It does not check for
46                    ; multiple col keys pressed at once.
47                    ; ********************************************
48                    ; Include files
49                            INCLUDE  portad.inc
60i                             INCLUDE "bits.inc"
146                   ; Define symbols used in the module
147                           XDEF  hex_key_scan
148                   ;********************************************
149                   ; Define constants
150     0000 000E ROW3:    EQU    %00001110 ; Row 3 scan code
151     0000 000D ROW2:    EQU    %00001101 ; Row 2 scan code
152     0000 000B ROW1:    EQU    %00001011 ; Row 1
153     0000 0007 ROW0:    EQU    %00000111 ; Row 0
154     0000 000F OUTPUTS:EQU    %00001111 ; Row outputs
155     0000 00F0 INPUTS: EQU    %11110000 ; Col inputs
156     0000 0070 COL3:    EQU    %01110000 ; Col 3 scan code
157     0000 00B0 COL2:    EQU    %10110000 ; Col 2
158     0000 00D0 COL1:    EQU    %11010000 ; Col 1
159     0000 00E0 COL0:    EQU    %11100000 ; Col 0
160     0000 00F0 KEY_MASK: EQU %11110000
161     0000 00F0 NO_KEYS:  EQU %11110000  ; Code for no
                                             keys pressed
162                   ; Code Section
```

```
163                      SubCode:  SECTION
164                      hex_key_scan:
165                      ; Save registers used
166 000000 36               psha
167 000001 34               pshx
168 000002 35               pshy
169                      ; Check to see if the port has been set up
170                      ;   IF the data direction register is not set
171                      ;   to output on the ROW_OUT bits
172 000003 B602 72          ldaa   DDRAD
173 000006 810F             cmpa   #OUTPUTS
174 000008 270B             beq    endif
175                      ;   THEN initialize the port
176 00000A 1C02 720F        bset   DDRAD,OUTPUTS  ; Set DDR outputs
177 00000E 1C02 74F0        bset   PERAD,INPUTS   ; Enable input
                                                    pull-ups
178 000012 4C8D F0          bset   ATDDIEN,INPUTS ; Enable ATD
                                                    input bits
179                      ;   ENDIF
180                      endif:
181                      ; Init counter and pointer to scan all cols
182 000015 8604             ldaa   #NUM_ROWS
183 000017 CExx xx          ldx    #Row_Codes
184                      ; DO
185                      ;   Output each row code and
186                      ;   read the col code each time.
187                      scan_cols:
188 00001A 180D 0002    movb  0,x,PTAD       ; Assert row scan
   00001E 70
189 00001F A7               nop            ; Delay a little
190 000020 A7               nop            ; before reading
191 000021 F602 70          ldab   PTAD    ; Read it
192 000024 180B 0F02    movb #OUTPUTS,PTAD ; Reset the row output
   000028 70
193 000029 37               pshb
194                      ;   IF Key pressed
195 00002A C4F0             andb   #KEY_MASK  ; Mask off row code
196 00002C C1F0             cmpb   #NO_KEYS   ; Compare no keys
197 00002E 33               pulb
198                      ;     THEN process the key to output the code
```

```
199 00002F 2607               bne    key_pressed
200                    ;    ELSE no key hit, continue
201 000031 08                 inx              ; Point to next Col
202                    ;    END_IF
203                    ; WHILE more Cols to scan
204 000032 0430 E5            dbne   a,scan_cols
205                    ; If here, must be no keys pressed,
206                    ; just return with scan code $00
207 000035 C7                 clrb             ; Clears carry too
208 000036 2018               bra    all_done
209                    ;*******************************************
210                    key_pressed:
211                    ; Now have to find the key that was pressed.
212                    ; The variable scan_code has the col and row
213                    ; code. Just scan through a look up table to
214                    ; find a match and return the user defined code.
215 000038 CExx xx            ldx    #Good_Codes  ; Point to table
216 00003B CDxx xx            ldy    #Key_Codes   ; Point to key codes
217 00003E 8610               ldaa   #NUM_KEYS
218                    ; DO
219                    ;    Scan through the table of good codes.
220                    check_codes:
221                    ;    IF B = the col & row code
222 000040 E100               cmpb   0,x
223 000042 2606               bne    continue
224                    ;    THEN have found a match.
225                    ;    Scan_code retrieved from Key_Code table
226 000044 E640               ldab   0,y          ;scan_code
227 000046 1401               sec              ; Set the carry bit
228 000048 2006               bra    all_done
229                    continue:
230                    ;    ELSE continue looking at the table
231 00004A 08                 inx              ; Point to next code
232 00004B 02                 iny
233                    ; END_IF
234                    ; WHILE not reached end of table
235 00004C 0430 F1            dbne   a,check_codes
236                    ; If reach here, must be a faulty col code,
237                    ; i.e. more than one key pressed in a row
```

```
238                      ; Return the no key pressed code
239 00004F C7                    clrb              ; B = 0, C = 0
240                      all_done:
241                      ; Note: Some debouncing should be done here
242                      ; Restore registers and return
243 000050 31                    puly
244 000051 30                    pulx
245 000052 32                    pula
246 000053 3D                    rts
247                      ;*******************************************
248                      ; Constant data section
249                      subConst: SECTION
250 000000 0E    Row_Codes:  DC.B   ROW3    ; Row 3 scan code
251 000001 0D                DC.B   ROW2    ; Row 2 scan code
252 000002 0B                DC.B   ROW1    ; Row 1
253 000003 07                DC.B   ROW0    ; Row 0
254              Good_Codes:
255 000004 7E                DC.B   COL3 | ROW3 ; "1"
256 000005 BE                DC.B   COL2 | ROW3 ; "2"
257 000006 DE                DC.B   COL1 | ROW3 ; "3"
258 000007 EE                DC.B   COL0 | ROW3 ; "A"
259 000008 7D                DC.B   COL3 | ROW2 ; "4"
260 000009 BD                DC.B   COL2 | ROW2 ; "5"
261 00000A DD                DC.B   COL1 | ROW2 ; "6"
262 00000B ED                DC.B   COL0 | ROW2 ; "B"
263 00000C 7B                DC.B   COL3 | ROW1 ; "7"
264 00000D BB                DC.B   COL2 | ROW1 ; "8"
265 00000E DB                DC.B   COL1 | ROW1 ; "9"
266 00000F EB                DC.B   COL0 | ROW1 ; "C"
267 000010 77                DC.B   COL3 | ROW0 ; "*"
268 000011 B7                DC.B   COL2 | ROW0 ; "0"
269 000012 D7                DC.B   COL1 | ROW0 ; "#"
270 000013 E7                DC.B   COL0 | ROW0 ; "D"
271                      ; User defined key codes. These are ASCII.
272 000014 3132 3341 Key_Codes:  DC.B   "123A456B789C*0#D"
    000018 3435 3642
    00001C 3738 3943
    000020 2A30 2344
273                      ;*******************************************
```

EXPLANATION OF EXAMPLE 18-8

Although Example 18-8 looks pretty complicated, the key to understanding how it works lies in seeing how the row and column scan codes map into the key code that is returned. The hardware connections are shown in Figure 18-4. Port AD bits 3–0 are outputs to scan each row. The column codes are read by port AD bits 7–4. Constant definitions define the code to output for each row (*lines 150–153*) and the code that will be read for each column (*lines 156–159*). Each row code is output in sequence and the column code is read. If no keys are pressed in that row, the column code will be %1111xxxx, and so the next row code is output. If all rows have been scanned and no keys pressed, the module returns a $00 indicating no keys (*line 207*). If a key is pressed, the row and column codes are used to find the correct key code to return.

Let us say the "9" key is pressed. This key is in row 1 and column 1. When the row 1 code (%1011) is output, reading port AD will return %11011011 where the most significant nibble (%1101) is the column code and the least significant nibble (%1011) is the row code (when reading a port the output bits are read too). This byte is used with two lookup tables to find the key code. A Good_Codes table (*lines 255–270*) gives all 16 possible column–row combinations. The Key_Codes table (*line 272*) contains the ASCII codes for all the keys. Two pointers are initialized to the start of each table (*lines 215 and 216*) and a counter is initialized in *line 217*. The program scans through the two tables, by incrementing both pointers (*lines 231 and 232*), until it finds a match for the column and row code. At this point the Y register is pointing to the key code to be returned.

Example 18-9 C Routine to Get a Key Code

```
/******************************************************************
 * Sample program for debouncing the keys
 * on a 16 key Grayhill keypad
 * Calling:
 * unsigned char keypad_debounce(unsigned int bounce_length);
 * where
 *    bounce_length is the bounce duration in milliseconds
 *    It returns the key valid at the end of the debounce time.
 ******************************************************************/
unsigned char hex_key_scan( void );  /* 4x4 keypad scanner */
void delay_X_ms( int );              /* Variable millisecond
                                        delay */
/******************************************************************/
/* Debouncer for the keypad */
unsigned char keypad_debounce(unsigned int bounce_length){
    int count;
    unsigned char get_key,new_key;
/******************************************************************/
    /* Get a key from the keypad */
    get_key = hex_key_scan();
    /* Debounce the keypad in case we caught it bouncing */
```

```
count = bounce_length/2;     /* initialize count */
while ( (count > 0) && (count < bounce_length) ){
    delay_X_ms(1);                /* Delay a millisecond */
    new_key = hex_key_scan(); /* Read the keyboard again */
    if (new_key == get_key) ++count;
    else --count;
}
/* IF count == bounce_length then get_key has been read
 * long enough to quit bouncing so return it */
if (count == bounce_length) return(get_key);
/* Otherwise, the new_key was read so return it */
else return(new_key);
}
```

18.3 Simple Display Devices

The most simple display device is a single light-emitting diode (LED). An LED lights when current of 10–20 milliamperes is passed in the forward direction. Figure 18-5 shows how to interface a single LED. When designing an LED driver you must determine the output current (sourcing or sinking) of the device turning on the LED. In Figure 18-5(a), a 74LS04 can sink up to 16 mA but can source only 400 μA. Therefore, by using the inverter, a logic 1 at the inverter's input will turn the LED on. The current limiting resistor, R, is designed to limit the current through the diode. In Figure 18-5(b) a latch is used as an *output device* to latch 1's and 0's to keep the LED on or off. In Figure 18-5(c) a 74ACT125 three-state buffer, which can source up to 24 mA, is designed to drive the LED connected to the ground. To use an LED in this configuration you must make sure the device can source sufficient current to turn on the LED.

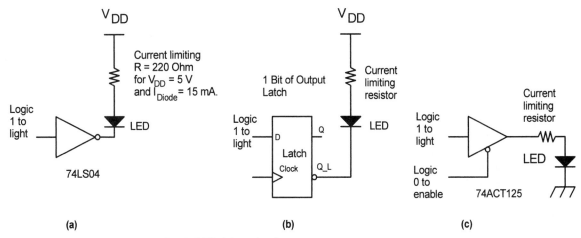

Figure 18-5 Single LED driver circuits.

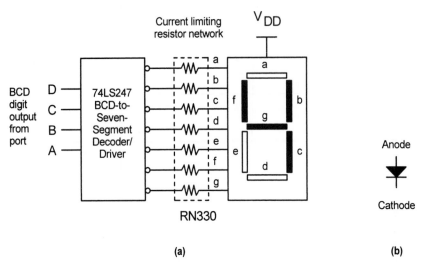

Figure 18-6 (a) Common anode seven-segment display; (b) LED showing anode and cathode.

A seven-segment LED display shows numeric characters. LED displays come in two varieties—common anode and common cathode. Figure 18-6(a) is a common anode display using a 74LS247 BCD-to-Seven-Segment Decoder/Driver. A BCD number is output by the CPU to the 74LS247 and its active low outputs turn on the appropriate segments to display the number.

Sometimes more than one display is required. Figure 18-7 shows how to multiplex a four-digit display using only one decoder/driver and common cathode LEDs. Each of the four

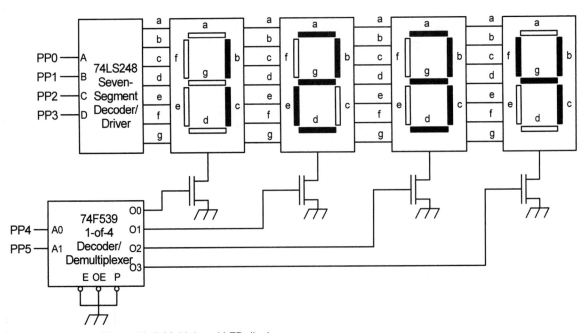

Figure 18-7 Multiplexed LED display.

digits is illuminated in turn by 4 bits from an output port or a decoder. The information on each display is output on the seven segment lines from a port or an active-high, seven-segment decoder such as a 74LS248. This is called a refreshed display, and if each display is turned on at a greater rate than about 20 Hz, our eyes will not detect any flickering. See Examples 18-10 and 18-11.

If your microcontroller has sufficient I/O lines the 74LS247 in Figure 18-6(a) and the 74LS248 and 74F539 chips in Figure 18-7 can be eliminated.

Example 18-10 Multiplexed LED Display Driver

```
Metrowerks HC12-Assembler
(c) COPYRIGHT METROWERKS 1987-2003


Abs. Obj. code Source line
---- --------- -----------
    1             ; *****************************************
    2             ; LED multiplexer example
    3             ; Input Registers: None
    4             ; Output Registers: None
    5             ; Registers modified: CCR
    6             ; Input variable data:
    7             ;   Thous = 1000's BCD digit
    8             ;   Huns = 100's BCD digit
    9             ;   Tens = 10's BCD digit
   10             ;   Ones = 1's BCD digit
   11             ; *****************************************
   12             ; Each time this routine is called the
   13             ; four BCD digits in the variable data
   14             ; are displayed.
   15             ; Assumptions:
   16             ; Port P, bits 3 - 0 are connected to a
   17             ; seven-segment decoder/driver and
   18             ; bits 5 - 4 to a 2-to-4 decoder.
   19             ; Port P is assumed to be initialized to
   20             ; be an output port.
   21             ; *****************************************
   22                        INCLUDE  portp.inc
   48 0000 0000 D1000:  EQU   %000000 ;Code to display 1000's
   49 0000 0010 D100:   EQU   %010000 ;Code to display 100's
   50 0000 0020 D10:    EQU   %100000 ;Code to display 10's
```

```
51 0000 0030 D1:     EQU   %110000 ;Code to display 1's
52            ; ****************************************
53            ; Define symbols used in the subroutine
54                      XDEF  led_mux, Thous, Huns,
55                      XDEF  Tens, Ones
56                      XREF  delay
57            ; ****************************************
58            ; Code Section
59            SubCode:  SECTION
60            led_mux:
61 36                  psha         ; Save the A reg
62            ; Display each digit in turn
63 B6xx xx              ldaa  Thous ; 1000's digit
64 8A00                oraa  #D1000
65 7A02 58             staa  PTP
66 16xx xx             jsr   delay ; Delay for 10 ms
67 B6xx xx             ldaa  Huns  ; 100's digit
68 8A10                oraa  #D100
69 7A02 58             staa  PTP
70 16xx xx             jsr   delay
71 B6xx xx             ldaa  Tens  ; 10's digit
72 8A20                oraa  #D10
73 7A02 58             staa  PTP
74 16xx xx             jsr   delay
75 B6xx xx             ldaa  Ones  ; 1's digit
76 8A30                oraa  #D1
77 7A02 58             staa  PTP
78 16xx xx             jsr   delay
79            ; Done now, restore A reg and return
80 32                  pula
81 3D                  rts
82
83            ; ****************************************
84            ; Variable Data Section
85            SubData:  SECTION
86            Thous:    DS.B  1   ; Thousands digit data
87            Huns:     DS.B  1   ; Hundreds digit data
88            Tens:     DS.B  1   ; Tens digit data
89            Ones:     DS.B  1   ; Ones digit data
```

Example 18-11 C Multiplexed LED Display Driver

```c
/******************************************************************
 * This sample program drives a four digit multiplexed
 * BCD LED display.
 * Port P, bits 0, 1, 2, and 3 are the BCD digit output
 * to the 74LS248 seven-segment decoder driver.
 * Port P, bits 4 and 5 is a two-bit code to multiplex
 * the display.
 * The input to the display routine is the 4 digit BCD
 * number - thousands, hundreds, tens, ones.
 * This function needs to be called repetitively to
 * refresh the display.
 ******************************************************************/
#include <mc9s12c32.h>                    /* derivative information */
typedef unsigned char BCD;
/* Define the decoder inputs */
#define DISP_1000 0x00  /* Display the 1000's digit */
#define DISP_100  0x10  /* Display the 100's digit */
#define DISP_10   0x20  /* Display the 10's digit */
#define DISP_1    0x30  /* Display the 1's digit */
void delay_X_ms( unsigned int X);     /* X millisecond delay */
/******************************************************************/
void led_mux( BCD thousands, BCD hundreds, BCD tens, BCD ones) {
/******************************************************************/
 /* Check to see if Port P, bits 0 - 5 are set to be output */
    if ((DDRP & 0x3F) != 0x3F) {
    /* THEN set the data direction register bits for output */
     DDRP = DDRP | 0x3F;  /* Set bits 5 - 0 and leave others alone */
  }
  /* Output the thousands digit */
    PTP = thousands + DISP_1000;
    delay_X_ms( 50 );
  /* Output the hundreds digit */
    PTP = hundreds + DISP_100;
    delay_X_ms( 50 );
  /* etc for the rest of the digits */
    PTP = tens + DISP_10;
    delay_X_ms( 50);
    PTP = ones + DISP_1;
    delay_X_ms( 50 );
}
```

18.4 Parallel I/O Expansion

Although microcontrollers have many parallel I/O lines, you may have an application that requires more I/O bits or some specialized I/O devices. One solution is to implement an expanded mode system with external I/O devices, or you might choose a different microcontroller with more I/O. Another solution, useful for simpler systems without a requirement for high-speed I/O, is shown in Figure 18-8.

One bidirectional port, such as port AD, can be used to emulate an external bidirectional data bus. Device selection, normally done by decoding an address, can be done by bits from a second, output port, such as port T. We saw in Chapter 11 that the fundamental component of an output port is a latch while for an input port it is a three-state gate. Figure 18-8 shows two 8-bit output latches and two 8-bit input three-state gates. You can expand the number of these input and output interfaces depending on the number of select signals available in port T. If more devices are needed, a decoder can be added to the output on port T. Example 18-14 shows modules to use for inputting and outputting data. (See Examples 18-12 and 18-13.)

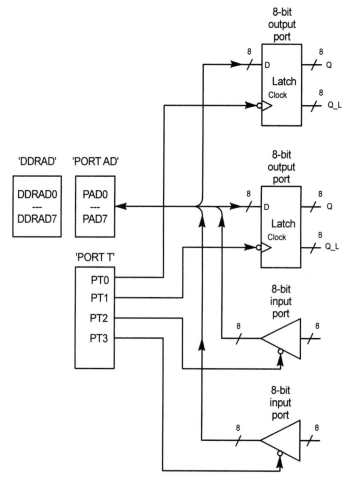

Figure 18-8 Parallel I/O expansion.

Example 18-12 Initialize Parallel I/O

The parallel I/O expansion shown in Figure 18-8 must have port T initialized. Write a short subroutine that enables port T bits 0–3 to be outputs.

Solution:

```
Metrowerks HC12-Assembler
(c) COPYRIGHT METROWERKS 1987-2003


Rel. Loc    Obj. code Source line
---- ---    --------- -----------
    1                 ; ****************************************
    2                 ; Initialize Port T output bits
    3                 ; Assembly calling:
    4                 ;        jsr  io_setup  ; Initialize I/O
    5                 ; C calling:
    6                 ; void io_setup( void );
    7                 ; ****************************************
    8                 ; Include files to define ports
    9                         INCLUDE bits.inc
   10                         INCLUDE base.inc
   11                         INCLUDE portt.inc
   12         0000 0001 PORT0_CK: EQU   BIT0 ; Latch Clock
   13         0000 0002 PORT1_CK: EQU   BIT1 ; Latch Clock
   14         0000 0004 PORT2_EN: EQU   BIT2 ; Three-state enable
   15         0000 0008 PORT3_EN: EQU   BIT3 ; Three-state enable
   16                 ; ****************************************
   17                 ; Define symbols used in the subroutine
   18                         XDEF  io_set_up
   19                 ; ****************************************
   20                 ; Code Section
   21                 SubCode:  SECTION
   22                 ; ****************************************
   23                 ; io_set_up:
   24                 ; This module sets Port T bits 0 - 3 to be
   25                 ; outputs and initializes them high
                        because the
   26                 ; enables are active low.
   27                 ; Input registers: None
   28                 ; Output registers: None
   29                 ; Registers modified: CCR
```

```
30                         ; Data used: None
31                         ; ****************************************
32                         io_set_up:
33                         ; Set the Port T bits high first
34 000000 1C02 400F             bset   PTT,PORT0_CK|PORT1_CK|
                                       PORT2_EN|PORT3_EN
35                         ; Now enable the output
36 000004 1C02 420F             bset   DDRT,PORT0_CK|PORT1_CK|
                                       PORT2_EN|PORT3_EN
37 000008 3D                    rts
38                         ; ****************************************
```

Example 18-13 Parallel I/O Input

Write a subroutine to input data from either of the two input interfaces shown in Figure 18-8.

Solution:

```
Metrowerks HC12-Assembler
(c) COPYRIGHT METROWERKS 1987-2003


Rel. Loc    Obj. code Source line
---- ---    --------- -----------

   1                  ; ****************************************
   2                  ; Get data from expanded parallel I/O port
   3                  ; Assembly calling:
   4                  ;         ldab   #Port_Num
   5                  ;         jsr    get_port  ; Get 8=bit value
   6                  ;   8-bit data returned in B
   7                  ; C calling:
   8                  ; char get_port( unsigned char Port_Num );
   9                  ; ****************************************
  10                  ; Include files to define ports
  11                          INCLUDE bits.inc
  12                          INCLUDE base.inc
  13                          INCLUDE portad.inc
  5i                            INCLUDE "bits.inc"
  14                          INCLUDE portt.inc
```

```
15        0000 0001 PORT0_CK: EQU   BIT0 ; Latch Clock
16        0000 0002 PORT1_CK: EQU   BIT1 ; Latch Clock
17        0000 0004 PORT2_EN: EQU   BIT2 ; Three-state enable
18        0000 0008 PORT3_EN: EQU   BIT3 ; Three-state enable
19        0000 0000 PORT0:    EQU   0    ; Port numbers
20        0000 0001 PORT1:    EQU   1
21        0000 0002 PORT2:    EQU   2
22        0000 0003 PORT3:    EQU   3
23                           ; ****************************************
24                           ; Define symbols used in the subroutine
25                                    XDEF  get_port
26                           ; ****************************************
27                           ; Code Section
28                           SubCode:  SECTION
29                           ; ****************************************
30                           ; get_port:
31                           ; This module reads the 8-bit input device
32                           ; and returns the data in the B register
33                           ; Input: B = Port Address
34                           ; Output: B = 8-bit data
35                           ; Registers modified: B, CCR
36                           ; Data used: None
37                           ; ****************************************
38                           get_port:
39                           ; Make Port AD input
40 000000 1D02 72FF          bclr  DDRAD,ALLBITS
41                           ; Enable the three-state gates
42                           ; IF B = PORT2
43 000004 C102               cmpb  #PORT2
44 000006 2606               bne   check_port3
45                           ; THEN enable PORT2
46 000008 1D02 4004          bclr  PTT,PORT2_EN
47 00000C 2009               bra   get_data
48                           ; ELSE check for PORT3
49                           check_port3:
50 00000E C103               cmpb  #PORT3
51 000010 2608               bne   endif_get ; Port 2 or 3 not
                                   specified
52                           ; Enable PORT3
```

```
53 000012 1D02 4008          bclr  PTT,PORT3_EN
54 000016 A7                 nop
55                     ; Read the switches
56                     get_data:
57 000017 F602 70            ldab  PTAD
58                     ; Disable the three-state gates
59                     endif_get:
60 00001A 1C02 4003          bset  PTT,PORT2|PORT3
61 00001E 3D                 rts
62                     ; ****************************************
```

Example 18-14 Parallel I/O Output

Write a subroutine to output data to either of the two output interfaces shown in Figure 18-8.

Solution:

```
Metrowerks HC12-Assembler
(c) COPYRIGHT METROWERKS 1987-2003


Rel. Loc    Obj. code Source line
---- ---     --------- -----------
   1                  ; ****************************************
   2                  ; Expanded Parallel I/O Example
   3                  ; Put data to expanded parallel I/O port
   4                  ; Assembly calling:
   5                  ;    ldab  Data
   6                  ;    pshb              ; Transfer data on stack
   7                  ;    ldab #Port_Num
   8                  ;    jsr   put_port  ; Output 8-bit value
   9                  ;    pulb              ; Restore stack pointer
  10                  ; C calling:
  11                  ; void put_port( char data, unsigned char
                        Port_Num );
  12                  ; ****************************************
  13                  ; put_port:
  14                  ; This module outputs the B register to the
  15                  ; 8-bit output latch
  16                  ; Input: B = port for output
  17                  ;        Data to output on the stack
```

```
18                      ;           (CodeWarrior calling convention)
19                    ; Output: None
20                    ; Registers modified: CCR
21                    ; Data used: None
22                    ; ****************************************
23                    ; Include files to define ports
24                            INCLUDE bits.inc
25                            INCLUDE base.inc
26                            INCLUDE portad.inc
 5i                             INCLUDE "bits.inc"
27                            INCLUDE portt.inc
28        0000 0001 PORT0_CK: EQU   BIT0 ; Latch Clock
29        0000 0002 PORT1_CK: EQU   BIT1 ; Latch Clock
30        0000 0004 PORT2_EN: EQU   BIT2 ; Three-state enable
31        0000 0008 PORT3_EN: EQU   BIT3 ; Three-state enable
32        0000 0000 PORT0:   EQU   0    ; Port numbers
33        0000 0001 PORT1:   EQU   1
34        0000 0002 PORT2:   EQU   2
35        0000 0003 PORT3:   EQU   3
36                    ; ****************************************
37                    ; Define symbols used in the subroutine
38                            XDEF   put_port
39                    ; ****************************************
40                    ; Code Section
41                    SubCode:  SECTION
42                    ; ****************************************
43                    put_port:
44                    ; Make Port AD output
45 000000 1C02 72FF        bset   DDRAD,ALLBITS
46                    ; Output the data from the stack
47 000004 180D 8202        movb   2,sp,PTAD
   000008 70
48                    ; Strobe the latch active low clock
49                    ; IF B = PORT0
50 000009 C100            cmpb   #PORT0
51 00000B 260B            bne    check_port1
52                    ; THEN strobe the latch signal low
53 00000D 1D02 4001        bclr   PTT,PORT0_CK
54 000011 A7              nop
55 000012 1C02 4001        bset   PTT,PORT0_CK
56 000016 200D            bra    endif_put
```

```
57                          ; ELSE check for PORT1
58                          check_port1:
59 000018 C101                   cmpb   #PORT1
60 00001A 2609                   bne    endif_put ; If not PORT1 error
61                          ; Strobe the latch signal low
62 00001C 1D02 4002              bclr   PTT,PORT1_CK
63 000020 A7                     nop
64 000021 1C02 4002              bset   PTT,PORT1_CK
65                          ; Make Port AD input
66                          endif_put:
67 000025 1D02 72FF              bclr   DDRAD,ALLBITS
68 000029 3D                     rts
69                          ; *************************************
```

EXPANDED I/O ROUTINES IN C

These routines are presented in Examples 18-15 to 18-18.

Example 18-15 Initialize Expanded I/O in C
The parallel I/O expansion shown in Figure 18-8 must have port T initialized. Write a short function in C that enables port T bits 0–3 to be outputs.

Solution:

```c
/****************************************************************
 * Initialize Port T output bits
 * void io_setup( unsigned char PTTout );
 * where PTTout is a byte with 1's in the output bit positions
 ***************************************************************/
#include <mc9s12c32.h>              /* derivative information */
/***************************************************************/
void io_setup( unsigned char PTTout ) {
/***************************************************************/
  /* Set the Port T bits high first before setting the direction */
  PTT = PTTout;
  /* Set the direction bits in DDRT */
  DDRT = PTTout;
}
```

Example 18-16 Expanded I/O Input in C

Write a function to input data from either of the two input interfaces shown in Figure 18-8.

Solution:

```
/*****************************************************************
 * Get data from the expanded I/O port
 * char get_port( unsigned char Port_Num );
 * This function gets the 8-bit value from the expanded
 I/O port.
 * the valid port numbers (Port_Num) are 2 and 3
 *****************************************************************/
#include <mc9s12c32.h>                /* derivative information */
/*****************************************************************/
char get_port( unsigned char Port_Num ) {
volatile char port_data;
/*****************************************************************/
  /* Make sure Port AD is an input port */
    DDRAD = 0;
    ATDDIEN = 0xFF;
  /* Make sure a valid port number is given */
    switch ( Port_Num ) {
      case 2: {
        /* Enable the three-state gate and get the data */
        PTT_PTT2 = 0;           /* Active low enable */
        port_data = PTAD;     /* Read the data */
      }
      case 3: {
        PTT_PTT3 = 0;                    /* Active low enable */
        port_data = PTAD;         /* Read the data */
      }
    }
  /* Disable the three-state gates */
    PTT_PTT2 = 1;
    PTT_PTT3 = 1;

    return( port_data );
}
```

Example 18-17 Expanded I/O Output in C

Write a function to output data to either of the two output interfaces shown in Figure 18-8.

Solution:

```
/*****************************************************************
* Put data to the expanded I/O port
* void put_port( char data, unsigned char Port_Num);
* This function puts an 8-bit value to the expanded
I/O port.
* the valid port numbers (Port_Num) are 0 and 1
*****************************************************************/
#include <mc9s12c32.h>              /* derivative information */
/*****************************************************************/
void put_port( char data, unsigned char Port_Num ) {
/*****************************************************************/
  /* Make sure Port AD is an output port */
    DDRAD = 0xFF;
  /* Make sure a valid port number is given */
    switch ( Port_Num ) {
      case 0: {
        /* output the data to the port */
        PTAD = data;
        /* Strobe the latch signal active low and then high */
        PTT_PTT0 = 0;
        asm ( nop );    /* Do a little time */
        PTT_PTT0 = 1;
      }
      case 1: {
        PTAD = data;
        /* Strobe the latch signal active low and then high */
        PTT_PTT1 = 0;
        asm ( nop );    /* Do a little time */
        PTT_PTT1 = 1;
      }
    }
    /* Make PTAD an input again */
    DDRAD = 0;
  }
```

Example 18-18 Expanded I/O Test Program
Write a short program to test the expanded I/O routines in Examples 18-15 to 18-17.

Solution:

```
/****************************************************************
 * Expanded I/O port test software
 * Gets a byte from port 3 and outputs it to port 1
 ****************************************************************/
#include <mc9s12c32.h>                 /* derivative information */
void put_port( char data, unsigned char Port_Num);
void io_setup( unsigned char PTTout );
char get_port( unsigned char Port_Num );
/****************************************************************/
void main ( void ) {
char ByteData;
/****************************************************************/
  /* Initialize the I/O */
  io_setup( 0x0F );
  for ( ; ;) {
    ByteData = get_port( 3 );
    put_port( ByteData, 1 );
  }
}
```

Parallel I/O Electronics

Some additional design is needed to interface our somewhat fragile electronics to the real, sometimes cruel, world. We must take care to protect our electronics from overvoltages and static discharges and to provide signal levels compatible with the logic circuits we are using. Although good printed circuit board design, shielding, and power supply design are outside the scope of this text, here are some simple interface circuits for digital input and output.

Figure 18-9 shows a simple input interface. The two 1N4001 (or similar) diodes limit the voltage excursion on the digital input signal to a maximum of a diode drop higher than V_{DD}

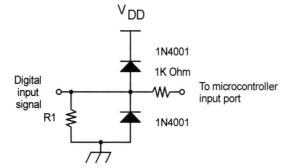

Figure 18-9 Digital input.

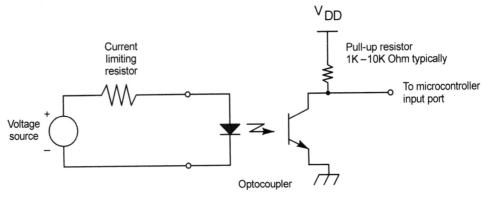

Figure 18-10 Optocoupler digital input.

and to a minimum of one diode drop below ground. If the input signal is high frequency with fast rise and fall times, resistor R1 can provide an impedance match to the driving circuit. It may be eliminated for low-frequency signals. The 1 kΩ series resistor provides some current limiting and further protection for the microcontroller's input pin.

Figures 18-10 is a circuit that can be used when the input signal is derived from a high voltage source. The optocoupler is a light-emitting diode with a photo transistor. The current limiting resistor is designed to provide the correct current for the LED. Notice that there is no physical connection between the signal voltage source and the microcontroller. This is a big advantage when interfacing to high voltage, high power, noisy circuits. The optocoupler can be used as an output interface as well by simply turning it around and connecting the LED to the microcontroller output port.

Figures 18-11 and 18-12 show how to use a bipolar junction transistor (2N2222) and a field effect transistor (2N7000) to output digital values. In each of these cases the output is an open collector or open drain. If you wish to have a logic level at this output you will have to add a pull-up resistor.

Microcontrollers often have to drive relays. Figure 18-13 shows a relay driver using the 2N2222 transistor. Relays often have multiple contacts, some of them normally open (NO) and some normally closed (NC). Is this case, "normally" refers to the relay not being energized. An important addition to this circuit is the 1N4001 clamp diode across the relay coil. This limits the voltage spike produced by the coil when the relay is de-energized to one diode drop greater than V.

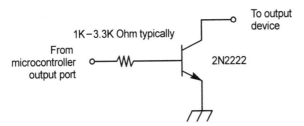

Figure 18-11 Transistor output buffer.

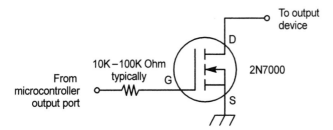

Figure 18-12 FET output buffer.

18.5 Serial I/O Devices

The serial I/O interfaces described in Chapter 15 (Serial Communications Interface—SCI— and Serial Peripheral Interface—SPI) and the Controller Area Network (MSCAN) interface discussed in Chapter 16 are useful I/O interfaces. They are attractive because only a few wires are needed to input and output data compared to the parallel I/O interfaces.

Serial Communications Interface

The SCI port described in Chapter 15 produces an asynchronous serial communications data stream. In the personal computer world this is called the *com* port and many PCs have one or more com ports, although the Universal Serial Bus (USB) is taking over many of the jobs the serial com port used to do. Nonetheless, the SCI is a useful method of transporting data over long distances using only three wires (at a minimum).

Often the serial port is called an RS-232-C interface and this refers to a standard interface specification that defines, among other element, the logic levels defined for one and zero,

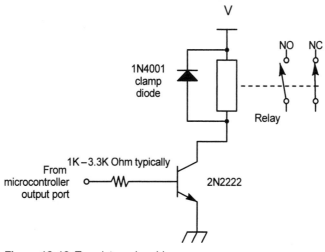

Figure 18-13 Transistor relay driver.

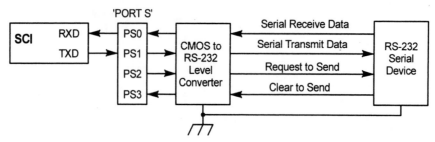

Figure 18-14 HCS12 Serial Communications Interface (SCI).

signal names for data and handshaking signals, and connectors to be used (Figure 18-14). The standard defines two kinds of devices—*Data Terminal Equipment (DTE)* and *Data Communications Equipment (DCE)*.

DTE AND DCE

The signal flow directions defined in Table 18-2 are based on the signal flow defined for a DTE device and computers often are configured as Data Terminal Equipment devices. For example, the transmitted data pin—TxD—is being sourced by a DTE device. Data Communications Equipment (DCE) devices include modems and some printers. The signal flow and signal names defined for DCE devices are often misapplied. For example, the signal Transmitted

TABLE 18-2 RS-232-C Signal Definitions

DE9	DB25	Signal	Purpose
	1	PG	*Protective Ground.* This is usually the shield in a shielded cable. It is designed to be connected to the equipment frame and may be connected to external grounds.
3	2	TxD	*Transmitted Data.* **Sourced** by DTE and **received** by DCE. Data terminal equipment cannot send unless RTS, CTS, DSR, and DTR are asserted.
2	3	RxD	*Received Data.* **Received** by DTE, **sourced** by DCE.
7	4	RTS	*Request to Send.* **Sourced** by DTE, **received** by DCE. RTS is asserted by the DTE when it wants to send data. The DCE responds by asserting CTS.
8	5	CTS	*Clear to Send.* **Sourced** by DCE, **received** by DTE. CTS must be asserted before the DTE can transmit data.
6	6	DSR	*Data Set Ready.* **Sourced** by DCE, **received** by DTE. Indicates that the DCE has made a connection on the telephone line and is ready to receive data from the terminal. The DTE must see this asserted before it can transmit data.
5	7	SG	*Signal Ground.* Ground reference for the signal is separate from pin 1, protective ground.
1	8	DCD	*Data Carrier Detect.* **Sourced** by DCE, **received** by DTE. Indicates that a DCE has detected the carrier on the telephone line. Originally it was used in half-duplex systems but can be used in full-duplex systems too.
4	20	DTR	*Data Terminal Ready.* **Sourced** by DTE, **received** by DCE. Indicates the DTE is ready for sending or receiving.
9	22	RI	*Ring Indicator.* **Sourced** by DCE, **received** by DTE. Indicates that a ringing signal is detected.

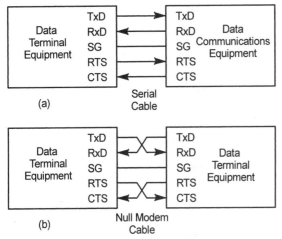

Figure 18-15 Serial communications: (a) DTE–DCE and (b) DTE–DTE.

Figure 18-16 RS-232-C tester.

Data (TxD) is actually *received* by the DCE device. When connecting one device to another, we must be sure what kind of devices are being used and we must select the proper cable. See Figure 18-15.

A very useful tool to have when working with RS-232-C interface devices is the RS-232-C tester shown in Figure 18-16. This tester shows what serial lines are active and allows us to easily determine if we are connecting to a DTE or DCE device.

SERIAL INTERFACE ELECTRICAL SPECIFICATIONS

There are four electrical specifications in use for interconnecting serial interfaces. These are shown in Table 18-3. The two most widely used are RS-232-C and RS-485. The latter offers much higher data rates over longer distances than the RS-232-C standard. It is less widespread in use, however.

TABLE 18-3 Summary of RS-232-C, RS-423, RS-422, and RS-485 Standards

Specification	RS-232-C	RS-423	RS-422	RS-485
Receiver input voltage	±3 to ±15 V	±200 mV to ±12 V	±200 mV to ±7 V	±200 mV to −7 to +12 V
Driver output signal	±5 to ±15 V	±3.6 to ±6 V	±2 to ±5 V	±1.5 to ±5 V
Maximum data rate	20 kb/s	100 kb/s	10 Mb/s	10 Mb/s
Maximum cable length	50 feet	4000 feet	4000 feet	4000 feet
Driver source impedance	3–7 kΩ	450 Ω min	100 Ω	54 Ω
Receiver input resistance	3 kΩ	4 kΩ	4 kΩ min	12 kΩ
Mode	Single-ended	Single-ended	Differential	Differential
Number of drivers and receivers allowed on one line	1 Driver 1 Receiver	1 Driver 10 Receivers	1 Driver 10 Receivers	32 Drivers 32 Receivers

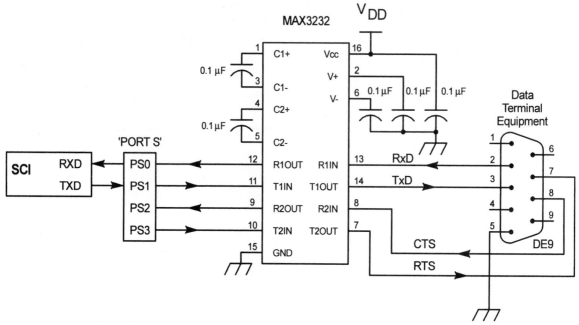

Figure 18-17 SCI with RS-232-C interface.

Each of the electrical standards shown in Table 18-3 requires a level converter to translate the TTL or CMOS logic levels of the microcontroller's serial data input and output lines to the voltages specified by the standard. Figure 18-17 shows the SCI connected to a MAX3232 CMOS-to-RS-232 level converter. This chip provides two conversion channels for both input and output, allowing the RTS and CTS handshaking signals to be implemented on PS-2 and PS-3 if needed in the application. Return to Chapter 15 to inspect C callable serial I/O routines.

LOW VOLTAGE, DIFFERENTIAL SIGNALING (LVDS)

Another electrical interface being used increasingly for higher speed serial data networks in both on-board and off-board applications is *Low Voltage, Differential Signaling (LVDS)*. This interface is similar to RS-485 in that differential transmitters and receivers are used. Differential line drivers can operate at much higher speeds because the differential line pair is relatively immune to common mode noise. Data rates up to 2 gigabits/second are possible with this technology.

The LVDS is a standard promoted as ANSI/TIA/EIA-644-A.[2] Unlike RS-232-C, the standard does not include functional specifications and protocol or cable characteristics. It does specify a differential line driver and receiver configuration as shown in Figure 18-18.

Serial Peripheral Interface

The *Serial Peripheral Interface (SPI)* discussed in Chapter 15 is a convenient way to provide additional I/O capabilities without using scarce microcontroller parallel I/O resources.

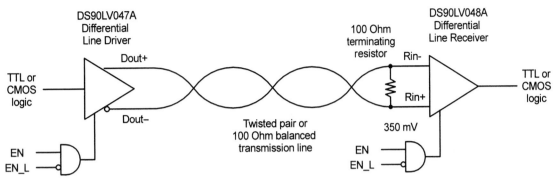

Figure 18-18 Low Voltage, Differential Signaling (LVDS) interface.

DIGITAL-TO-ANALOG OUTPUT

While the MC9S12C32 microcontroller, like many microcontrollers, has an analog-to-digital converter input port, it does not have a corresponding digital-to-analog converter for analog output signals. Figure 18-19 shows a MAX512,[3] three-channel digital-to-analog converter. It interfaces to the SPI and, as shown, outputs two analog channels on OUTA and OUTB. The third channel, OUTC, and the latched digital output, LOUT, are not used in this application. (See Examples 18-19 and 18-20.)

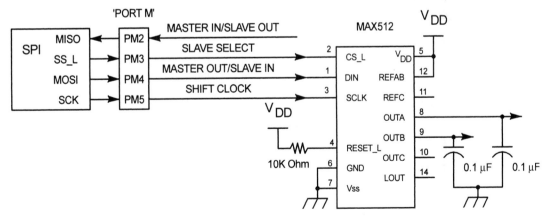

Figure 18-19 SPI with Maxim 512 D/A converter.

---

[2] ANSI = American National Standards Institute; TIA = Telecommunications Industry Association; EIA = Electronic Industries Association.

[3] *Maxim MAX512/MAX513 Low-Cost, Triple, 8-bit Voltage-Output DACs with Serial Interface.* http://www.maxim-ic.com.

Example 18-19 Digital-to-Analog Output Using the SPI

```
Metrowerks HC12-Assembler
(c) COPYRIGHT METROWERKS 1987-2003


Rel. Loc    Obj. code Source line
---- ---     --------- -----------
    1                   ;******************************************
    2                   ; Serial Peripheral Interface Example
    3                   ; This program continuously outputs a sawtooth
    4                   ; wave to the SPI port, which is connected
    5                   ; to a Maxim MAX512 serial D/A converter.
    6                   ;******************************************
    7                   ;******************************************
    8                   ; Define the entry point for the main program
    9                         XDEF   Entry, main
   10                         XREF   __SEG_END_SSTACK ; Note double
                                                      underbar
   11                   ;******************************************
   12                   ; Include files
   13                         INCLUDE bits.inc
   14                         INCLUDE portm.inc
   15                         INCLUDE spi.inc
   45i
   16                   ;******************************************
   17                   ; Constants definitions
   18                   ; Maxim D/A control byte
   19         0000 00B1 DASETUP0:EQU  %10110001  ; Output A enabled
   20                   ;******************************************
   21                   ; Code Section
   22                   MyCode: SECTION
   23                   Entry:
   24                   main:
   25                   ;******************************************
   26                   ; Initialize stack pointer register
   27 000000 CFxx xx         lds    #__SEG_END_SSTACK
   28                   ; Initialize I/O
   29                   ; DO
   30                   ;   SPI Setup
   31                   ;   Set Port M direction to control SS_L output
   32 000003 1C02 5208       bset   DDRM,SS
```

```
33                        ;    Enable the SPI in master mode
34 000007 8650                     ldaa    #SPE|MSTR
35 000009 5AD8                     staa    SPICR1
36                        ;    Set the SCLK to 4 MHz
37 00000B 8600                     ldaa    #$00
38 00000D 5ADA                     staa    SPIBR
39                        ;    Initialize the output data
40 00000F 79xx xx                    clr    sawtooth
41                   ; ENDO Initialize
42                   ; Main Loop
43                   ; Continuously output a sawtooth wave
44                   main_loop:
45                   ; Set SS low to select the D/A
46 000012 1D02 5008                bclr    PTM,SS
47                   ; Send the first byte to the D/A
48                        ;    Write an 8-bit control word to the serial
49                        ;    D/A converter before writing the data
50 000016 86B1                      ldaa    #DASETUP0  ; Write control
                                                          word to D/A
51 000018 5ADD                      staa   SPIDR
52                        ;    Spin loop to do nothing until the data
                              is shifted
53                        ;    out onto the serial line (MOSI).  Wait
                              for SPIF.
54 00001A 4FDB 80FC SPIN:   brclr SPISR,SPIF,SPIN
55                        ;    Clear the flag by reading the status
56                        ;    register and the data register
57 00001E 96DB                     ldaa    SPISR       ; Status register
58 000020 96DD                     ldaa    SPIDR       ; Data register
59                   ; Send the second byte to the D/A
60                        ;    Get the current sawtooth data value
61 000022 F6xx xx                    ldab    sawtooth
62                        ;    and store it into the serial shift
                              register
63 000025 5BDD                     stab    SPIDR
64                        ;    Increment the sawtooth value for the
                              next time
65 000027 72xx xx                    inc    sawtooth
66                        ;    Spin loop to do nothing until the data
                              is shifted
67                        ;    out onto the serial line (MOSI)
68 00002A 4FDB 80FC SPIN2:  brclr SPISR,SPIF,SPIN2
```

```
69 00002E 96DB                 ldaa   SPISR     ; Clear the serial
                                                   intr flag
70 000030 96DD                 ldaa   SPIDR
71                    ; Raise the SS line to tell the D/A converter
72                    ; to output the data
73 000032 1C02 5008            bset   PTM,SS
74
75                    ; FOREVER
76
77 000036 20DA                 bra    main_loop
78                    ;*****************************************
79                    MyData: SECTION
80                    ; Place variable data here
81 000000            sawtooth: ds.b   1
```

Example 18-20 Digital-to-Analog Converter with C

```c
/*******************************************************************
 * Sample Serial Peripheral Interface Example
 * This program continuously outputs a sawtooth wave to the SPI
 * port, which is connected to a Maxim MAX512 serial D/A converter
 ******************************************************************/
#include <mc9s12c32.h>                /* derivative information */
#define DA_SETUP 0b10110001           /* Output A enabled */
/******************************************************************/
void main(void) {
volatile char temp;
unsigned char sawtooth;
/******************************************************************/
  /* Initialize I/O */
  /* Set Port M direction to be able to control SS_L output */
    DDRM_DDRM3 = 1;
  /* Enable SPI and set master mode */
    SPICR1 =  SPICR1_SPE_MASK | SPICR1_MSTR_MASK;
  /* Set the SCK to 4 MHz */
    SPIBR = 0;
  /* Initialize the sawtooth data */
```

```
        sawtooth = 0;
    /* DO */
      for(;;) {
      /* Set SS_L low to select the D/A */
        PTM_PTM3 = 0;
      /* Send the first byte to the D/A
         * Write an 8-bit control word to the serial D/A
         * converter before writing the data */
        SPIDR = DA_SETUP;
      /* Wait until the byte is shifted out by waiting for SPIF */
        while ( SPISR_SPIF == 0 ) { };
      /* Send the second byte to the D/A and increment the value */
        SPIDR = sawtooth++;
      /* Wait until the byte is shifted out by waiting for SPIF */
        while ( SPISR_SPIF == 0 ) { };
      /* Clear the SPI flags */
        temp = SPISR;
        temp = SPIDR;
      /* Raise the SS_L line to tell the D/A to output the data */
        PTM_PTM3 = 1;

      } /* wait forever */
    }
```

EXPANDING PARALLEL I/O WITH THE SPI

Section 18.4 showed how to expand the parallel I/O capabilities using a bidirectional port as a pseudo-data bus and another port for enabling input ports or latching output ports. With serial-in/parallel-out and parallel-in/serial-out shift registers you can add more parallel input and output bits almost indefinitely. Figure 18-20 shows a 74HC595 8-bit serial-in/serial-or-parallel-out shift register in use for additional output lines. Although this examples shows only 8 bits, the serial-out pin (Q7′) can be used as the serial input for another, cascaded 8-bit port. Figure 18-21 shows adding eight input bits with a 74HC165 parallel-in/serial-out 8-bit shift register. It too can be expanded to provide more input bits by cascading additional chips.

LIQUID CRYSTAL DISPLAY

There are many inexpensive liquid crystal displays that make excellent display devices for embedded systems. Most have at least two ways to interface to the microcontroller including 4-bit and 8-bit parallel connections.

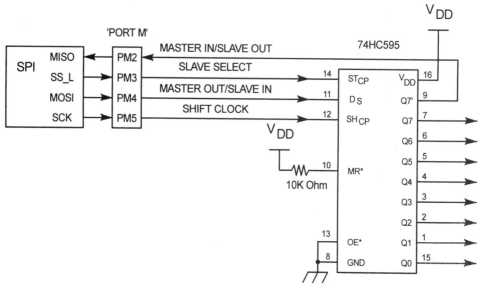

Figure 18-20 Adding parallel output with the SPI.

Figure 18-22 shows an LCD module connected to the Student Learning Kit. To reduce the parallel I/O bits needed to drive the LCD, a 74HC595 serial-in/parallel-out shift register is connected to the SPI port on the microcontroller. Figure 18-23 illustrates the hardware design and Example 18-21 shows software to display characters on the LCD.

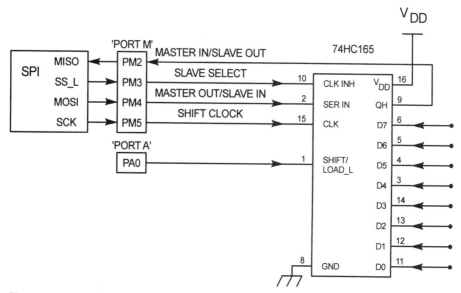

Figure 18-21 Adding parallel input with the SPI.

Figure 18-22 LCD module and Student Learning Kit.

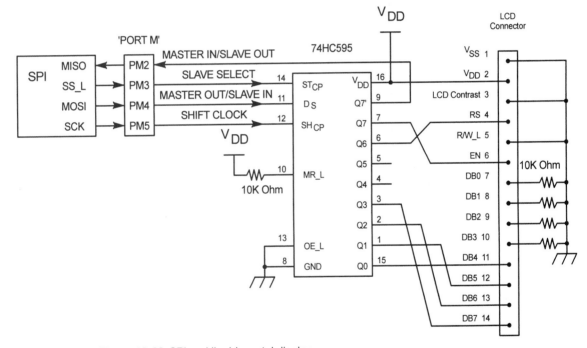

Figure 18-23 SPI and liquid crystal display.

Example 18-21 Liquid Crystal Display Drivers

```
/*****************************************************************
 * LCD Display Program
 *****************************************************************/
#include <mc9s12c32.h>    /* derivative information */
/*****************************************************************
 * Define the commands to be sent to the LCD
```

```
    ****************************************************************/
#define EN 0x80
#define RS 0x40
#define DB7 0x80
#define FSET_8_BIT 0x03
#define FSET_4_BIT 0x02
#define FSET_4_LINE  0x28
#define FSET_D_OFF   0x08
#define FSET_CLEAR   0x01
#define FSET_ENTRY   0x06
#define FSET_D_ON    0x0E
#define FSET_CUR_OFF 0x0C
#define HOME   0x02
/*****************************************************************
 * Define an array to be used for addressing each line
 *****************************************************************/
unsigned char LINE[4][20] = {
{0,1,2,3,4,5,6,7,8,9,10,11,12,13,14,15,16,17,18,19},
{64,65,66,67,68,69,70,71,72,73,74,75,76,77,78,79,80,81,82,83},
{20,21,22,23,24,25,26,27,28,29,30,31,32,33,34,35,36,37,38,39},
{84,85,86,87,88,89,90,91,92,93,94,95,96,97,98,99,100,101,102,103}};
/*****************************************************************/
void delay_X_ms( unsigned int );       /* Variable Delay */
void spi_send_byte( unsigned char );   /* Send a byte */
void lcd_put_command( unsigned char ); /* Send a command */
void init_spi( void );                 /* Initialize the SPI */
void lcd_init( void );                 /* Initialize the LCD */
void lcd_print( char *str_pointer );   /* Print a string */
void lcd_move_cursor( unsigned char, unsigned char ); /* Move
                                                       cursor */

/*****************************************************************
 * Test program printing a message on the LCD
 *****************************************************************/

void main(void) {
  /* Initialize the SPI and the LCD */
    lcd_init();
  /* Print the message below */
    lcd_move_cursor( 1,1 );                  /* Line 1 */
    lcd_print( "**** MC9S12C32 ****\0");
```

```
      lcd_move_cursor( 2,1 );                 /* Line 2 */
      lcd_print( "* Microcontrollers *\0" );
      lcd_move_cursor( 3,1 );                 /* Line 3 */
      lcd_print( "*  ROCK at Montana **\0" );
      lcd_move_cursor( 4,1 );                 /* Line 4 */
      lcd_print( "* State University *\0");
  /* Turn cursor off */
      lcd_put_command( FSET_CUR_OFF );
      for(;;) {
        } /* wait forever */
}


/******************************************************************
 * Initialize the SPI
 ******************************************************************/
void init_spi( void ) {
/******************************************************************/
  /* Initialize the SPI */
  /* Enable SPI in master mode */
    SPICR1 = SPICR1_SPE_MASK|SPICR1_MSTR_MASK|SPICR1_SSOE_MASK;
    SPICR2 = SPICR2_MODFEN_MASK;
  /* Set the SCK to 4 MHz */
    SPIBR = 0;
}


/******************************************************************
 * Initialize the LCD
 ******************************************************************/
void lcd_init( void ){
/******************************************************************/
  /* Initilize the SPI */
    init_spi();
  /* Initialize the LCD */
  /* Delay 15 ms in case the power just came on */
    delay_X_ms( 15 );
  /* Send the first command */
    lcd_put_command( FSET_8_BIT );
  /* Delay 5 ms */
    delay_X_ms( 5 );
  /* Send the first command again */
    lcd_put_command( FSET_8_BIT );
```

```
        /* Delay > 100 us */
          delay_X_ms( 1 );
        /* Send the first command again */
          lcd_put_command( FSET_8_BIT );
        /* Set the interface to 4 bits */
          lcd_put_command( FSET_4_BIT );
        /* Set interface to 4 line, 5x7 chars */
          lcd_put_command( FSET_4_LINE );
        /* Set display off */
          lcd_put_command( FSET_D_OFF );
        /* Clear display */
          lcd_put_command( FSET_CLEAR );
        /* Set entry mode */
          lcd_put_command( FSET_ENTRY );
        /* Turn display on */
          lcd_put_command( FSET_D_ON );
}


/******************************************************************
 * Send a byte to the SPI
 ******************************************************************/
void spi_send_byte( unsigned char spidata ){
  volatile unsigned char status, data;
/******************************************************************/
      SPIDR = spidata;
    /* Wait until the data has shifted out */
      while( SPISR_SPIF == 0 );
    /* Clear the flag by reading the status register
     * and then the data register */
      status = SPISR;
      data = SPIDR;
}


/******************************************************************
 * Send byte to display to the LCD
 ******************************************************************/
void lcd_put_data( unsigned char character ){
  static unsigned char msn, lsn;
/******************************************************************/
    /* Input is an ASCII character to display */
    /* Split into two nibbles and send the ms nibble first */
```

```
    msn = (character >> 4) | RS;    /* Setting the RS high */
    lsn = (character & 0x0f) | RS;
  /* Send the most significant nibble */
    spi_send_byte( msn  );
    spi_send_byte( msn | EN );    /* Set the enable high */
    spi_send_byte( msn );
    delay_X_ms( 1 );
  /* Send the least significant nibble */
    spi_send_byte( lsn );
    spi_send_byte( lsn | EN );
    spi_send_byte( lsn );
    delay_X_ms( 1 );
}


/****************************************************************
 * Send command byte to the LCD
 ****************************************************************/
void lcd_put_command( unsigned char character ){
  static unsigned char msn, lsn;
/****************************************************************/
  /* Input is an ASCII character to display */
  /* Split into two nibbles and send the ms nibble first */
    msn = (character >> 4);
    lsn = (character & 0x0f);
  /* Send the most significant nibble */
    spi_send_byte( msn  );
    spi_send_byte( msn | EN );    /* Set the enable high */
    spi_send_byte( msn );
    delay_X_ms( 2 );
  /* Send the least significant nibble */
    spi_send_byte( lsn );
    spi_send_byte( lsn | EN );
    spi_send_byte( lsn );
    delay_X_ms( 2 );
}
/****************************************************************
 * Print a null terminated string on the LCD
 ****************************************************************/
void lcd_print( char *str_pointer ){
/****************************************************************/
  /* Print a null terminated string on the display */
```

```
        while (*str_pointer != 0){
        lcd_put_data( *str_pointer++ );
    }
    return;
}


/***************************************************************
 * Move cursor to a line, column.
 * Input is the line number 1 - 4, column number 1 - 20
 * If line is not 1 - 4, line is set to 1.
 * If column is not 1 - 20, column is set to 1
 ***************************************************************/
void lcd_move_cursor( unsigned char line, unsigned char column ) {
    /***************************************************************/
    /* Check for line 1-4 */
    if (line < 1 || line > 4) line = 0;
    else line = line - 1;
    /* Check for column 1 - 20 */
    if (column < 1 || column > 20) column = 0;
    else column = column - 1;
    lcd_put_command(LINE[line][column] | DB7);
}
```

SPI DEVICES

There are a wide variety of devices that can be interfaced to your microcontroller using the SPI bus. Table 18-4 shows a sample of what you can find just by searching a few manufacturers.

18.6 The Analog-to-Digital Converter Interface

Figure 18-24 shows a typical A/D interface circuit. This particular circuit is for a sensor that produces a voltage in the range of 0–2.9 volts. The Texas Instruments OPA4344 is a quad (there are four operational amplifiers in one package), rail-to-rail amplifier. This means it can operate on a single 5-volt power supply and still achieve an output voltage of nearly 5 volts. The gain of this noninverting amplifier is given by

$$A_V = \frac{R_1 + R_2}{R_1} = 2.8$$

TABLE 18-4 SPI Devices

Device	Company
CPU Supervisor with 4K–64K bit EEPROM	Intersil
Digitally Controlled Pot with Voltage Comparator	Intersil
Digital-to-Analog Converter	Intersil
CMOS Serial 8-bit I/O Port	Intersil
CMOS Serial 10-bit A/D Converter	Intersil
24-Bit, High Precision, Sigma Delta A/D Converter	Intersil
CMOS Serial Real-Time Clock with RAM and Power Sense/Control	Intersil
8-Channel, 16-bit, High Precision, Sigma-Delta A/D Subsystem	Intersil
3 Cell Li-Ion Battery Protection and Monitor IC	Intersil
18-Channel TFT-LCD Reference Voltage Generator	Intersil
256-Bit, 16 × 16 bit; Serial AUTOSTORE™ NOVRAM	Intersil
Quad Power Drivers with Serial Diagnostic Interface	Intersil
Dual Voltage Monitor with Integrated System Battery Switch and EEPROM	Intersil
HSP50016 Digital Down Converter, 75MSPS, 16-bit Input, Low Pass I/Q Filter	Intersil
20-Port and 28-Port I/O Expander	Maxim
LED Display Drivers	Maxim
Real-Time Clock	Maxim
Vacuum Fluorescent Display Controller	Maxim
Digital Temperature Sensor	TI
Dual 12-bit Digital-to-Analog Converter	TI
FLEX™ Decoder	TI
Analog Monitoring and Control	TI
Memory Stick Interconnect Extender	TI
24-Bit Analog-to-Digital Converter	TI
Touch Screen Controller	TI
16-Output Switch	Freescale
Octal Serial Switch	Freescale
Dual Motor Drive H-Bridge	Freescale
Multiple Switch Detection Interface	Freescale
Power Management and Audio Component	Freescale

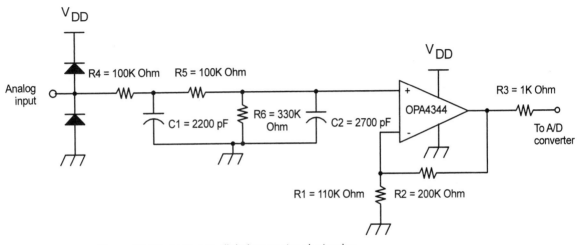

Figure 18-24 Analog-to-digital converter electronics.

The gain of the input voltage divider is

$$\frac{R_6}{R_4 + R_5 + R_6} = 0.62$$

giving an overall gain of 1.7 for DC. A two-pole lowpass filter is given by R_4, C_1 and R_5, R_6, C_2. We can treat these two filter sections independently because the coupling resistor, R_5, is sufficiently high to isolate the second filter from the first.

The cutoff frequency of the R_4, C_1 is

$$f_{C_1} = \frac{1}{2\pi R_4 C_1} = 723 \text{ Hz}$$

The cutoff frequency of R_5, R_6, C_2 is

$$f_{C_2} = \frac{1}{2\pi (R_5 // R_6) C_2} = 775 \text{ Hz}$$

18.7 I/O Software

I/O Software has an *initialization* part and a data *input/output* part and must be *synchronized* with the I/O device.

There are three major parts in your I/O software. First, as shown in previous chapters, is an *initialization* part to set up the function of the ports and the direction of data flow. Second, there are *data input and output* sections that simply read from or write to the appropriate I/O register. There is a third element, namely, *software synchronization*. I/O software must synchronize the reading and writing of the data with the timing requirements of the I/O device. Typically, microprocessors are much faster than the I/O devices they serve and must be synchronized using software and/or hardware techniques. Hardware handshaking techniques for the microcontroller are discussed next, and there are two software I/O synchronization methods. Interrupts are used also for I/O synchronization, and we discussed those in Chapter 12.

Real-Time Synchronization

Real-time synchronization uses a software delay to match the timing requirements of the software and hardware. For example, consider outputting characters to a parallel port (say, port AD) at a rate no faster than 1000 characters per second. If we assume negligible time is spent in getting and outputting each character, a delay of 1 millisecond is required between each output operation. This could be done with a pair of subroutines; one gets and outputs the characters, and one delays 1 millisecond before returning.

Real-time synchronization has its problems. It is dependent on the CPU's clock frequency and it usually has some overhead cycles that cause errors so the timing is not exact. Thus, depending on the requirements of the application, software timing loops may not be accurate enough. In Chapter 14 we saw how to generate highly accurate timing delays using the microcontroller timer system. These are far better than software delays, although they too depend on the clock frequency.

Polled I/O

Polled I/O software uses additional I/O bits as *status bits* for external I/O devices. An external device receiving data from the microcontroller via port P could use port J0 as a status bit. PJ0 will be asserted by the external device when it is ready for new data and deasserted when it is not. See Figure 18-25(a) for the hardware used. Obviously, hardware logic is required in the external device to assert and deassert this bit. The polling software monitors the status bit and outputs data only when the external device is ready. Example 18-22 shows a program that polls port J bit-0 to determine when it is safe to output more data to bits 3–0 of port P.

Polled input software (and hardware) is similar. When the software is ready to input new data from the external input device it checks the RDY_IN signal on PJ1 and waits until it is asserted by the external device before inputting the data. The polling software can be doing other things while it is waiting for the external device to supply new data. Figure 18-25(b) shows the signals needed for input and output polling and Example 18-23 gives a sample of program code.

A question you might reasonably ask at this point is: "In the polled input scenario, how does the external device know when the CPU has taken its data?" The RDY_IN bit is information *from* the external device *to* the microcontroller. There is no corresponding timing information going in the other direction to let the external device know that the microcontroller has taken the current data and that it is safe to supply new data. The solution to this problem may have two forms. First, RDY_IN could activate an *interrupt* bit in port J and generate an interrupt to ensure the CPU takes the data in a timely fashion. We have discussed this procedure more completely in Chapter 12. Second, *handshaking I/O* can be used as discussed in the next section.

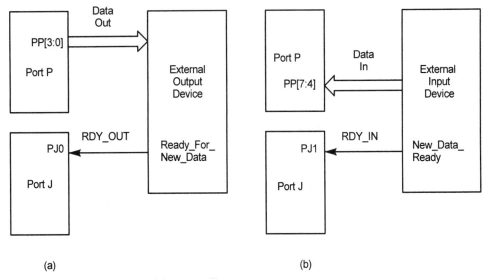

(a)   (b)

Figure 18-25 (a) Output and (b) input polling.

Example 18-22 Using a Status Bit for Output Polling

```
Metrowerks HC12-Assembler
(c) COPYRIGHT METROWERKS 1987-2003

Rel. Loc    Obj. code Source line
---- ---    --------- -----------
   1                  ; Program example showing how to use a status
   2                  ; bit to determine when an output device is
   3                  ; ready to accept data.
   4                  ; ****************************************
   5                  ; Include Files
   6                          INCLUDE  bits.inc
   7                          INCLUDE  portj.inc
   8                          INCLUDE  portp.inc
   9                  ; ****************************************
  10     0000 0001 RDY_OUT:  EQU %00000001   ; Bit 0
  11     0000 000F O_BITS:   EQU %00001111   ; Bits to be output
  12
  13                  ;       . . .
  14                  ; I/O Initialization
  15                  ; Set up PORTP[3:0] to be output
  16 000000 1C02 5A0F         bset  DDRP,O_BITS
  17                  ;       . . .
  18                  ; Output data to Port P bits 3 - 0
  19                  ; Wait until status bit, Port J, bit 0 is a 1
  20                    spin2:
  21 000004 1F02 6801         brclr PTJ,RDY_OUT,spin2
     000008 FB
  22                  ; Now can output the data from Accum A
  23 000009 7A02 58           staa  PTP
  24                  ;       . . .
```

Example 18-23 Using a Status Bit for Input Polling

```
Metrowerks HC12-Assembler
(c) COPYRIGHT METROWERKS 1987-2003

Rel. Loc    Obj. code Source line
---- ---    --------- -----------
   1                  ; Program example showing how to use a status
   2                  ; bit to determine when an output device is
```

```
  3                         ; ready to accept data.
  4                         ; ****************************************
  5                         ; Include Files
  6                                   INCLUDE   bits.inc
  7                                   INCLUDE   portj.inc
  8                                   INCLUDE   portp.inc
  9                         ; ****************************************
 10      0000 0002 RDY_IN:   EQU %00000010    ; Bit 1
 11      0000 000F O_BITS:   EQU %00001111    ; Bits to be output
 12                         ;       . . .
 13                         ; I/O Initialization
 14                         ; Set up PORTP[3:0] to be output
 15 000000 1C02 5A0F              bset   DDRP,O_BITS
 16                         ;       . . .
 17                         ; Input data from Port P bits 7 - 4
 18                         ; Wait until status bit, Port J, bit 1 is a 1
 19                         spin2:
 20 000004 1F02 6802              brclr PTJ,RDY_IN,spin2
    000008 FB
 21                         ; Now can input the data
 22 000009 B602 58               ldaa   PTP
 23                         ;       . . .
```

MICROCONTROLLER INTERNAL POLLED I/O

The HCS12 internal I/O devices all have status bits, called *flags*, that allow polling or interrupt I/O synchronization. For example, when using the SCI transmitter, one must not output any new data before the last data byte has been sent. The SCI sets a *transmit data register empty flag* (*TDRE*) when all bits have cleared the transmit data register. Our polling software monitors this flag to tell when it can output new data. See the *putchar* serial I/O software in Chapter 15 for an example.

Handshaking I/O

The HCS12 does not have dedicated hardware circuitry for handshaking I/O. Instead, general purpose I/O bits plus software similar to polling in Examples 18-22 and 18-23 are used. Figure 18-26 shows the hardware picture for output and input handshaking. There are a variety of schemes to accomplish handshaking depending on the timing requirements of each device and the microcontroller. One such scheme is shown Figure 18-27.

The I/O handshaking software, Example 18-24, consists of three parts—the initialization, output handshaking, and input handshaking.

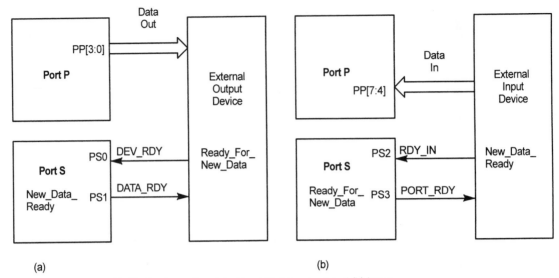

Figure 18-26 Hardware handshaking I/O: (a) output and (b) input.

I/O Initialization: The initialization code must set the data direction register for port P[3:0] and port S-1 and -3 to enable the output direction. It then clears the handshaking signals DATA_RDY and PORT_RDY (*lines 16–22*).

Output Handshaking: When the program is ready to output new data it starts the handshaking process by asserting the DATA_RDY signal on port S bit 1 (*line 26*). It then waits until the external device is ready for new data by polling the DEV_RDY bit on port S bit 0 (*line 29*). When DEV_RDY is asserted (there must be hardware in the external output device that asserts this signal when it is ready to receive data), the program knows that the external device has processed the last data and is ready for new. It then outputs the new data to port P[3:0] (*line 31*) and deasserts DATA_RDY signal (*line 33*). The external device then deasserts DEV_RDY until it is ready for new data again. Figure 18-27(a) shows the timing diagram associated with this output handshaking transfer of data.

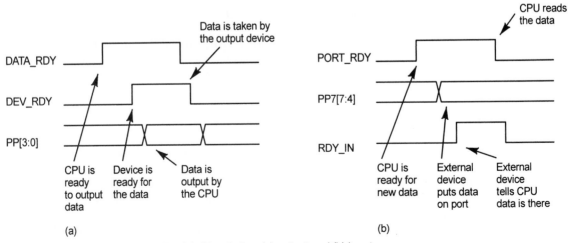

Figure 18-27 Handshaking timing: (a) output and (b) input.

Input Handshaking: Input handshaking shown in Figure 18-26(b) is very similar to output handshaking. When the microcontroller is ready to receive new data, it asserts the PORT_RDY control signal on port S bit 3 (*line 38*) and begins polling RDY_IN on port S bit 2 (*line 42*). When the external device has new data ready it places it on port S[7:4] and asserts RDY_IN. The microcontroller takes the data (*line 44*)and deasserts its handshaking signal PORT_RDY (*line 46*). The external device then deasserts RDY_IN and the cycle repeats. Figure 18-27(b) shows a timing diagram for this process.

A common variation on both input and output handshaking themes is that the DEV_RDY and RDY_IN status bits generate interrupts in the microcontroller. This is a good scheme to implement if there are timing issues to be resolved, and it also allows the program to go about other business while waiting for the data to be ready.

Example 18-24 Handshaking I/O Software

```
Metrowerks HC12-Assembler
(c) COPYRIGHT METROWERKS 1987-2003


Rel. Loc    Obj. code Source line
---- ---     --------- -----------
   1                    ;*****************************************
   2                    ; Hardware Input and Output Handshaking
   3                    ;*****************************************
   4                    ;
   5                              INCLUDE  ports.inc
   6                              INCLUDE  portp.inc
   7                    ;*****************************************
   8      0000 0001 DEV_RDY:  EQU %00000001   ; Device Ready
   9      0000 0002 DATA_RDY: EQU %00000010   ; Data Ready signal
  10      0000 0004 RDY_IN:   EQU %00000100   ; Ready Input
  11      0000 0008 PORT_RDY: EQU %00001000   ; Port Ready signal
  12      0000 000F O_BITS:   EQU %00001111   ; Bits to be output
  13                    ;        . . .
  14                    ;*****************************************
  15                    ; I/O Initialization
  16                    ; Set up PORTP[3:0] to be output
  17 000000 1C02 5A0F          bset  DDRP,O_BITS
  18                    ; Set up Port S bits 1 and 3 to be output
  19 000004 1C02 4A0A          bset  DDRS,DATA_RDY | PORT_RDY
  20                    ; Clear the handshaking signals
  21 000008 1D02 480A          bclr  PTS,DATA_RDY | PORT_RDY
  22                    ;        . . .
```

```
23                    ;*****************************************
24                    ; Handshaking output to Port P[3:0]
25                    ; Assert the New Data Ready signal
26 00000C 1C02 4802           bset  PTS,DATA_RDY
27                    ; Wait until the external device is ready
28                    spin1:
29 000010 1F02 4801           brclr PTS,DEV_RDY,spin1
   000014 FB
30                    ; Now we can output the data
31 000015 7A02 58            staa  PTP
32                    ; Now we de-assert the DATA_RDY signal
33 000018 1D02 4802           bclr  PTS,DATA_RDY
34                    ;         . . .
35                    ;*****************************************
36                    ; Handshaking to input data from Port P[7:4]
37                    ; Assert the Ready for New Data Signal
38 00001C 1C02 4808           bset  PTS,PORT_RDY
39                    ; Wait until the external device asserts
40                    ; New Data Ready
41                    spin2:
42 000020 1F02 4804           brclr PTS,RDY_IN,spin2
   000024 FB
43                    ; Now can read the data
44 000025 B602 58            ldaa  PTP
45                    ; and can de-assert the PORT_RDY signal
46 000028 1D02 4808           bclr  PTS,PORT_RDY
47                    ;         . . .
```

18.8 Conclusion and Chapter Summary Points

In this chapter we showed a wide variety of application examples using a microcontroller.

- Switches are used for inputting binary information.
- Mechanical switches bounce when contact is made and in some applications software or hardware debouncing must be used.
- Pull-up (or pull-down) resistors must be used with switch inputs to avoid floating input.
- Internal pull-up (or pull-down) resistors may be enabled in the microcontroller's I/O port.

- Keypads and keyboards are switches and diodes that connect an output from an I/O port to an input line. The keypad is scanned to determine what key is being pressed.
- When an LED is interfaced, the circuit designer must make sure the output device can supply enough current to light the LED.
- LEDs require several milliamperes to light.
- Parallel I/O expansion can be done with a bidirectional port acting as a data bus.
- Parallel I/O expansion can be done also with the SPI and serial/parallel shift registers.
- Serial I/O such as the SCI usually requires level translation to convert CMOS logic levels to another device electronic standard such as RS-232-C.
- The SPI can provide easy connection to a wide variety of external I/O devices.
- I/O software must be synchronized with software delays, polling, handshaking, or interrupts.

18.9 Bibliography and Further Reading

Cady, F. M., *Microcontrollers and Microprocessors*, Oxford University Press, New York, 1997.

Ganssle, J. G., *A Guide to Debouncing*, http://www.ganssle.com/debouncing.pdf, 2004.

18.10 Problems

Basic

18.1 Design an output circuit with eight LEDs connected to port B. The LEDs are to be on when bits in a byte stored in location DATA1 are 1's. Show the hardware and software required. **[c]**

18.2 Design an input circuit to input the states of eight switches to the HCS12. **[c]**

18.3 Why is the RS-232 voltage specification for mark and space logic levels used for serial communications instead of TTL voltage levels? **[a]**

18.4 Two computers are to be connected using their COM ports. **[c]**

  a. For this to work, what operational parameters need to be specified?

  b. In this application, what is meant by "data flow" synchronization?

  c. What are two ways in which data flow synchronization can be achieved?

18.5 A system is to be designed to transfer asynchronous serial data over a distance of 500 m. What interface standard could you use and what data rates are possible? **[a, c]**

Intermediate

18.6 A mythical microprocessor has two 8-bit output ports (P and Q) and two 8-bit input ports (R and S). Assume that a set of eight switches is connected to port S and a set of eight LEDs is connected to port P. Describe (a diagram would be nice) how you would implement a scheme using these resources (plus any others you would like, i.e., more switches, buffers, latches, etc.) that would allow you to input data from the switches only after the user has completed entering new data, and then to display the 8-bit data on the LEDs. The hardware is to be as simple and cheap as possible. Describe the operation of your system to be able to input data from the switches and output to the LEDs. **[c, k]**

18.7 Now, assuming the hardware you have proposed in Problem 18.6, describe, from a high-level, using pseudocode, how you would do the following: **[c]**

  a. Input data from the switches.

  b. Output data to the LEDs.

Mark

Space

18.8 A serial I/O port sends the following waveform: **[a]**

  a. What is the ASCII character being sent?

  b. What type of parity is being used?

18.9 A system is to be designed to transfer serial data from one place to another over a distance of 200 feet. Data is to be transferred in one direction only and there is no data flow problem. Data transfer rate is to be a minimum of 100 kilobits/second. You are to compare an asynchronous serial port approach (SCI) with a synchronous serial port (SPI) approach. **[b, c]**

  a. How many wires will be needed to connect the two systems (including the ground wire)?

  b. For this distance and data rate, what signaling interface standard would you propose?

18.10 You are to define a serial cable to connect two PCs configured as RS-232 DTE devices. Each PC has a DE9P connector on its back panel. The software used in each PC for file transfer uses hardware (RTS/CTS) flow control. Draw an appropriate cable using the *minimum* number of wires. Be sure to show each connection, what the signal name is, what the data flow direction is, and what connectors are to be used on each end of the cable. **[c]**

18.11 You are to define a serial cable to connect a PC configured as RS-232 DTE device to a microcontroller system configured as a DCE device. The PC has a DE9P connector on its back panel and the embedded system uses a DE9S connector. There is no flow control for the data transfer between the two computers. Draw an appropriate cable using the *minimum* number of wires. Be sure to show each connection, what the signal name is, what the data flow direction is, and what connectors are to be used on each end of the cable. **[c]**

Advanced

18.12 An eight-digit LED display is multiplexed with each digit being refreshed at 100 Hz by an interrupt service routine. The ISR changes the display to the next digit and requires 8 μs to execute. **[b]**

  a. If the interrupt service routine is started by an interrupt from the timer system, what is the interrupt rate to be able to refresh each digit in the display at 100 Hz?

  b. What percentage of the processor's time is spent refreshing the eight-digit display?

19 HCS12 Fuzzy Logic

OBJECTIVES

In this chapter we investigate the fundamental principles of fuzzy logic and then see how to implement these ideas as microcontroller control programs.

19.1 Introduction

Control systems take inputs and produce outputs to control some process such as the fan speed of a furnace.

The goal of any control system is to provide appropriate output drive signals for every possible combination of input signals. In simple cases for which there are two inputs and one output, the system can be graphically represented by a three-dimensional *control surface* where X and Y axes represent the inputs and the Z axis represents the output.

An easy way to implement such a control system in a microcontroller is with a table lookup algorithm. For a one-input, one-output system you could simply build a two-dimensional table in which the value stored at each address represents the desired output drive value. The "control surface" in this case is a line as shown in Figure 19-1. You can easily see there is one output for each input. If you tried to build such a table for a system with two 8-bit resolution inputs and one 8-bit output, you would quickly see the problem with using a table. Now the control surface is actually a surface as shown in Figure 19-2. Since each input has 256 possible values, the table would need to have 256 times 256 elements or 65,536 data points. Offsetting this size disadvantage, tables are computationally fast and simple. A 16-bit offset pointer can be created by using one input as the upper order eight bits and the other input as the lower order eight bits. For microcontrollers such as the HCS12, the CPU cannot handle such a large table directly. Over the years, programmers have found ways to compromise by reducing the input resolution or by limiting the range of interest to get tables that are more manageable.

In some systems, the input to output relationships can be expressed mathematically. Instead of building a very large table, you can compute the needed results for current values of system inputs. Proportional-Integral-Differential (PID) systems are an example of this technique. In a PID system the inputs are the error from the desired output (P), the integral of the error (I), and the first derivative of the error (D). Separate constant multipliers are provided for each of these terms, and the output is computed as the sum of these three product terms. Control is achieved by adjusting the three constant multipliers and analyzing the resulting system behavior.

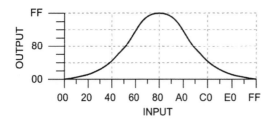

Figure 19-1 A one-input, one-output control surface.

Control surfaces for PID controllers are planes because the equations are linear. Since there are three inputs, the control system is now a family of control surfaces rather than a single three-dimensional plot so it becomes more difficult to visualize the system. The linear nature of PID systems points to a significant limitation. While many natural systems are fundamentally linear, they often involve secondary effects that interfere with ideal linear behavior. In these cases the resulting control system is only as good as the mathematical model. If you can develop a more accurate model to take into account additional factors, it will involve more complex computations to implement the control system.

19.2 Our Digital Heritage

Digital controllers are based on *Boolean* (binary) logic principles.

Modern microcontrollers are digital devices based on Boolean logic. This maps well onto traditional (Western) ideas of logic dating back at least to Aristotle. In this Boolean world, everything is yes or no, true or false, black or white. For many practical applications this is fine. Very useful work is being done every day using these ideas. Fuzzy logic allows us to get beyond this binary limitation so we can deal effectively with problems that involve shades of gray.

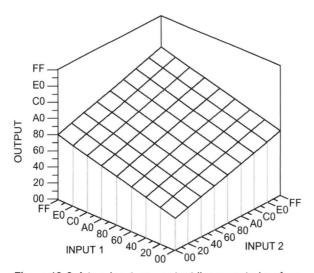

Figure 19-2 A two-input, one-output linear control surface.

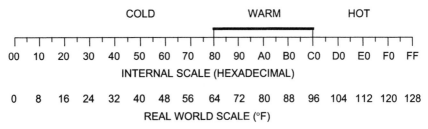

Figure 19-3 Traditional representation of abstract concepts.

Although microcontrollers are fundamentally digital devices, we have learned to represent analog quantities to arbitrarily chosen levels of resolution using A/D converters. We also know how to use digital programs to perform floating point computations. We try to limit the use of floating point computations because they tend to be slow and expensive (in terms of time and computational resources).

Fuzzy logic allows *subjective* concepts such as "kind of warm" to be used for control.

Until now we have had relatively little success getting digital computers to understand abstract or subjective concepts such as "warm." You could select a range of actual temperatures and say that if temperature falls in that range it is "warm. " Any temperature below this could be "cold" and any temperature above the range could be "hot" (see Figure 19-3). This still leaves a problem at the boundaries. If you said the lower bound of the warm range was 64 degrees Fahrenheit, that implies that 63.5 is cold while 64.00 is warm. This is definitely not what humans think of when you say "warm."

As remarkable as human senses are, they do not resolve environmental temperatures to small parts of a degree (at least not consciously). In the decision process, people also do not perform complex mathematical computations to support their decisions. Yet ordinary humans routinely make complex decisions that cannot be made by traditional computing methods.

19.3 How Is Fuzzy Logic Different?

Fuzzy logic does not mean imprecise logic.

Fuzzy logic introduces new concepts that allow a user to attach an unambiguous numerical meaning to linguistic terms such as "warm." In turn, this allows the programmer to perform computations with these linguistic inputs to perform useful control and decision algorithms. While the approach is quite different from traditional control methods, the result can still be mapped onto a control surface. For every possible set of input conditions there is a precise and repeatable control output. The common misconception, that the control output from a fuzzy logic system is approximate or imprecise, is simply incorrect.

The usual starting point for any control program is with an application expert. If you are trying to control a furnace in a steel plant, you start by understanding what the furnace is supposed to do. With traditional methods you would try to model the operation of the furnace system. Since an expert furnace operator is not likely to be familiar with mathematical modeling techniques, there is a problem already. Either the operator has to explain to the model designer what he/she does or the designer/programmer has to learn what the operator does. Often there is very little common ground for good communication.

In fuzzy logic systems, the control program is written in terms the application expert understands rather than some abstract mathematical or programming language. When it comes time to debug and fine-tune the system, the human expert can actively participate.

19.4 What Is Fuzzy About Fuzzy Logic?

Since fuzzy logic involves new concepts, some new terminology is needed. Some of the terms sound so strange to traditional control system engineers that some have dismissed fuzzy logic as frivolous nonsense. The following extreme example illustrates how this new terminology can cause problems. A major Japanese camera manufacturer introduced a new camera in the United States with the phrase "fuzzy auto focus"! Would you buy a camera that automatically produced fuzzy pictures? Fuzzy just doesn't sound precise at all. It sounds too much like guessing or approximation—not anything a scientist or mathematician would want to use in place of traditional control methods. The problem here is that fuzzy refers to a way of describing sets and is not an adjective describing the quality of the control system outputs.

Traditional set theory is binary in that something either is or is not a member of a particular set. There is no way to express "somewhat warm." The first powerful contribution from fuzzy logic is the idea that this traditional view of sets is a very limited subset of what fuzzy logic calls "fuzzy sets." To understand this concept see Figure 19-4, which illustrates three different ways to express the meaning of cold, warm, and hot.

Figure 19-4(a) shows how traditional digital computing methods might quantify the meaning of these three terms. The X-axis represents all possible values of temperature from some temperature sensor input. Any specific temperature either is or is not "warm."

Figure 19-4(b) tries to show what a human might mean by the terms cold, warm, and hot. In this case there are areas of uncertainty along the X-axis. It isn't that the human doesn't know what the temperature is. Rather, the human isn't quite sure what to call these temperatures. In the band between warm and cold, you might say the temperature is somewhat cold and/or somewhat warm. There has been no good way to code this into traditional control programs.

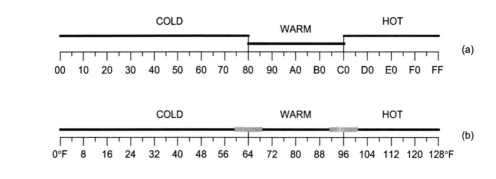

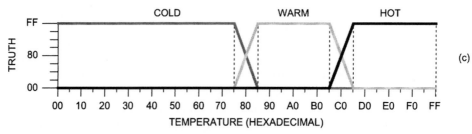

Figure 19-4 Different ways to represent abstract concepts.

Figure 19-4(c) shows how fuzzy logic deals with this problem. In this graphical representation, the X-axis still represents all possible values of the temperature input. The Y-axis represents truth, where zero means completely false and $FF means completely true. (In fuzzy logic theory, the Y-axis would range from zero to one. The zero to $FF range is just more convenient for small microcontrollers.) This is much closer to what a human means by the terms cold, warm, and hot. As temperature changes from lower values within the cold range to higher values in the warm range, there is a gradual transition from cold to warm.

> Fuzzy logic allows unambiguous numerical meaning to be given to abstract concepts such as "somewhat."

The ideas demonstrated in Figure 19-4(c) are so important that they deserve additional discussion. In a digital microcontroller you can now store an unambiguous numerical representation of the meaning of an abstract concept such as "warm." This graphical representation is called a "membership function" and in this figure trapezoids, that can easily be stored in a microcontroller as four 8-bit values (two points and two slopes), are used. This means a programmer can now write programs that perform mathematical computations with linguistic terms such as "warm."

Figure 19-4(c) also demonstrates that the boundaries of sets can be gradual or "fuzzy." This is a major improvement over the traditional methods where a hundredth of a degree could mean the difference between cold and warm. It is also interesting that sets can (and almost always do) overlap. As temperature rises from $78 to $88, cold gradually goes from completely true to false at the same time warm goes from false to completely true. In between you would say that it is both cold and warm at the same time (to different degrees of truth). It should be easy to imagine from this that a controller using this information would gradually change from doing what is expected when temperature is cold to doing what it should when temperature is warm.

19.5 Structure of a Fuzzy Logic Inference Program

> The three main operations in a fuzzy inference program are *fuzzification, rule evaluation*, and *defuzzification*.

Figure 19-5 is a block diagram of a fuzzy logic inference program. System inputs enter from the left and outputs exit to the right. Major blocks in the diagram show the three main operations performed by the inference program. *Fuzzification* transforms digital input values into fuzzy truth values for linguistic labels such as cold, cool, warm, and so on. The *rule evaluation* stage then uses these fuzzy input values to compute fuzzy output values for the fuzzy outputs. Similar to fuzzy input values, fuzzy output values represent the truth of linguistic output labels such as slow, mid, fast, and so on. The final *defuzzification* block transforms the set of fuzzy outputs into a specific digital output value that is suitable to drive some output device.

In Figure 19-5, temperature and pressure are system inputs and they each have a digital value between $00 and $FF when the fuzzy inference program starts to execute. Temperature has five labels in this example, COLD, COOL, NORMAL, WARM, and HOT. During the fuzzification step, a truth value is computed for each of these five labels. These results are called fuzzy inputs and each is represented by an 8-bit value in RAM. A $00 in the fuzzy input for cold means the linguistic expression "temperature is cold" is false. An $80 in the fuzzy input for warm means the expression "temperature is warm" is half true. We will see exactly how fuzzy inputs are computed a little later in this section, but for now you can see that subjective-sounding linguistic expressions such as "temperature is warm" can be expressed as concrete values and further precise calculations are possible.

> Rule evaluation sets fuzzy output labels to a truth value.

In the *MIN-MAX rule evaluation* step, rules are processed using current input conditions (as represented by the current values in the fuzzy inputs) to produce fuzzy outputs. Similar to fuzzy inputs, fuzzy outputs are 8-bit values in RAM that represent the truth

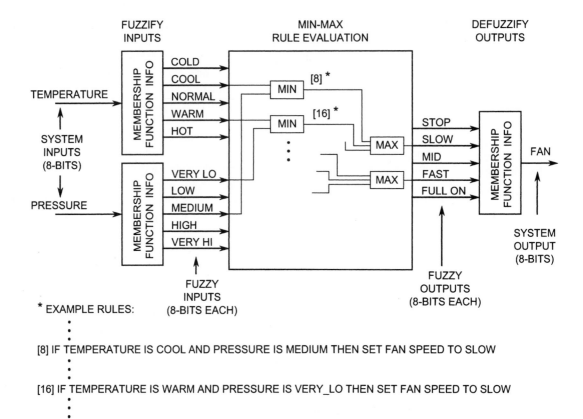

* EXAMPLE RULES:

⋮

[8] IF TEMPERATURE IS COOL AND PRESSURE IS MEDIUM THEN SET FAN SPEED TO SLOW

⋮

[16] IF TEMPERATURE IS WARM AND PRESSURE IS VERY_LO THEN SET FAN SPEED TO SLOW

⋮

Figure 19-5 Fuzzy logic inference block diagram.

of a linguistic expression such as "set fan to medium." After rule evaluation, there can be more than one fuzzy output that is true to some degree. It would be ambiguous to attempt to drive a typical output device with these seemingly contradictory signals.

Rule evaluation consists of finding the smallest (minimum) fuzzy input associated with a rule and then applying that value to the associated fuzzy output, subject to a maximum computation. The maximum computation finds the rule that is most true associated with each fuzzy output and applies that truth value to the corresponding fuzzy output.

> Defuzzification computes the final system output values.

The final step, called defuzzification, performs a weighted average computation to resolve the fuzzy outputs into a single specific output drive signal. Later in this section we will discuss differences between the kinds of membership functions used for inputs compared to those used for outputs.

Fuzzy logic is inherently a parallel computing technology because all fuzzy inputs could be computed simultaneously and independently if the hardware or computing resources were available. Similarly, all rules could be processed independently and simultaneously (rules are not considered to have any sequential importance). In practice, these tasks are performed sequentially because the CPU can do only one task at a time.

Figure 19-6 is another view of the fuzzy logic block diagram. In this figure, the system is broken into two sections, and each of these sections has a component associated with each of the three steps in the fuzzy inference process. The knowledge base is a set of data representing the knowledge that is specific to a particular application. This information is typically

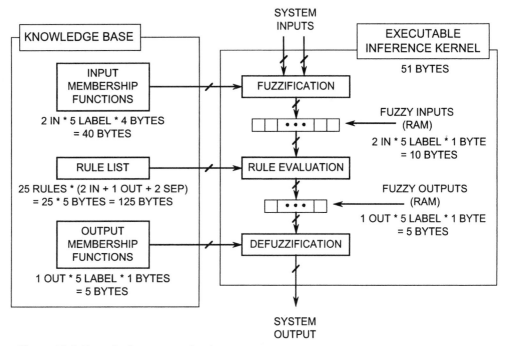

Figure 19-6 Fuzzy logic program structure.

provided by an application expert. The fuzzy executable inference kernel is the processing portion of the fuzzy inference program. This section is somewhat independent of any specific application and in fact may be used without modification to solve many different application problems simply by changing the information in the knowledge base.

To explain the program and process, we will assume the specific example illustrated in the original fuzzy block diagram, Figure 19-5. There are two system inputs—Temperature and Pressure. Temperature has the labels Cold, Cool, Normal, Warm, and Hot. Pressure has the labels Very Lo, Low, Medium, High, and Very Hi. Each of these input labels corresponds to a fuzzy input in the system. There is one system output named Fan that has the labels Stop, Slow, Mid, Fast, and Full On, each corresponding to one fuzzy output in the system.

19.6 Fuzzification

The inputs to the fuzzification step are the system inputs (8-bit data values indicating current system conditions), and input membership functions from the knowledge base. To be compatible with the membership function evaluation instruction (MEM) in the HCS12, the input membership functions are *trapezoids*. These trapezoids are specified with the x-positions of the two endpoints of the base and the 8-bit values representing the slopes of the sides of the trapezoid.

Creation of membership functions is a job we tackle with the help of the application expert. For example, our expert might say that temperatures in the range of 44 °F to 56 °F ($58–$70) are definitely Cool. Below 32 °F ($40) or above 68 °F are definitely not Cool. The transition areas from 32 °F to 44 °F and 56 °F to 68 °F are still considered part of the

Membership functions to be evaluated by the MEM instruction are trapezoidal.

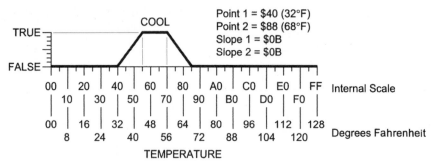

Figure 19-7 A trapezoidal membership function.

Cool set, but the degree of truth varies between $00 and $FF. (The hexadecimal values are what the A/D converter gives us when we read the temperature.) A trapezoidal membership function for Cool is shown in Figure 19-7 and is defined by the two x-values of the base ($40 and $88) and the slopes of the two sides ($FF/$18 = $(255/25)_{10}$ = 10.6_{10} = $0B).[1]

Figure 19-8 illustrates what is done in the fuzzification step of the fuzzy inference program. The trapezoidal membership functions for each label of the temperature input and the pressure input are shown. All five membership functions for temperature share a common scale along their X-axis. Similarly, all five membership functions for pressure share a common scale along their X-axis. The vertical lines through all five temperature, and all five pressure, membership functions represent a current specific temperature and pressure.

To the right of each graphical membership function is the 4-byte representation of that membership function as it appears in the knowledge base memory. This memory would typically be ROM or some other nonvolatile memory because the membership functions would not usually change during normal operation of the application program.

During the fuzzification step, the program compares the current system input value (temperature or pressure) to each membership function and stores the resulting Y-intercept into the corresponding fuzzy input location in RAM. These values are shown to the left of each graphic membership function. Note that these values would change if the temperature or pressure changed. These fuzzy input values represent an instantaneous snapshot of current input conditions.

In HCS12 assembly language, the MEM instruction does most of this work for us. When a MEM instruction is executed, the X-index register points at a membership function data structure, the Y-index register points at the RAM location in which the fuzzy input result will be stored, and the A accumulator holds the current value of the system input. The MEM instruction reads the 4-byte membership function definition from memory, computes the Y-value corresponding to the current value of the system input, stores this result in the RAM location pointed to by Y, and updates X = X + 4 and Y = Y + 1 so the index pointers are ready for the next membership function computation. The contents of the A accumulator are not disturbed. By planning the arrangement of membership functions and fuzzy inputs in memory, the whole fuzzification step becomes a simple matter of setting up X and Y to initial positions and executing one MEM instruction for each label of each input (usually in a loop).

---

[1] The magnitude of the slope is given. There are no negative values for the slope.

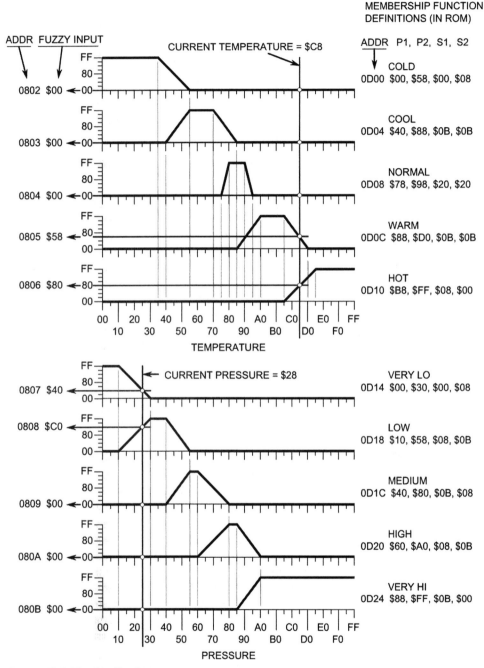

Figure 19-8 The fuzzification process.

A *crisp* input is a value from a sensor.

The assembly language code segment in Example 19-1 performs the entire fuzzification step for a system with two system inputs, each having five labels. For this example we will name the system inputs Temperature and Pressure. We will also assume any preprocessing of these inputs to adjust for gain or offset of the sensors was done prior

Example 19-1 Fuzzification Program Segment

```
8000  CE0D00  132  Fuzzify:    LDX   #Input_MFs   ;Point at MF
                                                   definitions
8003  CD0802  133              LDY   #FuzzyIns    ;Point  at  fuzzy
                                                   ins in RAM
8006  B60800  134              LDAA  Temperature  ;Get first system
                                                   input
8009  C605    135              LDAB  #5           ;Temperature  has
                                                   5 labels
800B  01      136  Fuz_loop:   MEM                ;Evaluate  one  MF
800C  0431FC  137              DBNE  B,Fuz_loop   ;For  5  labels  of
                                                   1 input
800F  B60801  138              LDAA  Pressure     ;Get second
                                                   system input
8012  C605    139              LDAB  #5           ;Pressure has
                                                   5 labels
8014  01      140  Fuz_loop1:  MEM                ;Evaluate  one  MF
8015  0431FC  141              DBNE  B,Fuz_loop1  ;For  5  labels  of
                                                   1 input
```

to calling the fuzzy kernel program so the current (compensated) values of these inputs are simply stored in RAM locations named Temperature and Pressure. Because these system inputs are *real* values, not fuzzy, they are sometimes called *crisp* inputs.

19.7 Rule Evaluation

The REV or REVW instruction does the rule evaluation.

From the subjective sound of a rule such as "If temperature is warm and pressure is high then fan should be slow," you might think there could be an infinite number of rules, but this is not the case. The number of inputs and the number of labels per input force a limit to the maximum number of rules possible in a system. In our example, we have two inputs and each has five labels so there could not be more than five times five or twenty-five rules.

A common way to be sure you have specified a rule for all possible input combinations is to draw a rule matrix similar to Table 19-1, listing the labels for one input down the side and the labels for the other input across the bottom. If you have a third input, you would need as many of these matrices as there are labels for the third input. You can think of this rule matrix as a crude control surface viewed from above. This matrix only shows cases in which a single input label is true at a time. Later we will discuss what happens when the input falls in a range where more than one label is partially true at the same time.

Next, we need to code the rules into a compact form that can be processed by the REV or REVW instructions. The input expressions, such as "If temperature is warm," correspond to a

TABLE 19-1 Rule Matrix

Pressure

	Cold	Cool	Normal	Warm	Hot
Very Hi	Stop [5]	Slow [10]	Mid [15]	Fast [20]	Full On [25]
High	Stop [4]	Slow [9]	Mid [14]	Fast [19]	Full On [24]
Med	Stop [3]	Slow [8]	Mid [13]	Fast [18]	Fast [23]
Low	Stop [2]	Stop [7]	Slow [12]	Mid [17]	Fast [22]
Very Lo	Stop [1]	Stop [6]	Slow [11]	Slow [16]	Mid [21]

Temperature

[1] IF Temperature IS Cold AND Pressure IS Very Lo THEN SET Fan TO Stop.

•

• OTHER RULES

•

[13] IF Temperature IS Normal AND Pressure IS Medium THEN SET Fan TO Mid

•

• OTHER RULES

•

[25] IF Temperature IS Hot AND Pressure IS Very Hi THEN SET Fan TO Full On

fuzzy input so you can replace the subjective-sounding expression with a pointer to the fuzzy input. Each output expression, such as "fan should be fast," corresponds to a fuzzy output so you can replace the subjective-sounding expression with a pointer to the fuzzy output. Again, we call on our application expert to help us fill in the matrix and write the rules. It is the expert furnace operator who can tell us, for example, that when the temperature is warm and the pressure is low the fan ought to be at its mid setting. For the HCS12, the rule evaluation instructions use separator characters to tell when the rule input expressions stop and the rule output expressions start. This allows the instructions to work with rules that have a variable number of inputs and outputs.

All rules are in the following form:

> IF *system_input_1* is [equal to] *fuzzy_input_label_m* AND *system_input_2* is [equal to] *fuzzy_input_label_n* AND ... THEN set *system_output_1* to *fuzzy_output_label-p*, *system_output_2* to *fuzzy_output_label_q*, . . .

For example, rule number [8] is

IF Temperature IS Cool AND Pressure IS Medium THEN SET Fan Speed TO Slow

Min-max rule evaluation uses the fuzzy AND operator to connect rule inputs and the (implied) fuzzy OR operator to connect successive rules. The fuzzy AND is equivalent to the mathematical MINIMUM operator and the fuzzy OR is equivalent to the mathematical MAXIMUM operator. Other types of rules and operators are possible in fuzzy logic but they are beyond the scope of this text.

In our example program, each rule has two inputs, a separator character, one output, and another separator before the next rule. A rule occupies 5 bytes in memory and is encoded as follows:

```
DB      Pointer_to_temperature_fuzzy_input_cool
DB      Pointer_to_pressure_fuzzy_input_medium
DB      $FE        ; Separates inputs from output
DB      Pointer_to_fan_fuzzy_output_slow
DB      $FE        ; Separates this rule from the next
```

The pointers to fuzzy inputs and fuzzy outputs are 8-bit offsets from a base address so a complete rule takes 5 bytes in the knowledge base. A special separator character is used to mark the end of the last rule.

During rule evaluation, the MCU simply looks up each fuzzy input and finds the smallest one (minimum). This value is the "truth value" for the rule. Before any rules are processed, all fuzzy outputs are set to zero (not true). As the rules are processed the truth value for a rule is compared against the referenced fuzzy output. If the fuzzy output is not already bigger (maximum), the truth value is stored to the fuzzy output location. After all rules are processed, each fuzzy output holds the truth value for the rule that was most true (and referenced this fuzzy output).

The HCS12 has two rule evaluation instructions. REV is used for unweighted min-max rule evaluation, while REVW is used for weighted min-max rule evaluation. In our example we will use the unweighted REV instruction. These are basically list processing instructions that process a complete set of rules. In our example there are 25 rules.

Since the list of rules can be long, the REV and REVW instructions were designed so they can be interrupted and they will resume when you return from the interrupt. Everything the REV and REVW instructions need to resume is held in HCS12 CPU registers so the normal stacking and unstacking for the interrupt takes care of the save and resume functions. Even the information about whether the instruction is currently processing inputs or outputs is held in the CCR V bit so processing of rules can be interrupted anywhere in a rule without losing your place. The interrupt could even contain other REV or REVW instructions without interfering with the interrupted rule evaluation instruction.

The rule evaluation instruction does most of the work, but there is a little bit of setup before REV can start. First, all fuzzy outputs need to be cleared. This is part of the mechanism for doing the maximum computations. Say the first two rules, for example, [1] and [2] in Table 19-1, both refer to the same fuzzy output (Stop). We clear all fuzzy outputs before starting the REV instruction. REV finds the truth for the first rule and compares this with the current contents of the fuzzy output (zero because it has not been changed since it was cleared before processing any rules). The truth value for the first rule is then

---

Example 19-2 Rule Evaluation Code Segment

```
              142   * Here X points at rule list, Y at fuzzy
                                                      outputs
8018 C605     143              LDAB  #5          ;5 fuzzy outputs
                                                  in RAM
801A 6970     144   Rule_eval: CLR   1,Y+        ;Clr a fuz out &
                                                  inc pntr
801C 0431FB   145              DBNE B,Rule_eval  ;Loop to clr all
                                                  fuz outs
801F CD0802   146              LDY #FuzzyIns     ;Pointer to fuz
                                                  ins & outs
              147   * X already pointing at top of rule list
8022 86FF     148              LDAA #$FF         ;Init A & clears
                                                  V-bit
8024 183A     149              REV               ;Process rule
                                                  list
```

---

written into the fuzzy output. The truth value for the second rule is now found, and this value is compared to the fuzzy output (which now holds the result of the first rule). The larger value (comparing the fuzzy output set by the last rule and the truth value from the second) is now written to the fuzzy output. In this way the fuzzy output will eventually hold the truth value from the rule that was most true (and that referred to this fuzzy output). The assembly language code segment in Example 19-2 performs the entire rule evaluation step.

Figure 19-9 is similar to the rule matrix in Table 19-1, but it shows more detail so we can see what happens when more than one rule is active at the same time. The areas in which only one rule is active are marked by the number of the rule in square brackets. For example, in the rectangular area marked [1], only the rule that states "IF Temperature IS Cold AND Pressure IS Very Lo" will be active. In order for any other rule to become active, temperature would have to get higher than $40 and/or pressure would have to get higher than $10.

The five rules corresponding to "Temperature IS Normal" ([11]–[15]) correspond to vertical lines at temperature $88 rather than rectangular areas because any temperature slightly above or below this value would cause "Temperature IS Warm" or "Temperature IS Cool" to be true to a small degree so other rules would become partially active.

In areas in which only one rule is active, the output drive level is simply the level stated in the corresponding rule because there is nothing to modify the conclusion proposed by that rule. Things get more interesting when more than one rule is active at the same time because the proposed output drive level from each of the active rules could be different (and somewhat contradictory). Lightly shaded areas indicate areas in which two rules are active to some degree at the same time. Heavily shaded areas indicate four rules are active at the same time. The defuzzification step will resolve these apparent conflicts.

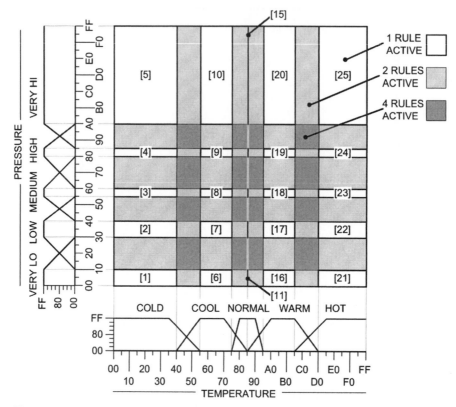

Figure 19-9 More detailed rule matrix.

19.8 Defuzzification

The WAV instruction is used to defuzzify the output.

Control system output devices are typically things such as variable-speed motors or heating elements. Only one control value can drive such a device at a time but rule evaluation can result in more than one nonzero fuzzy output value. Defuzzification combines these fuzzy output values into a single value that is suitable to drive the control system output device. Although there are other methods of defuzzification, the most common is the weighted average method that is supported directly by the WAV instruction.

Output membership functions are a little different from input membership functions. Both input and output membership functions provide the meaning of linguistic labels in an application system. We saw earlier how a trapezoidal membership function can be used to express the meaning of an input label such as "warm." For system outputs, we can use a simpler type of membership function called a singleton.

Singleton membership functions can be represented graphically as shown in Figure 19-10. In this figure, the X-axis represents all possible values of the output signal. Each singleton membership function is a vertical line of zero width whose height is equal to the fuzzy output (weight). The position of these membership functions is specified by a single 8-bit value in the knowledge base (the singleton position).

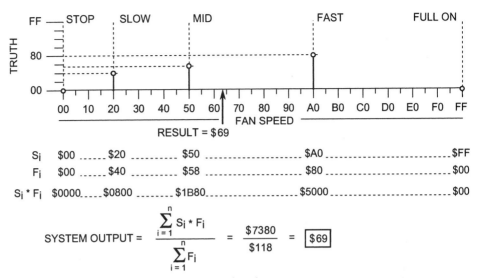

Figure 19-10 Singleton output membership function.

The fan speed is actually continuously variable. The five labels arise because these linguistic labels occur in the rules provided by the application expert. The fuzzy logic program needs to know what these linguistic terms mean in unambiguous numerical terms. You could ask the application expert to set the fan speed to "Fast" and then measure this speed. After discussing this with the expert you would add the output membership functions to the knowledge base.

```
DB      $00     ;   MF for Fan Stop
DB      $20     ;   MF for Fan Slow
DB      $50     ;   MF for Fan Mid
DB      $A0     ;   MF for Fan Fast
DB      $FF     ;   MF for Fan Full on
```

When more than one fuzzy output has a nonzero value, we compute a weighted average to combine these separate output instructions into a single output drive level. The formula used to compute the weighted average is

$$\text{System Output } = \frac{\sum_{i=1}^{n} S_i \times F_i}{\sum_{i=1}^{n} F_i} \tag{19-1}$$

where n is the number of labels for this system output (5 in our example system), S_i are the singleton positions from the knowledge base, and F_i are the corresponding fuzzy outputs that resulted from the rule evaluation step (weights).

The weighted average instruction (WAV) in the HCS12 computes the sum-of-products and the sum-of-weights. Before executing the WAV instruction, you must set the number of iterations, n, in the B accumulator. You also set the X index register to point at the first element in the S_i list, and the Y index register to point at the first element in the F_i list. Since

Example 19-3 Defuzzification Code Segment

```
              150   * Here X points at output MFs, Y at fuzzy inputs
8026 CD080C   151   Defuz:    LDY  #FuzzyOuts ;Point at fuzzy outputs
              152   * X already pointing at singleton MFs
8029 C605     153             LDAB #5          ;5 fuzzy outs per sys out
802B 183C     154             WAV              ;Calc sums for wtd ave
802D 11       155             EDIV             ;Final divide for wtd ave
802E B764     156             TFR  Y,D         ;Move result to A:B
8030 7B0811   157             STAB FanSpeed    ;Store system output
```

this instruction could take more CPU cycles than normal instructions, it is designed so that it can be interrupted. When the WAV instruction is completed, the sum-of-products is returned in the 32-bit register pair Y:D and the sum-of-weights is returned in X. These are the correct registers so that the EDIV instruction can be used to complete the final division for the weighted average computation, returning the final result in the lower 8 bits of the Y index register. The assembly language code segment in Example 19-3 performs the entire defuzzification step.

19.9 Putting It All Together

We now have all the information necessary to see how the fuzzy inference engine works. Before we look at a complete program, let us work through an example by hand to make sure we know what is going on. Figure 19-8 shows the membership functions for the temperature and pressure and let us assume the current temperature is \$C8 and pressure is \$28. The MEM operator calculates the "truth" values for each of the fuzzy inputs. These are shown down the left side of Figure 19-8 and are repeated here in Table 19-2. Table 19-3 shows the rule matrix of Table 19-1 with the truth values for each of the fuzzy inputs added.

TABLE 19-2 MEM Calculations

System Input	Fuzzy Input Label	Truth Value
Temperature	Cold	\$00
	Cool	\$00
	Normal	\$00
	Warm	\$58
	Hot	\$80
Pressure	Very Lo	\$40
	Low	\$C0
	Med	\$00
	High	\$00
	Very Hi	\$00

TABLE 19-3 Rule Matrix

Pressure	Cold	Cool	Normal	Warm	Hot
Very Hi	Stop	Slow	Mid	Fast	Full On
$00	[5]	[10]	[15]	[20]	[25]
High	Stop	Slow	Mid	Fast	Full On
$00	[4]	[9]	[14]	[19]	[24]
Med	Stop	Slow	Mid	Fast	Fast
$00	[3]	[8]	[13]	[18]	[23]
Low	Stop	Stop	Slow	Mid	Fast
$C0	[2]	[7]	[12]	[17]	[22]
Very Lo	Stop	Stop	Slow	Slow	Mid
$40	[1]	[6]	[11]	[16]	[21]
	Cold	Cool	Normal	Warm	Hot
	$00	$00	$00	$58	$80

Temperature

The rule evaluation process (REV) takes each rule and, first, finds the minimum of the two fuzzy inputs associated with that rule. It then sets the specified fuzzy output to the maximum of this calculated minimum and the current fuzzy output. This is what is known as min-max rule evaluation. Table 19-4 shows this process for all 25 rules.

The rule evaluation step gives us the following fuzzy output (weights):

Stop = $00

Slow = $40

Mid = $58

Fast = $80

Full On = $00

Figure 19-10 shows the singleton positions that we defined with our application expert (the S_i) and the weights assigned by the rule evaluation step (the F_i). The system output is calculated by Equation (19-1) and gives us the following:

$$Fan\ Speed = \frac{\$00 \times \$00 + \$20 \times \$40 + \$50 \times \$58 + \$A0 \times \$80 + \$FF \times \$00}{\$00 + \$40 + \$58 + \$80 + \$00}$$

$$Fan\ Speed = \frac{\$7380}{\$118} = \$69$$

TABLE 19-4 Rule Evaluation

Rule	Fuzzy Input #1	Fuzzy Input #2	Min(#1:#2)	Current Fuzzy Output	New Fuzzy Output Max(Min:Current)
1	Cold = $00	Very Lo = $40	$00	Stop = $00	Stop = $00
2	Cold = $00	Low = $C0	$00	Stop = $00	Stop = $00
3	Cold = $00	Medium = $00	$00	Stop = $00	Stop = $00
4	Cold = $00	High = $00	$00	Stop = $00	Stop = $00
5	Cold = $00	Very Hi = $00	$00	Stop = $00	Stop = $00
6	Cool = $00	Very Lo = $40	$00	Stop = $00	Stop = $00
7	Cool = $00	Low = $C0	$00	Stop = $00	Stop = $00
8	Cool = $00	Medium = $00	$00	Slow = $00	Slow = $00
9	Cool = $00	High = $00	$00	Slow = $00	Slow = $00
10	Cool = $00	Very Hi = $00	$00	Slow = $00	Slow = $00
11	Normal = $00	Very Lo = $40	$00	Slow = $00	Slow = $00
12	Normal = $00	Low = $C0	$00	Slow = $00	Slow = $00
13	Normal = $00	Medium = $00	$00	Mid = $00	Mid = $00
14	Normal = $00	High = $00	$00	Mid = $00	Mid = $00
15	Normal = $00	Very Hi = $00	$00	Mid = $00	Mid = $00
16	Warm = $58	Very Lo = $40	$40	Slow = $00	Slow = $40
17	Warm = $58	Low = $C0	$58	Mid = $00	Mid = $58
18	Warm = $58	Medium = $00	$00	Fast = $00	Fast = $00
19	Warm = $58	High = $00	$00	Fast = $00	Fast = $00
20	Warm = $58	Very Hi = $00	$00	Fast = $00	Fast = $00
21	Hot = $80	Very Lo = $40	40	Mid = $58	Mid = $58
22	Hot = $80	Low = $C0	$80	Fast = $00	Fast = $80
23	Hot = $80	Medium = $00	$00	Fast = $80	Fast = $80
24	Hot = $80	High = $00	$00	Full On = $00	Full On = $00
25	Hot = $80	Very Hi = $00	$00	Full On = $00	Full On = $00

19.10 The Complete Fuzzy Inference System

Figure 19-11 shows a three-dimensional graphical representation of the entire process for our example system that has two system inputs and one system output. The fuzzy logic process is also applicable to much more complex systems, but it becomes very difficult to visualize what is happening when there are more than two inputs.

As in Table 19-1 and Figure 19-9, numbers in square brackets mark places at which a single rule is active. These areas correspond to flat areas on the control surface that are parallel to the base of the drawing and are located at a height (z) corresponding to a label of the system output. For example, the area marked [10] corresponds to rule 10, "If Temperature is Cool and Pressure is Very Hi then Fan Speed should be Slow," so the flat area marked [10] is at height $20 that corresponds to slow on the Z-axis.

The sloping area between [5] and [10] corresponds to the range where Cold and Cool overlap on the temperature scale and pressure is Very Hi. In this area both rules [5] and [10] are partially active at the same time. Defuzzification performs a surface interpolation to identify a point somewhere on the sloping surface.

In actual use, the fuzzy inference program only needs to compute one point on this control surface (the point corresponding to the current specific values of temperature and pressure at the time of the calculation). The complete control surface is shown here as a convenience to illustrate the input-to-output relationship for any combination of input values.

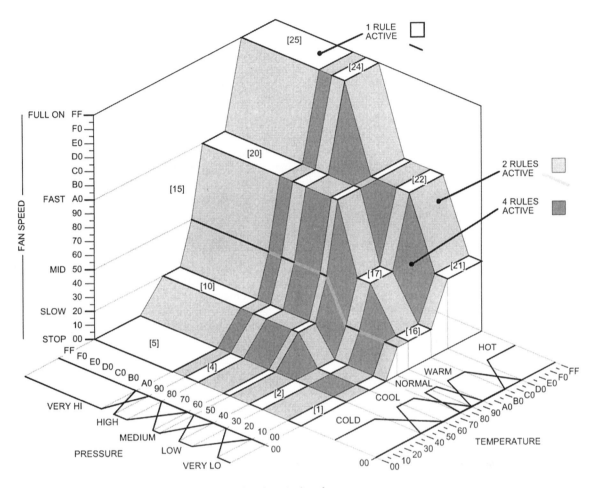

Figure 19-11 Three-dimensional control surface.

Although Figure 19-11 shows the transition from [5] to [10] as a sloping plane, this is not absolutely accurate because of the way the trapezoidal membership functions for Cold and Cool intersect. Usually adjacent membership functions would intersect like those for Pressure. For example, Low pressure starts to taper down from completely true at $40 just as Medium begins to rise from completely false. Low completes the transition to false at $58 just as Medium completes its transition to true. Cold and Cool do not quite follow this pattern. Cold begins its decent at $38 but Cool doesn't begin its rise until $40. Even though Cold is less than completely true in the range $38 to $40, it is the only label that is true at all so the sloping transition to [10] doesn't start until $40. Thus the nonsymmetrical overlap between the Cold and Cool membership functions causes a small amount of curvature in the sloping transition from [5] to [10].

EXPLANATION OF EXAMPLE 19-4

Example 19-4 is a complete listing for the fuzzy logic system that was used as an example throughout this chapter. *Lines 12* and *13* allocate 1 byte of RAM each for the current system temperature and pressure input values. Five bytes are allocated for fuzzy inputs for the five labels of the input temperature (*lines 18–22*) and 5 bytes are allocated for the five labels for the input pressure (*lines 24–28*). Next, five locations are allocated for the five labels of the output fan speed (*lines 32–36*). Finally, 1 byte is allocated for the defuzzified output value FanSpeed (*line 39*).

The next section defines the knowledge base for the example problem. This knowledge base includes input membership functions for each label of each system input (*lines 47–58*), the list of control rules (*lines 89–118*), and output membership functions for each label of the system output (*lines 121–125*). *Lines 60–80* attached mnemonic names to offsets from a reference pointer at the start of the fuzzy inputs and outputs (FuzzyIns on *line 16*) to each fuzzy input and fuzzy output. These new program labels make the encoded rules in *lines 89–118* easier to read in the listing. Two special separator characters, Sep and Rule_end, used by the REV instruction for encoding rules are defined in *lines 82* and *84*.

The actual executable fuzzy inference program is located in *lines 132–158*. *Lines 132–135* set up registers for the loop Fuz_loop. The two-instruction loop in *lines 136* and *137* performs fuzzification of the five labels of the temperature input. *Lines 138* and *139* load the current pressure value and reset the loop counter. The two-instruction loop in *lines 140* and *141* performs fuzzification of the five labels of the pressure input. At this point, the fuzzy inputs reflect current conditions of temperature and pressure in terms that can be used in the rule evaluation step to follow. Note that this program example does not show the steps that input the crisp values of temperature and pressure from the A/D converter. The two-instruction loop in *lines 144* and *145* clears all fuzzy outputs in preparation for rule evaluation. *Lines 146* and *148* complete the setup for rule evaluation and the REV instruction in *line 149* processes the entire list of 25 rules. At this point, the values in the fuzzy outputs reflect the results of rule evaluation for current system conditions. *Lines 151* and *153* set up registers for the defuzzification step, and *lines 154* and *155* actually perform the defuzzification computation. Finally, *lines 156* and *157* move the 8-bit result into the RAM location FanSpeed. The RTS in *line 158* makes it possible to call the executable fuzzy inference program as a subroutine.

Example 19-4 ORGs the variables in RAM, the knowledge base in EEPROM, and the executable portion in the Flash (for an MC68HC912B32). For experimentation with this program in an SLK board, you may wish to change these ORG statements (*lines 7, 41,* and *128*) to other, more convenient, locations.

Example 19-4 Complete Fuzzy Inference Program

```
Metrowerks HC12-Assembler
(c) COPYRIGHT METROWERKS 1987-2003


Rel. Loc   Obj. code Source line

---- ---    --------- ----------

   1                  ;**********
   2                  ;* Fuzzy Logic Inference Program
   3                  ;*  a small instructive example
   4                  ;*  2 system inputs with 5 labels
   5                  ;*  1 system output with 5 labels
   6                  ;**********
```

```
 7                        ORG  $800    ;Data memory start (RAM)
 8                  ;*********
 9                  ;* Following are the runtime RAM data storage
                    locations
10                  ;*********
11                  ;* System inputs (1 byte each in RAM)
12 000800          Temperature DS.B  1
13 000801          Pressure    DS.B  1
14
15                  ;* Fuzzy Inputs (1 byte of RAM per label of a
                    system input)
16     0000 0802 FuzzyIns:   EQU  * ;Used to ref fuz ins & outs
17                  ;* Temperature
18 000802          Cold        DS.B  1
19 000803          Cool        DS.B  1
20 000804          Normal      DS.B  1
21 000805          Warm        DS.B  1
22 000806          Hot         DS.B  1
23                  ;* Pressure
24 000807          VeryLo      DS.B  1
25 000808          PLow        DS.B  1
26 000809          Medium      DS.B  1
27 00080A          PHigh       DS.B  1
28 00080B          VeryHi      DS.B  1
29                  ;* Fuzzy Outputs (1 byte of RAM per label of
                    a system output)
30     0000 080C FuzzyOuts:   EQU  *
31                  ;* Fan Speed
32 00080C          FStop       DS.B  1
33 00080D          Slow        DS.B  1
34 00080E          Mid         DS.B  1
35 00080F          Fast        DS.B  1
36 000810          FullOn      DS.B  1
37
38                  ;* System output (1 byte of RAM)
39 000811          FanSpeed:   DS.B  1
40
41                        ORG  $D00    ;Knowledge  base
                                       memory start
42                  ;*********
43                  ;* Following is the application specific
                    knowledge base
```

```
44                    ;**********
45                    Input_MFs:
46                    ;* Each trapezoidal MF def by 4 bytes -
                      pt1,pt2,slope1,slope2
47                    ;* Temperature
48 000D00 0058 0008           DC.B    $00,$58,$00,$08   ;MF  for
                                                         temp Cold
49 000D04 4088 0B0B           DC.B    $40,$88,$0B,$0B   ;MF  for
                                                         temp Cool
50 000D08 7898 2020           DC.B    $78,$98,$20,$20   ;MF  for
                                                         temp Normal
51 000D0C 88D0 0B0B           DC.B    $88,$D0,$0B,$0B   ;MF  for
                                                         temp Warm
52 000D10 B8FF 0800           DC.B    $B8,$FF,$08,$00   ;MF  for
                                                         temp Hot
53                    ;* Pressure
54 000D14 0030 0008           DC.B    $00,$30,$00,$08   ;MF  for
                                                         pressure VeryLo
55 000D18 1058 080B           DC.B    $10,$58,$08,$0B   ;MF  for
                                                         pressure Low
56 000D1C 4080 0B08           DC.B    $40,$80,$0B,$08   ;MF  for
                                                         pressure Medium
57 000D20 60A0 080B           DC.B    $60,$A0,$08,$0B   ;MF  for
                                                         pressure High
58 000D24 88FF 0B00           DC.B    $88,$FF,$0B,$00   ;MF  for
                                                         pressure VeryHi
59
60                    ;*****
61                    ;* Setup offsets so rules more understandable
62                    ;*  all offsets relative to start of Fuzzy_ins
63                    ;*****
64     0000 0000 T_Cold      EQU    (Cold-FuzzyIns) ;offset to
                                                     fuzzy input
65     0000 0001 T_Cool      EQU    (Cool-FuzzyIns)
66     0000 0002 T_Normal    EQU    (Normal-FuzzyIns)
67     0000 0003 T_Warm      EQU    (Warm-FuzzyIns)
68     0000 0004 T_Hot       EQU    (Hot-FuzzyIns)
69
70     0000 0005 P_VeryLo    EQU    (VeryLo-FuzzyIns)
71     0000 0006 P_Low       EQU    (PLow-FuzzyIns)
72     0000 0007 P_Medium    EQU    (Medium-FuzzyIns)
73     0000 0008 P_High      EQU    (PHigh-FuzzyIns)
74     0000 0009 P_VeryHi    EQU    (VeryHi-FuzzyIns)
75
```

```
76      0000 000A  F_Stop      EQU  (FStop-FuzzyIns)
77      0000 000B  F_Slow      EQU  (Slow-FuzzyIns)
78      0000 000C  F_Mid       EQU  (Mid-FuzzyIns)
79      0000 000D  F_Fast      EQU  (Fast-FuzzyIns)
80      0000 000E  F_FullOn    EQU  (FullOn-FuzzyIns)
81
82      0000 00FE  Sep         EQU  $FE       ;Seperator between
                                              ins/outs
83              ;* Same $FE is used to separate successive
                rules
84      0000 00FF  Rule_end    EQU  $FF       ;Rule list
                                              terminator
85
86              Rule_list:
87              ;* This comment shows the first rule as an
                example.
88              ;* If Temp is Cold & Pressure is Very_lo then
                set Fan to Stop.
89 000D28 0005 FE0A        DC.B T_Cold,P_VeryLo,Sep,F_Stop,Sep
                           ;Rule 1
   000D2C FE
90 000D2D 0006 FE0A        DC.B T_Cold,P_Low,Sep,F_Stop,Sep
                           ;Rule 2
   000D31 FE
91 000D32 0007 FE0A        DC.B T_Cold,P_Medium,Sep,F_Stop,Sep
                           ;Rule 3
   000D36 FE
92 000D37 0008 FE0A        DC.B T_Cold,P_High,Sep,F_Stop,Sep
                           ;Rule 4
   000D3B FE
93 000D3C 0009 FE0A        DC.B T_Cold,P_VeryHi,Sep,F_Stop,Sep
                           ;Rule 5
   000D40 FE
94
95 000D41 0105 FE0A        DC.B T_Cool,P_VeryLo,Sep,F_Stop,Sep
                           ;Rule 6
   000D45 FE
96 000D46 0106 FE0A        DC.BL T_Cool,P_Low,Sep,F_Stop,Sep
                           ;Rule 7
   000D4A FE
97 000D4B 0107 FE0B        DC.B T_Cool,P_Medium,Sep,F_Slow,Sep
                           ;Rule 8
   000D4F FE
```

```
 98 000D50 0108 FE0B        DC.B T_Cool,P_High,Sep,F_Slow,Sep
                            ;Rule 9

    000D54 FE
 99 000D55 0109 FE0B        DC.B T_Cool,P_VeryHi,Sep,F_Slow,Sep
                            ;Rule 10

    000D59 FE
100
101 000D5A 0205 FE0B        DC.B T_Normal,P_VeryLo,Sep,F_Slow,Sep
                            ;Rule 11

    000D5E FE
102 000D5F 0206 FE0B        DC.B T_Normal,P_Low,Sep,F_Slow,Sep
                            ;Rule 12

    000D63 FE
103 000D64 0207 FE0C        DC.B T_Normal,P_Medium,Sep,F_Mid,Sep
                            ;Rule 13

    000D68 FE
104 000D69 0208 FE0C        DC.B T_Normal,P_High,Sep,F_Mid,Sep
                            ;Rule 14

    000D6D FE
105 000D6E 0209 FE0C        DC.B T_Normal,P_VeryHi,Sep,F_Mid,Sep
                            ;Rule 15

    000D72 FE
106
107 000D73 0305 FE0B        DC.B T_Warm,P_VeryLo,Sep,F_Slow,Sep
                            ;Rule 16

    000D77 FE
108 000D78 0306 FE0C        DC.B T_Warm,P_Low,Sep,F_Mid,Sep
                            ;Rule 17

    000D7C FE
109 000D7D 0307 FE0D        DC.B T_Warm,P_Medium,Sep,F_Fast,Sep
                            ;Rule 18

    000D81 FE
110 000D82 0308 FE0D        DC.B T_Warm,P_High,Sep,F_Fast,Sep
                            ;Rule 19

    000D86 FE
111 000D87 0309 FE0D        DC.B T_Warm,P_VeryHi,Sep,F_Fast,Sep
                            ;Rule 20

    000D8B FE
112
113 000D8C 0405 FE0C        DC.B T_Hot,P_VeryLo,Sep,F_Mid,Sep
                            ;Rule 21

    000D90 FE
```

```
114 000D91 0406 FE0D        DC.B  T_Hot,P_Low,Sep,F_Fast,Sep
                            ;Rule 22
    000D95 FE

115 000D96 0407 FE0D        DC.B  T_Hot,P_Medium,Sep,F_Fast,Sep
                            ;Rule 23
    000D9A FE

116 000D9B 0408 FE0E        DC.B  T_Hot,P_High,Sep,F_FullOn,Sep
                            ;Rule 24
    000D9F FE

117 000DA0 0409 FE0E        DC.B  T_Hot,P_VeryHi,Sep,F_FullOn
                            ;Rule 25

118 000DA4 FF               DC.B  Rule_end
119
120              Output_MFs:
121 000DA5 00               DC.B  $00       ;MF for Fan Stop
122 000DA6 20               DC.B  $20       ;MF for Fan Slow
123 000DA7 50               DC.B  $50       ;MF for Fan Mid
124 000DA8 A0               DC.B  $A0       ;MF for Fan Fast
125 000DA9 FF               DC.B  $FF       ;MF for Fan FullOn
126              ;* End of Knowledge Base
127
128                  ORG   $8000      ;Program memory start
129              ;**********
130              ;* Following is the executable part of the
                 fuzzy program
131              ;**********
132 008000 CE0D 00   Fuzzify: LDX  #Input_MFs  ;Point    at    MF
                                                 definitions
133 008003 CD08 02            LDY  #FuzzyIns    ;Point at fuzzy
                                                 ins in RAM
134 008006 B608 00            LDAA Temperature  ;Get first sys-
                                                 tem input
135 008009 C605               LDAB #5           ;Temperature
                                                 has 5 labels
136 00800B 01     Fuz_loop:   MEM               ;Evaluate one MF
137 00800C 0431 FC            DBNE B,Fuz_loop   ;For 5 labels
                                                 of 1 input
138 00800F B608 01            LDAA Pressure     ;Get second
                                                 system input
139 008012 C605               LDAB #5           ;Pressure has 5
                                                 labels
140 008014 01    Fuz_loop1: MEM                 ;Evaluate one MF
```

```
141 008015 0431 FC            DBNE B,Fuz_loop1   ;For 5 labels of
                                                  1 input
142                    ;* Here X points at rule list, Y at fuzzy
                       outputs
143 008018 C605             LDAB #5              ;5 fuzzy outputs
                                                  in RAM
144 00801A 6970    Rule_eval: CLR  1,Y+          ;Clr a fuz out &
                                                  inc pntr
145 00801C 0431 FB          DBNE B,Rule_eval     ;Loop to clr all
                                                  fuz outs
146 00801F CD08 02          LDY  #FuzzyIns       ;Pointer to fuz
                                                  ins & outs
147                    ;* X already pointing at top of rule list
148 008022 86FF             LDAA #$FF            ;Init A & clears
                                                  V-bit
149 008024 183A             REV                  ;Process rule
                                                  list
150                    ;* Here X points at output MFs, Y at fuzzy
                       inputs
151 008026 CD08 0C    Defuz:   LDY #FuzzyOuts    ;Point at fuzzy
                                                  outputs
152                    ;* X already pointing at singleton MFs
153 008029 C605             LDAB #5              ;5 fuzzy outs per
                                                  sys out
154 00802B 183C             WAV                  ;Calc sums for
                                                  wtd ave
155 00802D 11               EDIV                 ;Final divide for
                                                  wtd ave
156 00802E B764             TFR  Y,D             ;Move result to
                                                  A:B
157 008030 7B08 11          STAB FanSpeed        ;Store system
                                                  output
158 008033 3D               RTS
159                    ;**********
```

19.11 Fuzzy Logic

Fuzzy Logic Instructions

The fuzzy logic instructions are the *Membership Function (MEM), Rule* and *Weighted Rule Evaluation (REV* and *REVW), and *Weighted Average (WAV).*

MEMBERSHIP FUNCTION

MEM fuzzifies the inputs. A current system input value is compared against a stored input membership function to determine the degree to which a label of the system input is true. MEM finds the y-value of the current input on a trapezoidal membership function. Before MEM is executed, the A, X, and Y registers must be set up as follows:

A must hold the current, 8-bit, crisp value of a system input variable.

X must point to a 4-byte data structure that describes the trapezoidal membership function for a system input label.

Y must point to the fuzzy input (RAM location) in which the resulting grade of membership (the truth value) is to be stored.

The data structure that describes the trapezoidal membership function is defined by the following 4 bytes, in order, starting at the address contained in X:

Point_1: The starting point for the leading edge of the trapezoid.

Point_2: The position for the rightmost point of the trapezoid.

Slope_1: The slope of the leading edge of the trapezoid.

Slope_2: The magnitude of the trailing edge slope.

A slope of $00 indicates a special case of infinite slope. Slope_1 = $00 indicates the membership function starts with a value of $FF at Point_1; Slope_2 = $00 is for a trapezoid whose ending value is $FF at Point_2.

After MEM is executed, A is unchanged, the memory location to which Y pointed contains the grade of membership calculated by MEM, Y is incremented (by one) to point to the next fuzzy input location, and X is incremented (by four) to point to the next membership function data structure.

Example 19-4 shows the use of the membership functions and the MEM instruction. *Lines 48–58* show the trapezoidal membership functions for each of the five labels for the two input variables. The membership grade (the fuzzy input truth value) for each label is evaluated by the MEM instructions on *lines 136* and *140*.

RULE EVALUATION

REV (*Rule Evaluation*) and *REVW* (*Weighted Rule Evaluation*) are similar and perform unweighted and weighted evaluation of a list of rules. They use fuzzy inputs (produced by MEM) to produce fuzzy outputs. REV and REVW can be interrupted so they do not adversely affect interrupt latency. The rule evaluation step first finds the minimum of the fuzzy inputs specified by the rule. It then sets the fuzzy output(s) to the maximum of this calculated minimum and the current fuzzy output.

Before executing the REV instruction, the following setup operations must be done:

X must point to the first 8-bit element in the rule list.

Y must point to the base address for fuzzy inputs and outputs.

A must contain the value $FF and the CCR V-bit must be 0. (LDAA #$FF places the correct value in A and clears V.)

All current fuzzy outputs must be set to zero.

Each rule in the knowledge base consists of a table of 8-bit offsets (from the base address of the table contained in the Y register) defining the fuzzy inputs associated with the rule.

These are called the antecedent offsets. Terminating this list is the separator character $FE. Following the separator character is a list of 8-bit offsets pointing the fuzzy output(s) to be set. This is called the consequent list. Following this list is either a $FE separator or, if the rule is the last in the knowledge base, an end-of-list indicator $FF.

During execution X is incremented to point to the next rule. When the rule evaluation is complete, X points to the address following the $FF and A contains the truth value for the last rule. The value in Y does not change.

In Example 19-4 the rule list is defined in *lines 89–118* and the rule setup and evaluation is carried out by the code in *lines 142–149*.

REVW is similar to REV except that the minimum calculated result can be multiplied by a weighting factor before finding the maximum for the fuzzy output. The knowledge base is similar also except the antecedent and consequent pointers are 16 bits each, $FFFE is the separator code, and $FFFF signifies the end of the list.

The multiplication by the weighting factor is optional and controlled by the state of the C (carry) bit in the CCR. Setting the C bit enables weighted rule evaluation.
The following setup operations must be done:

X must point to the first 16-bit element in the rule list.

A must contain the value $FF and the CCR V bit must be 0. (LDAA #$FF places the correct value in A and clears V.)

All current fuzzy outputs must be set to zero.

Set or clear the carry bit (to enable or disable) weighted evaluation. When weighted evaluation is to be done, the Y register must point to the first item in a table of weighting factors. These factors are 8 bits each and there must be one for each rule to be evaluated.

WEIGHTED AVERAGE

After the rule evaluation step gives us the fuzzy output values, the final step is to defuzzify the outputs to give a final value to each system output. The *Weighted Average (WAV)* does this. WAV computes a sum-of-products and a sum-of-weights that are needed to find a final weighted average. The knowledge base includes a list of singleton output membership functions (8-bit values relating each output label to a specific output value). Before WAV executes, the X register points to the first singleton membership function in the knowledge base, the Y register points to the fuzzy outputs, and the B register contains the number of elements to be included in the calculation. WAV calculates a 24-bit sum-of-products and returns it in Y:D. The sum of the weights is calculated and returned in X. The final weighted average is calculated by following WAV with an EDIV instruction.

In Example 19-4 the output singleton membership functions are in *lines 121–125*. The defuzzification step is accomplished in *lines 151–157*.

Minimum and Maximum Instructions

The minimum and maximum instructions were implemented for general purpose applications but they can be useful also for custom fuzzy logic programs. They are shown in Table 19-5 and find the minimum or maximum of two operands. The instructions perform an unsigned subtraction, setting the condition code register bits appropriately, and replacing the specified operand with the minimum (or maximum).

TABLE 19-5 Fuzzy Logic and Maximum/Minimum Instructions

			Addressing Mode						Condition Codes			
Function	**Opcode**	**Symbolic Operation**	**IMM**	**DIR**	**EXT**	**IDX**	**IDR**	**INH**	**N**	**Z**	**V**	**C**
Membership Function	MEM								?	?	?	?
Rule Evaluation	REV								?	?	↕	?
Weighted Rule Evaluation	REVW								?	?	↕	↕
Weighted Average	WAV								?	1	?	?
Minimum → D	EMIND	MIN(D,(M:M + 1)) → D				x	x		↕	↕	↕	↕
Minimum → (M)	EMINM	MIN(D,(M:M + 1)) → M				x	x		↕	↕	↕	↕
Minimum → A	MINA	MIN(A,(M)) → A				x	x		↕	↕	↕	↕
Minimum → (M)	MINM	MIN(A,(M)) → (M)				x	x		↕	↕	↕	↕
Maximum → D	EMAXD	MAX(D,(M:M + 1)) → D				x	x		↕	↕	↕	↕
Maximum → (M)	EMAXM	MAX(D,(M:M + 1)) → M				x	x		↕	↕	↕	↕
Maximum → A	MAXA	MAX(A,(M)) → A				x	x		↕	↕	↕	↕
Maximum → (M)	MAXM	MAX(A,(M)) → (M)				x	x		↕	↕	↕	↕
Ext Mult and Accum	EMACS								↕	↕	↕	↕
Table Lookup	ETBL					x			↕	↕	—	↕
Table Lookup	TBL					x			↕	↕	—	↕

Table Lookup Instructions

The table lookup instructions, TBL and ETBL, linearly interpolate to one of 256 result values that fall between pairs of data entries in a lookup table stored in memory. TBL uses 8-bit table entries and returns an 8-bit result in A. ETBL uses 16-bit table entries and provides a 16-bit result in D. To use either instruction, initialize the X register pointing to the first data point (P1) and the B register with the binary fraction (radix point to the left of the most significant bit) representing the interpolation point for which a value is needed. The value returned in A or D is the unrounded result:

$$A\ (\text{or } D) = P1 + [(B) \times (P2 - P1)]$$

where P1 is the first data point and P2 is the second data point immediately following P1. See Example 19-5.

Example 19-5

An 8-bit data table with two entries is defined by the following:

```
P1:   DB   2
P2:   DB   8
```

Show how to initialize the registers to use the TBL instruction and specify the result returned when you wish to find the data value midway between P1 and P2. Specify the result for three-quarters of the way between P1 and P2.

Solution:

```
; Initialize X and B
        ldx     #P1                 ; Point to the first value
        ldab    #%10000000          ; Binary fraction for 0.5
        tbl     0,x
; On return, A contains the value 5
        ldab    #%11000000          ; Binary fraction for 0.75
        tbl     0,x
; On return, A contains the value 6
```

19.12 Conclusion and Chapter Summary Points

Like other control system methodologies, fuzzy logic produces a precise repeatable output for every combination of input conditions. Although the term "fuzzy" seems to suggest the control output is approximate, this is simply not true. Instead, fuzzy refers to a new way of representing sets that describe linguistic concepts such as warm. The boundaries of fuzzy sets are more like dimmer controls than digital on–off switches. As temperature decreases from a value we might agree is "warm," it eventually stops being warm and becomes cool or even cold. This transition is not sudden at some specific temperature—rather the degree to which we agree that the temperature is warm gradually decreases until we would no longer consider it to be warm. Since fuzzy logic offers a way to attach an unambiguous numerical meaning to these abstract concepts such as warm, we can now perform computations on these linguistic variables to design automatic control systems.

There are three processing steps in a fuzzy logic program:

- *Fuzzification* compares the current value of a system input against the membership functions of each label of that input to determine fuzzy input values.
- *Rule evaluation* plugs fuzzy input values into control rules to determine fuzzy output values.
- *Defuzzification* combines all fuzzy outputs into a single composite control output value.

The executable fuzzy logic program is generally not specific to any application.

For each of the three fuzzy program steps there is a data structure in the knowledge base that provides application-specific information. Input membership functions provide the meaning of each input label for the fuzzification step. A list of rules is encoded as a list of pointers to fuzzy inputs and fuzzy outputs. Output membership functions provide the meaning of each output label for the defuzzification step.

Although it seems as if there could be a large number of linguistic rules in a control system, the number of rules is actually limited to the product of the number of labels in each of the system inputs. For example, a system with two inputs with 5 labels each would have a maximum of 5 × 5 or 25 rules.

19.13 Bibliography and Further Reading

AN1295: Demonstration Model of fuzzyTECH® Implementation on M68HC12, Motorola Semiconductor Application Note, Austin, TX, 1996.

M68HC12 Reference Manual, Motorola, 1996.

von Altrock, C., *Fuzzy Logic & NeuroFuzzy Applications Explained*, Prentice Hall PTR, Englewood Cliffs, NJ, 1995.

19.14 Problems

Basic

19.1 Using table lookup (instead of a fuzzy logic system), how many bytes would be needed for the table in a system with three inputs with 8-bit resolution on each input and one output with 8-bit resolution? **[k]**

19.2 Referring to Figure 19-4, for what temperature or temperatures are the expressions "temperature is cold" and "temperature is warm" true to the same degree? **[k]**

Intermediate

19.3 Sketch graphical representations of the following input membership functions. **[k]**

 a. LOW FCB $20,$60,$08,$10 ;pt1,pt2,slope1,slope2

 b. MID FCB $60,$A0,$10,$0B ;pt1,pt2,slope1,slope2

 c. HI   FCB $A0,$E0,$0B,$08 ;pt1,pt2,slope1,slope2

19.4 For the fuzzy logic system in Example 19-4, how many bytes of memory in the knowledge base did it take for all 25 rules? **[k]**

19.5 In Figure 19-9, how many rules would be active if temperature was $C0 and pressure was $90? Which rule(s) would be active? **[k]**

19.6 Using Figure 19-11, what value should be in FanSpeed if temperature is $B0 and pressure is $38? **[k]**

Advanced

19.7 What is the maximum number of rules that would be needed in a fuzzy logic system that has three inputs with three labels each? How many fuzzy inputs would there be in such a system and how many bytes of RAM would that require? **[k]**

20 Debugging Systems

OBJECTIVES

In this chapter we investigate the on-chip systems that support application development and debugging. The Background Debug™ Module (BDM) allows access to internal operation of the microcontroller through a single interface pin. An on-chip *in-circuit emulator* (ICE) system allows a developer to set up complex trigger conditions and capture real-time bus activity in an on-chip capture buffer for later examination. An instruction tagging mechanism allows external development hardware to stop the system on specific instruction boundaries.

20.1 Introduction

As silicon processing has progressed, it has become possible to build more and more complex processing systems into a single chip. The rate of change has been so rapid that the tools that engineers use to debug systems based on these microcontrollers have been unable to keep pace. In some cases, entirely new debugging strategies were enabled. The background debug system in the HCS12 is an example of such a new strategy. The BDM allows a developer to see and modify internal memory and registers as well as control breakpoints and single stepping of the application program. A single microcontroller pin (BKGD) is used for the BDM in order to minimize any disturbance of the application system.

One of the most common types of development tool in the past was the in-circuit emulator (ICE). In an external ICE system, you would remove the actual microcontroller chip from the target application and connect an emulator in its place. The emulator attempted to duplicate the functions of the original microcontroller while allowing direct access to internal data and control signals. This allowed the user to stop the system at any time and examine, or even modify, the contents of internal registers before continuing with the original program. The user could also establish complex sets of trigger conditions and cause the system to stop if or when these conditions ever occurred.

It is difficult, if not impossible, to develop an in-circuit emulator for modern microcontrollers.

Traditional ICE systems have run into two major problems. First, the newer microcontrollers have many more pins and smaller pin spacing than older microcontrollers. The smallest leadless packages such as QFN packages even place the connection points under the device to reduce the overall size of the package. This makes the mechanical connection of an ICE much more difficult or even impossible.

The second challenge is that modern microcontrollers have fast internal signals, internal pipelines, and multiple data buses so there can be more activity going on within the microcontroller than you can report on the pins of the chip. To complicate the situation even further, the tiny propagation delays for signals to pass through interface pin buffers can be so long compared to signals inside the chip, that it is almost impossible to duplicate on-chip processes with external logic. Bundling dozens of such fast signals into an umbilical cable to connect the emulator to the target application can make the problem completely unsolvable.

The newest HCS12 microcontrollers address this problem by building the circuits for an ICE inside the microcontroller to form an on-chip ICE system. This logic is just as fast as the other parts of the microcontroller system. In the past, this would not have been practical because it would have increased the size (and therefore the cost) of the microcontroller even though this logic is primarily used only during application development. The size of individual transistors on a modern microcontroller has decreased so dramatically and the total number of transistors in the microcontroller has increased so much, that debug logic can now be included at negligible cost. As silicon technology continues to advance, this trend will continue until arbitrarily complex development logic can be included on the silicon die with the microcontroller.

20.2 Software and Hardware Breakpoints

We discussed how to use breakpoints to stop a program's flow and showed the CodeWarrior debugger in Chapter 9. We will now compare three types of breakpoints that are useful in different situations. SWI-based software breakpoints are established by replacing an instruction opcode in an application program with an SWI opcode. BGND-based software breakpoints are used with a background debug pod and are similar to the SWI breakpoints except a BGND opcode is used rather than the SWI opcode. Hardware breakpoints use comparator logic to detect specific address or data patterns and then directly force the microcontroller to active background mode. Since hardware breakpoints do not require a SWI or BGND opcode to be written into the application program, they work in ROM and Flash memory areas where software breakpoints cannot be used. Hardware breakpoints also work to stop a program when some data location is accessed rather than being limited to instruction opcode addresses.

SWI-BASED SOFTWARE BREAKPOINTS

This type of breakpoint has been used in debug monitor programs for many years. The user specifies the address(s) where a breakpoint is needed. When a GO or CONTINUE command is entered to start execution of the application program, the monitor replaces the application opcode at each breakpoint address with an SWI opcode. The application opcodes are saved so they can be restored when the monitor program regains control. The application program executes normally until one of these SWI opcodes is encountered, at which time the user's CPU register values are saved on the application program's stack and the monitor regains control. The monitor then restores the original application opcodes at each breakpoint address. This assures that the user can examine the application program opcodes while the monitor program is in control.

The monitor program can allow as many software breakpoints as it likes by choosing how much monitor storage memory it wishes to dedicate for breakpoints. The monitor needs 3 bytes of RAM for each breakpoint—2 bytes for the address and another byte to save the original application opcode while the application program (as opposed to the monitor program) has control. The D-Bug12 monitor program allows ten SWI-based breakpoints while it is configured for EVB mode. Since SWI-based breakpoints execute an SWI instruction to save user registers when a breakpoint is encountered, nine extra bytes are needed on the user's stack. Usually this is not a problem, but if the user's program causes the stack pointer to be corrupted, the monitor can lose track of the contents of user registers (they will be reported incorrectly after the breakpoint). There could also be unintended writes to unexpected application memory areas if the user program accidentally sets the stack pointer to point into a register or variable area of memory.

Since this breakpoint mechanism uses SWI, it interferes with any application use of SWI. There are techniques to allow both SWI breakpoints and user SWI instructions, but it causes significant complications and may have trouble in some unusual circumstances such as when a breakpoint is set at the address of a user SWI instruction. This breakpoint mechanism also requires the program being debugged to be located in RAM so the SWI opcode switch can be made by the monitor program.

BGND-BASED SOFTWARE BREAKPOINTS

The D-Bug12 monitor program, used with some evaluation boards with M68HC12 microcontrollers, can operate in two distinct modes. In the EVB mode, it behaves as a traditional ROM monitor. In the POD mode, D-Bug12 uses a BDM interface to control a target M68HC12 or HCS12, so the monitor program is not using any memory resources in the target system. In EVB mode, software breakpoints are based on SWI opcodes as described earlier. In the POD mode, BGND opcodes are used instead of SWI opcodes. BGND can only be used for breakpoints when the monitor is controlling a target system through BDM.

The software that manages breakpoints is more or less the same for SWI-based or BGND-based breakpoints. You still need to store a 16-bit address and the user opcode from that address for each breakpoint. Since the monitor is not using memory from the user's memory map for this, it is less intrusive to add more breakpoints. Breakpoints are still entered into the application memory as control changes from the monitor to the user program (as in a GO command), and they are removed when control goes back to the monitor program (as in when a breakpoint is encountered). When the user application is running and a BGND-based breakpoint is encountered, the BGND opcode causes the target microcontroller to enter active background mode, but user registers are not stacked on the user's stack as they were for SWI-based breakpoints. This is significantly less intrusive than the older SWI-based breakpoints. BGND-based breakpoints also work correctly even if the application program fails to maintain the stack pointer correctly because the BGND instruction does not use the application stack to save registers.

BGND-based software breakpoints do not interfere with any application use of the SWI instruction. There are no application uses for BGND instructions so there is no danger of any interference between monitor and application use of BGND instructions.

HARDWARE BREAKPOINTS

Hardware breakpoints provide a way to stop normal execution of an application program without adding external hardware. Successful breakpoint match events can cause one of two actions depending on the selected breakpoint mode. A breakpoint can cause the microcontroller to go into active BDM mode, or it can cause the CPU to execute an SWI instruction instead of the opcode at the breakpoint address. In addition, there are two different mechanisms for handling a breakpoint. A tagging mechanism causes the CPU to execute a BGND or an SWI instruction rather than the opcode at the breakpoint address. In addition to the tagging mechanism, an interrupt-like mechanism can be used which causes the CPU to enter active BDM mode at the next instruction boundary after the match event occurred. When this interrupt mechanism is used, R/W_L and data can be included in the match comparisons and the breakpoint need not be set to the address of an opcode. In all cases where the breakpoint causes the microcontroller to enter active BDM mode, it is assumed that the ENBDM control bit in the BDM STATUS register has been previously set to allow active BDM mode (or the microcontroller will just return to the application program).

In the MC9S12C32, hardware breakpoints are supported by comparators in the on-chip ICE system. This system can be configured to operate in a legacy breakpoint compatibility mode. BKP mode configures the comparators to operate like the breakpoint module in older MC68HC12 devices. The newer on-chip ICE, or DBG module, provides better support for breakpoints in expanded PPAGE memory space.

Hardware breakpoints are different from software breakpoints in several important ways. Hardware breakpoints do not write anything into application program memory so they can be used to set breakpoints in memory that cannot be modified during a debug session (such as ROM or Flash memory). Since hardware breakpoints are based on real-time comparison against address and data buses rather than opcode substitution, hardware breakpoints are not limited to addresses that correspond to instruction opcodes. This means you can set a break to occur when a certain control register or memory location is written (something you cannot do with a software breakpoint).

Hardware breakpoints require comparators and control logic in the target system. Since this logic takes up valuable space on every microcontroller, there is a limit to the number of such breakpoints you can allow. The silicon area for this logic is getting smaller as silicon processing shrinks. In the MC9S12C32, the on-chip ICE module supports three address breakpoints or one address + data breakpoint and one address-only breakpoint.

20.3 Development Related Features of the MC9S12C32

There are three systems in the MC9S12C32 that support development and debugging of application programs and systems. After a brief introduction to these systems, we will take a closer look at each one.

- **BDM:** The background debug system is the most obvious debug feature in the HCS12. Serial commands are provided to the target microcontroller and data are sent-to and received-from the target microcontroller through a special single-wire interface. This system allows you to read from or write to any system memory location, read from or write to CPU registers, stop processing application programs, trace application program instructions one at a time, or go to any address to start executing application program instructions.

- **On-Chip ICE:** This system is like a traditional external emulator, but it is built entirely within the microcontroller die and uses no microcontroller pins. The system includes comparators, trigger logic, and a capture buffer, which captures address and/or data information in real time without affecting normal program execution or timing. When the on-chip ICE system is disabled, these comparators can function as hardware breakpoints to set three address breakpoints or one full address/data/control breakpoint plus an address-only breakpoint. Although this system is used primarily for debugging, in a ROM microcontroller it can be used to provide for a few patches to correct software errors in the mask-programmed ROM.

- **Instruction Tagging:** This is a way of making the CPU stop execution of an application program when you reach a specific instruction. In older microprocessor systems, interrupts were used for this function, but the HCS12 has an instruction queue (or pipe), which means instructions are not executed immediately after they are fetched from memory. In the case of jumps or branches, it is even possible that an instruction will be fetched, but never executed. Instruction tags let a developer mark instructions as they are fetched from memory, and then if or when that instruction reaches the CPU, the background debug mode is entered rather than executing the tagged instruction.

20.4 Background Debug Module (BDM)

> The HCS12 BDM system uses a single wire (plus ground) for all debugging.

The background debug system in the HCS12 allows access to the internal operation of the target system and does not interfere with the application hardware or software. You can even read or write to target system memory locations without stopping the running application program. It uses a separate serial I/O interface (BKGD pin) and no user memory. With the HCS12 BDM, you need only provide a way to connect to ground and the BKGD pin. The HCS12 also includes logic for tracing a single instruction at a time in the target system.

As its name implies, the background debug system is primarily intended for system development and debugging. However, the BDM system is also useful for other applications. The BDM can be used to load or reload an application program into a target system after the product has been completely assembled. It can be used also to calibrate finished systems or perform field upgrades of operation software. In a data logging application, the BDM can be used to retrieve logged data, thus making it unnecessary to add this function to the application software. We will explore some of these nondebug uses later in this section after a more detailed explanation of the BDM system.

As Figure 20-1 shows, a typical HCS12 BDM system consists of a desktop PC, an intelligent interface BDM pod, and the target HCS12 system. The link from the PC to the intelligent pod is typically a USB connection. The link from the pod to the target system is a custom serial interface to the single-wire BDM system of the HCS12 target.

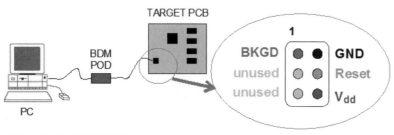

Figure 20-1 HCS12 BDM system.

A standard six-pin connector allows a pod to be connected to any target HCS12 system. The connection can be as simple as the BKGD and ground pins. It is more common to include reset so the PC can remotely force a target system reset. A V$_{dd}$ connection would allow the pod to steal power from the target system.

Single-Wire Physical Interface

The single background interface pin (BKGD) is used for three functions in the HCS12 microcontroller. During reset, this pin is a mode select input that selects between normal and special modes of operation. After reset, this pin becomes the dedicated serial interface pin for the background debug mode. If an appropriate serial command is received, this pin can be used to tag selected program instructions as they are fetched into the instruction queue.

Special Mode Select Function

The background debug system is always available regardless of operating mode. Memory access does not require background mode to be "active," so these debug commands may be used at any time and in any operating mode. Other debug commands require the microcontroller to be in "active" background mode, where the application program is stopped and BDM firmware is being executed instead. One way to get into active background mode is to reset the microcontroller into special single-chip mode. This means you could design a system that operates in normal single-chip mode unless a debug system is connected to the BKGD pin. When the debug system is connected, it could hold BKGD low (selecting special vs. normal mode) while providing an active-low reset signal.

This could be especially useful in a system where the main program is in Flash. A programmer or debug system could be connected to the BKGD pin, and the Flash memory could be programmed before attempting to reset in normal mode. When the programmer is removed from the BKGD pin, the BKGD pin is pulled up through a simple pull-up resistor and the target system resets in normal single-chip mode.

Tagging Function

Because of the HCS12's instruction queue (or pipe), instructions are fetched from memory several cycles before they are executed. In the case of change-of-flow instructions such as jump and branch, some instructions that were fetched into the pipe may be flushed from the pipe without ever being executed. If you want the CPU to stop at a certain instruction, you would tag the instruction as it was fetched into the instruction queue. This tag signal follows along with the program information in the instruction queue. If this tag is set as the CPU12 is about to start an instruction, an exception causes the microcontroller to enter active background mode instead of executing the tagged instruction.

There are two tagging inputs so that you can tag the even byte, the odd byte, or both bytes of the program word as it is fetched into the instruction queue. The BKGD pin is used to tag the even (high) byte if tagging is enabled. The low strobe (LSTRB_L) pin is used to tag the odd (low) byte if tagging is enabled. Tag signals are captured at the falling edge of Bus Clock where the fetched program word is also valid.

Tagging is enabled when a serial debug command (TAG_GO) is issued to resume an application program from active background mode. While tagging is enabled, the BKGD pin cannot be used for serial debug commands because the interface would not be able to distinguish a serial command from a tag signal. If a debugger, connected to the BKGD pin, wanted to regain control, it could simply drive the BKGD pin low and hold it. Within a few instructions, this would be seen as a tag request, and the microcontroller would enter active background mode. While in active background mode, tagging is disabled and the BKGD pin is dedicated for serial BDM communications.

Background Communication Function

The primary use of the BKGD pin is for bidirectional communications between an external host, such as a PC, and the target HCS12 microcontroller. To keep this to a single pin, a custom serial protocol was devised. This protocol is more tolerant of speed variations than other asynchronous protocols like that used for ordinary RS-232 terminals. A debugging pod, as shown in Figure 20-1, converts high-level commands—for example, setting a breakpoint—into the custom serial bit stream. A debugging program, such as CodeWarrior, provides the high-level user interface for the background debug applications discussed next and for the in-circuit emulator shown in Section 20.6.

20.5 Application Uses for BDM

This section discusses applications of the BDM system including several non debug applications.

Nonvolatile Memory Programming

> The BDM can program Flash memory, EPROM, and EEPROM in a target application.

One of the most valuable uses for the BDM system is to program system memory including internal Flash memory and EEPROM as well as external nonvolatile memory from other manufacturers. This is possible because the BDM system can gain direct control of a target system and can access any location the CPU can access through a simple BDM connector with only two to six pins. This is independent of any user software in the target system so the user does not have to plan for BDM access other than to provide the physical BDM connector. With so few connections needed, it is often possible to tack wires into a system even if a BDM connector was not provided.

Target System Calibration

Suppose you have a target system with sensors that need to be calibrated. Due to sensor-to-sensor variations and sensitive analog circuitry, it is best to perform calibration after final assembly of the application system so all parasitic effects can be taken into account during the calibration. BDM allows access to the final system through a minimum two-wire interface. This is small enough for even the smallest embedded control application systems. As an added advantage, the calibration code can be downloaded into the application system through BDM at the time of calibration so it need not take up any memory space in the normal application program.

Field Diagnostics and Code Changes

Debugging problems in the field can be done with the BDM.

It would be nice if you could write complex controller programs and just forget about them. Unfortunately, it is all too common to discover problems in the field after system delivery. Often, the problems disappear as soon as the controller is returned to a factory service center (because the subtle conditions that caused the problem are only present at the field location where it failed). The BDM system allows access to the internal operation of the product while it is executing the application program. If the problem cannot be located unobtrusively, the BDM allows you to load pieces of diagnostic code to help isolate the application bugs. Once a bug is found, the BDM can be used to erase and reprogram the system Flash memory or EEPROM. The same technique can be used to load maintenance updates to application products in the field without completely disassembling the product.

This technique was used at a major automotive manufacturer to locate a bug in the BDLC module of an early version of the 68HC912B32. Because of the simplicity of a BDM connection, the debugging could be performed on the module without removing it from the vehicle. While looking for the bug, it was discovered that the BDLC communications system in the vehicle was experiencing thousands of corrupted bus communications (it was surprising the bus could even continue to function with so many errors). The silicon bug would not have been detected if it were not for the bus errors, but the bus errors were also a serious problem that needed to be solved. If the controller module had been removed from the vehicle to diagnose the errors, they would not have been found because the errors had to do with the interaction of the module with the noisy BDLC communication bus.

This debug session also relied heavily on the use of the on-chip hardware breakpoint module in the MC68HC912B32. The system used the COP watchdog system and that interfered with debugging. Every time the debugger pod tried to gain control through BDM, a COP timeout would trigger, causing the system to reset. To get past this, a hardware breakpoint was inserted before the initialization code that enabled the COP. In response to this breakpoint, the debugger inserted a substitute initialization routine that left the COP disabled. This allowed the debug session to proceed.

Data Logging Applications

In a typical data logging application you would have a program to collect and record the desired data. You would also include some utility program to allow the collected data to be transferred from the data collection system into some other system for analysis and recordkeeping. The BDM system can eliminate the need for this extra utility program, leaving more memory space for the primary data collection software. After data has been collected, the BDM system can be used to read the recorded information out of the data collection system. BDM commands can then be used to erase the logging memory so the data collection system is ready for its next use.

20.6 On-Chip Real-Time ICE System

The on-chip in-circuit emulator allows a program to run at full speed to a breakpoint while capturing essential information for debugging.

The on-chip in-circuit emulator (ICE) system essentially duplicates the functions of a traditional external emulator system with on-chip circuits. As the name implies, an emulator emulates the system under development but also provides access to internal bus information and controls that allow a developer to eavesdrop on running programs and stop the system under development at critical times during execution. These systems also provide complex trigger logic, which allows the developer to specify much more subtle combinations of conditions leading to a trigger or breakpoint than standard breakpoints allow.

The problems associated with traditional emulator systems have made them difficult to use for general debugging. As a result, they are normally used only in the most difficult debugging situations. The on-chip ICE system overcomes these problems so this style of debugging is now available to any designer. We will discuss some of the problems with traditional emulation and explain how the on-chip version solves these problems.

Real-Time Bus Capture Versus Breakpoints and Single-Instruction Trace

Prior to the availability of the on-chip ICE system, debugging was usually done with hardware breakpoints and single-instruction trace techniques. The idea was to place a breakpoint near the area of interest in the program under development. When this breakpoint was reached, the target system would stop executing the application program and wait for further commands. The developer would use background commands to examine CPU registers and memory contents and would then continue executing the application program one instruction at a time to watch for unexpected behavior. This is a perfectly good way to debug many programs. The MC9S12C32 supports this style of debugging through its single-wire background debug system and hardware breakpoint comparators.

In some applications it can be bad to stop the program, interrupt it with a breakpoint, or slow it down by single stepping. Imagine a program that controls a high-current load such as a motor H-bridge or a fuel injector. The normal program may be designed to activate such loads for very short periods. If the program was stopped while one of these loads was being driven, it could cause damage to the device. In this sort of application, the bus capture approach of an emulator is much better than using breakpoints and single stepping. The program can run through the process at normal speed while the emulator records what happened on a cycle-by-cycle basis. Later the developer can study the captured information to see exactly what the program was doing.

Real-time bus capture allows developers to see what their programs are doing at every bus cycle without having to stop the program. If you captured every bus cycle from the start of a program there would be too much information to search to find the area of interest, so emulators use comparators similar to hardware breakpoints to control when the emulator captures bus information. Additional logic is provided so that more complex trigger conditions can be set up. For example, two breakpoints can be combined such that the breakpoints must both be reached in the proper sequence before a trigger event is recognized. In other cases one comparator can monitor addresses while the other monitors data. In still other cases, two address comparators can set the endpoints of an address range and a trigger can be set up to occur if the address falls within or outside this range. These more complex trigger conditions allow developers to focus their examination in very specific areas of the program under development.

Traditional External ICE

An external emulator replaces the target system's microcontroller with an umbilical cord and attaches it to the development system.

Traditional external emulators replace the microcontroller on the target printed circuit board with the emulator through an umbilical cable and connection system. Inside the emulator there is usually a microcontroller similar to the one that is being emulated, except it operates this microcontroller in an expanded memory mode and disables internal nonvolatile memory systems such as Flash and EEPROM. These memories are replaced by emulation memory in the emulator, which is typically fast static RAM. Since the emulator is using many microcontroller pins to support the expansion bus, it also has extra circuitry to rebuild the normal pin functions for these pins. Finally, the emulator contains trigger logic and a large bus capture buffer.

It is not a simple matter to remove the microcontroller from a target system so you can connect the emulator. Modern microcontroller packages are often very fine pitch surface mount devices, so it can be a major challenge to connect the emulator umbilical cable. There are significant differences in the capacitive loading of the emulator cable compared to the original microcontroller. This extra capacitance affects timing and edge characteristics on microcontroller signals. The umbilical cable is also awkward and may interfere with access to other target system components.

The logic that rebuilds the normal pin functions is generally good at duplicating logical pin functions, but there are often differences in timing and drive characteristics. These differences can result in differences in operation with an emulator compared to operation with the actual microcontroller.

Because of the difficulty in designing external emulators and the very small production volumes, they often cost tens of thousands of dollars. As microcontrollers become faster and physically smaller, these costs continue to rise. Higher speed microcontroller buses force the emulators to use even faster static RAM, which is also very expensive. Address and data buses are getting wider so more pins are required to get information out of the emulation microcontroller.

Benefits of On-Chip ICE

Many of the difficulties associated with external emulators disappear when the emulation circuitry is built inside the microcontroller. The umbilical and connection systems are eliminated. It is no longer necessary to rebuild normal pin functions because no microcontroller pins are needed to connect to external emulator components. Speed and timing are no longer a problem because the transistors for the emulation circuits are identical to other logic transistors in the microcontroller. In fact, now the emulation can operate over the same temperature and supply voltage ranges as the microcontroller (because it is the microcontroller and not an emulated version of it).

Building the emulator on-chip does limit the number of separate trigger comparators and the amount of capture buffer memory that can be included without starting to increase the die cost for the microcontroller. The relative cost of this circuitry is getting smaller continuously as silicon processing technology continues to shrink and improve. Thus the next generation of on-chip emulators will continue to get bigger and better.

In the MC9S12C32 there are three comparators and the capture buffer is sixty-four 16-bit words. As we will see later in this section, information filtering can be used to make sure only important information is captured so this 64-word buffer can be equivalent to a much larger capture buffer in a traditional external emulator.

The biggest advantage to the on-chip emulation approach is that it eliminates the expensive external emulator box, thus saving thousands of dollars. This means that emulation-style debugging is now readily accessible to any HCS12 user. In fact, a student working in a dormitory room with free software tools and an inexpensive evaluation board can now perform the same kind of sophisticated development tasks for which major corporations pay tens or hundreds of thousands of dollars.

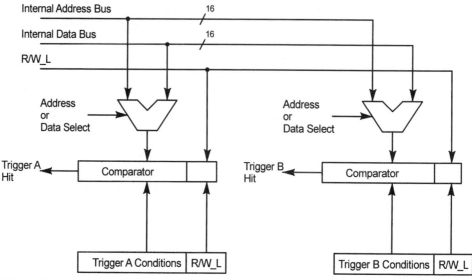

Figure 20-2 On-chip ICE.

Features of HCS12 On-Chip ICE

The on-chip ICE system, or DBG module, in the MC9S12C32 consists of comparator circuits to detect specific addresses and data, trigger logic, and capture mode logic. See Figure 20-2. There is a first in–first out (FIFO) capture buffer to hold the address and data bus information captured during a trace run. After listing the major features, we will discuss important in-circuit emulator concepts.

- Three 16-bit comparators for breakpoints and trace triggers.
  - Compare to address or data (high byte, low byte, or both).
- Nine trigger modes to control trace runs.
  - **A-Only:** Basic trace run with one trigger point.
  - **A-OR-B:** Basic trace run with two trigger points.
  - **A-then-B:** Sequence trace run.
  - **Event-Only-B:** Capture data associated with address in B.
  - **A-then-Event-Only-B:** Sequence-qualified data capture.
  - **A-AND-B:** Address matches A; data matches B.
  - **A-AND-NOT-B:** Address matches A; data does not match B.
  - **Inside-Range:** Address $\geq$ A AND address $\leq$ B.
  - **Outside-Range:** Address $<$ A OR address $>$ B.
- Four information capture modes.
  - **Normal:** Capture change-of-flow or data values.
  - **LOOP1:** Capture only first change-of-flow in repeating program loops.
  - **DETAIL:** Capture address and data in consecutive buffer words for every significant bus cycle.
  - **Profile:** Return current PC value each time trace buffer is read.

- A 64-word FIFO capture buffer.
  - Filtering increases the effective buffer depth.
- Accessible from CPU bus or through BDM.
  - BDM provides full nonintrusive access.
  - CPU access enables resident monitor programs like the serial monitor.
- Stop in active background mode after a trace run, or continue to run the application program.

On-Chip ICE Concepts

This section explains several important terms and concepts that are used in the discussions about the on-chip ICE system.

CHANGE-OF-FLOW

A *change-of-flow* is a branch or jump to some other place in the program.

For most instructions, the program simply progresses to the next sequential memory location after each instruction is finished. It is not necessary to record the location of every instruction in order to track the flow of a program. Interrupts and a few instructions cause the program to vector, branch, or jump to some nonsequential address rather than continuing to the next sequential address. These are called "changes-of-flow." If you were trying to record the flow of a program, you would need to record an address for each change-of-flow. When an interrupt occurs, the address of the interrupt service routine is recorded as the change-of-flow (COF) address. When an RTS, RTC, or RTI return instruction is executed, the return address is fetched from the stack so this address is recorded as the COF address. For branch (and long-branch) instructions, if the branch is not taken, the next instruction is completely predictable. However, if the branch is taken, the destination address is recorded as the COF address. BSR and most JSR and JMP instructions can be reconstructed by the host after a trace run, but if an indexed addressing mode is used, the host development system cannot compute the jump destination because it does not (usually) know what is in the index register. For indexed JMP or JSR instructions, the destination address is recorded as the COF address.

TRIGGER MODES

Trigger modes control the initiation or termination of trace runs. They also control the operation of breakpoints when the on-chip ICE system is operating in DBG mode. Trace runs can be set up to start collecting information when the trigger occurs (begin-type trace) or to capture information continuously in the FIFO capture buffer until the trigger occurs (end-type trace). If the trace buffer is ignored, triggers are effectively hardware breakpoints. The following paragraphs describe the types of trigger conditions that can be specified with comparators A and B.

Basic: These modes include A-Only and A-OR-B trigger modes. The A, or A and B comparators are compared against address signals and each comparator can be further qualified by the R/W_L signal and/or instruction tagging. In these modes the comparators operate like basic hardware breakpoints.

Event-Only: These trigger modes are used to record data values associated with the address stored in the B comparator. There is a simple Event-Only-B mode and a sequence version (A-then-Event-Only-B). In the sequence version, comparator A is used to identify an address that must be detected before data associated with the address in B is recorded. Event-Only triggers are only used for begin-type trace runs.

Sequence: In the sequence trigger modes, the first address is used to identify a prequalifying event that must occur before the second event will be recognized. A-then-B mode is used to specify two addresses or instructions. A-then-Event-Only-B is used to specify a qualifying event before starting to record data values for accesses to address B.

Full Address Plus Data: In the A-AND-B mode, the address must match the value in comparator A and the data must match the value in comparator B during the same bus cycle. This allows you to set a breakpoint that will trigger when a very specific data value is read-from or written-to a specific address. The second variation, A-AND-NOT-B, triggers when the address matches the value in comparator A but the data does not match the value in comparator B. This can be useful to detect when a status register returns a value that does not match the expected value.

Range: To detect when execution reaches a specific subroutine, you can store the address of the first instruction of a subroutine in comparator A, the address of the last instruction in comparator B, and then specify Inside-Range trigger mode. Outside-Range can be used to detect code runaway by specifying the address of the first program location in comparator A and the last program address in comparator B. If code runaway occurs due to some program error, the program is likely to go outside the range of the normal program instructions, which would cause a trigger. If an end-type trace run were used, the FIFO would contain the last 64 changes-of-flow that led to the erroneous access outside the range of normal memory.

TAG VERSUS FORCE TRIGGER POINTS

Because of the instruction queue mechanism in the HCS12 , it is not always sufficient to specify an address for a breakpoint. Often an instruction opcode is fetched into the instruction queue, but before it propagates to the head of the queue where it would be executed, some interrupt or other change-of-flow causes it to be flushed out of the queue without ever being executed. Imagine setting a breakpoint at the first instruction of a subroutine (B), and further assume that another subroutine (A) happens to be located just before this in the Flash memory. The last instruction in subroutine (A) would be an RTS or RTC instruction, which causes a change-of-flow. Every time subroutine (A) is executed, the first opcode of subroutine (B) is fetched into the instruction queue, shortly before the RTS or RTC is executed. Instead of executing that instruction, the queue is flushed due to the change-of-flow. If the address alone was used as a breakpoint, you would get a false trigger before subroutine (B) was ever actually executed.

A tag mechanism is provided to overcome this potential problem. If the breakpoint or trigger is specified as a "tag-type" trigger or breakpoint, a tag signal is latched into an extra bit in the instruction queue. As the instructions in the queue are advanced, the tag signal advances with them. If the instruction reaches the head of the queue and is about to be executed, the breakpoint or trigger is executed rather than the tagged instruction.

When the tag mechanism is not used, it is called a "force-type" trigger or breakpoint. With a force-type breakpoint, the trigger or breakpoint is requested as soon as the address matches, and then it takes effect at the next instruction boundary.

CAPTURE MODES

In begin-type trace runs, information is captured into the trace buffer after the trigger condition is satisfied. In end-type trace runs, information is continuously captured into the trace buffer in circular FIFO fashion, and capture stops or ends when the trigger condition is satisfied. Four basic capture modes control what is captured and how it is captured.

Normal: This is the most common capture mode. In this mode, an address is recorded at each change-of-flow, such as a branch (if taken), and interrupt, an indexed jump, or a return instruction. The host debug system can easily reconstruct all of the other instructions that execute sequentially between these changes-of-flow.

In the Event-Only-B trigger modes, a data value is captured into the trace buffer each time the address in comparator B is accessed. A control bit allows you to specify read accesses, write accesses, or both.

LOOP1: In this capture mode, the address in comparator C is updated at each change-of-flow. If the new COF address is the same as the address that was stored in comparator C at the last COF event, trace buffer capture is inhibited to save space in the capture buffer. After the trace run, you can tell that you executed the loop at least once, but you cannot tell (and typically do not care) how many times you repeated the loop.

It does not matter how many instructions are inside the loop as long as there is only one change-of-flow in the loop. Two common places where this is useful are delay counter loops and loops that are waiting for some peripheral flag such as waiting for RDRF to indicate a new serial character has arrived in the SCI receiver.

When the LOOP1 capture mode is used, comparator C is not available as a third breakpoint.

DETAIL: This mode causes two buffer captures per significant bus cycle. The first capture records the address value and the second records the data value. Program-fetch (P) cycles and free (f) cycles are not recorded because the host can easily reconstruct these after the trace run. This mode is useful for monitoring the data transactions as well as the program flow. The extra data bus information allows a developer to reconstruct what was in memory and CPU registers when the program was executing.

Profile: Technically this is not a capture mode because nothing is actually stored in the trace buffer. When this mode is selected, each read of the 16-bit DBGTB trace buffer data port register returns the value of the PC for the most-recently-executed instruction. By reading DBGTB periodically over a relatively long time, the host debugger can build up a histogram of program addresses that were executed. The Debugger can then display the percentage of time that was spent in various areas of the application program.

FILTERING CAPTURE INFORMATION

One of the disadvantages of the current on-chip ICE system is that the capture buffer is smaller than it would be in a traditional external emulator. In an external emulator, additional buffer memory is just a matter of buying more fast static RAMs. Since the emulator already costs thousands of dollars, a few more RAM chips do not add that much to the price. With the on-chip ICE system, it is a different matter. Internal logic devices are very small, so adding a few thousand devices to implement the on-chip ICE is not that significant, but buffer memory quickly adds up to a measurable additional silicon cost. For this reason, the size of the capture buffer needs to be limited to avoid adding cost to every device for circuitry that is needed only during development.

The capture buffer in the MC9S12C32 is sixty-four 16-bit words. In order to make the best use of this limited memory, the on-chip ICE filters the information that is captured. One example of filtering is to store only the addresses corresponding to changes-of-flow (COF). It is a trivial matter for the host development system to fill in the entire program flow, using only the COF addresses, so it is not necessary to store the address for every bus cycle.

The Event-Only-B trigger modes capture data values associated only with accesses to the address in comparator B. Logic analyzers tend to capture everything for every bus cycle and then let the user try to sort out the useful data from the massive amount of captured data. While the on-chip ICE approach does have some limitations, it often speeds up the debugging process to capture selectively only the most significant pieces of data.

LOOP1 capture mode is another form of filtering. Many programs have program loops that execute repeatedly by counting down a delay constant or waiting for a flag from a peripheral module. If you blindly capture all changes-of-flow, you would capture a new COF value each time through such a loop, and you would quickly fill up the capture buffer. When you are debugging such a program, you need to know that the loop is there, but typically it is not useful to see it execute a thousand times. Instead, you usually want to know what happens when it finally times out or gets a flag from the peripheral module. LOOP1 mode takes care of this by capturing the COF address into comparator C on each new COF event. If the new COF event is the same as the previous one (as it would be in a loop), the capture is inhibited. Therefore, when you get to a loop, only the first COF to the beginning of the loop gets captured no matter how many times the loop is repeated.

DETAIL mode captures both the address and the data for every significant bus cycle, which fills up the capture buffer very quickly. However, even in this mode, filtering is taking place. The terminology "every significant bus cycle" is a clue that not "every" bus cycle is recorded in the capture buffer. There are two very common types of bus cycles, which are easily recognized, and the debugger host easily reconstructs them after the trace run. Program-fetch cycles (P cycles) can be reconstructed by the host because the host is aware of the object code that makes up the program. These P cycles are simply 16-bit read cycles of the object code. Free bus cycles (f cycles) are cycles where the CPU is not using the bus so it does not matter what was on the bus during these cycles. This filtering of uninteresting bus cycles makes it possible to store information for much more than 32 bus cycles at two words per cycle.

CodeWarrior User Interface to the On-Chip ICE System

Studying the internal details of the on-chip ICE system can help you learn what you can do with it, but most users will experience the system through a host debugger tool such as CodeWarrior. As a user and programmer, you think of your applications in terms of the source programs and listing files. In a debugging session, you think of CPU register contents, memory component windows, source and disassembly windows, and variable windows. All of these components of the debugger GUI have meaning to you and help you visualize what your program is doing.

You do not think about the registers and control bits inside the on-chip ICE system (at least not consciously and directly). To use the on-chip ICE system effectively, you need to understand how to make it show you what you want to know about the program that you are debugging. Ideally, the setup controls should be as intuitive as possible. CodeWarrior tries to do this by using right clicks and pop-up menus that show the options that are appropriate for the current context. Some menus are hierarchical so that as you make initial selections, the additional options you need to specify a trace run are presented at the logical time.

In this section, we will examine the basics of the user interface to the on-chip ICE system by going through a practical application example. You can find more detailed information in the CodeWarrior documentation, the "Getting Started" application note, and other references that are shown in Section 20.8.

A Simple Trace Example

Example 20-1 shows a simple program that we will use to demonstrate some of the debugging features of the in-circuit emulator. This program has a bug in the subroutine. The stack use is unbalanced and so the subroutine will not return to the main program when the rts is executed. As we discussed in Chapter 9, this bug is relatively easy to find with breakpoints, but let us now see what the program execution looks like with the trace feature in the ICE.

Example 20-1 Example Program for Trace

```
Metrowerks HC12-Assembler
(c) COPYRIGHT METROWERKS 1987-2003


Rel. Loc    Obj. code Source line
---  ----   --------- -----------
  1                   ;*******************************
  2                   ; Example program to demonstrate the ICE
                        debugger
  3                   ;*******************************
  4                   ; Define the entry point for the main
                        program
  5                         XDEF  Entry, main
  6                         XREF  __SEG_END_SSTACK ; Note double
                                                    underbar
  7                   ;*******************************
  8                   ; Code Section
  9                   MyCode: SECTION
 10                   Entry:
 11                   main:
 12                   ;*******************************
 13                   ; Initialize stack pointer register
```

```
14 000000 CFxx xx                lds    #__SEG_END_SSTACK
15                        ;*******************************
16                        main_loop:
17                        ; DO
18 000003 CDxx xx                ldy    #DATA2
19 000006 CExx xx                ldx    #DATA1
20 000009 16xx xx                jsr    sub1
21 00000C A7                     nop
22 00000D A7                     nop
23                        ; FOREVER
24 00000E 20F3                   bra    main_loop
25                        ;*******************************
26                        ; Subroutine with a stack problem
27                        sub1:
28 000010 34                     pshx   ; Save the registers
29 000011 35                     pshy
30 000012 7Axx xx                staa   DATA2 ; Code to test memory
                                                 access
31 000015 A7                     nop
32 000016 A7                     nop
33 000017 A7                     nop
34 000018 31                     puly         ; Restore the
                                                 registers (bug!)
35 000019 3D                     rts
36                        ;*******************************
37                        MyConst:SECTION
38                        ; Place constant data here
39 000000 5468 6973 DATA0:  DC.B     "This is a string"
   000004 2069 7320
   000008 6120 7374
   00000C 7269 6E67
40 000010 A7A7      DATA1:  DC.B    $A7, $A7
41                        ;*******************************
42                        MyData: SECTION
43                        ; Place variable data here
44 000000           DATA2: DS.B    2
```

SET UP THE TRACE RUN

Before starting to set up the trace run, remember to open a trace component window on the debugger screen. To do this, select *Component → Open . . .* , scroll down to the *Trace* component (it has an icon of a bear's footprint) and double-click it. Right click in the Trace window

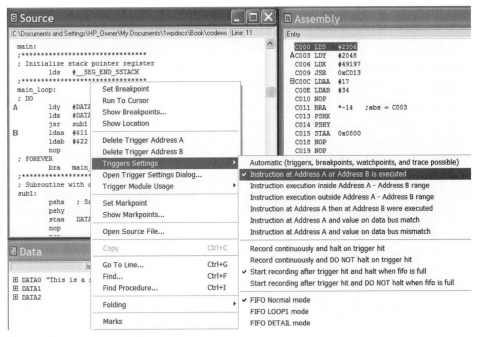

Figure 20-3 Using hierarchical pop-up menus.

and select *Instructions.* (Textual display is another way to show more information.) Resize and reposition this trace window on the debugger screen. After you have opened this trace component window, you do not need to do it again no matter how many trace runs you perform.

Scroll down in the source window to set both A and B triggers. Position the cursor over the `ldy #DATA2` instruction and right click on that source line and then select *Set Trigger Address A*. A bold red letter A at the left end of the line indicates the location of the trigger point. Position the cursor over the `ldaa #$11` following the `jsr sub`, right click and select *Set Trigger Address B*. Right click anywhere in the source window, drag down to *Trigger Settings* ▶ and use the hierarchical pop-up menus shown in Figure 20-3 to select the following trigger settings:

  ✓ Instruction at Address A or Address B is executed
  ✓ Start recording after trigger hit and halt when fifo is full
  ✓ FIFO Normal mode

Notice that the menus present the choices in the language of the programmer and not in terms of microcontroller register control bits. If a trigger address for only Address A or Address B had been set, the menus would have offered different choices that are appropriate for that context.

You are now ready to start a trace run. Click the *Run* button (green arrow). The program runs until it hits the instruction at the Trigger A address. The on-chip ICE will begin capturing change-of-flow addresses into its capture buffer as the program continues. When the 64-word capture buffer (FIFO) gets full, the program stops in active background mode and the results are displayed in the trace window, Figure 20-4.

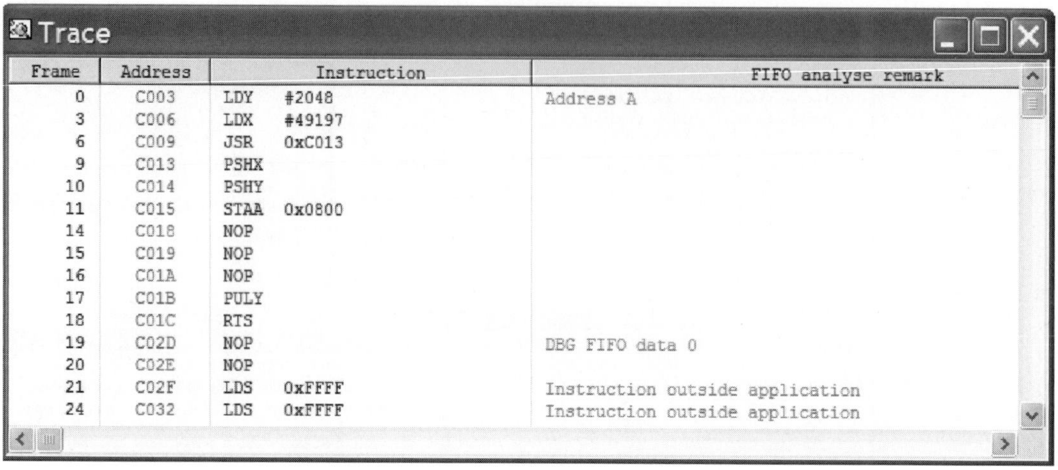

Frame	Address	Instruction	FIFO analyse remark
0	C003	LDY #2048	Address A
3	C006	LDX #49197	
6	C009	JSR 0xC013	
9	C013	PSHX	
10	C014	PSHY	
11	C015	STAA 0x0800	
14	C018	NOP	
15	C019	NOP	
16	C01A	NOP	
17	C01B	PULY	
18	C01C	RTS	
19	C02D	NOP	DBG FIFO data 0
20	C02E	NOP	
21	C02F	LDS 0xFFFF	Instruction outside application
24	C032	LDS 0xFFFF	Instruction outside application

Figure 20-4 Trace results.

INTERPRETING THE RESULTS

Figure 20-4 shows the result of this trace run. The Trace window shows a *Frame* of information that includes the bus address, the instruction, and any FIFO analysis remarks. We see all of the instructions that were executed following the Trigger A event and everything looks OK up to the `RTS` instruction. At that point we see that instead of returning to the Trigger B address ($C00C) the program returns to $C02D and starts to execute code from that point, eventually filling the FIFO buffer and stopping. This gives us our clue that there is something wrong with the stack operations in the subroutine.

SETTING A TRACE FOR MEMORY ACCESSES

To illustrate another feature of the ICE that you cannot do with other breakpoints, right click in the source window and *Delete Trigger Address A* and *Trigger Address B*. Then in the Data window click on the + next to DATA2 and expand the array. Position the cursor on the first element, right click, and select *Set Trigger Address A → Read/Write Access* and then on the last data element, right click and select *Set Trigger Address B → Read/Write Access*. Then select the following trigger settings:

✓ Memory access inside Address A–Address B range
✓ Record continuously and halt on trigger hit
✓ FIFO Normal mode

Push the Reset button and then the Run button. The Trace screen, Figure 20-5, shows the instructions executed up until a trigger was hit for the `STAA 0x0800` instruction in Frame 14. Frames 17 and 18 show two more instructions that were in the instruction pipe at the time of the trigger hit.

This tool is very useful if you suspect your program is writing into areas of memory that it should not, for example, a stack overflow.

See Example 20-2.

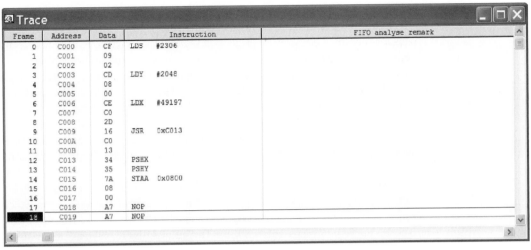

Figure 20-5 Trace up to a memory write.

A Practical Example of a Trace Run

As another example, we will trace an unusual program that can be called to program Flash memory locations from a program that is running in the same Flash memory. During Flash programming, the Flash memory is temporarily disabled from the memory map of the microcontroller and so, during that critical time, you cannot execute instructions out of the Flash memory. This makes it impossible to use traditional breakpoints and single stepping to debug this routine. The on-chip ICE can be used, because it just records COF addresses in real time during the trace run. With traditional breakpoints and single stepping, the debugger tries to read values out of the Flash during debugging, but since the Flash is disabled during the programming operation, these accesses would fail.

BRIEF DESCRIPTION OF THE EXAMPLE PROGRAM

The DoOnStack subroutine is the basis for using one or more blocks of Flash as EEPROM. What makes this program unusual is that it copies a subroutine onto the stack and calls that routine to perform a critical portion of the Flash programming operation. This program is included in the serial monitor that is preprogrammed in the Flash memory of the CSM12C32 microcontroller module that comes with the microcontroller SLK student learning kit. This

---

Example 20-2

The program in Example 20-1 has an extra push in the subroutine, causing it to crash because it does not return from the subroutine to the main program. Describe how you could use the in-circuit emulator to detect when the CPU starts to execute an instruction outside the valid program.

Solution: Set the Trigger A address at the first instruction and Trigger B at the last instruction in the program. Then choose Trigger Setting → Instruction execution outside Address A–Address B range.

serial monitor is described in application note AN2548/D and the CodeWarrior project files (AN2548SW1) are available at the Freescale website.

After loading the CodeWarrior project and starting the IDE, open the source file named S12SerMon2r1.def and modify the first three code blocks to configure the serial monitor for the MC9S12C32. You do this by commenting out the old lines of code and removing the semicolons on the lines for the MC9S12C32 on the CSM12C32 microcontroller module. Only one set of options in each of these code blocks should be enabled at a time. In the first block, the C32 lines should be active. In the second block, the lines for the 16-MHz oscillator should be active. In the third block, the lines for PTP5 should be active because SW2 on the CSM12C32 microcontroller module is configured to be the Run/Load switch for the serial monitor. After making these changes, save and close the file.

Select the C32 in the drop-down menu at the top of the project window. Click the *Make* button to compile the serial monitor program. Three warnings will appear (only the first time you run Make), but they can be ignored. You can close the errors and warnings window. Use *Find . . .* in the *Search* menu to search for DoOnStack. Next, select *Find Definition* in the Search menu. From here, you can examine the DoOnStack routine, the SpSub routine, and any of the nearby routines such as WriteD2IX. Some of the code around WriteD2IX will look strange because the serial monitor is designed to work on any HCS12 including devices with multiple Flash memory modules and extended paging through the PPAGE register. The MC9S12C32 does not have extended memory space or multiple Flash modules.

The main program in which we are interested is the DoOnStack routine, which moves the SpSub subroutine from Flash onto the stack and calls it with an indexed JSR. The DoOnStack source code is shown in Figure 20-6. SpSub is said to be "position independent" because it still operates normally even after it has been moved to some arbitrary location in stack memory. With the HCS12, it is easy to write a position-independent routine. You just need to avoid using instructions that jump to an absolute memory address. Branches are inherently position independent.

The initial steps of a Flash programming sequence can be executed from Flash because the Flash is not disabled until near the end of the sequence. The Flash is out of the map so you cannot execute out of Flash during the last two steps where you write 1 to FCBEF in the FSTAT register to register the command and then wait for the FCCF flag to indicate the operation is complete. These two steps are done in the SpSub subroutine, which executes out of stack RAM.

This is a fairly complicated and tricky sequence. It is just the type of program that would need to be carefully traced and debugged to be sure it was executing exactly like you expected it to. We will trace this program with the on-chip ICE.

Rather than write our own example program to demonstrate the on-chip ICE, we will run the serial monitor program from the CodeWarrior True Time Simulator & Real Time Debugger and trace the EraseAll function, which erases all of the EEPROM and Flash except the serial monitor itself. Because it is not erasing the entire Flash, EraseAll erases one block at a time from the beginning of Flash until just before the serial monitor object code starts. We will perform two trace runs. One will use normal capture mode while the other will use LOOP1 capture mode to demonstrate how that filters capture information.

SET UP THE TRACE RUN

Before starting to set up the trace run, remember to open a trace component window on the debugger screen. To do this, select *Component → Open . . .* from the menu, scroll down to the *Trace* component (it has an icon of a bear's footprint) and double-click it. Resize and reposition this trace window on the debugger screen.

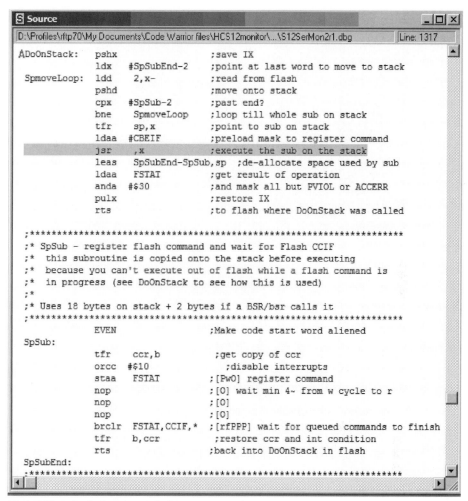

```
S Source                                                                    _ □ ×
D:\Profiles\rftp70\My Documents\Code Warrior files\HCS12monitor\...\S12SerMon2r1.dbg    Line: 1317

ADoOnStack:   pshx                      ;save IX
              ldx     #SpSubEnd-2       ;point at last word to move to stack
SpmoveLoop:   ldd     2,x-              ;read from flash
              pshd                      ;move onto stack
              cpx     #SpSub-2          ;past end?
              bne     SpmoveLoop        ;loop till whole sub on stack
              tfr     sp,x              ;point to sub on stack
              ldaa    #CBEIF            ;preload mask to register command
              jsr     ,x                ;execute the sub on the stack
              leas    SpSubEnd-SpSub,sp ;de-allocate space used by sub
              ldaa    FSTAT             ;get result of operation
              anda    #$30              ;and mask all but PVIOL or ACCERR
              pulx                      ;restore IX
              rts                       ;to flash where DoOnStack was called


;******************************************************************
;* SpSub - register flash command and wait for Flash CCIF
;*  this subroutine is copied onto the stack before executing
;*  because you can't execute out of flash while a flash command is
;*  in progress (see DoOnStack to see how this is used)
;*
;* Uses 18 bytes on stack + 2 bytes if a BSR/bsr calls it
;******************************************************************
              EVEN                      ;Make code start word aliened
SpSub:
              tfr     ccr,b             ;get copy of ccr
              orcc    #$10               ;disable interrupts
              staa    FSTAT             ;[PwO] register command
              nop                       ;[O] wait min 4~ from w cycle to r
              nop                       ;[O]
              nop                       ;[O]
              brclr   FSTAT,CCIF,*      ;[rfPPP] wait for queued commands to finish
              tfr     b,ccr              ;restore ccr and int condition
              rts                       ;back into DoOnStack in flash
SpSubEnd:
;******************************************************************
```

Figure 20-6 DoOnStack and SpSub source code.

Reset the target and then click the *green Run* arrow. The serial monitor will run through its initialization and stick in a loop waiting for a <cr> (0x0D) character. Click the red Stop button. The PC (IP for Instruction Pointer in the debugger) should be 0xFC82 and accumulator A should be 0x00. Manually set IP = 0xFC8A and A = 0x0D and click the Run button again. This time the monitor will run until it gets to Prompt1, where it is waiting for a command character. Click Stop, change IP from 0xFC82 to 0xFC8A, and change A from 0x3E to 0xB6 (this is the code for EraseAll).

Scroll down in the source window until you find DoOnStack. Right click on that source line, and select *Set Trigger Address A*. A bold red letter A at the left end of the line indicates the location of the trigger point. Right click anywhere in the source window and use the hierarchical pop-up menus to select the following trigger settings:

✓ Instruction at Address A is executed
✓ Start Recording after trigger hit and halt when fifo is full
✓ FIFO Normal mode

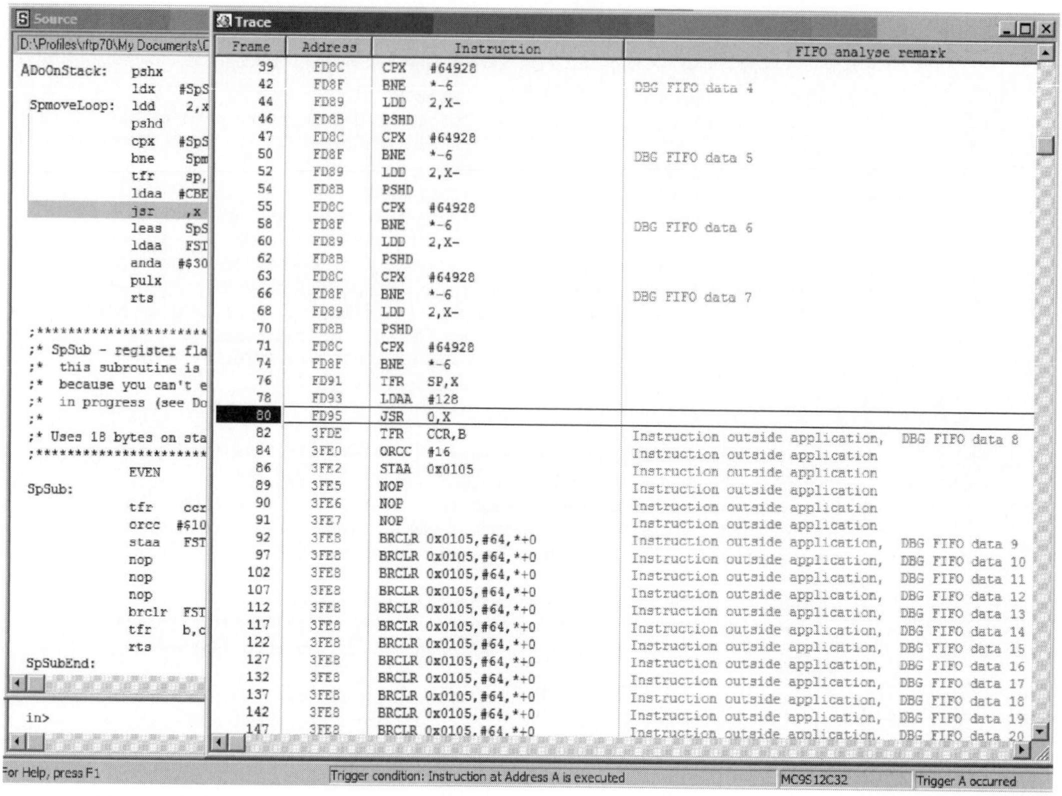

Figure 20-7 Trace results (Normal capture mode).

You are now ready to perform a trace run. Click the *Run* button (green arrow). The serial monitor program will resume from address 0xFC8A, decode the command, and start to execute the EraseAll command program. When it starts to erase the first block of Flash, trigger A will fire as the first instruction in DoOnStack is executed. This causes the on-chip ICE to begin capturing change-of-flow addresses into its capture buffer as the program continues. When the 64-word capture buffer (FIFO) gets full, the program stops in active background mode and the results are displayed in the trace window of the CodeWarrior debugger. The results of this trace run are shown in Figure 20-7 and then we will do another trace run in LOOP1 capture mode to get a better picture of what happened in our program.

To repeat the trace run in LOOP1 capture mode, reset the target microcontroller and follow the same steps as before, except this time select LOOP1 capture mode in the trace settings hierarchical menu before clicking the Go button. The results of the LOOP1 trace run are shown in Figure 20-8 and discussed after the Normal capture trace run.

INTERPRET THE RESULTS

Figure 20-7 shows the trace window after the trace run with part of the source window showing in the background to the left. This trace window is set to show disassembled instructions. For other types of trace runs and in other debug situations, the trace window can be set up to display different information such as data values, object code, or even waveforms. The numbers in the *Frame* column correspond to bytes of object code so, for example, frames 39 and 42 in the figure indicate that the CPX #64928 instruction takes 3 bytes of object code. These

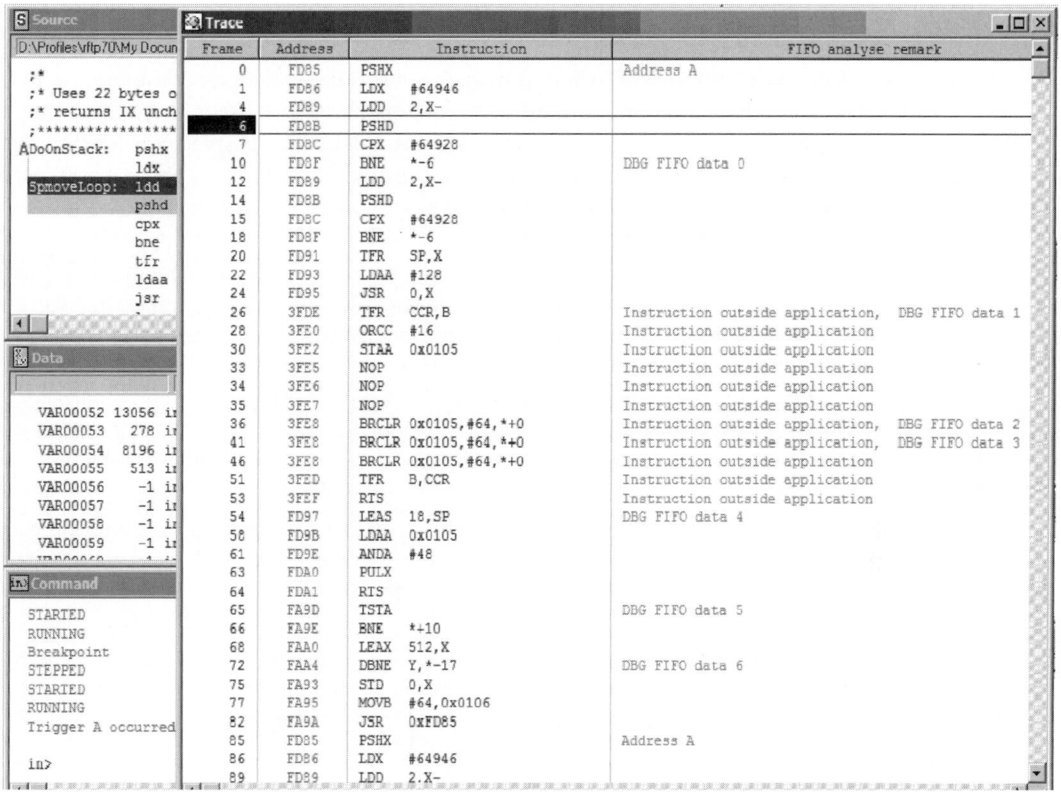

Frame	Address	Instruction		FIFO analyse remark
0	FD85	PSHX		Address A
1	FD86	LDX	#64946	
4	FD89	LDD	2,X-	
6	FD8B	PSHD		
7	FD8C	CPX	#64928	
10	FD8F	BNE	*-6	DBG FIFO data 0
12	FD89	LDD	2,X-	
14	FD8B	PSHD		
15	FD8C	CPX	#64928	
18	FD8F	BNE	*-6	
20	FD91	TFR	SP,X	
22	FD93	LDAA	#128	
24	FD95	JSR	0,X	
26	3FDE	TFR	CCR,B	Instruction outside application, DBG FIFO data 1
28	3FE0	ORCC	#16	Instruction outside application
30	3FE2	STAA	0x0105	Instruction outside application
33	3FE5	NOP		Instruction outside application
34	3FE6	NOP		Instruction outside application
35	3FE7	NOP		Instruction outside application
36	3FE8	BRCLR	0x0105,#64,*+0	Instruction outside application, DBG FIFO data 2
41	3FE8	BRCLR	0x0105,#64,*+0	Instruction outside application, DBG FIFO data 3
46	3FE8	BRCLR	0x0105,#64,*+0	Instruction outside application
51	3FED	TFR	B,CCR	Instruction outside application
53	3FEF	RTS		Instruction outside application
54	FD97	LEAS	18,SP	DBG FIFO data 4
58	FD9B	LDAA	0x0105	
61	FD9E	ANDA	#48	
63	FDA0	PULX		
64	FDA1	RTS		
65	FA9D	TSTA		DBG FIFO data 5
66	FA9E	BNE	*+10	
68	FAA0	LEAX	512,X	
72	FAA4	DBNE	Y,*-17	DBG FIFO data 6
75	FA93	STD	0,X	
77	FA95	MOVB	#64,0x0106	
82	FA9A	JSR	0xFD85	
85	FD85	PSHX		Address A
86	FD86	LDX	#64946	
89	FD89	LDD	2,X-	

Figure 20-8 Trace results (LOOP1 capture mode).

numbers start at zero at the top of the scrolling window and are used primarily as line number references in this example. The *Address* and *Instruction* columns show a disassembled version of the instructions that were executed during the trace run.

The column labeled *FIFO analyse remark* includes comments for some frames. The comment, DBG FIFO data 4, at frame 42 indicates this frame corresponds to a change-of-flow that was captured into the trace buffer during the trace run. The instruction that caused the capture was the BNE *-6 instruction and the address that was captured into the buffer was FD8F. We also know the branch was "taken" because otherwise it would not have caused a change-of-flow.

Frame 82 also includes the remark, "Instruction outside application," because it corresponds to the first instruction that is executed in the SpSub subroutine that is currently located in stack RAM. Notice the JSR 0,X instruction in frame 80 that jumped to the relocated SpSub subroutine on the stack. Frame 80 is highlighted by clicking and dragging in the trace window. As you drag over instructions, the source window automatically scrolls to that area of the source program and highlights the corresponding source instruction in light gray so you can associate the frames in the trace window to the instructions in the source program. Because the SpSub program is a relocated version, scrolling over these lines will not highlight lines in the source program.

Although frames 0–38 are not shown in Figure 20-7, they show the beginning of the sequence that copies SpSub onto the stack. The last four iterations of that loop are shown in frames 39–74. Each loop pass consists of a load D (indexed with auto decrement), push D, compare X (immediate), and a BNE *-6. This copies the 18-byte SpSub subroutine onto the stack starting with the last two bytes and ending with the first two bytes of the subroutine.

Frame 76 shows the then-current stack pointer value being transferred to X in preparation for the indexed JSR in frame 80.

Frames 82 through 92 show the stack subroutine executing. Frame 92 is the BRCLR instruction that is waiting for the Flash erase operation to finish and the FCCF flag to be set. The block erase operation takes 20 ms and because the bus frequency is 8 MHz, the BRCLR instruction only takes $\frac{5}{8}$ μs. Thus the BRCLR instruction will repeat about 32,000 times before it finds FCCF set. If you scroll down in the trace window, the rest of the buffer is filled with these repeated BRCLR instructions.

Figure 20-8 shows the trace window after the LOOP1 trace run. This will filter out any repeating loops and just show the first iterations. The value of doing this is that we will be able to see more of the interesting execution activity than we could in the previous trace. For example, it's important to know that the BRCLR instruction branched back to be executed again, but we usually are not interested in seeing it happen 32,000 times.

Frames 4–10 show the first iteration of the loop to move SpSub onto the stack. Even though the BNE *-6 repeats seven more times, these extra changes-of-flow are filtered out to save space in the capture buffer for more interesting activity. Frame 24 is the next captured value corresponding to the JSR 0,X change-of-flow. Frames 26–53 show the complete execution of the stack subroutine including the RTS return to DoOnStack.

Frames 36–46 need additional explanation because we see two FIFO captures at 36 and 41 when we only expected to see one due to LOOP1 capture mode. This happened because the loop is so short and it is aligned in the instruction pipe such that the first byte of object code for the second iteration of the loop is already in the pipe when the first BRCLR executes. There is not enough time to update comparator C with the new COF address before the second COF event starts so a second capture sneaks through in this unusual case. The longer loop related to copying SpSub onto the stack is captured once only, as we expected, because there was enough time to update comparator C. The third BRCLR in frame 46 is there because the COF event in frame 41 indicated that the branch was taken and the BRCLR executed after frame 41. Since the BRCLR in frame 46 is not marked as a change-of-flow, it means this branch was not taken (because FCCF was finally detected as being set).

Frames 54–64 correspond to the last five instructions in DoOnStack. Frames 65–82 correspond to instructions in the EraseAll command program. Frames 65 and 66 check for errors in the previous page-erase operation, and frames 68 and 72 update the Flash pointer to point at the next 512-byte Flash page. Frame 75 performs the first step in a new page-erase operation by writing any data to an address in the page to be erased. Frame 77 performs the second step by writing the page-erase command code to the Flash command register, and frame 82 calls DoOnStack to complete the page-erase operation.

Figure 20–8 is showing only the first iteration of a larger page-erase loop where frames 0–82 show the first iteration. Much more trace information is shown if you scroll down in this window. The information from this trace run goes on to show more than eight full iterations of this page-erase loop.

20.7 Conclusion and Chapter Summary Points

The MC9S12C32 offers extensive resources for developing and debugging application programs. The single-wire background debug interface (BDM) is useful for things other than debugging. The on-chip ICE system offers a level of debugging support that previously required external equipment costing thousands of dollars.

In this chapter we discussed the following points:

- The single-wire background debug system provides a convenient nonintrusive way to access the internal systems of the HCS12 microcontroller. This system can be used to program the on-chip Flash, even when it is soldered into the application system. The finished application system can be calibrated and adjustment factors can be saved in internal nonvolatile memory. It is also a great debugging tool allowing microcontroller memory and registers to be read or written while the application continues to run normally. This system allows the application program to be stopped or single stepped. Finally, this is the most common way to access the separate on-chip ICE system.
- The on-chip in-circuit emulator (ICE) system was discussed in detail. This system replaces older external emulators and offers many important improvements over those systems. The on-chip ICE is used to capture address and data bus information without disturbing the operation or timing of the application program. This approach can be used in some places where breakpoints and single stepping do not work. We walked through a practical example of using the ICE system to trace a Flash erase program to demonstrate some of the basic principles of this system and style of debugging.
- We explained hardware and software breakpoints. Hardware breakpoints use the comparators that are located in the on-chip ICE system. You can think of these breakpoints as being a small but important subset of the functions that are supported by the on-chip ICE.
- We discussed the hardware tagging mechanism that is used in large external emulator systems. While most programmers will use the internal ICE system, the traditional external emulators still have a few advantages such as a very large capture buffer, which can be useful in very unusual debugging situations.

20.8 Bibliography and Further Reading

Cady, F. M. and J. M. Sibigtroth, *Software and Hardware Engineering Motorola M668HC12*, Chapter 14 and Appendix A, Oxford University Press, New York, 2000.

Debug (DBG) Module V1 Block User Guide, S12DBGV1/D, Freescale Semiconductor, Inc., Austin, TX, 2003.

Debugger HCS12 Onchip DBG Module User Interface, Metrowerks, 2003.

MC9S12C128 Data Sheet: Covers MC9S12C Family and MC9S12GC Family, Freescale Semiconductor, Inc., Austin, TX, 2005.

Montañez, E. and S. Ruggles, MCUSLK_CSM12C32, *Getting Started with the Microcontroller Student Learning Kit (MCUSLK)*, Freescale Semiconductor, Inc., Austin, TX, 2005.

Montañez, E., AN2596/D, *Using the HCS08 Family On-Chip In-Circuit Emulator (ICE)*, Freescale Semiconductor, Inc., Austin, TX, 2004.

Sibigtroth, J. M., CPU12RM/AD Rev. 3, *CPU12 Reference Manual*, Freescale Semiconductor, Inc., Austin, TX, 2002.

Williams, J., AN2548/D, *Serial Monitor Program for HCS12 Microcontrollers*, Freescale Semiconductor, Inc., Austin, TX, 2003.

Williams, J., AN2548SW1, *CodeWarrior Project Software Files for AN2548/D*, Freescale Semiconductor, Inc., Austin, TX, 2003.

21 Advanced HCS12 Hardware

OBJECTIVES

In this chapter we finish our discussion of the HCS12 by filling in some of the gaps left in previous chapters.

21.1 Clocks and Reset Generator

The CRG block provides the system clocks and reset features such as the external reset and COP. Table 21-1 shows the main features of the block.

TABLE 21-1 Features of the Clocks and Reset Generator Block

Feature	Includes
System clock generator	Clock quality check.
	Clock switch for either oscillator-based or PLL-based system clocks.
	User selectable disabling of clocks during wait mode for reduced power consumption.
Phase locked loop frequency multiplier	Reference divider.
	Automatic bandwidth control mode for low-jitter operation.
	Automatic frequency lock detector.
	CPU interrupt on entry or exit from locked condition.
	Self-clock mode in absence of reference clock.
Computer operating properly (COP)	Watchdog timer with time-out clear window. See Chapter 12.
System reset generation from the following sources	Power-on reset.
	Low-voltage reset.
	COP reset.
	Loss of clock reset.
	External pin reset.
Real-time interrupt (RTI)	Real-time interrupts. See Chapter 14.

21.2 System Clock Generator

The clock generator shown in Figure 21-1 generates all clocks used by various subsystems in the HCS12. The fundamental input to the clock generator is the external oscillator, usually a crystal or ceramic resonator for high-frequency stability, or an internal phase locked loop (PLL) oscillator. The crystal oscillator produces the OSCCLK signal and the phase locked loop the PLLCLK signal. A clock monitor circuit monitors OSCCLK to ensure it continues to run properly. If the clock monitor detects a failure of the oscillator, the *Self Clock Mode Status (SCM)* bit in the CRGFLG register is set and all clocks are then switched to the PLL running at its lowest frequency. This allows the HCS12 system to keep running even if the external crystal oscillator fails. Two multiplexers select one of these signals for various clock outputs to be used in the rest of the system. The SCM bit is normally reset and all system clocks are derived from the OSCCLK (if the PLL is not selected as described later).

The main CRG outputs are the *Bus Clock*, the *Core Clock*, and the *Oscillator Clock*. The CRG also supplies the clocks for the COP and the real-time interrupt. Peripheral modules,

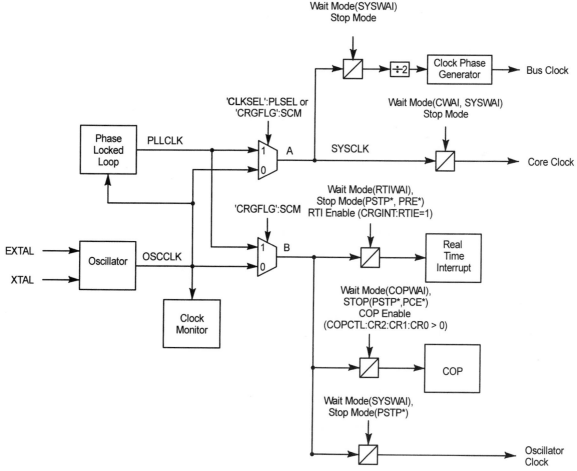

Figure 21-1 System clocks generator.

TABLE 21-2 Clock Generator Control Bits

Clock Control Bit	Register	Definition
SCM	CRGFLG	Self Clock Mode Status.
PLLSEL	CLKSEL	PLL Select.
PSTP	CLKSEL	Pseudo Stop Bit.
SYSWAI	CLKSEL	System Clocks Stop in Wait Mode.
ROAWAI	CLKSEL	Reduced Oscillator Amplitude in Wait Mode.
PLLWAI	CLKSEL	PLL Stops in Wait Mode.
CWAI	CLKSEL	Core Stops in Wait Mode.
RTIWAI	CLKSEL	RTI Stops in Wait Mode.
COPWAI	CLKSEL	COP Stops in Wait Mode.
PRE	PLLCTL	RTI Enable during Pseudo Stop Mode.
PCE	PLLCTL	COP Enable during Pseudo Stop Mode.

such as the timer, A/D, and serial ports, use the Bus Clock, and its frequency is one-half the external oscillator or PLL frequency. The Core Clock is used by the CPU core to generate timing for memory and I/O accesses and other CPU activities. The Oscillator Clock is used by some peripheral modules, such as the MSCAN. We should note that its frequency is the same as the external oscillator or phase locked loop oscillator.

A clock gate, shown by a box with a diagonal line, is enabled by a combination of control bits in registers and CPU operating modes such as Wait and Stop. The clock gate controls each of the output clocks in Figure 21-1. For example, the Clock Phase Generator, generating the Bus Clock, receives its input through a clock gate controlled by the Wait and Stop modes. If our program executes a STOP instruction and we enter Stop Mode, the Bus Clock is stopped. The Bus Clock is stopped also if we execute a WAI instruction and if the *System Clocks Stop in Wait Mode* bit (SYSWAI) is set. Table 21-2 gives brief definitions of these signals with more complete explanations in the sections that follow.

System Clocks

OSCCLK: Output from the external oscillator module and used normally as the basis for all other system clocks.

PLLCLK: Output from the internal phase locked loop oscillator. This clock is used as the basis for all other system clocks if the Clock Monitor detects a failure in OSCCLK. In addition, the user can select this clock to generate SYSCLK by setting PLLSEL = 1 in CLKSEL.

SYSCLK: The SYSCLK is derived from either the OSCCLK or the PLLCLK. See Table 21-3.

TABLE 21-3 SYSCLK Generation

SCM	PLLSEL	SYSCLK	Explanation
0	0	OSCCLK	Normal operation.
0	1	PLLCLK	Oscillator operating normally but user selects the PLL clock.
1	x	PLLCLK	Oscillator failure detected and Self Clock Mode has been entered.

Bus Clock: The Bus Clock is used by the peripheral modules and is one-half the frequency of the SYSCLK. See Table 21-4.

Core Clock: The Core Clock is equal to SYSCLK and is used by the CPU core to generate basic internal timing sequences. The Core Clock is stopped in Wait and Stop Modes as shown in Table 21-5.

Real-Time Interrupt Clock: The Real-Time Interrupt Clock is derived from either the OSCCLK (normally) or the PLLCLK if a clock failure has occurred. See Table 21-6.

COP Clock: The COP Clock is derived from either the OSCCLK (normally) or the PLLCLK if a clock failure has occurred. See Table 21-7.

Oscillator Clock: The CPU core, Flash memory, and the MSCAN device use the Oscillator Clock. Table 21-8 shows how it is controlled.

TABLE 21-4 Bus Clock Generation

Wait Mode	SYSWAI	Stop Mode	Bus Clock	Explanation
No	x	No	SYSCLK/2	Normal operation.
x	x	Yes	Stopped	In Stop Mode.
Yes	0	No	SYSCLK/2	In Wait Mode but SYSWAI is not set.
Yes	1	No	Stopped	In Wait Mode and SYSWAI is set.

TABLE 21-5 Core Clock Generation

Wait Mode	CWAI	SYSWAI	Stop Mode	Core Clock	Explanation
No	x	x	No	SYSCLK	Normal operation.
No	x	x	Yes	Stopped	In Stop Mode.
Yes	x	1	No	Stopped	In Wait Mode and SYSWAI is set.
Yes	1	x	No	Stopped	In Wait Mode and CWAI is set.

TABLE 21-6 RTI Clock Generation

SCM	Wait Mode	RTIWAI	Stop Mode	PSTP	PRE	RTI Clock	Explanation
0	No	x	No	x	x	OSCCLK	Normal operation.
1	No	x	No	x	x	PLLCLK	Clock failure occurred; in Self Clock Mode
x	x	x	Yes	0	x	Stopped	In Stop Mode.
x	x	x	Yes	1	0	Stopped	In Pseudo Stop Mode and RTI stops in Pseudo Stop Mode.
0 or 1	x	x	Yes	1	1	OSCCLK or PLLCLK	In Pseudo Stop Mode and RTI runs in Pseudo Stop Mode.
0 or 1	Yes	0	No	x	x	OSCCLK or PLLCLK	In Wait Mode but RTI runs in Wait Mode.
x	Yes	1	No	x	x	Stopped	In Wait Mode and RTI stops in Wait Mode.

TABLE 21-7 COP Clock Generation

SCM	Wait Mode	COPWAI	Stop Mode	PSTP	PCE	COP Clock	Explanation
0	No	x	No	x	x	OSCCLK	Normal operation.
1	No	x	No	x	x	PLLCLK	Clock failure occurred; in Self Clock Mode
x	x	x	Yes	0	x	Stopped	In Stop Mode.
x	x	x	Yes	1	0	Stopped	In Pseudo Stop Mode and COP stops in Pseudo Stop Mode.
0 or 1	x	x	Yes	1	1	OSCCLK or PLLCLK	In Pseudo Stop Mode and COP runs in Pseudo Stop Mode.
0 or 1	Yes	0	No	x	x	OSCCLK or PLLCLK	In Wait Mode but COP runs in Wait Mode.
x	Yes	1	No	x	x	Stopped	In Wait Mode and COP stops in Wait Mode.

TABLE 21-8 Oscillator Clock Generation

SCM	Wait Mode	SYSWAI	Stop Mode	PSTP	Oscillator Clock	Explanation
0	No	x	No	x	OSCCLK	Normal operation.
1	No	x	No	x	PLLCLK	Clock failure occurred; in Self Clock Mode.
x	x	x	Yes	0	Stopped	In Stop Mode.
0 or 1	x	x	Yes	1	OSCCLK or PLLCLK	In Pseudo Stop Mode and Clock runs in Pseudo Stop Mode.
0 or 1	Yes	0	No	x	OSCCLK or PLLCLK	In Wait Mode but Clock runs in Wait Mode.
x	Yes	1	No	x	Stopped	In Wait Mode and Clock stops in Wait Mode.

Clock Generator Control Registers

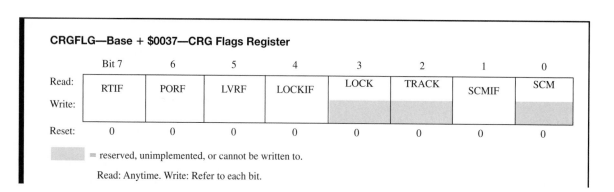

CRGFLG—Base + $0037—CRG Flags Register

	Bit 7	6	5	4	3	2	1	0
Read:	RTIF	PORF	LVRF	LOCKIF	LOCK	TRACK	SCMIF	SCM
Write:								
Reset:	0	0	0	0	0	0	0	0

▓ = reserved, unimplemented, or cannot be written to.

Read: Anytime. Write: Refer to each bit.

RTIF

Real Time Interrupt Flag

0 = RTI time-out has not yet occurred (default).
1 = RTI time-out has occurred.

RTIF is set at the end of an RTI period. Writing a one to the bit clears this flag. If enabled (RTIE = 1), RTIF causes an interrupt request. See Chapter 14.

PORF

Power On Reset Flag

0 = Power on reset has not occurred (default).
1 = Power on reset has occurred.

PORF is set to 1 when a power on reset occurs. This flag can be cleared by writing a one to the bit.

LVRF

Low Voltage Reset Flag

0 = Low voltage reset has not occurred (default).
1 = Low voltage reset has occurred.

LVRF is set when a low voltage reset occurs. This flag can be cleared by writing a one to the bit. Some versions of the HCS12 do not have this feature. For these, LVRF always reads zero.

LOCKIF

PLL Lock Interrupt Flag

0 = No change in LOCK bit (default).
1 = LOCK bit has changed.

LOCKIF is set when the LOCK status bit changes. This flag can be cleared by writing a one to the bit. If enabled (LOCKIE = 1), LOCKIF causes an interrupt request.

LOCK

Lock Status Bit

0 = PLL VCO is not within the desired tolerance of the target frequency (default).
1 = PLL VCO is within the desired tolerance of the target frequency.

LOCK reflects the current state of the PLL track condition. The bit is cleared in Self Clock Mode and writes to this bit have no effect.

TRACK

Track Status Bits

0 = Acquisition mode status (default).
1 = Track mode status.

TRACK reflects the current state of the PLL track condition. The bit is cleared in Self Clock Mode and writes to this bit have no effect.

SCMIF

Self Clock Mode Interrupt Flag

0 = No change in SCM bit (default).
1 = SCM bit has changed.

SCMIF is set when the SCM status bit changes. This flag can be cleared by writing a one to the bit. If enabled (SCMIE = 1), LOCKIF causes an interrupt request.

SCM

Self Clock Mode Status Bit

0 = MCU is operating normally with OSCCLK available (default).
1 = MCU is operating in Self Clock Mode with OSCCLK in an unknown state. All clocks are derived from PLLCLK running at its minimum frequency.

CRGINT—Base + $0038—CRG Interrupt Enable Register

	Bit 7	6	5	4	3	2	1	0
Read:	RTIE	0	0	LOCKIE	0	0	SCMIE	0
Write:								
Reset:	0	0	0	0	0	0	0	0

    = reserved, unimplemented, or cannot be written to.

Read: Anytime. Write: Anytime.

RTIE

Real Time Interrupt Enable

0 = Interrupt requests from the RTI are disabled (default).
1 = Enables operation of the RTI. Interrupts will be requested when RTIF is set.

See Chapter 14 and Table 21-9.

LOCKIE

Lock Interrupt Enable

0 = LOCK interrupt requests are disabled (default).
1 = LOCK interrupts will be generated when LOCKIF is set.

SCMIE

Self Clock Mode Interrupt

0 = SCM interrupt requests are disabled (default).
1 = Interrupts will be generated when SCM is set.

TABLE 21-9 Interrupt Vector Assignments

Priority[a]	Vector Address	Interrupt Source	Local Enable Bit	See Register	HPRIO Value to Promote
29	$FFC4:FFC5	CRG Self Clock Mode	SCMIE	CRGINT	$C4
28	$FFC6:FFC7	CRG PLL Lock	LOCKIE	CRGINT	$C6
7	$FFF0:FFF1	Real Time Interrupt	RTIE	CRGINT	$F0

[a] The numbers given for the priority show zero as the highest priority. This numbering scheme is also used by the CodeWarrior linker to locate the vector in the proper place in memory. See Chapter 12.

CLKSEL—Base + $0039—CRG Clock Select Register

	Bit 7	6	5	4	3	2	1	0
Read: Write:	PLLSEL	PSTP	SYSWAI	ROAWAI	PLLWAI	CWAI	RTIWAI	COPWAI
Reset:	0	0	0	0	0	0	0	0

Read: Anytime. Write: Anytime.

PLLSEL

PLL Select Bit

0 = System clock derived from OSCCLK (Bus Clock = OSCCLK/2) (default).
1 = System clock derived from PLLCLK (Bus Clock = PLLCLK/2).

Writing a one when LOCK = 0 and AUTO = 1, or TRACK = 0 and AUTO = 0 has no effect. This prevents the selection of an unstable PLLCLK as SYSCLK. PLLSEL bit is cleared when the MCU enters Self Clock Mode, Stop Mode, or Wait Mode with PLLWAI bit set.

PSTP

Pseudo Stop Bit

0 = Oscillator is disabled in Stop Mode (default).
1 = Oscillator continues to run in Stop Mode (Pseudo Stop). This allows for faster STOP recovery. It also reduces the mechanical stress and aging of the resonator in case of frequent STOP conditions but with slightly increased power consumption.

SYSWAI

System Clocks Stop in Wait Mode

0 = In Wait Mode the system clocks continue to run (default).
1 = System clocks are stopped in Wait Mode.

RTI and COP are not affected by SYSWAI.

ROAWAI

Reduced Oscillator Amplitude in Wait Mode

0 = Normal oscillator amplitude in Wait Mode (default).

1 = Reduced amplitude in Wait Mode.

PLLWAI

PLL Stops in Wait Mode

0 = PLL keeps running in Wait Mode (default).

1 = PLL stops in Wait Mode.

If PLLWAI is set, the CRG will clear the PLLSEL bit before entering Wait Mode. The PLLON bit remains set during Wait Mode but the PLL is powered down. Upon exiting Wait Mode, the PLLSEL bit has to be set manually (by your program) if PLL clock is required.

While the PLLWAI bit is set, the AUTO bit is set to one in order to allow the PLL to lock automatically on the selected target frequency after exiting Wait Mode.

CWAI

Core Stops In Wait Mode

0 = Core clock keeps running in Wait Mode (default).

1 = Core clock stops in Wait Mode.

RTIWAI

RTI Stops in Wait Mode

0 = RTI keeps running in Wait Mode (default).

1 = RTI stops and initializes the RTI dividers when in Wait Mode.

COPWAI

COP Stops in Wait Mode

0 = COP keeps running in Wait Mode (default).

1 = COP stops and initializes the COP dividers when in Wait Mode.

21.3 External Oscillator

The oscillator block in Figure 21-1 uses an external crystal, ceramic resonator, or CMOS compatible external oscillator. At reset, port E pin 7 acts as an XCLKS_L input signal, which controls whether a crystal or ceramic oscillator is used with an internal Colpitts (low-power) oscillator (XCLKS_L = 1) or whether an external Pierce oscillator or CMOS oscillator is used (XCLKS_L = 0). Each of these circuits is shown in Figure 21-2. There may be other oscillator circuits suggested by the crystal or resonator manufacturer. Please refer to the oscillator block user guide, *OSC Block User Guide*, and AN2727 *Designing Hardware for the HCS12 D-Family* by Gallop for more in-depth design information.

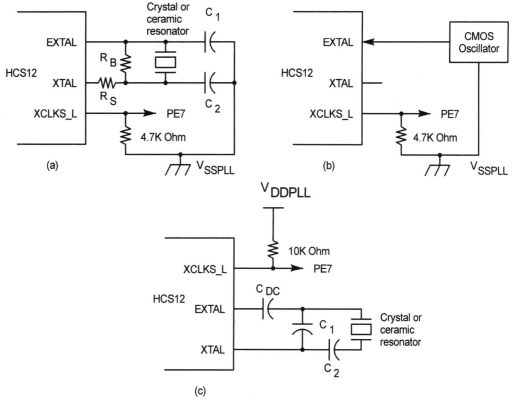

Figure 21-2 Oscillators: (a) Pierce, (b) CMOS compatible, and (c) Colpitts.

21.4 Phase Locked Loop Oscillator

The phase locked loop clock oscillator can be used to generate the SYSCLK and run the CPU at a different frequency than the incoming OSCCLK. In normal operation, the OSCCLK can be divided by 1 to 16 to provide the reference frequency. The PLL can multiply this by a factor ranging from 2 to 128 based on the SYNR register.

PLL External Connections

V_{DDPLL} AND V_{SSPLL}

The PLL requires an operating voltage (V_{DDPLL}) and ground (V_{SSPLL}). These allow the PLL operating voltage to be independently bypassed for better noise performance. If the PLL is not being used, these pins must be connected to proper voltage levels.

XFC

A PLL requires a second-order, lowpass filter in the feedback loop connected to the XFC pin as shown in Figure 21-3. If the PLL is not used, this pin must be connected to V_{DDPLL}. For

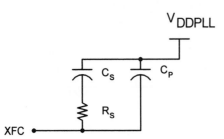

V_{DDPLL}

XFC

Figure 21-3 PLL filter.

more details, see Gallop's application note AN2727 *Designing Hardware for the HCS12 D-Family*. A PLL component calculator is available from Freescale as well.[1]

PLL Programming

As described in Section 21.2, the PLL can be chosen to serve as the SYSCLK by setting the PLLON bit in the CLKSEL register. The PLL has a variety of controlling bits in the SYNR, REFDV, and PLLCTL registers. Example 21-1 shows how to use the PLL to provide a higher frequency SYSCLK. Please refer to *CRG Block User Guide* or *HCS12 PLL Component Calculator User Guide* for more in-depth information on using the phase locked loop oscillator.

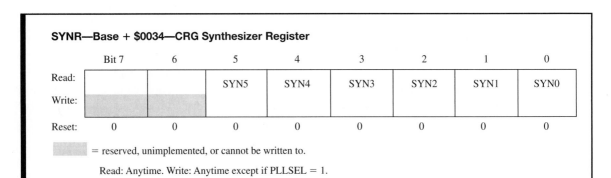

SYNR—Base + $0034—CRG Synthesizer Register

	Bit 7	6	5	4	3	2	1	0
Read:			SYN5	SYN4	SYN3	SYN2	SYN1	SYN0
Write:								
Reset:	0	0	0	0	0	0	0	0

= reserved, unimplemented, or cannot be written to.

Read: Anytime. Write: Anytime except if PLLSEL = 1.

SYN5:SYN0

Multiplication Factor of the PLL

The SYNR register controls the multiplication factor of the PLL. If the PLL is on, the count in the loop divider register (SYNR) effectively multiplies up the PLL clock from the reference frequency by a factor of $2 \times (SYNR + 1)$. Writes to this register initialize the lock detector and track detector bits.

[1] *HCS12 PLL Component Calculator User Guide*, HCS12PLLCALUG/D, http://www.freescale.com/files/microcontrollers/software_tools/initialization/boot_code_generation/HCS12PLLCALSW.zip?srch=1.

REFDV—Base + $0035—CRG Reference Divider Register

	Bit 7	6	5	4	3	2	1	0
Read:	0	0	0	0	REFDV3	REFDV2	REFDV1	REFDV0
Write:								
Reset:	0	0	0	0	0	0	0	0

░ = reserved, unimplemented, or cannot be written to.

Read: Anytime. Write: Anytime except if PLLSEL = 1.

REFDV3:REFDV0

Reference Divider Bits

The REFDV register provides a finer granularity for the PLL multiplier steps. The count in the reference divider divides OSCCLK frequency by REFDV + 1. Writes to this register initialize the lock detector and track detector bits.

$$PLLCLK = 2 \times OSCCLK \times \frac{SYNR + 1}{REFDV + 1}$$

PLLCTL—Base + $003A—PLL Control Register

	Bit 7	6	5	4	3	2	1	0
Read:	CME	PLLON	AUTO	ACQ	0	PRE	PCE	SCME
Write:								
Reset:	1	1	1	1	0	0	0	1

░ = reserved, unimplemented, or cannot be written to.

Read: Anytime. Write: Anytime.

CME

Clock Monitor Enable

0 = Clock monitor is disabled.
1 = Clock monitor is enabled. Slow or stopped clocks will cause a clock monitor reset sequence of Self Clock Mode (default).

PLLON

Phase Locked Loop On

0 = PLL is turned off.
1 = PLL is turned on. If AUTO bit is set, the PLL will lock automatically (default).

AUTO

Automatic Bandwidth Control

0 = Automatic Mode Control is disabled and the PLL is under software control using the ACQ bit.
1 = Automatic Mode Control is enabled and the ACQ bit has no effect (default).

ACQ

Acquisition Bit

0 = Low bandwidth filter is selected.

1 = High bandwidth filter is selected (default).

PRE

RTI Enable During Pseudo Stop

0 = RTI stops running during Pseudo Stop Mode (default).

1 = RTI continues running during Pseudo Stop Mode.

If the PRE bit is cleared the RTI dividers will remain static while Pseudo Stop Mode is active. They are not initialized like in Wait Mode with RTIWAI bit set.

PCE

COP Enable During Pseudo Stop

0 = COP stops running during Pseudo Stop Mode (default).

1 = COP continues running during Pseudo Stop Mode.

If the PCE bit is cleared the COP dividers will remain static while Pseudo Stop Mode is active. They are not initialized like in Wait Mode with COPWAI bit set.

SCME

Self Clock Mode Enable

0 = Detection of crystal clock failure causes clock monitor reset (default).

1 = Detection of crystal clock failure forces the microcontroller into Self Clock Mode. This bit can be written once only in normal modes.

Example 21-1 Programming the PLL

```
Metrowerks HC12-Assembler
(c) COPYRIGHT METROWERKS 1987-2003


Rel. Loc  Obj. code Source line
---- ---  --------- -----------
    1                 ; This example shows how to speed up the Bus
                        Clock
    2                 ; by switching to the PLL. The PLL frequency is
    3                 ; set to 24 MHz. This example uses the timer
    4                 ; overflow flag to generate a square wave
    5                 ; by toggling port A bit 0 every time the
                        timer overflows.
```

```
 6                     ;********************************
 7                     ; Define the entry point for the main program
 8                             XDEF      Entry, main
 9                             XREF      __SEG_END_SSTACK
10                             INCLUDE   timer.inc   ; Timer defns
11                             INCLUDE   base.inc    ; Reg base defn
12                             INCLUDE   bits.inc    ; Bit defns
13                             INCLUDE   porta.inc   ; Port adr defns
14                             INCLUDE   crg.inc
15                     ;********************************
16                     ; Constants
17                     ; Set the SYNR and REFDV factors for a 16 MHz
18                     ; OSCCLK
19                     ;     Multiplier (SYNR+1) = 3
20   0000 0003 SYN_MULT: EQU   3
21                     ;     Divider (REFDV+1) = 4
22   0000 0004 REF_DIV:  EQU   4
23                     ;********************************
24                     ; Code Section
25                     MyCode: SECTION
26                     Entry:
27                     main:
28 000000 CFxx xx              lds     #__SEG_END_SSTACK
29                     ; Program the PLL to make a 24 MHz clock
30                     ;   Make sure SYSCLK = OSCCLK
31 000003 4D39 80              bclr  CLKSEL,PLLSEL
32                     ;   Turn on the PLL
33 000006 4C3A 40              bset  PLLCTL,PLLON
34                     ;   Set the SYNR and REFDV registers
35 000009 180B 0300            movb  #SYN_MULT,SYNR
   00000D 34
36 00000E 180B 0400            movb  #REF_DIV,REFDV
   000012 35
37 000013 A7                   nop              ; Give it some time
38 000014 A7                   nop
39                     ;   Wait for PLL to stabilize and lock
40                     spin:
41 000015 4F37 08FC            brclr CRGFLG,LOCK,spin
42                     ;   Turn on the PLL now
43 000019 4C39 80              bset  CLKSEL,PLLSEL
44                     ; Initialize the I/O
```

```
45 00001C 4C02 01            bset   DDRA,BIT0 ; Make PA-0 output
46                        ; Clear the TOF flag
47 00001F 8680              ldaa   #TOF
48 000021 5A4F              staa   TFLG2
49 000023 8601              ldaa   #BIT0     ; Initial LED
                                                display
50 000025 5A00              staa   PTA
51                        ; Enable the timer
52 000027 4C46 80            bset   TSCR1,TEN
53                        main_loop:
54                        ; DO
55                        ; DO
56                        ;   Wait until the TOF occurs
57                        spin1:
58 00002A F700 4F            tst    TFLG2
59 00002D 2AFB              bpl    spin1
60                        ; Reset the TOF
61 00002F 8680              ldaa   #TOF
62 000031 5A4F              staa   TFLG2
63                        ; Toggle the bit
64 000033 7100 00            com    PTA
65                        ; ENDO
66
67                        ; FOREVER
68 000036 20F2              bra    main_loop
69                        ;*******************************
```

21.5 HCS12 Operating Modes

Clock Generator Modes

We have referred to a variety of *operating modes* in this and in previous chapters. We summarize them here and refer you to other documentation for more in-depth information.

Run Mode: This is our normal operating mode with all clocks running.

Wait Mode: The Wait Mode is entered by executing a WAI instruction. The system goes into a medium power-down mode with some clocks stopping and others continuing to run. The user can control what clocks are stopped by setting or resetting bits in the CLKSEL and other registers. The microcontroller exits from a Wait Mode when it receives an external reset, clock monitor reset, COP reset, self clock mode interrupt, or a real-time interrupt.

Stop Mode: The Stop Mode is entered by executing a STOP instruction. A majority of clocks are stopped resulting in significant power saving. There are two varieties of Stop Mode–*Full Stop Mode* and *Pseudo Stop Mode.*

In Full Stop Mode the oscillator is disabled and all system and core clocks are stopped. The COP and the RTI are stopped also and remain frozen in their current state. The microcontroller exits from the Stop Mode by receiving an external interrupt or external reset.

In Pseudo Stop Mode (PSTP in CLKSEL is set) the oscillator continues to run but most of the system and core clocks are stopped. Pseudo Stop Mode consumes higher power than Full Stop Mode but recovery from this mode when an interrupt or reset occurs is faster.

Self Clock Mode: The Self Clock Mode is entered if the Clock Monitor Enable bit (CME) and the Self-Clock Mode Enable bit (SCME) are both set and the clock monitor in the oscillator block detects a loss of clock. When Self Clock Mode is entered the clock generator starts to perform a clock quality check. The microcontroller stays in Self Clock Mode until the incoming clock signal meets approval of the clock monitor. Self Clock Mode provides reduced functionality in case of a loss of clock.

See *CRG Block User Guide* for more in-depth information about these clock modes.

Modes at Reset

The operating modes entered at reset were described in Chapter 4 and include the modes shown in Table 21-10. The mode is established by the states of the MODC, MODB, and MODA pins on the rising edge of the reset signal.

Many MC9S12C microcontroller family applications will be in normal single-chip mode. Two expanded modes are available for those applications requiring more resources, particularly for more RAM. In expanded modes, port A, port B, and port E are pressed into service

TABLE 21-10 HCS12 Operating Modes

Operating Mode	Mode Description
Normal Single-Chip	Many embedded systems use this mode. The processor operates entirely within its internal memory and I/O resources because there is no expansion bus in this mode. Ports A, B, K (in some version), and most of port E are available as general purpose I/O.
Normal Expanded Wide	Ports A and B are used as a 16-bit multiplexed address and data bus. Port E provides bus controls and status signals. In this mode, 16-bit external memory and I/O devices can be interfaced to the processor.
Normal Expanded Narrow	Ports A and B are configured as a 16-bit address bus and port A is used as a multiplexed 8-bit data bus. Port E provides bus control and status signals.
Special Single-Chip	This mode may be used for debugging single-chip operation, bootstrapping, or security-related operations. The active background mode controls the CPU and BDM firmware is waiting for additional serial commands through the BKGD pin. There is no expansion bus in this mode.
Emulation Wide and Emulation Narrow	Developers use these two modes for emulating systems in which the target application is Normal Expanded Wide or Narrow Modes.
Special Test	Ports A and B are used as a 16-bit multiplexed address and data bus and port E provides bus control and status signals. This mode is used for factory testing.
Special Peripheral	This mode is used for factory testing. The CPU is inactive and an external bus master drives the address and data buses and the control pins.

TABLE 21-11 Power Supply Current

Operating Mode	Typical Current at 27 °C
Single-Chip Run Mode	45 mA
Wait Mode all modules enabled	33 mA
Wait Mode only RTI enabled	8 mA
Pseudo Stop Mode, RTI and COP enabled	500 μA
Pseudo Stop Mode, RTI and COP disabled	200 μA
Full Stop Mode	25 μA

for address, data, and control buses. A number of registers and control and status bits are available for use in these modes. Please refer to *Multiplexed External Bus Interface Block User Guide* and the application notes by Gallop and by Williams for more details.

Wait and Stop Modes

Many applications require us to be conservative with power, and like any CMOS circuit, the strategy to reduce power consumption is to reduce the frequency of, or even stop, clocked circuits. There are two power-saving operation modes—Wait Mode and Stop Mode—and Table 21-11 shows the reduced power supply current achieved by using these modes.

WAIT MODE

Wait Mode is entered when a WAI instruction is executed. A variety of clocks can be stopped when in Wait Mode as shown in Table 21-12 and there are five ways to restart your program after entering a Wait Mode as shown in Table 21-13.

PSEUDO STOP MODE

Pseudo Stop Mode is entered when a STOP instruction is executed and the Pseudo Stop bit (PSTP) is set. Pseudo Stop Mode differs from Full Stop Mode in that the oscillator continues to run, allowing a faster recovery from the stopped clocks condition. Pseudo Stop Mode is

TABLE 21-12 Clocks and Devices Stopped During Wait Mode

Register	Bit	Effect	Default
CLKSEL	SYSWAI	Bus Clock, Core Clock, and Oscillator Clock stopped	Not stopped
CLKSEL	ROAWAI	Oscillator amplitude reduced	Normal amplitude
CLKSEL	PLLWAI	PLL stopped	Not stopped
CLKSEL	CWAI	Core Clock stopped	Not stopped
CLKSEL	RTIWAI	Real-time interrupt clock stopped	Not stopped
CLKSEL	COPWAI	COP clock stopped	Not stopped
PWMCTL	PSWAI	PWM stopped	Not stopped
SPICR2	SPIWAI	SPI stopped	Not stopped
SCICR1	SCIWAI	SCI stopped	Not stopped
CANCTL0	CSWAI	CAN stopped	Not stopped
ATDCTL2	AWAI	ATD stopped	Not stopped
TSCR1	TSWAI	Timer stopped	Not stopped

TABLE 21-13 Wait Mode Exit Options

Wait Mode Restart	Comments
External Reset	An external reset acts as any normal reset. All configuration bits are restored to their default and the microcontroller enters the run mode.
Clock Monitor Reset or Self Clock Mode Interrupt	If the clock monitor is enabled (CME = 1) and the clock has failed, the microcontroller can leave wait mode by two paths. If the self clock mode is enabled (SCME = 1) and if the self clock mode interrupt is enabled, then a self clock mode interrupt is generated. The microcontroller exits wait mode, enters self clock mode, and continues with normal operation. If SCME = 0, self clock mode is not entered and a clock monitor reset is performed. The reset vector is fetched from the clock monitor failure reset vector.
COP Reset	A COP time-out resets the system. The COPWAI bit must be reset for this to work. The reset vector is fetched from the COP failure reset vector.
Other Interrupts	Any other interrupt request (if interrupts are unmasked and enabled) will exit wait mode. If the PLL was switched off when wait mode was entered (PLLWAI = 1), it will remain off and the software must manually set the PLLON bit again.

essentially the same as Wait Mode in that the user can choose to allow the RTI and COP clocks to run by setting the PRE and/or the PCE bits. Pseudo Stop Mode does request other microcontroller units to enter their individual power-saving modes. Recovering from Pseudo Stop Mode is the same as from Wait Mode. See Table 21-14.

FULL STOP MODE

Full Stop Mode is entered when a STOP instruction is executed and PSTP = 0. All clocks are stopped and the PLL, if in use, is switched off. An external reset or interrupt request is required to return to run mode.

21.6 Designing a System Using an MC9S12C Family Microcontroller

We have given you most of the operational details needed to operate the MC9S12C Family microcontroller in a variety of applications. More information than what we have covered here is needed to be able to design your own microcontroller printed circuit board. Engineers at Freescale have produced a variety of documents that will help you in this task. See Section 21.8 for a list of some of these useful resources.

TABLE 21-14 RTI and COP Operation During Pseudo Stop Mode

Bit	Effect	Default
PRE = 1	RTI runs during Pseudo Stop Mode.	Stopped
PCE = 1	COP runs during Pseudo Stop Mode.	Stopped

TABLE 21-15 CRG System Registers

Name	Register	Address Base +
SYNR	CRG Synthesizer Register	$0034
REFDV	CRG Reference Divider Register	$0035
CRGFLG	CRG Flags Register	$0037
CRGINT	CRG Interrupt Enable Register	$0038
CLKSEL	CRG Clock Select Register	$0039
PLLCTL	PLL Control Register	$003A

21.7 Clock Generator System Register Address Summary

This summary is given in Table 21-15.

21.8 Bibliography and Further Reading

CRG Block User Guide V04.05, S12CRGV4/D, Freescale Semiconductor, Inc., Austin, TX, 2002.

Gallop, M., *Designing Hardware for the HCS12 D-Family*, AN2727, Rev. 0, Freescale Semiconductor, Inc., Austin, TX, December 2004.

MC9S12C Family Device User Guide V01.05, 9S12C128DGV1/D, Freescale Semiconductor, Inc., Austin, TX, 2004.

MC9S12C128 Data Sheet Covers MC9S12C Family and MC9S12GC Family, MC9S12C128 Rev 1.1.6, mc9s128_v1.pdf, Freescale Semiconductor, Inc., Austin, TX, October 2005.

Multiplexed External Bus Interface Block User Guide V03.0, S12MEBIV3/D, Freescale Semiconductor, Inc., Austin, TX, 2003.

OSC Block User Guide V02.03, S12OSCV2/D, Freescale Semiconductor, Inc., Austin, TX, 2003.

Rob, S., *HCS12 PLL Component Calculator User Guide*, HCS12PLLCALUG/D, Rev. 0 , Freescale Semiconductor, Inc., Austin, TX, December 2002. http://www.freescale.com/files/microcontrollers/software_tools/initialization/boot_code_generation/HCS12PLLCALSW.zip?srch=1

Williams, J., *HCS12 External Bus Design*, AN2287/D, Rev. 1, Freescale Semiconductor, Inc., Austin, TX, August 2004.

Appendix A: Binary Codes

A.1 Binary Codes Review

Coding is a two-part process consisting of encoding and decoding. Encoding means converting information into a form that can be used in the microcontroller, generally into a binary code. Decoding allows us to convert the coded information back to the information. Whenever a binary code is chosen, we consider the following:

The type of information to be encoded: Is the information numerical? Are there negative and positive numbers? Are there fractional or just integer numbers? If the information is not numerical, is there a standard code to be used? How much information is there? What is its range of values? To what resolution do we need to know and encode the information?

The number of bits needed to represent the information: The number of bits needed depends on the amount of information to be encoded and the resolution to which we need to know the information.

$$Number\ of\ Bits \geq log_2\ (Number\ of\ Information\ Elements)$$

or

$$Number\ of\ Bits \geq log_2\ (Full\ Scale\ Value/Resolution\ Value)$$

When we know the number of bits required, we can calculate the number of code words available.

$$Number\ of\ Code\ Words = 2^{Number\ of\ Bits}$$

(See Examples A-1 and A-2.)

Example A-1

A binary code is needed to identify each of the 83 students in a class. How many bits are required?

Solution:

$$N \geq \log_2 \text{ (Number of Codes)}$$
$$\log_2 83 = 6.375; \text{ therefore, } N = 7$$

How much larger can the class grow before another bit is needed?

Solution:

$$2^7 = 128$$

The class can grow by 45 students.

Example A-2

A binary code is needed to encode a voltage converted to a digital value by an analog-to-digital converter. The maximum voltage is 5.0 volts and the resolution required is 0.01 volt. How many bits are required?

Solution:

$$N \geq \log_2 \text{ (Full Scale Value/Resolution Required)}$$
$$= \log_2 (5.0/0.1) = 5.6; \quad \text{therefore, } N = 6$$

Binary Codes for Numerical Information

There are several codes that we use for numerical information. The five that are most important to microcontroller users are (1) unsigned-binary, (2) signed/magnitude, (3) ones'-complement, (4) two's complement, and (5) binary coded decimal.

UNSIGNED-BINARY CODE

The unsigned-binary code is a positive weighted code where each bit in the code word has a weight (or value) according to its position. Each digit is assigned a position starting from the right (at the binary point) with zero, increasing to the left, and decreasing to the right. The weight of each position is the base raised to the power of the digit position. The leftmost bit is the *most significant bit* (MSB) and the rightmost bit is the *least significant bit* (LSB). See Table A-1.

TABLE A-1 Binary Word Bit Positions

Bits	...	b_3	b_2	b_1	b_0	.	b_{-1}	b_{-2}	b_{-3}	...
Bit Position	...	3	2	1	0	.	-1	-2	-3	...
Bit Weight	...	2^3	2^2	2^1	2^0	.	2^{-1}	2^{-2}	2^{-3}	...
Weights	...	8	4	2	1	.	0.5	0.25	0.125	...
	MSB									LSB

The unsigned-binary code uses all positive weights and represents only positive information. The number of bits, and therefore the number of codes, determines how much information can be encoded. In a code word with p + q bits, the number of codes is

$$Number\ of\ Codes = 2^{p+q}$$

The range of numerical information that can be represented in a code word with p-integer and q-fractional bits is

$$Range = 0\ to\ 2^p - 2^{-q}$$

The resolution is the value of the least significant bit. In this case

$$Resolution = 2^{-q}$$

(See Examples A-3 to A-5.)

The unsigned-binary code can represent only positive information, but there are several other codes used for negative information. The three used most commonly in the microcomputer world are the signed/magnitude, radix-1-complement (ones'-complement) and radix-complement (two's complement) codes.

Example A-3

An unsigned-binary code has four integer bits (p = 4) and two fractional bits (q = 2). How many codes are there? What is the range of numbers that can be encoded? What is the smallest number that can be encoded?

Solution:

The number of codes is $2^{p+q} = 2^6 = 64$.

The range of numbers is from 0 to $2^p - 2^{-q} = 16 - 0.25 = 15.75$.

The smallest number that can be encoded is the resolution $= 2^{-q} = 2^{-2} = 0.25$.

Example A-4

What are the weights of each of the bits p–w in an unsigned-binary code word pqrst.uvw?

Solution:

Code:	p	q	r	s	t	.	u	v	w
Weights:	2^4	2^3	2^2	2^1	2^0	.	2^{-1}	2^{-2}	2^{-3}
	16	8	4	2	1	.	0.5	0.25	0.125

Example A-5

How many codes are there, what is the range of numbers that can be represented, and what is the resolution of an unsigned-binary code word pqrst.uvw?

Solution:

$$\text{Number of Codes} = 2^8 = 256$$
$$\text{Range} = 0 \text{ to } 2^5 - 2^{-3} = 0 \text{ to } 31.875$$
$$\text{Resolution} = 2^{-3} = 0.125$$

SIGNED/MAGNITUDE BINARY CODE

Signed/magnitude, ones'-complement, and *two's complement* binary codes are used for positive and negative numbers.

The *signed/magnitude binary code* is similar to our decimal number system. The decimal code word for plus ten is written $+10$ or just 10. Minus ten is encoded -10. Two additional symbols, $+$ and $-$, are added to the front of the digits used for the magnitude. These symbols double the number of code words to be able to represent both positive and negative numbers. Notice that there are two codes for zero, $+0$ and -0. By convention, we never use the code for minus zero.

In the binary system, an additional bit to encode the sign is added to the binary digits encoding the magnitude. A zero is positive and one negative. Table A-2 shows the layout for a signed/magnitude binary code. Example A-6 shows a 7-bit binary code with one bit used as a sign and six bits to encode the magnitude.

The range of information that can be represented with p-integer bits (including the sign bit) plus q-fractional bits is

$$-(2^{p-1} - 2^{-q}) \ to \ +(2^{p-1} - 2^{-q})$$

For Example A-6, the range is -15.75 to $+15.75$. Again, there is a code for plus and minus zero (see Example A-7).

ONES'-COMPLEMENT CODE

The definition of the *radix-1* or *ones'-complement* of a number X is

$$Ones'\text{-}Complement = 2^p - X - 2^{-q}$$

where p is the number of integer bits and q the number of fractional bits.

Example A-8 shows how to form the ones'-complement code for ± 6.25. The leftmost bit is an *indicator* (called the *sign bit*) for the sign of the number with 0 representing positive and 1 negative. The range and resolution of the ones'-complement code are the same as the signed/magnitude code and, again, there are two codes for zero. The ones'-complement code is not a weighted code.

TABLE A-2 Signed/Magnitude Code Bits

Bits	b_{p-1}	...	b_2	b_1	b_0	.	b_{-1}	b_{-2}	...	b_{-q}
Bit Position	$p-1$	...	2	1	0	.	-1	-2	...	$-q$
Bit Weight	Sign	...	2^2	2^1	2^0	.	2^{-1}	2^{-2}	...	2^{-q}
Weights	$0 = +$	...	4	2	1	.	0.5	0.25	...	
	$1 = -$									

Example A-6 Signed/Magnitude Binary Code Examples

$$0 \quad 1 \quad 0 \quad 1 \quad 1 \quad . \quad 1 \quad 1 \quad = \quad +11.75$$
$$1 \quad 1 \quad 0 \quad 1 \quad 1 \quad . \quad 1 \quad 1 \quad = \quad -11.75$$
$$\underbrace{}_{\text{Sign}} \underbrace{}_{\text{Magnitude Code}}$$

The leftmost bit is the sign bit, and the magnitude is encoded with a 6-bit unsigned-binary code.

Example A-7

How many codes are there, what is the range of numbers that can be represented, and what is the resolution of a signed/magnitude binary code word pqrst.uvw, where p is the sign bit?

Solution:

$$\text{Number of Codes} = 2^8 = 256 \text{ (but two are used for zero)}$$
$$\text{Range} = -(2^4 - 2^{-3}) \text{ to } (2^4 - 2^{-3}) = -15.875 \text{ to } 15.875$$
$$\text{Resolution} = 2^{-3} = 0.125$$

Example A-8

Find the ones'-complement code for -6.25 assuming a code of the form pqrst.uvw.

Solution: Find the unsigned-binary code for $+6.25$ and add a sign bit in the most significant bit position. Then, to find the code for -6.25, complement all bits.

6.25		1	1	0	.	0	1	0		Unsigned-binary code for 6.25
+6.25	= 0	1	1	0	.	0	1	0		Ones'-complement binary code for +6.25
−6.25	= 1	0	0	1	.	1	0	1		Ones'-complement binary code for −6.25

TWO'S COMPLEMENT CODE

Two's complement binary codes are used for negative numbers in microcontroller systems.

In the binary number system, the radix-complement is the *two's complement* binary code. The definition of p-integer bit, two's complement of number X is

$$Two's\ Complement = 2^p - X$$

This is a *negatively* weighted code because the most significant bit has a negative weight as shown in Table A-3 and Example A-9.

There is only one code for zero in the two's complement scheme. The extra code represents the most negative number. We can see this by looking at the range of the two's complement binary code. For a number with p-integer and q-fractional bits, the range is

$$Range = -(2^{p-1})\ to\ +(2^{p-1} - 2^{-q})$$

TABLE A-3 Two's Complement Code Bits

Bits, 0 or 1	b_{p-1}	$\ldots$	b_2	b_1	b_0	.	b_{-1}	b_{-2}	$\ldots$	b_{-q}
Bit Position	$p-1$	$\ldots$	2	1	0	.	-1	-2	$\ldots$	$-q$
Bit Weight	-2^{p-1}	$\ldots$	2^2	2^1	2^0	.	2^{-1}	2^{-2}	$\ldots$	2^{-q}
Weights	$0 = +$	$\ldots$	4	2	1	.	0.5	0.25	$\ldots$	
	$1 = -$									

The range of the binary number in Example A-9 is -8.00 to $+7.875$. The resolution is $2^{-q} = 0.125$.

THE SIGN OF THE NUMBER

In the three codes shown in Examples A-9 to A-11, the most significant bit gives the sign of the number, although the sign bit for signed/magnitude code could be placed anywhere in the code word. In two's complement codes the sign bit carries a negative weight. In ones'-complement codes the sign bit does not carry a weight; it indicates the sign.

Example A-9

Show the weights of a two's complement binary number 1 0 1 1 . 0 1 1.

Solution:

$$1\,0\,1\,1\,.\,0\,1\,1 = 1 \times (-2^3) + 0 \times 2^2 + 1 \times 2^1 + 1 \times 2^0 + 0 \times 2^{-1} + 1 \times 2^{-2} + 1 \times 2^{-3}$$
$$= -8 + 2 + 1 + 0.25 + 0.125 = -4.625$$

Example A-10

What are the weights of each of the bits in a two's complement binary code word pqrst.uvw?

Solution:

Code:	p	q	r	s	t	.	u	v	w
Weights:	-2^4	2^3	2^2	2^1	2^0	.	2^{-1}	2^{-2}	2^{-3}
	-16	8	4	2	1	.	0.5	0.25	0.125

Example A-11

How many codes are there, what is the range of numbers that can be represented, and what is the resolution of a two's complement binary code word pqrst.uvw?

Solution:

$$\text{Number of Codes} = 2^8 = 256 \text{ (but two are used for zero)}$$
$$\text{Range} = -(2^4) \text{ to } (2^4 - 2^{-3}) = -16.000 \text{ to } 15.875$$
$$\text{Resolution} = 2^{-3} = 0.125$$

FINDING THE CODE FOR THE NEGATIVE

In decimal, when we want the code for the negative of a number, we "take the negative of it" by simply changing the sign. For a signed/magnitude binary code, the same is true. The sign bit is *complemented* to change a positive to a negative and vice versa. See Examples A-6 and A-8.

The ones'-complement code for a negative number is found by complementing each of the bits in the code for the positive number. This process is called *ones'-complementing* or just *complementing* the bits. See Example A-12.

Finding the code for the negative in a two's complement number system involves an extra step. We find the code for the negative by *taking the two's complement* of the code for the positive. This is analogous to "taking the negative" of a signed/magnitude code. The two's complement of any number is found by

$$Two's\ Complement\ Code = Ones'\text{-}Complement\ Code + 2^{-q}$$

Taking the two's complement to find the negative is a three-step process:

1. Find the two's complement code for the positive number.
2. Complement each of the bits (ones'-complement).
3. Add one to the least significant bit position.

This procedure is shown in Examples A-13 to A-15.

Example A-12 Ones'-Complement Binary Code Examples

$$0\ 1\ 0\ 1\ 1\ .\ 1\ 1 = +11.75$$

Complement each bit to find the code for -11.75.

$$1\ 0\ 1\ 0\ 0\ .\ 0\ 0 = -11.75$$

Example A-13 Taking the Two's Complement

3.25 =	0 0 1 1 . 0 1 0 0	Two's complement code for $+3.25$
Ones'-complement =	1 1 0 0 . 1 0 1 1	
Add 2^{-4}	0 0 0 0 . 0 0 0 1	
-3.25 =	1 1 0 0 . 1 1 0 0	Two's complement code for -3.25

Example A-14

Find the two's complement binary code for -6.25 assuming a code of the form pqrst.uvw.

Solution:

$+6.25$ =	0 0 1 1 0 . 0 1 0	Two's complement code for $+6.25$
Ones'-complement =	1 1 0 0 1 . 1 0 1	
Add 2^{-4}	0 0 0 0 0 . 0 0 1	
-6.25 =	1 1 0 0 1 . 1 1 0	Two's complement code for -6.25

Example A-15

Take the two's complement of the code for -6.25 to find the code for $+6.25$.

Solution:

$$
\begin{array}{rcccccccccc}
-6.25 &=& 1 & 1 & 0 & 0 & 1 & . & 1 & 1 & 0 \\
\text{Ones'-complement} &=& 0 & 0 & 1 & 1 & 0 & . & 0 & 0 & 1 \\
\text{Add } 2^{-4} & & 0 & 0 & 0 & 0 & 0 & . & 0 & 0 & 1 \\
+6.25 &=& 0 & 0 & 1 & 1 & 0 & . & 0 & 1 & 0 \\
\end{array}
$$

UNSIGNED-BINARY ARITHMETIC

Adding and subtracting unsigned-binary numbers is done just like we add and subtract the magnitudes of decimal numbers. In each case we keep track of carries into or borrows from the next-most significant digit position. See Examples A-16 and A-17.

An overflow occurs if the two numbers being added result in a number outside the allowable range. In Example A-18 we try to add 9 and 10 using a 4-bit unsigned-binary code. The expected result, 19, requires 5 bits. The carry bit out of the most significant bit indicates an overflow of the available bits. In microcontroller systems this error is detected with a special flag called the carry flag.

TWO'S COMPLEMENT BINARY ARITHMETIC

The beauty of using the two's complement code for signed numbers is that the hardware to do addition and subtraction is the same as the hardware for unsigned-binary coded arithmetic.

Example A-16 Unsigned-Binary Addition

Add the 4-bit binary codes for 6 and 3.

Solution:

$$
\begin{array}{rcccc}
\text{Carries} & & 1 & 1 & 0 & \\
6 &=& 0 & 1 & 1 & 0 \\
\underline{3} &=& \underline{0} & \underline{0} & \underline{1} & \underline{1} \\
9 &=& 1 & 0 & 0 & 1 \\
\end{array}
$$

Example A-17 Unsigned-Binary Subtraction

Subtract the 4-bit binary code 3 from 6.

Solution:

$$
\begin{array}{rcccc}
6 &=& 0 & 1 & 1 & 0 \\
\underline{-3} &=& \underline{0} & \underline{0} & \underline{1} & \underline{1} \\
\text{Borrows} & & 0 & 1 & 1 & \\
3 &=& 0 & 0 & 1 & 1 \\
\end{array}
$$

Example A-18 Unsigned-Binary Overflow

Add the 4-bit binary codes for 9 and 10. The result musts be 4 bits also.

Solution:

```
Carries        1   0   0   0
    +9    =    1   0   0   1
   +10    =    1   0   1   0
    3?    =    0   0   1   1
```

Furthermore, one can easily subtract two numbers by adding the two's complement of the subtrahend to the minuend. This is shown in Example A-19 for a 6-bit two's complement code and in Example A-20 for an 8-bit code.

A two's complement overflow occurs when the result of an addition or subtraction is outside the allowable range of numbers for the number of bits available. When two's complement numbers are added or subtracted, a carry out of the most significant bit position does not indicate an

Example A-19 Subtraction by the Addition of the Two's Complement

	Binary Subtraction	Subtraction by Adding the Two's Complement
$+5 =$	0 0 0 ! 0 1	$+5 = 0\ 0\ 0\ 1\ 0\ 1$
$-3 =$	0 0 0 0 1 1	$+(-3) = 1\ 1\ 1\ 1\ 0\ 1$
$+2 =$	0 0 0 0 1 0	$+2 = 0\ 0\ 0\ 0\ 1\ 0$

Example A-20

Using 8-bit, two's complement binary codes (5 integer and 3 fractional bits) compute $8.75 - 10.5$.

Solution:

```
   8.75 = 0 1 0 0 0 . 1 1 0   Two's complement binary code for 8.75
  10.50 = 0 1 0 1 0 . 1 0 0   Two's complement binary code for 10.5
 -10.50 = 1 0 1 0 1 . 1 0 0   Two's complement binary code for -10.75
```

Therefore

```
     8.75 = 0 1 0 0 0 . 1 1 0
 +(-10.50) = 1 0 1 0 1 . 1 0 0
            1 1 1 1 0 . 0 1 0
```

The result is negative. To find the magnitude of the result, take the two's complement.

$$0\ 0\ 0\ 0\ 1\ .\ 0\ 1\ 0\ = 1.75$$

and so the answer is -1.75.

Example A-21 Two's Complement Overflow

Add the 4-bit, two's complement numbers +6 and +3 and detect if an overflow occurs. The result is to be 4 bits.

Solution:

$$
\begin{array}{rcl}
\text{Carries} & & 0 \quad 1 \quad 1 \quad 0 \\
+6 & = & 0 \quad 1 \quad 1 \quad 0 \\
+3 & = & 0 \quad 0 \quad 1 \quad 1 \\
\hline
-6? & = & 1 \quad 0 \quad 0 \quad 1
\end{array}
$$

Two's complement overflow has occurred because the sign of the result is different from the sign of the two numbers.

overflow as it does in unsigned-binary arithmetic. There are various algorithms for detecting a two's complement overflow (see Example A-21). One of the easiest to understand is the following:

A two's complement overflow occurs if when adding or subtracting two numbers of the same sign gives a result with a different sign.

Two's complement overflow cannot occur when adding or subtracting two numbers of opposite sign.

BINARY CODED DECIMAL

A 4-bit, unsigned-binary code is sometimes used to encode the ten decimal digits 0–9. This is called *natural binary coded decimal* and is used so frequently that it is usually just called *binary coded decimal* or *BCD*. Table A-4 shows the natural BCD code. A packed BCD code uses 8 bits to encode two binary coded decimal digits. See Example A-22.

TABLE A-4 Four-Bit Binary Code Comparison

Code Word	Unsigned Binary	Ones'-Complement	Two's Complement	Signed/Magnitude	BCD	Hexadecimal
0000	0	0	0	0	0	0
0001	1	1	1	1	1	1
0010	2	2	2	2	2	2
0011	3	3	3	3	3	3
0100	4	4	4	4	4	4
0101	5	5	5	5	5	5
0110	6	6	6	6	6	6
0111	7	7	7	7	7	7
1000	8	−7	−8	−0	8	8
1001	9	−6	−7	−1	9	9
1010	10	−5	−6	−2	NA	A
1011	11	−4	−5	−3	NA	B
1100	12	−3	−4	−4	NA	C
1101	13	−2	−3	−5	NA	D
1110	14	−1	−2	−6	NA	E
1111	15	−0	−1	−7	NA	F
Range	0 to 15	−7 to +7	−8 to +7	−7 to +7	0 to 9	0 to F

Example A-22 Packed BCD

Using an 8-bit packed BCD code, give the code for the decimal numbers 23, 45, and 99.

Solution: A packed BCD code has 4 bits for each decimal digit in one-half of each byte. The most significant nibble has the most significant digit's code.

$$
\begin{array}{rcccc|cccc}
23 & = & 0 & 0 & 1 & 0 & 0 & 0 & 1 & 1 \\
45 & = & 0 & 1 & 0 & 0 & 0 & 1 & 0 & 1 \\
99 & = & 1 & 0 & 0 & 1 & 1 & 0 & 0 & 1 \\
\end{array}
$$

HEXADECIMAL CODES

The hexadecimal, or base 16, number system is a shorthand for strings of binary digits. Like the BCD code, the 16 hexadecimal digits, 0–9, A–F, are encoded using an unsigned-binary code (see Example A-23). The hexadecimal digits and their binary codes are shown in Table A-4.

YOU HAVE TO KNOW THE CODE

If given a binary number and asked what it means, you can't answer unless you know what code is being used. Table A-4 shows the different information that is decoded from a 4-bit code word using the different codes covered in this section.

Binary Codes for Nonnumerical Information

Sometimes encoded (or decoded) information isn't a number. A common example is the alphanumeric information sent from a keyboard to a computer or from a computer to a display. Codes used for this application are called unweighted codes because, unlike the numerical codes, there isn't a weight associated with a bit's position. To find out what a code means, you must look it up in a table.

THE ASCII CODE

The *American Standard Code for Information Interchange* (*ASCII*) is used to encode alphanumeric information, for example, keys on a keyboard or letters displayed on a terminal. The ASCII codes for alphanumeric information are shown in Table A-5.

Example A-23 Hexadecimal Codes

Covert the binary number 1 0 1 1 0 1 0 1 to hexadecimal.

Solution: Start with the four least significant bits (0 1 0 1) = 5; the most significant four bits (1 0 1 1) = B. The hexadecimal number is B5.

TABLE A-5 ASCII 7-bit Codes for Alphanumeric Characters

LS Digit	MS Digit							
	0	**1**	**2**	**3**	**4**	**5**	**6**	**7**
0	NUL	DLE	SP	0	@	P	`	p
1	SOH	DC1	!	1	A	Q	a	q
2	STX	DC2	"	2	B	R	b	r
3	ETX	DC3	#	3	C	S	c	s
4	EOT	DC4	$	4	D	T	d	t
5	ENQ	NAK	%	5	E	U	e	u
6	ACK	SYN	&	6	F	V	f	v
7	BEL	ETB	'	7	G	W	g	w
8	BS	CAN	(	8	H	X	h	x
9	HT	EM	)	9	I	Y	i	y
A	LF	SUB	*	:	J	Z	j	z
B	VT	ESC	+	;	K	[	k	{
C	FF	FS	,	<	L	\	l		
D	CR	GS	-	=	M	]	m	}
E	SO	RS	.	>	N	^	n	~
F	SI	US	/	?	O	_	o	DEL

The two leftmost columns (MS Digit = 0 and 1) are control codes that have been defined for serial data communications. (See Table A-6.)

TABLE A-6 ASCII Control Codes

00	NUL	Null	Character with all zeros.
01	SOH	Start of Header	Used at the beginning of a sequence of characters which constitutes a machine-readable address of routing information. The header is terminated by the STX character.
02	STX	Start of Text	A character that precedes a sequence of characters to be treated as an entity. STX may be used to terminate a sequence of characters started by SOH.
03	ETX	End of Text	Character used to terminate a sequence of characters started with STX.
04	EOT	End of Transmission	Indicates the conclusion of a transmission.
05	ENQ	Enquiry	Used as a request for a response from a remote station.
06	ACK	Acknowledge	Character transmitted by a receiver as an affirmative response to the sending station.
07	BEL	Bell	Character used to control an alarm or attention device.
08	BS	Back Space	Controls the movement of the printing mechanism back one space.
09	HT	Horizontal Tab	Controls the movement of the printing mechanism to the next predefined tab position.
0A	LF	Line Feed	Moves the printing mechanism to the next line. In some systems this may be interpreted as a "New Line" (NL), where the print mechanism moves to the beginning of the next line.
0B	VT	Vertical Tab	Controls the movement of the printing mechanism to the next predefined printing line position.
0C	FF	Form Feed	Moves the printing mechanism to the start of the next page.
0D	CR	Carriage Return	Moves the printing mechanism to the start of the line.

TABLE A-6 Continued

0E	SO	Shift Out	Indicates that the code combinations following are outside the character set of the standard ASCII table until a Shift In character is received.
0F	SI	Shift In	Indicates that the code characters following are to be interpreted according to the standard ASCII table.
10	DLE	Data Link Escape	Changes the meaning of a limited number of following characters. DLE is usually terminated by a Shift In character.
11	DC1	Device Controls	These characters are used to control ancillary devices associated with data processing.
12	DC2		
13	DC3		
14	DC4		
15	NAK	Negative Acknowledge	Transmitted by a receiver as a negative response to the sender.
16	SYN	Synchronous Idle	Character used by a synchronous transmission system in the absence of any other characters to maintain synchronism between the transmitter and receiver.
17	ETB	End of Transmission Block	Used to indicate the end of a block of data.
18	CAN	Cancel	Indicates that the data with which it is sent is in error or is to be disregarded.
19	EM	End of Medium	Sent with data to represent the physical end of the medium.
1A	SUB	Substitute	A character that may be substituted for a character that is invalid or in error.
1B	ESC	Escape	A control character intended to provide code extension. It is usually a prefix affecting the interpretation of a limited number of contiguously following characters.
1C	FS	File Separator	These information separators may be used within data.
1D	GS	Group Separator	
1E	RS	Record Separator	
1F	US	Unit Separator	

A.2 Problems

Basic

A.1 Encode your name using the ASCII code. [a]

A.2 Decode the ASCII message 44 65 73 69 67 6e 69 6e 67 20 77 69 74 68 20 6d 69 63 72 6f 70 72 6f 63 65 73 73 6f 72 73 20 69 73 20 46 55 4e 21. [a]

A.3 Give the decimal value of the following binary code words assuming (i) unsigned-binary, (ii) two's-complement binary, and (iii) signed/magnitude codes. [a]

a. 10101010

b. 01010101

c. 11001100

d. 00110011

e. 10000000

f. 01111111

A.4 Find the two's complement binary code for the following decimal numbers: [a]

a. 26

b. −26

c. 32.125

d. −32.125

A.5 Find the decimal equivalent of the following two's complement numbers: **[a]**

   a. 0101101.1

   b. 1010010.1

   c. 1000

   d. 1010.1101

A.6 A 6-bit, two's complement binary code is to be used for integer numbers. What is the range of information, what is the resolution, and how many codes are there? **[a]**

A.7 How many bits are required to encode the decimal number 238 using a BCD code? How many using an unsigned-binary code? How many using a two's complement binary code? **[a]**

A.8 Find the binary code words for the following hexadecimal numbers: **[a]**

   a. BEEF

   b. FEED

   c. C0FFEE

   d. F00D

A.9 Find the hexadecimal code words for the following binary code words: **[a]**

   a. 01011010

   b. 11110101

   c. 110101

   d. 101

Intermediate

A.10 Prove that two's complement overflow cannot occur when two numbers of different signs are added. **[a]**

A.11 For 4-bit number two's complement addition, choose four examples to demonstrate the following: **[a, b]**

   a. Addition with no carry out and no two's complement overflow.

   b. Addition with no carry out and two's complement overflow.

   c. Addition with carry out and no two's complement overflow.

   d. Addition with carry out and two's complement overflow.

Appendix B: HCS12 Instruction Set

HCS12 Instruction Set[a]

Mnemonic	Operation	Mnemonic	Operation
Load Registers			
LDAA	$(M) \rightarrow A$	LDAB	$(M) \rightarrow B$
LDD	$(M:M + 1) \rightarrow D$	LDS	$(M:M + 1) \rightarrow SP$
LDX	$(M:M + 1) \rightarrow X$	LDY	$(M:M + 1) \rightarrow Y$
LEAS	$EA \rightarrow SP$	LEAX	$EA \rightarrow X$
LEAY	$EA \rightarrow Y$		
PULA	$(SP) \rightarrow A$	PULB	$(SP) \rightarrow B$
PULD	$(SP:SP + 1) \rightarrow D$	PULC	$(SP) \rightarrow CCR$
PULX	$(SP:SP + 1) \rightarrow X$	PULY	$(SP:SP + 1) \rightarrow Y$
Store Registers			
STAA	$A \rightarrow (M)$	STAB	$B \rightarrow (M)$
STD	$D \rightarrow (M:M + 1)$	STS	$SP \rightarrow (M:M + 1)$
STX	$X \rightarrow (M:M + 1)$	STY	$Y \rightarrow (M:M + 1)$
PSHA	$A \rightarrow (SP)$	PSHB	$B \rightarrow (SP)$
PSHD	$D \rightarrow (SP:SP + 1)$	PSHC	$CCR \rightarrow (SP)$
PSHY	$Y \rightarrow (SP:SP + 1)$	PSHX	$X \rightarrow (SP:SP + 1)$
Transfer/Exchange Registers			
TFR	Any Reg $\rightarrow$ Any Reg	EXG	Any Reg $\leftarrow \rightarrow$ Any Reg
Move Memory Contents			
MOVB	$(M1) \rightarrow (M2)$	MOVW	$(M1:M1 + 1) \rightarrow (M2:M2 + 1)$
Decrement/Increment			
DEC	$(M) - 1 \rightarrow (M)$	DECA	$A - 1 \rightarrow A$
DECB	$B - 1 \rightarrow B$	DES	$SP - 1 \rightarrow SP$
DEX	$X - 1 \rightarrow X$	DEY	$Y - 1 \rightarrow Y$
INC	$(M) + 1 \rightarrow (M)$	INCA	$A + 1 \rightarrow A$
INCB	$B + 1 \rightarrow B$	INS	$SP + 1 \rightarrow SP$
INX	$X + 1 \rightarrow X$	INY	$Y + 1 \rightarrow Y$
Clear/Set			
CLR	$0 \rightarrow (M)$	CLRA	$0 \rightarrow A$
CLRB	$0 \rightarrow B$		
BCLR	$0 \rightarrow (M \text{ bits})$	BSET	$1 \rightarrow (M \text{ bits})$

HCS12 Instruction Set (Continued)

Mnemonic	Operation	Mnemonic	Operation
Arithmetic			
ABA	$A + B \rightarrow A$	ABX	$B + X \rightarrow X$ (see LEAX)
ABY	$B + Y \rightarrow Y$ (see LEAY)	ADDA	$A + (M) \rightarrow A$
ADDB	$B + (M) \rightarrow B$	ADDD	$D + (M:M + 1) \rightarrow D$
ADCA	$A + (M) + C \rightarrow A$	ADCB	$B + (M) + C \rightarrow B$
DAA	Decimal adjust		
SUBA	$A - (M) \rightarrow A$	SBA	$A - B \rightarrow A$
SUBD	$D - (M:M + 1) \rightarrow D$	SUBB	$B - (M) \rightarrow B$
SBCB	$B - (M) - C \rightarrow B$	SBCA	$A - (M) - C \rightarrow A$
NEG	Two's Complement (M)	NEGA	Two's Complement $\rightarrow A$
NEGB	Two's Complement B	SEX	Sign extend A,B,CCR
MUL	Unsigned $A*B \rightarrow D$	EMUL	Unsigned $D*Y \rightarrow Y:D$
EMULS	Signed $D*Y \rightarrow Y:D$		
IDIV	Unsigned $D/X \rightarrow X,D$	EDIV	Unsigned $Y:D/X \rightarrow Y,D$
EDIVS	Signed $Y:D/X \rightarrow Y,D$	IDIVS	Signed $D/X \rightarrow X,D$
FDIV	Fractional $D/X \rightarrow X,D$		
Logic			
ANDA	$A \cdot (M) \rightarrow A$	ANDB	$B \cdot (M) \rightarrow B$
ANDCC	$CCR \cdot (M) \rightarrow CCR$		
EORB	$B \, EOR \, (M) \rightarrow B$	EORA	$A \, EOR \, (M) \rightarrow A$
ORAB	$B \, OR \, (M) \rightarrow B$	ORAA	$A \, OR \, (M) \rightarrow A$
ORCC	$CCR \, OR \, (M) \rightarrow CCR$		
COM	Ones'-complement (M)	COMA	Ones'-complement A
COMB	Ones'-complement B		
Rotates and Shifts			
ROL	Rotate left (M)	ROLA	Rotate left A
ROLB	Rotate left B	ROR	Rotate right (M)
RORA	Rotate right A	RORB	Rotate right B
ASL	Arith shift left (M)	ASLA	Arith shift left A
ASLB	Arith shift left B	ASLD	Arith shift left D
ASR	Arith shift right (M)	ASRA	Arith shift right A
ASRB	Arith shift right B		
LSLA	Logic shift left A	LSL	Logic shift left (M)
LSLD	Logic shift left D	LSLB	Logic shift left B
LSRA	Logic shift right A	LSR	Logic shift right (M)
LSRD	Logic shift right D	LSRB	Logic shift right B
Data Test			
BITA	Test bits in A	BITB	Test bits in B
CBA	$A - B$	CMPA	$A - (M)$
CMPB	$B - (M)$	CPD	$D - (M:M + 1)$
CPX	$X - (M:M + 1)$	CPY	$Y - (M:M + 1)$
CPS	$SP - (M:M + 1)$		
TST	Test $(M) = 0$ or negative	TSTA	Test $A = 0$ or negative
TSTB	Test $B = 0$ or negative		
Fuzzy Logic and Specialized Math			
MEM	Membership function	REV	MIN-MAX rule evaluation
REVW	Weighted rule evaluation	WAV	Weighted average
EMINM	$MIN(D, (M:M + 1)) \rightarrow (M:M + 1)$	EMIND	$MIN(D, (M:M + 1)) \rightarrow D$
MINM	$MIN(A, (M)) \rightarrow (M)$	MINA	$MIN(A, (M)) \rightarrow A$
EMAXM	$MAX(D, (M:M + 1)) \rightarrow (M:M + 1)$	EMAXD	$MAX(D, (M:M + 1)) \rightarrow D$

HCS12 Instruction Set (Continued)

Mnemonic	Operation	Mnemonic	Operation
Fuzzy Logic and Specialized Math			
MAXM	MAX(A, (M)) $\to$ (M)	MAXA	MAX(A, (M)) $\to$ A
ETBL	16-Bit table interpolate	EMACS	Multiply and accumulate
TBL	8-Bit table interpolate		
Conditional Branch			
BMI	Short branch minus	LBMI	Long branch minus
BPL	Short branch plus	LBPL	Long branch plus
BVS	Short branch two's complement overflow set	LBVS	Long branch two's complement overflow set
BVC	Short branch two's complement overflow clear	LBVC	Long branch two's complement overflow clear
BLT	Short branch two's complement less than	LBLT	Long branch two's complement less than
BGE	Short branch two's complement greater than or equal	LBGE	Long branch two's complement greater than or equal
BLE	Short branch two's complement less than or equal	LBLE	Long branch two's complement less than or equal
BGT	Short branch two's complement greater than	LBGT	Long branch two's complement greater than
BEQ	Short branch equal	LBEQ	Long branch equal
BNE	Short branch not equal	LBNE	Long branch not equal
BHI	Short branch higher	LBHI	Long branch higher
BLS	Short branch lower or same	LBLS	Long branch lower or same
BHS	Short branch higher or same	LBHS	Long branch higher or same
BLO	Short branch lower	LBLO	Long branch lower
BCC	Short branch carry clear	LBCC	Long branch carry clear
BCS	Short branch carry set	LBCS	Long branch carry set
Loop Primitive			
DBEQ	Decrement and branch = 0	DBNE	Decrement and branch <> 0
IBEQ	Increment and branch = 0	IBNE	Increment and branch <> 0
TBEQ	Test and branch = 0	TBNE	Test and branch <> 0
Jump and Branch			
JMP	Jump to address		
JSR	Jump to subroutine	Call	Call subroutine
RTS	Return to subroutine	RTC	Return from CALL
BSR	Branch to subroutine		
BRN	Short branch never	LBRN	Long branch never
BRA	Short branch always	LBRA	Long branch always
BRSET	Branch bits set		
BRCLR	Branch bits clear		
Condition Code			
ANDCC	Clear CCR bits	ORCC	Set CCR bits
Interrupt			
CLI	Clear interrupt mask	SEI	Set interrupt mask
SWI	S/W interrupt	RTI	Return from interrupt
WAI	Wait for interrupt	TRAP	S/W interrupt
Miscellaneous			
NOP	No operation	STOP	Stop clocks
BGND	Background debug mode		

[a] (M) indicates the instruction addresses memory using immediate, direct, extended, or index addressing. Register Name (A, B, D, X, Y, SP, PC) indicates the contents of that register. (SP) means on the stack. C denotes the contents of carry flag. CCR denotes the contents of the condition code register. EA means Effective Address.

Appendix C: CodeWarrior Assembler

CodeWarrior HCS12 Assembler Directives

Section Definition

ORG	Set program counter to the origin of the program in an absolute assembler mode.
SECTION	Define a relocatable section.
OFFSET	Define an offset section.

Constant Definition

EQU	Equate symbol to an expression (cannot be redefined).
SET	Assign a name to an expression (can be redefined).

Reserving or Allocating Memory Locations

DS	Define storage.

Defining Constants in Memory

DC.B	Define byte constant.
DC.W	Define word constant.
DCB	Define a constant block.
RAD50	RAD50 encoded string constants.

Export or Import Global Symbols

ABSENTRY	Specify the entry point in an absolute assembly file.
XDEF	Make a symbol public (visible to some other file).
XREF	Import reference to an external symbol.
XREFB	Import reference to an external symbol located on the direct page.

Assembly Control

ALIGN	Define alignment constraint.
BASE	Specify default base for constants.
END	End of assembly unit.
EVEN	Define two-byte alignment constraint.
FAIL	Generate user defined error or warning messages.
INCLUDE	Include text from another file.
LONGEVEN	Define four-byte alignment constraint.

Repetitive Assembly Control

FOR	Repeat assembly blocks.
ENDFOR	End of FOR block.

Listing Control

CLIST	Include conditional assembly block.
LIST	Specify that all following assembly lines are in the list file.
LLEN	Define line length.

CodeWarrior HCS12 Assembler Directives (Continued)

MLIST	Include macro expansions.
NOLIST	Specify that all following assembly lines are not in the list file.
NOPAGE	Disable pagination in the list file.
PAGE	Insert page break.
PLEN	Define page length.
SPC	Insert empty or blank line.
TABS	Define number of characters to insert for the <tab>.
TITLE	User defined title.

Macro Definition

ENDM	End of user defined macro.
MACRO	Start of user defined macro.
MEXIT	Exit from macro expansion.

Conditional Assembly

ELSE	Alternate block, code included if IF statement not true.
ENDIF	End of conditional block.
IF	Start of conditional block.
IFC	Test if two string expressions are equal.
IFDEF	Test if a symbol is defined.
IFEQ	Test if an expression is null.
IFGE	Test if an expression is greater than or equal to 0.
IFGT	Test if an expression is greater than 0.
IFLE	Test if an expression is less than or equal to 0.
IFLT	Test if an expression is less than 0.
IFNC	Test if two string expressions are different.
IDNDEF	Test if a symbol is undefined.
IFNE	Test if an expression is not null.

Base Designators for Constants

Base	Prefix
Binary (2)	%
Decimal (10)	none
Hexadecimal (16)	$ (Default)
Octal	@

Assembler Expressions

+	Addition	−	Subtraction
*	Multiplication		
/	Division produces truncated result	%	Modulo division
>>	Shift right	<<	Shift left
&	Bitwise AND	|	Bitwise OR
^	Bitwise Exclusive OR (XOR)	~	Ones' complement
!	Logical NOT		
!= or <>	Not equal	= or ==	Equal
<=	Less than or equal	<	Less than
>=	Greater than or equal	>	Greater than
HIGH	High byte of an address	LOW	Low byte of an address
PAGE	Page byte of an address		

Appendix D: HCS12 I/O Registers

Parallel I/O Registers

See Chapter 11 for operational details for the parallel I/O.

Name	Register	Address Base+
PORTA	Port A I/O Register	$0000
PORTB	Port B I/I Register	$0001
DDRA	Data Direction Register Port A	$0002
DDRB	Data Direction Register Port B	$0003
PTE	Port E I/O Register	$0008
DDRE	Data Direction Register Port E	$0009
PEAR	Port E Assignment Register	$000A
PUCR	Pull Device Enable Register	$000C
RDRIV	Reduced Drive Control	$000D
INITRM	Initialize RAM Location	$0010
INITRG	Initialize Register Location	$0011
INITEE	Initialize EEPROM Location	$0012
ATDDIEN	ATD Digital Input Enable Register	$008D
PORTAD0	ATD Digital Input Register	$008F
PTT	Port T I/O Register	$0240
PTIT	Port T Input Register	$0241
DDRT	Data Direction Register Port T	$0242
RDRT	Reduced Drive Register T	$0243
PERT	Pull-up or Pull-down Enable Port T	$0244
PPST	Port T Polarity Select	$0245
MODRR	Port T Module Routine Register	$0247
PTS	Port S I/O Register	$0248
PTIS	Port S Input Register	$0249
DDRS	Data Direction Register Port S	$024A
RDRS	Reduced Drive Register S	$024B
PERS	Pull-up or Pull-down Enable Port S	$024C
PPSS	Port S Polarity Select	$024D
WOMS	Wired-OR Mode	$024E
PTM	Port M I/O Register	$0250
PTIM	Port M Input Register	$0251
DDRM	Data Direction Register Port M	$0252
RDRM	Reduced Drive Register M	$0253

Name	Register	Address Base+
PERM	Pull-up or Pull-down Enable Port M	$0254
PPSM	Port M Polarity Select	$0255
WOMM	Wired-OR Mode	$0256
PTP	Port P I/O Register	$0258
PTIP	Port P Input Register	$0259
DDRP	Data Direction Register Port P	$025A
RDRP	Reduced Drive Register P	$025B
PERP	Pull-up or Pull-down Enable Port P	$025C
PPSP	Port P Polarity Select	$025D
PIEP	Port P Interrupt Enable Register	$025E
PIFP	Port P Interrupt Flag Register	$025F
PTJ	Port J I/O Register	$0268
PTIJ	Port J Input Register	$0269
DDRJ	Data Direction Register Port J	$026A
RDRJ	Reduced Drive Register J	$026B
PERJ	Pull-up or Pull-down Enable Port J	$026C
PPSJ	Port J Polarity Select	$026D
PIEJ	Port J Interrupt Enable Register	$026E
PIFJ	Port J Interrupt Flag Register	$026F
PTAD	Port AD I/O Register	$0270
PTIAD	Port AD Input Register	$0271
DDRAD	Data Direction Register Port AD	$0272
RDRAD	Reduced Drive Register Port AD	$0273
PERAD	Pull-up or Pull-down Enable Port AD	$0274
PPSAD	Port AD Polarity Select	$0275

Interrupt System Registers

See Chapter 12 for details.

Name	Register	Address Base+
ARMCOP	Arm COP Register	$003F
COPCTL	COP Control Register	$003C
HPRIO	Highest Priority Interrupt Register	$001F
INTCR	Interrupt Control Register	$001E
ITCR	Interrupt Test Control Register	$0015
ITEST	Interrupt Test Registers	$0016
PERJ	Pull-up or Pull-down Enable Port J	$026C
PERP	Pull-up or Pull-down Enable Port P	$025C
PIEJ	Port J Interrupt Enable Register	$026E
PIEP	Port P Interrupt Enable Register	$025E
PIFJ	Port J Interrupt Flag Register	$026F
PIFP	Port P Interrupt Flag Register	$025F
PLLCTL	CRG PLL Control Register	$003A
PPSJ	Port J Polarity Select	$026D
PPSP	Port P Polarity Select	$025D

Memory System Registers

See Chapter 13 for details.

Name	Register	Address Base+
INITRM	Initialize RAM Location	$0010
INITRG	Initialize Register Location	$0011
INITEE	Initialize EEPROM Location	$0012
MISC	Miscellaneous System Control	$0013
MEMSIZ0	Memory Size Register 0	$001C
MEMSIZ1	Memory Size Register 1	$001D
PPAGE	Program Page Index Register	$0030

Timer Module Registers

See Chapter 14 for details.

Name	Register	Address Base+
TIOS	Timer Input Capture/Output Compare Select Register	$0040
CFORC	Timer Compare Force Register	$0041
OC7M	Output Compare 7 Mask Register	$0042
OC7D	Output Compare 7 Data Register	$0043
TCNT (hi)	Timer Count Register	$0044:0045
TSCR1	Timer System Control Register 1	$0046
TTOV	Timer Toggle On Overflow Register	$0047
TCTL1	Timer Control Register 1	$0048
TCTL2	Timer Control Register 2	$0049
TCTL3	Timer Control Register 3	$004A
TCTL4	Timer Control Register 4	$004B
TIE	Timer Interrupt Enable Register	$004C
TSCR2	Timer System Control Register 2	$004D
TFLG1	Main Timer Interrupt Flag 1	$004E
TFLG2	Main Timer Interrupt Flag 2	$004F
TC0	Timer Input Capture/Output Compare Register 0	$0050:0051
TC1	Timer Input Capture/Output Compare Register 1	$0052:0053
TC2	Timer Input Capture/Output Compare Register 2	$0054:0055
TC3	Timer Input Capture/Output Compare Register 3	$0056:0057
TC4	Timer Input Capture/Output Compare Register 4	$0058:0059
TC5	Timer Input Capture/Output Compare Register 5	$005A:005B
TC6	Timer Input Capture/Output Compare Register 6	$005C:005D
TC7	Timer Input Capture/Output Compare Register 7	$005E:005F
PACTL	16-Bit Pulse Accumulator Control Register	$0060
PAFLG	Pulse Accumulator Flag Register	$0061
PACNT	Pulse Accumulator Count Registers	$0062, 63
Reserved		$0064–$006F

Pulse-Width Modulation Module Registers

See Chapter 14 for details.

Name	Register	Address Base+
PWME	PWM Enable	$00E0
PWMPOL	PWM Polarity	$00E1
PWMCLK	PWM Clock Select	$00E2
PWMPRCLK	PWM Prescale Clock Select	$00E3
PWMCAE	PWM Center Align	$00E4
PWMCTL	PWM Control	$00E5
PWMTST	PWM Test	$00E6
PWMPRSC	PWM Prescale Counter	$00E7
PWMSCLA	PWM Scale A	$00E8
PWMSCLB	PWM Scale B	$00E9
PWMSCNTA	PWM Scale A Counter	$00EA
PWMSCNTB	PWM Scale B Counter	$00EB
PWMCNT0	PWM Channel 0 Counter	$00EC
PWMCNT1	PWM Channel 1 Counter	$00ED
PWMCNT2	PWM Channel 2 Counter	$00EE
PWMCNT3	PWM Channel 3 Counter	$00EF
PWMCNT4	PWM Channel 4 Counter	$00F0
PWMCNT5	PWM Channel 5 Counter	$00F1
PWMPER0	PWM Channel 0 Period	$00F2
PWMPER1	PWM Channel 1 Period	$00F3
PWMPER2	PWM Channel 2 Period	$00F4
PWMPER3	PWM Channel 3 Period	$00F5
PWMPER4	PWM Channel 4 Period	$00F6
PWMPER5	PWM Channel 5 Period	$00F7
PWMDTY0	PWM Channel 0 Duty	$00F8
PWMDTY1	PWM Channel 1 Duty	$00F9
PWMDTY2	PWM Channel 2 Duty	$00FA
PWMDTY3	PWM Channel 3 Duty	$00FB
PWMDTY4	PWM Channel 4 Duty	$00FC
PWMDTY5	PWM Channel 5 Duty	$00FD

SCI and SPI System Registers

See Chapter 15 for details.

Name	Register	Address Base+
SCIBDH	SCI Baud Rate Register High	$00C8
SCIBDL	SCI Baud Rate Register Low	$00C9
SCICR1	SCI Control Register 1	$00CA
SCICR2	SCI Control Register 2	$00CB
SCISR1	SCI Status Register 1	$00CC
SCISR2	SCI Status Register 2	$00CD
SCIDRH	SCI Data Register High	$00CE
SCIDRL	SCI Data Register Low	$00CF

Name	Register	Address Base+
SPICR1	SPI Control Register 1	$00D8
SPICR2	SPI Control Register 2	$00D9
SPIBR	SPI Baud Rate Register	$00DA
SPISR	SPI Status Register	$00DB
SPIDR	SPI Data Register	$00DD
PTS	Port S I/O Register	$0248

MSCAN Registers

See Chapter 16 for details.

Name	Register	Address Base+
CANCTL0	Control Register 0	$0140
CANCTL1	Control Register 1	$0141
CANBTR0	Bus Timing Register 0	$0142
CANBTR1	Bus Timing Register 1	$0143
CANRFLG	Receiver Flag Register	$0144
CANRIER	Receiver Interrupt Enable Register	$0145
CANTFLG	Transmitter Flag Register	$0146
CANTIER	Transmitter Interrupt Enable Register	$0147
CANTARQ	Transmitter Message Abort Control Register	$0148
CANTAAK	Transmitter Message Abort Acknowledge Register	$0149
CANTBSEL	Transmit Buffer Selection Register	$014A
CANIDAC	Identifier Acceptance Control Register	$014B
CANRXERR	Receive Error Counter Register	$014E
CANTXERR	Transmit Error Counter Register	$014F
CANIDAR0–CANIDAR7	Identifier Acceptance Registers	$0150–$015B
CANIDMR0–CANIDMR7	Identifier Mask Registers	$0154–$015F
CANRIDR0–CANRIDR3	Receiver Identifier Registers	$0160–$0163
CANRDSR0–CANRDSR7	Receive Data Segment Registers	$0164–$016B
CANRDLR	Receive Data Length Register	$016C
CANRTSRH:CANRTSRL	Receiver Time Stamp Register	$016E:016F
CANTIDR0–CANTIDR3	Transmit Identifier Registers	$0170–$0173
CANTDSR0–CANTDSR7	Transmit Data Segment Registers	$0174–$017B
CANTDLR	Transmit Data Length Register	$017C
CANTBPR	Transmit Buffer Priority Register	$017D
CANTTSRH:CANTTSRL	Transmit Time Stamp Register	$017E:017F

ATD Registers

See Chapter 17 for details.

Name	Register	Address Base+
ATDCTL2	ATD Control Register 2	$0082
ATDCTL3	ATD Control Register 3	$0083
ATDCTL4	ATD Control Register 4	$0084
ATDCTL5	ATD Control Register 5	$0085

Name	Register	Address Base+
ATDDR0H	ATD Results Registers	$0090
ATDDR0L		$0091
ATDDR1H		$0092
ATDDR1L		$0093
ATDDR2H		$0094
ATDDR2L		$0095
ATDDR3H		$0096
ATDDR3L		$0097
ATDDR4H		$0098
ATDDR4L		$0099
ATDDR5H		$009A
ATDDR5L		$009B
ATDDR6H		$009C
ATDDR6L		$009D
ATDDR7H		$009E
ATDDR7L		$009F
ATDSTAT0	ATD Status Register 0	$0086
ATDSTAT1	ATD Status Register 1	$008B
ATDCTL2	ATD Control Register 2	$0082
ATDCTL2	ATD Control Register 2	$0082
ATDDIEN	ATD Digital Input Enable Register	$008D
PORTAD	Port AD Data Register	$008F
PTAD	Port AD I/O Register	$0270
ATDTEST1	ATD Test Register 1	$0009

CRG System Registers

See Chapter 21 for details.

Name	Register	Address Base+
SYNR	CRG Synthesizer Register	$0034
REFDV	CRG Reference Divider Register	$0035
CRGFLG	CRG Flags Register	$0037
CRGINT	CRG Interrupt Enable Register	$0038
CLKSEL	CRG Clock Select Register	$0039
PLLCTL	PLL Control Register	$003A
RTICTL	RTI Control Register	$003B

Appendix E: Include File Listings

ATD.INC

```
            NOLIST          ; Turn listing off
; ATD 9S12C32 Equates Include File
; atd.inc
            INCLUDE "bits.inc"
            NOLIST
            INCLUDE "base.inc"
            NOLIST
;*******************************
; Port AD Analog port
; A/D control registers
ATDCTL2:SET     BASE+$0082
ADPU:   SET     BIT7        ; A/D power up bit
AFFC:   SET     BIT6        ; ATD Fast Flag Clear
AWAI:   SET     BIT5        ; A/D wait mode
ETRIGLE:SET     BIT4        ; External trigger level
ETRIGP: SET     BIT3        ; External trigger polarity
ETRIGE: SET     BIT2        ; External trigger enable
ASCIE:  SET     BIT1        ; Seq complete interrupt enable
ASCIF:  SET     BIT0        ; Seq complete interrupt flag
;
ATDCTL3:SET     BASE+$0083
S8C:    SET     BIT6        ; Seq length code
S4C:    SET     BIT5
S2C:    SET     BIT4
S1C:    SET     BIT3
FIFO:   SET     BIT2        ; First in first out mode
FRZ1:   SET     BIT1        ; Background DBG freeze
FRZ0:   SET     BIT0
;
ATDCTL4:SET     BASE+$0084
SRES8:  SET     BIT7        ; ATD resolution
SMP1:   SET     BIT6        ; Sample time select
```

```
SMP0:     SET    BIT5
PRS4:     SET    BIT4          ; ATD Prescaler
PRS3:     SET    BIT3
PRS2:     SET    BIT2
PRS1:     SET    BIT1
PRS0:     SET    BIT0
;
ATDCTL5:SET      BASE+$0085
DJM:      SET    BIT7          ; Data justification
DSGN:     SET    BIT6          ; Data signed or unsigned
SCAN:     SET    BIT5          ; Scan mode
MULT:     SET    BIT4          ; Multi-channel mode
CC:       SET    BIT2          ; Channel select code
CB:       SET    BIT1
CA:       SET    BIT0
;
ATDSTAT0: SET BASE+$0086
SCF:      SET    BIT7          ; Seq complete flag
ETORF:    SET    BIT5          ; Ext trigger overrun flag
FIFOR:    SET    BIT4          ; FIFO overrun flag
CC2:      SET    BIT2          ; Conversion counter
CC1:      SET    BIT1
CC0:      SET    BIT0
;
ATDSTAT1: SET BASE+$008B
CCF7:     SET    BIT7          ; Conversion complete flags
CCF6:     SET    BIT6
CCF5:     SET    BIT5
CCF4:     SET    BIT4
CCF3:     SET    BIT3
CCF2:     SET    BIT2
CCF1:     SET    BIT1
CCF0:     SET    BIT0
;
ATDDIEN:SET      BASE+$008D ; ATD Input Enable
PORTAD0:SET      BASE+$008F ; ATD Digital Input Port
;
; ATD result registers
ATDDR0H:  SET    BASE+$0090
ATDDR0L:  SET    BASE+$0091
ATDDR1H:  SET    BASE+$0092
ATDDR1L:  SET    BASE+$0093
ATDDR2H:  SET    BASE+$0094
ATDDR2L:  SET    BASE+$0095
ATDDR3H:  SET    BASE+$0096
ATDDR3L:  SET    BASE+$0097
ATDDR4H:  SET    BASE+$0098
ATDDR4L:  SET    BASE+$0099
```

```
ATDDR5H: SET   BASE+$009A
ATDDR5L: SET   BASE+$009B
ATDDR6H: SET   BASE+$009C
ATDDR6L: SET   BASE+$009D
ATDDR7H: SET   BASE+$009E
ATDDR7L: SET   BASE+$009F
         LIST                    ; Turn listing back on
```

BASE.INC

```
         NOLIST
; Include file defining the register base address
; This should be included in each project
;
BASE:    SET    $0
         LIST
```

BITS.INC

```
         NOLIST
; Include file for bit definitions
;
BIT7:    SET    %10000000
BIT6:    SET    %01000000
BIT5:    SET    %00100000
BIT4:    SET    %00010000
BIT3:    SET    %00001000
BIT2:    SET    %00000100
BIT1:    SET    %00000010
BIT0:    SET    %00000001
ALLBITS:SET     %11111111
         LIST
```

COP.INC

```
         NOLIST            ; Turn listing off
; COP 9S12C32 Equates Include File
; cop.inc
         INCLUDE "base.inc"
         NOLIST
         INCLUDE "bits.inc"
         NOLIST
;*******************************
; COP Control Register
COPCTL: SET   BASE+$003C
WCOP:   SET   BIT7         ; Windowed COP
RSBCK:  SET   BIT6         ; Stop in BDM Mode
; COP Timeout Bits CR2:CR1:CR0
```

```
COPMS1:  SET   %001        ; 1.024 milliseconds
COPMS4:  SET   %010        ; 4.096 ms
COPMS16:SET    %011        ; 16.384 ms
COPMS65:SET    %100        ; 65.536 ms
COPMS262:SET   %101        ; 262.144 ms
COPMS524:SET   %110        ; 524.288 ms
COPMS1000:SET  %111        ; 1.049 sec
;******************************
ARMCOP: SET   BASE+$003F
        LIST               ; Turn listing back on
```

CRG.INC

```
          NOLIST           ; Turn listing off
; Clock and Reset Generator 9S12C32 Equates Include File
; crg.inc
          INCLUDE "base.inc"
          NOLIST
          INCLUDE "bits.inc"
          NOLIST
;******************************
; CRG Flags Register
CRGFLG: SET   BASE+$0037
RTIF:   SET   BIT7   ; Real-time interrupt flag
PORF:   SET   BIT6   ; Power on reset flag
LVRF:   SET   BIT5   ; Low voltage reset flag
LOCKIF: SET   BIT4   ; PLL lock interrupt flag
LOCK:   SET   BIT3   ; Lock status bit
TRACK:  SET   BIT2   ; PLL Track status bit
SCMIF:  SET   BIT1   ; SCM interrupt flag
SCM:    SET   BIT0   ; Self clock mode status
; ***
; Clock Select Register
CLKSEL: SET   BASE+$0039
PLLSEL: SET   BIT7   ; PLL Select
PSTP:   SET   BIT6   ; Pseudo Stop bit
SYSWAI: SET   BIT5   ; Sys clocks stop in wait
ROAWAI: SET   BIT4   ; Reduced amplitude in wait
PLLWAI: SET   BIT3   ; PLL stops in wait
CWAI:   SET   BIT2   ; Core stops in wait
RTIWAI: SET   BIT1   ; RTI stops in wait
COPWAI: SET   BIT0   ; COP stops in wait
; ***
; Synthesizer Register
SYNR:   SET   BASE+$0034
; ***
; Reference Divider Register
REFDV:  SET   BASE+$0035
```

```
            ; ***
            ; PLL Control register
            PLLCTL: SET   BASE+$003A
            CME:    SET   BIT7    ; Clock monitor enable
            PLLON:  SET   BIT6    ; PLL on
            AUTO:   SET   BIT5    ; Auto bandwidth control
            ACQ:    SET   BIT4    ; Acquisition bit
            PRE:    SET   BIT3    ; RTI enable during pseudo-stop
            PCE:    SET   BIT2    ; COP stops in pseudo-stop
            SCME:   SET   BIT0    ; Self clock mode enable
                    LIST          ; Turn listing back on
```

MSCAN.INC

```
                    NOLIST          ; Turn listing off
            ; PWM 9S12C32 Equates Include File
            ; mscan.inc
                    INCLUDE "base.inc"
                    NOLIST
                    INCLUDE "bits.inc"
                    NOLIST
            ;********************************
            ; CANCTL0 MSCAN Control Register 0
            CANCTL0: SET  BASE+$0140
            RXFRM:  SET   BIT7
            RXACT:  SET   BIT6
            CSWAI:  SET   BIT5
            SYNCH:  SET   BIT4
            TIME:   SET   BIT3
            WUPE:   SET   BIT2
            SLPRQ:  SET   BIT1
            INITRQ: SET   BIT0

            ;********************************
            ;  MSCAN Control Register 1
            CANCTL1: SET  BASE+$0141
            CANE:   SET   BIT7
            CLKSRC: SET   BIT6
            LOOPB:  SET   BIT5
            LISTEN: SET   BIT4
            WUPM:   SET   BIT2
            SLPAK:  SET   BIT1
            INITAK: SET   BIT0

            ;********************************
            ; CANBTR0 MSCAN Bus Timing Register 0
            CANBTR0: SET  BASE+$0142
```

```
;********************************
; CANBTR1 MSCAN Bus Timing Register 1
CANBTR1: SET   BASE+$0143


;********************************
;
; CANRFLG MSCAN Receiver Flag Register
CANRFLG: SET   BASE+$0144
WUPIF:   SET   BIT7
CSCIF:   SET   BIT6
RSTAT1:  SET   BIT5
RSTAT0:  SET   BIT4
TSTAT1:  SET   BIT3
TSTAT0:  SET   BIT2
OVRIE:   SET   BIT1
RXF:     SET   BIT0


;********************************
; CANRIER MSCAN Receiver Interrupt Enable Register
CANRIER: SET   BASE+$0145
WUPIE:   SET   BIT7
CSCIE:   SET   BIT6
RSTATE1: SET   BIT5
RSTATE0: SET   BIT4
TSTATE1: SET   BIT3
TSTATE0: SET   BIT2
OVRIE:   SET   BIT1
RXFIE:   SET   BIT0


;********************************
; CANTFLG MSCAN Transmitter Flag Register
CANTFLG: SET   BASE+$0146
TXE2:    SET   BIT2
TXE1:    SET   BIT1
TXE0:    SET   BIT0


;********************************
; CANTIER MSCAN Transmitter Interrupt Enable Register
CANTIER: SET   BASE+$0147
TXEIE2:  SET   BIT2
TXEIE1:  SET   BIT1
TXEIE0:  SET   BIT0


;********************************
; CANTARQ MSCAN Transmitter Message Abort Control Register
CANTARG: SET   BASE+$0148
ABTRQ2:  SET   BIT2
```

```
ABTRQ1:   SET   BIT1
ABTRG0:   SET   BIT0


;********************************
; CANTAAK MSCAN Transmitter Abort Acknowledge Register
CANTAAK:  SET   BASE+$0149
ABTAK2:   SET   BIT2
ABTAK1:   SET   BIT1
ABTAK0:   SET   BIT0


;********************************
; CANTBSEL MSCAN Transmit Buffer Select Register
CANTBSEL:SET   BASE+$014A
TX2:      SET   BIT2
TX1:      SET   BIT1
TX0:      SET   BIT0


;********************************
; CANIDAC MSCAN Identifier Acceptance Control Register
CANIDAC:  SET   BASE+$014B
IDAM1:    SET   BIT5
IDAM0:    SET   BIT4
IDHIT2:   SET   BIT2
IDHIT1:   SET   BIT1
IDHIT0:   SET   BIT0


;********************************
; CANRXERR MSCAN Receive Error Counter Register
CANRXERR:SET   BASE+$014E


;********************************
; CANTXERR MSCAN Transmit Error Counter Register
CANTXERR:SET   BASE+$014F


;********************************
; MSCAN Identifier Acceptance Registers
CANIDAR0: SET BASE+$0150
CANIDAR1: SET BASE+$0151
CANIDAR2: SET BASE+$0152
CANIDAR3: SET BASE+$0153


;********************************
; MSCAN Identifier Mask Registers
CANIDMR0: SET BASE+$0154
CANIDMR1: SET BASE+$0155
CANIDMR2: SET BASE+$0156
CANIDMR3: SET BASE+$0157
```

```
;********************************
; MSCAN Identifier Acceptance Registers
CANIDAR4: SET BASE+$0158
CANIDAR5: SET BASE+$0159
CANIDAR6: SET BASE+$015A
CANIDAR7: SET BASE+$015B


;********************************
; MSCAN Identifier Mask Registers
CANIDMR4: SET BASE+$015C
CANIDMR5: SET BASE+$015D
CANIDMR6: SET BASE+$015E
CANIDMR7: SET BASE+$015F


;********************************
; MSCAN Foreground Receiver Buffer
CANRIDR0: SET BASE+$0160
CANRIDR1: SET BASE+$0161
CANRIDR2: SET BASE+$0162
CANRIDR3: SET BASE+$0163
CANRDSR0: SET BASE+$0164
CANRDSR1: SET BASE+$0165
CANRDSR2: SET BASE+$0166
CANRDSR3: SET BASE+$0167
CANRDSR4: SET BASE+$0168
CANRDSR5: SET BASE+$0169
CANRDSR6: SET BASE+$016A
CANRDSR7: SET BASE+$016B
CANRDLR:  SET BASE+$016C
CANRTSRH: SET BASE+$016E
CANRTSRL: SET BASE+$016F


;********************************
; MSCAN Foreground Transmit Buffer
CANTIDR0: SET BASE+$0170
CANTIDR1: SET BASE+$0171
SRR:      SET BIT4
RTR_STAND:SET BIT4       ; Standard frame RTR
IDE:      SET BIT3
CANTIDR2: SET BASE+$0172
CANTIDR3: SET BASE+$0173
RTR_EXT:  SET BIT0       ; Extended frame RTR
CANTDSR0: SET BASE+$0174
CANTDSR1: SET BASE+$0175
CANTDSR2: SET BASE+$0176
CANTDSR3: SET BASE+$0177
CANTDSR4: SET BASE+$0178
CANTDSR5: SET BASE+$0179
```

```
CANTDSR6: SET BASE+$017A
CANTDSR7: SET BASE+$017B
CANTDLR:  SET BASE+$017C
CANTBPR:  SET BASE+$017D
CANTTSRH: SET BASE+$017E
CANTTSRL: SET BASE+$017F
          LIST              ; Turn listing back on
```

PIM.INC

```
            NOLIST           ; Turn listing off
; PWM 9S12C32 Equates Include File
; pwm.inc
            INCLUDE "base.inc"
            NOLIST
            INCLUDE "bits.inc"
            NOLIST
;********************************
; PWM Prescale Clock Select Register
PWMPRCLK SET BASE+$E3
PCKB2:    SET  BIT6
PCKB1:    SET  BIT5
PCKB0:    SET  BIT4
PCKB2:    SET  BIT2
PCKB1:    SET  BIT1
PCKB0:    SET  BIT0
; Clock B Divider Definitions
CBDIV2:   SET  %00010000   ; Clock B=Bus/2
CBDIV4:   SET  %00100000   ; Clock B=Bus/4
CBDIV8:   SET  %00110000   ; Clock B=Bus/8
CBDIV16:  SET  %01000000   ; Clock B=Bus/16
CBDIV32:  SET  %01010000   ; Clock B=Bus/32
CBDIV64:  SET  %01100000   ; Clock B=Bus/64
CBDIV128:SET   %01110000   ; Clock B=Bus/128
; Clock A Divider Definitions
CADIV2:   SET  %00000001   ; Clock A=Bus/2
CADIV4:   SET  %00000010   ; Clock A=Bus/4
CADIV8:   SET  %00000011   ; Clock A=Bus/8
CADIV16:  SET  %00000100   ; Clock A=Bus/16
CADIV32:  SET  %00000101   ; Clock A=Bus/32
CADIV64:  SET  %00000110   ; Clock A=Bus/64
CADIV128:SET   %00000111   ; Clock A=Bus/128
;********************************
; Clock A and B Scale Registers
PWMSCLA: SET  BASE+$00E8
PWMSCLB: SET  BASE+$00E9
;********************************
```

```
; Clock Select Register
PWMCLK:   SET   BASE+$00E2
PCKL5:    SET   BIT5
PCKL4:    SET   BIT4
PCKL3:    SET   BIT3
PCKL2:    SET   BIT2
PCKL1:    SET   BIT1
PCKL0:    SET   BIT0
;*******************************
PWM Enable Register
PWME:     SET   BASE+$00E0
PWME5:    SET   BIT5
PWME4:    SET   BIT4
PWME3:    SET   BIT3
PWME2:    SET   BIT2
PWME1:    SET   BIT1
PWME0:    SET   BIT0
;*******************************
; PWM Polarity Control
PWMPOL:   SET   BASE+$00E1
PPOL5:    SET   BIT5
PPOL4:    SET   BIT4
PPOL3:    SET   BIT3
PPOL2:    SET   BIT2
PPOL1:    SET   BIT1
PPOL0:    SET   BIT0
;*******************************
; PWM Center Alignment Control
PWMCAE:   SET   BASE+$00E4
CAE5:     SET   BIT5
CAE4:     SET   BIT4
CAE3:     SET   BIT3
CAE2:     SET   BIT2
CAE1:     SET   BIT1
CAE0:     SET   BIT0
;*******************************
; PWM Concatenation Control
PWMCTL:   SET   BASE+$00E5
CON45:    SET   BIT6
CON23:    SET   BIT5
CON01:    SET   BIT4
PSWAI:    SET   BIT3
PFRZ:     SET   BIT2
;*******************************
; PWM Counter Registers
PWMCNT0:  SET   BASE+$00EC
PWMCNT1:  SET   BASE+$00ED
PWMCNT2:  SET   BASE+$00EE
```

```
PWMCNT3: SET   BASE+$00EF
PWMCNT4: SET   BASE+$00F0
PWMCNT5: SET   BASE+$00F1
;*******************************
; PWM Period Registers
PWMPER0: SET   BASE+$00F2
PWMPER1: SET   BASE+$00F3
PWMPER2: SET   BASE+$00F4
PWMPER3: SET   BASE+$00F5
PWMPER4: SET   BASE+$00F6
PWMPER5: SET   BASE+$00F7
;*******************************
; PWM Duty Registers
PWMDTY0: SET   BASE+$00F8
PWMDTY1: SET   BASE+$00F9
PWMDTY2: SET   BASE+$00FA
PWMDTY3: SET   BASE+$00FB
PWMDTY4: SET   BASE+$00FC

PWMDTY5: SET   BASE+$00FD
         LIST            ; Turn listing back on
```

PORTA.INC

```
         NOLIST          ; Turn listing off
; PORT A 9S12C32 Equates Include File
; porta.inc
         INCLUDE "base.inc"
         NOLIST
         INCLUDE "bits.inc"
         NOLIST
;*******************************
; Port A
PTA:    SET    BASE+$0000
DDRA:   SET    BASE+$0002 ; Data Direction
PUCR:   SET    BASE+$000C ; Pull-up or -down enable
RDRIV:  SET    BASE+$000D ; Reduced drive
         LIST            ; Turn listing back on
```

PORTAD.INC

```
         NOLIST          ; Turn listing off
; PORT AD 9S12C32 Equates Include File
; portad.inc
         INCLUDE "bits.inc"
         NOLIST
         INCLUDE "base.inc"
         NOLIST
```

```
;*******************************
; Port AD
PTAD:    SET    BASE+$0270
PTIAD:   SET    BASE+$0271
DDRAD:   SET    BASE+$0272
RDRAD:   SET    BASE+$0273 ; Reduced drive
PERAD:   SET    BASE+$0274 ; Pull-up or -down enable
PPSAD:   SET    BASE+$0275 ; Polarity Select
;*******************************
ATDCTL2:SET    BASE+$0082 ; ATD Control Reg 2
ADPU:    SET    BIT7      ; ATD Power Up
AFFC:    SET    BIT6      ; Fast Flag Clear
AWAI:    SET    BIT5      ; Power Down in Wait
ETRIGLE:SET    BIT4      ; Ext Trigger Level
ETRIGP: SET    BIT3      ; Ext Trigger Polarity
ETRIGE: SET    BIT2      ; Ext Trigger Enable
ASCIE:  SET    BIT1      ; Sequence Complete Interrupt Enable
ASCIF:  SET    BIT0      ; Sequence Complete Interrupt Flag
;*******************************
ATDCTL3:SET    BASE+$0083 ; ATD Control Reg 3
FIFO:    SET    BIT1      ; FIFO Mode
;*******************************
ATDCTL4:SET    BASE+$0084 ; ATD Control Reg 4
SRES8:   SET    BIT7      ; ATD Resolution Select
;*******************************
ATDCTL5:SET    BASE+$0085 ; ATD Control Reg 5
DJM:     SET    BIT7      ; Data Justification
DSGN:    SET    BIT6      ; Data Signed
SCAN:    SET    BIT5      ; Conversion Sequence Mode
MULT:    SET    BIT4      ; Multi-channel Mode
;*******************************
ATDSTAT0:SET   BASE+$0088 ; ATD Status Register 0
SCF:     SET    BIT7      ; Sequence Complete Flag
ETORF:   SET    BIT5      ; Ext Trigger Overrun Flag
FIFOR:   SET    BIT4      ; FIFO Overrun Flag
;*******************************
ATDSTAT1:SET   BASE+$008B ; ATD Status Register 1
CCF7:    SET    BIT7      ; Conversion Complete 7
CCF6:    SET    BIT6      ; Conversion Complete 6
CCF5:    SET    BIT5      ; Conversion Complete 5
CCF4:    SET    BIT4      ; Conversion Complete 4
CCF3:    SET    BIT3      ; Conversion Complete 3
CCF2:    SET    BIT2      ; Conversion Complete 2
CCF1:    SET    BIT1      ; Conversion Complete 1
CCF0:    SET    BIT0      ; Conversion Complete 0
;*******************************
ATDDIEN: SET   BASE+$008D ; ATD Digital Input Enable
;*******************************
```

```
               ; Port AD Analog Port
PORTAD0: SET   BASE+$008F ; ATD Digital Input Port
;*******************************
               ; Port ATD Data Registers
ATDDR0H: SET   BASE+$0090 ; High Byte
ATDDR0L: SET   BASE+$0091 ; Low Byte
ATDDR1H: SET   BASE+$0092 ; High Byte
ATDDR1L: SET   BASE+$0093 ; Low Byte
ATDDR2H: SET   BASE+$0094 ; High Byte
ATDDR2L: SET   BASE+$0095 ; Low Byte
ATDDR3H: SET   BASE+$0096 ; High Byte
ATDDR3L: SET   BASE+$0097 ; Low Byte
ATDDR4H: SET   BASE+$0098 ; High Byte
ATDDR4L: SET   BASE+$0099 ; Low Byte
ATDDR5H: SET   BASE+$009A ; High Byte
ATDDR5L: SET   BASE+$009B ; Low Byte
ATDDR6H: SET   BASE+$009C ; High Byte
ATDDR6L: SET   BASE+$009D ; Low Byte
ATDDR7H: SET   BASE+$009E ; High Byte
ATDDR7L: SET   BASE+$009F ; Low Byte
          LIST               ; Turn listing back on
```

PORTB.INC

```
               NOLIST          ; Turn listing off
; PORT B 9S12C32 Equates Include File
; portb.inc
          INCLUDE "base.inc"
          NOLIST
;*******************************
; Port B
PTB:    SET   BASE+$0001
DDRB:   SET   BASE+$0003
PUCR:   SET   BASE+$000C ; Pull-up or -down enable
RDRIV:  SET   BASE+$000D ; Reduced drive
          LIST               ; Turn listing back on
```

PORTE.INC

```
               NOLIST          ; Turn listing off
; PORT E 9S12C32 Equates Include File
; porte.inc
          INCLUDE "base.inc"
          NOLIST
;*******************************
; Port E
PTE:    SET   BASE+$0008
DDRE:   SET   BASE+$0009 ; Data Direction
```

```
PEAR:   SET    BASE+$000A ; Port E assignment
MODE:   SET    BASE+$000B ; Mode
PUCR:   SET    BASE+$000C ; Pull-up or -down enable
RDRIV:  SET    BASE+$000D ; Reduced drive
        LIST              ; Turn listing back on
```

PORTJ.INC

```
            NOLIST        ; Turn listing off
; PORT J 9S12C32 Equates Include File
; portj.inc
            INCLUDE "base.inc"
            NOLIST
;*******************************
; Port J
PTJ:    SET    BASE+$0268
PTIJ:   SET    BASE+$0269
DDRJ:   SET    BASE+$026A
RDRJ:   SET    BASE+$026B ; Reduced drive
PERJ:   SET    BASE+$026C ; Pull-up or -down enable
PPSJ:   SET    BASE+$026D ; Polarity Select
PIEJ:   SET    BASE+$026E ; Interrupt enable
PIFJ:   SET    BASE+$026F ; Interrupt flag
        LIST              ; Turn listing back on
```

PORTM.INC

```
            NOLIST        ; Turn listing off
; PORT M 9S12C32 Equates Include File
; portm.inc
            INCLUDE "base.inc"
            NOLIST
;*******************************
; Port M
PTM:    SET    BASE+$0250
PTIM:   SET    BASE+$0251
DDRM:   SET    BASE+$0252
RDRM:   SET    BASE+$0253 ; Reduced drive
PERM:   SET    BASE+$0254 ; Pull-up or -down enable
PPSM:   SET    BASE+$0255 ; Polarity Select
WOMM:   SET    BASE+$0256 ; Wired-or mode
        LIST              ; Turn listing back on
```

PORTP.INC

```
            NOLIST        ; Turn listing off
; PORT P 9S12C32 Equates Include File
; portp.inc
```

```
                INCLUDE "base.inc"
                NOLIST
;*******************************
; Port
PTP:    SET    BASE+$0258
PTIP:   SET    BASE+$0259
DDRP:   SET    BASE+$025A
RDRP:   SET    BASE+$025B ; Reduced drive
PERP:   SET    BASE+$025C ; Pull-up or -down enable
PPSP:   SET    BASE+$025D ; Polarity Select
PIEP:   SET    BASE+$025E ; Interrupt enable
PIFP:   SET    BASE+$025F ; Interrupt flag
MODRR:  SET    BASE+$0247 ; Module routine
        LIST              ; Turn listing back on
```

PORTS.INC

```
                NOLIST          ; Turn listing off
; PORT S 9S12C32 Equates Include File
; ports.inc
                INCLUDE "base.inc"
                NOLIST
;*******************************
; Port
PTS:    SET    BASE+$0248
PTIS:   SET    BASE+$0249
DDRS:   SET    BASE+$024A
RDRS:   SET    BASE+$024B ; Reduced drive
PERS:   SET    BASE+$024C ; Pull-up or -down enable
PPSS:   SET    BASE+$024D ; Polarity Select
WOMS:   SET    BASE+$024E ; Wired-OR mode
        LIST              ; Turn listing back on
```

PORTT.INC

```
                NOLIST          ; Turn listing off
; PORT T 9S12C32 Equates Include File
; portt.inc
                INCLUDE "base.inc"
                NOLIST
;*******************************
; Port
PTT:    SET    BASE+$0240
        LIST              ; Turn listing back on
PTIT:   SET    BASE+$0241
DDRT:   SET    BASE+$0242
```

```
RDRT:    SET    BASE+$0243 ; Reduced drive
PERT:    SET    BASE+$0244 ; Pull-up or -down enable
PPST:    SET    BASE+$0245 ; Polarity Select
MODRR:   SET    BASE+$0247 ; Module routing
```

PWM.INC

```
          NOLIST          ; Turn listing off
; PWM 9S12C32 Equates Include File
; pwm.inc
          INCLUDE "base.inc"
          NOLIST
          INCLUDE "bits.inc"
          NOLIST
;*******************************
; PWM Prescale Clock Select Register
PWMPRCLK SET BASE+$E3
PCKB2:    SET    BIT6
PCKB1:    SET    BIT5
PCKB0:    SET    BIT4
PCKB2:    SET    BIT2
PCKB1:    SET    BIT1
PCKB0:    SET    BIT0
; Clock B Divider Definitions
;CBDIV2:   SET    %00010000  ; Clock B=Bus/2
;CBDIV4:   SET    %00100000  ; Clock B=Bus/4
;CBDIV8:   SET    %00110000  ; Clock B=Bus/8
;CBDIV16:  SET    %01000000  ; Clock B=Bus/16
;CBDIV32:  SET    %01010000  ; Clock B=Bus/32
;CBDIV64:  SET    %01100000  ; Clock B=Bus/64
;CBDIV128: SET    %01110000  ; Clock B=Bus/128
; Clock A Divider Definitions
;CADIV2:   SET    %00000001  ; Clock A=Bus/2
;CADIV4:   SET    %00000010  ; Clock A=Bus/4
;CADIV8:   SET    %00000011  ; Clock A=Bus/8
;CADIV16:  SET    %00000100  ; Clock A=Bus/16
;CADIV32:  SET    %00000101  ; Clock A=Bus/32
;CADIV64:  SET    %00000110  ; Clock A=Bus/64
;CADIV128: SET    %00000111  ; Clock A=Bus/128
;*******************************
; Clock A and B Scale Registers
PWMSCLA:  SET    BASE+$00E8
PWMSCLB:  SET    BASE+$00E9
;*******************************
; Clock Select Register
PWMCLK:   SET    BASE+$00E2
PCLK5:    SET    BIT5
```

```
PCLK4:    SET   BIT4
PCLK3:    SET   BIT3
PCLK2:    SET   BIT2
PCLK1:    SET   BIT1
PCLK0:    SET   BIT0
;*******************************
; PWM Enable Register
PWME:     SET   BASE+$00E0
PWME5:    SET   BIT5
PWME4:    SET   BIT4
PWME3:    SET   BIT3
PWME2:    SET   BIT2
PWME1:    SET   BIT1
PWME0:    SET   BIT0
;*******************************
; PWM Polarity Control
PWMPOL:   SET   BASE+$00E1
PPOL5:    SET   BIT5
PPOL4:    SET   BIT4
PPOL3:    SET   BIT3
PPOL2:    SET   BIT2
PPOL1:    SET   BIT1
PPOL0:    SET   BIT0
;*******************************
; PWM Center Alignment Control
PWMCAE:   SET   BASE+$00E4
CAE5:     SET   BIT5
CAE4:     SET   BIT4
CAE3:     SET   BIT3
CAE2:     SET   BIT2
CAE1:     SET   BIT1
CAE0:     SET   BIT0
;*******************************
; PWM Concatenation Control
PWMCTL:   SET   BASE+$00E5
CON45:    SET   BIT6
CON23:    SET   BIT5
CON01:    SET   BIT4
PSWAI:    SET   BIT3
PFRZ:     SET   BIT2
;*******************************
; PWM Counter Registers
PWMCNT0:  SET   BASE+$00EC
PWMCNT1:  SET   BASE+$00ED
PWMCNT2:  SET   BASE+$00EE
PWMCNT3:  SET   BASE+$00EF
PWMCNT4:  SET   BASE+$00F0
PWMCNT5:  SET   BASE+$00F1
```

```
;*******************************
; PWM Period Registers
PWMPER0: SET   BASE+$00F2
PWMPER1: SET   BASE+$00F3
PWMPER2: SET   BASE+$00F4
PWMPER3: SET   BASE+$00F5
PWMPER4: SET   BASE+$00F6
PWMPER5: SET   BASE+$00F7
;*******************************
; PWM Duty Registers
PWMDTY0: SET   BASE+$00F8
PWMDTY1: SET   BASE+$00F9
PWMDTY2: SET   BASE+$00FA
PWMDTY3: SET   BASE+$00FB
PWMDTY4: SET   BASE+$00FC
PWMDTY5: SET   BASE+$00FD
;*******************************
; Module Routing Register
;  Used to route PWM to Port T
MODRR:   SET   BASE+$0247
         LIST              ; Turn listing back on
```

SCI.INC

```
            NOLIST          ; Turn listing off
; SCI 9S12C32 Equates Include File
; sci.inc
            INCLUDE "base.inc"
            NOLIST
            INCLUDE "bits.inc"
            NOLIST
;*******************************
; SCI Baud Rate Registers
SCIBDH:  SET   BASE+$00C8
SCIBDL:  SET   BASE+$00C9
;*******************************
; SCI Control Register 1
SCICR1:  SET   BASE+$00CA
LOOPS:   SET   BIT7
SCISWAI: SET   BIT6
RSRC:    SET   BIT5
M:       SET   BIT4
MODE:    SET   M
WAKE:    SET   BIT3
ILT:     SET   BIT2
PE:      SET   BIT1
```

```
PT:       SET   BIT0
;*******************************
; SCI Control Register 2
SCICR2:   SET   BASE+$00CB
TIE:      SET   BIT7
TCIE:     SET   BIT6
RIE:      SET   BIT5
ILIE:     SET   BIT4
TE:       SET   BIT3
RE:       SET   BIT2
RWU:      SET   BIT1
SBK:      SET   BIT0
;*******************************
; SCI Status Register 1
SCISR1:   SET   BASE+$00CC
TDRE:     SET   BIT7
TC:       SET   BIT6
RDRF:     SET   BIT5
IDLE:     SET   BIT4
OR:       SET   BIT3
NF:       SET   BIT2
FE:       SET   BIT1
PF:       SET   BIT0
;*******************************
; SCI Status Register 2
SCISR2:   SET   BASE+$00CD
BRK13:    SET   BIT2
TXDIR:    SET   BIT1
RAF:      SET   BIT0
;*******************************
; SCI Data Register High
SCIDRH:   SET   BASE+$00CE
T8:       SET   BIT6
;*******************************
; SCI Data Register Low
SCIDRL:   SET   BASE+$00CF
; 8 MHz Bus Clock Baud Rate Register Values
B38400:   SET   13
B19200:   SET   26
B9600:    SET   52
B4800:    SET   104
B2400:    SET   208
B1200:    SET   417
B600:     SET   833
B300:     SET   1667
B150:     SET   3333
B110:     SET   4545
          LIST           ; Turn listing back on
```

SPI.INC

```
                NOLIST          ; Turn listing off
        ; SPI 9S12C32 Equates Include File
        ; sci.inc
                INCLUDE "base.inc"
                NOLIST
                INCLUDE "bits.inc"
                NOLIST
        ;*******************************
        ; SPI Control Registers
        SPICR1:    SET    BASE+$00D8 ; Control Register 1
        SPIE:      SET    BIT7       ; SPI Interrupt Enable
        SPE:       SET    BIT6       ; SPI Enable
        SPTIE:     SET    BIT5       ; Transmit Interrupt Enable
        MSTR:      SET    BIT4       ; Master/Slave Select
        CPOL:      SET    BIT3       ; Clock Polarity
        CPHA:      SET    BIT2       ; Clock Phase
        SSOE:      SET    BIT1       ; Slave Select Output Enable
        LSBFE:     SET    BIT0       ; LSB-First Enable
        ;
        SPICR2:    SET    BASE+$00D9 ; Control Register 2
        MODFEN:    SET    BIT4       ; Mode Fault Enable
        BIDIROE:   SET    BIT3       ; Bidirectional Output Enable
        SPISWAI:   SET    BIT1       ; SPI Stop in Wait Mode
        SPCO:      SET    BIT0       ; Serial Pin Bit Control
        ;
        SPIBR:     SET    BASE+$00DA ; SPI Baud Rate Register
        ;
        SPISR:     SET    BASE+$00DB ; SPI Status Register
        SPIF:      SET    BIT7       ; SPI Interrupt Flag
        SPTEF:     SET    BIT5       ; Transmit Empty Interrupt Flag
        MODF:      SET    BIT4       ; Mode Fault Flag
        ;
        SPIDR:     SET    BASE+$00DD ; SPI Data Register
        ;
        ; SPI Bits on Port M
        PTM:    SET    BASE+$0250
        DDRM:   SET    BASE+$0252
        SS:     SET    BIT3          ; Port M-3
        MOSI:   SET    BIT4                            ; Port M-4
        SCK:    SET    BIT5          ; Port M-5
                LIST            ; Turn listing back on
```

TIMER.INC

```
                NOLIST          ; Turn listing off
        ; Timer 9S12C32 Equates Include File
```

```
; timer.inc
        INCLUDE "base.inc"
        NOLIST
        INCLUDE "bits.inc"
        NOLIST
;******************************
; Timer Input Capture/Output Compare Select
TIOS:   SET   BASE+$0040
IOS7:   SET   BIT7
IOS6:   SET   BIT6
IOS5:   SET   BIT5
IOS4:   SET   BIT4
IOS3:   SET   BIT3
IOS2:   SET   BIT2
IOS1:   SET   BIT1
IOS0:   SET   BIT0
; ***
; Timer Compare Force Register
CFORC:  SET   BASE+$0041
FOC7:   SET   BIT7
FOC6:   SET   BIT6
FOC5:   SET   BIT5
FOC4:   SET   BIT4
FOC3:   SET   BIT3
FOC2:   SET   BIT2
FOC1:   SET   BIT1
FOC0:   SET   BIT0
; ***
; Output Compare 7 Mask Register
OC7M:   SET   BASE+$0042
OC7M7:  SET   BIT7
OC7M6:  SET   BIT6
OC7M5:  SET   BIT5
OC7M4:  SET   BIT4
OC7M3:  SET   BIT3
OC7M2:  SET   BIT2
OC7M1:  SET   BIT1
OC7M0:  SET   BIT0
; Output Compare 7 Data Register
OC7D:   SET   BASE+$0043
OC7D7:  SET   BIT7
OC7D6:  SET   BIT6
OC7D5:  SET   BIT5
OC7D4:  SET   BIT4
OC7D3:  SET   BIT3
OC7D2:  SET   BIT2
OC7D1:  SET   BIT1
OC7D0:  SET   BIT0
```

```
; ***
; Timer Counter Register
TCNT:   SET   BASE+$0044
; ***
; Timer System Control Register 1
TSCR1:  SET   BASE+$0046
TEN:    SET   BIT7                        ; Timer enable
TSWAI:  SET   BIT6        ; Timer stops in wait
TSFRZ:  SET   BIT5        ; Timer stops in freeze
TTFCA:  SET   BIT4        ; Timer fast flag clear all
; ***
; Timer Toggle on Overflow Register
TTOV:   SET   BASE+$0047
TOV7:   SET   BIT7
TOV6:   SET   BIT6
TOV5:   SET   BIT5
TOV4:   SET   BIT4
TOV3:   SET   BIT3
TOV2:   SET   BIT2
TOV1:   SET   BIT1
TOV0:   SET   BIT0
; ***
; Timer Control Register 1
TCTL1:  SET   BASE+$0048
; Output compare bit controls
OM7:    SET   BIT7
OL7:    SET   BIT6
OM6:    SET   BIT5
OL6:    SET   BIT4
OM5:    SET   BIT3
OL5:    SET   BIT2
OM4:    SET   BIT1
OL4:    SET   BIT0
; ***
; Timer Control Register 2
TCTL2:  SET   BASE+$0049
OM3:    SET   BIT7
OL3:    SET   BIT6
OM2:    SET   BIT5
OL2:    SET   BIT4
OM1:    SET   BIT3
OL1:    SET   BIT2
OM0:    SET   BIT1
OL0:    SET   BIT0
; ***
; Timer Control Register 3
TCTL3:  SET   BASE+$004A
EDG7B:  SET   BIT7
```

```
EDG7A:   SET   BIT6
EDG6B:   SET   BIT5
EDG6A:   SET   BIT4
EDG5B:   SET   BIT3
EDG5A:   SET   BIT2
EDG4B:   SET   BIT1
EDG4A:   SET   BIT0
; Timer Control Register 4
TCTL4:   SET   BASE+$004B
EDG3B:   SET   BIT7
EDG3A:   SET   BIT6
EDG2B:   SET   BIT5
EDG2A:   SET   BIT4
EDG1B:   SET   BIT3
EDG1A:   SET   BIT2
EDG0B:   SET   BIT1
EDG0A:   SET   BIT0
; ***

; Timer Interrupt Enable Register
TIE:    SET   BASE+$004C
; TIE Register Bits - Timer Interrupt Bits
C7I:    SET   BIT7
C6I:    SET   BIT6
C5I:    SET   BIT5
C4I:    SET   BIT4
C3I:    SET   BIT3
C2I:    SET   BIT2
C1I:    SET   BIT1
C0I:    SET   BIT0
; ***
; Timer System Control Register 2
TSCR2:   SET   BASE+$004D
TOI:    SET   BIT7        ; Timer overflow interrupt
TCRE:   SET   BIT3        ; Timer counter reset enable
PR2:    SET   BIT2        ; Prescaler bit 2
PR1:    SET   BIT1        ; Prescaler bit 1
PR0:    SET   BIT0        ; Prescaler bit 0
; Timer Interrupt Flag Register 1
TFLG1:   SET   BASE+$4E
; TFLG1 Register bits - Timer Interrupt Flags
C7F:    SET   BIT7
C6F:    SET   BIT6
C5F:    SET   BIT5
C4F:    SET   BIT4
C3F:    SET   BIT3
C2F:    SET   BIT2
C1F:    SET   BIT1
```

```
C0F:    SET  BIT0
; ***
; Timer Interrupt Flag Register 2
TFLG2:  SET  BASE+$004F
TOF:    SET  BIT7                    ; Timer overflow flag
; ***
; Timer Input Capture/Output Compare Registers
TC0:    SET  BASE+$0050         ; Register 0
TC1:    SET  BASE+$0052         ; Register 1
TC2:    SET  BASE+$0054         ; Register 2
TC3:    SET  BASE+$0056         ; Register 3
TC4:    SET  BASE+$0058         ; Register 4
TC5:    SET  BASE+$005A         ; Register 5
TC6:    SET  BASE+$005C         ; Register 6
TC7:    SET  BASE+$005E         ; Register 7
; Pulse Accumulator Control Register
PACTL:  SET  BASE+$0060
PAEN:   SET  BIT6        ; Pulse accumulator enable
PAMOD:  SET  BIT5        ; Pulse accumulator mode
PEDGE:  SET  BIT4        ; Pulse accumulator edge control
CLK1:   SET  BIT3        ; Clock select bit 1
CLK2:   SET  BIT2        ; Clock select bit 2
PAOVI:  SET  BIT1        ; Pulse accumulator overflow interrupt enable
PAIF:   SET  BIT0        ; Pulse accumulator input interrupt enable
; ***
; Pulse Accumulator Flag Register
PAFLG:  SET  BASE+$0061
PAOVF:  SET  BIT1        ; Pulse accumulator overflow flag
PAIF:   SET  BIT0        ; Pulse accumulator input edge flag
; Pulse Accumulator  Counter Register
PACNT:  SET  BASE+$0062
; ***
; Timer Test Register
TIMTST: SET  BASE+$006D
        LIST            ; Turn listing back on
```

Appendix F: HCS12 Interrupt Vector Assignments

Interrupt Vector Assignments

Priority	Vector Address	Interrupt Source	Local Enable Bit	See Register	See Chapter	HPRIO Value to Promote
—	$FF80:FF89	Reserved	—	—	—	—
58	$FF8A:FF8B	VREG LVI	LVIE	CTRL0		$8A
57	$FF8C:FF8D	PWM Emergency Shutdown	PWMIE	PWMSDN	14	$8C
56	$FF8E:FF8F	Port P	PIEP[7:0]	PIEP	12	$8E
33	$FF90:FFAF	Reserved	—	—	—	—
39	$FFB0:FFB1	CAN Transmit	TXEIE[2:0]	CANTIER	16	$B0
38	$FFB2:FFB3	CAN Receive	RXFIE	CANRIER	16	$B2
37	$FFB4:FFB5	CAN Errors	CSCIE, OVRIE	CANRIER	16	$B4
36	$FFB6:FFB7	CAN Wake-up	WUPIE	CANRIER	16	$B6
35	$FFB8:FFB9	Flash	CCIE, CBEIE	FCNFG		$B8
—	$FFBA:FFC3	Reserved	—	—	—	—
29	$FFC4:FFC5	CRG Self Clock Mode	SCMIE	CRGINT	20	$C4
28	$FFC6:FFC7	CRG PLL Lock	LOCKIE	CRGINT	20	$C6
—	$FFC8:FFCD	Reserved	—	—	—	—
24	$FFCE:FFCF	Port J	PIEP[7-6]	PIEP	11	$CE
—	$FFD0:FFD1	Reserved	—	—	—	—
22	$FFD2:FFD3	A/D Converter	ASCIE	ATDCTL2	17	$D2
—	$FFD4:FFD5	Reserved	—	—	—	—
20	$FFD6:FFD7	SCI Serial System	TIE, TCIE, RIE, ILIE	SCICR2	15	$D6
19	$FFD8:FFD9	SPI Serial Peripheral System	SPIE, SPTIE	SPICR1	15	$D8
18	$FFDA:FFDB	Pulse Accumulator Input Edge	PAI	PACTL	14	$DA
17	$FFDC:FFDD	Pulse Accumulator Overflow	PAOVI	PACTL	14	$DC
16	$FFDE:FFDF	Timer Overflow	TOI	TSCR2	14	$DE
15	$FFE0:FFE1	Timer Channel 7	C7I	TIE	14	$E0
14	$FFE2:FFE3	Timer Channel 6	C6I	TIE	14	$E2
13	$FFE4:FFE5	Timer Channel 5	C5I	TIE	14	$E4
12	$FFE6:FFE7	Timer Channel 4	C4I	TIE	14	$E6
11	$FFE8:FFE9	Timer Channel 3	C3I	TIE	14	$E8
10	$FFEA:FFEB	Timer Channel 2	C2I	TIE	14	$EA
9	$FFEC:FFED	Timer Channel 1	C1I	TIE	14	$EC
8	$FFEE:FFEF	Timer Channel 0	C0I	TIE	14	$EE
7	$FFF0:FFF1	Real-Time Interrupt	RTIE	CRGINT	14	$F0
6	$FFF2:FFF3	IRQ_L Pin	IRQEN	INTCR	12	$F2
5	$FFF4:FFF5	XIRQ_L Pin	X-bit	CCR	12	—
4	$FFF6:FFF7	SWI	None	—	—	—

Interrupt Vector Assignments (Continued)

Priority	Vector Address	Interrupt Source	Local Enable Bit	See Register	See Chapter	HPRIO Value to Promote
3	$FFF8:FFF9	Unimplemented Instruction Trap	None	—	—	—
2	$FFFA:FFFB	COP Failure Reset	CR2:CR1:CR0	COPCTL	12	—
1	$FFFC:FFFD	Clock Monitor Fail Reset	CME, SCME	PLLCTL	12	—
0	$FFFE:FFFF	External Reset	None	—	4	—

Index